Polyhydroxyalkanoates: Kinetics, Bioengineering and Industrial Aspects

Polyhydroxyalkanoates: Kinetics, Bioengineering and Industrial Aspects

Editor: Barbara Moore

www.callistoreference.com

Callisto Reference,
118-35 Queens Blvd., Suite 400,
Forest Hills, NY 11375, USA

Visit us on the World Wide Web at:
www.callistoreference.com

ISBN: 978-1-64116-802-1 (Hardback)

Cataloging-in-Publication Data

Polyhydroxyalkanoates : kinetics, bioengineering and industrial aspects / edited by Barbara Moore.
p. cm.
Includes bibliographical references and index.
ISBN 978-1-64116-802-1
1. Poly-beta-hydroxyalkanoates. 2. Motion. 3. Bioengineering. 4. Poly-beta-hydroxyalkanoates--Industrial applications.
I. Moore, Barbara.

QR92.P58 P65 2023
660.63--dc23

Table of Contents

Preface

Polyhydroxyalkanoates (PHAs) belong to a category of polymers that comprise a group of natural biodegradable polyesters which are synthesized by microorganisms. The vital properties of PHAs such as biodegradability, biocompatibility and non-toxicity make them appropriate for applications as biomaterials. PHAs are used in the field of tissue engineering as bone graft substitutes. They are extensively being used for developing new stents for nerve repair, cartilage repair and cardiovascular patches. In implantology, PHAs have been used as sutures and valves. They have also been used as drug carriers for drug delivery systems because the breakdown products are not hazardous for the body due their biodegradability. The mechanical properties of PHAs can be modified by the process of blending to obtain better polymeric properties. This book includes some of the vital pieces of works being conducted across the world with respect to kinetics, bioengineering and industrial aspects of polyhydroxyalkanoates. It will also provide interesting topics for research, which interested readers can take up.

This book is the end result of constructive efforts and intensive research done by experts in this field. The aim of this book is to enlighten the readers with recent information in this area of research. The information provided in this profound book would serve as a valuable reference to students and researchers in this field.

At the end, I would like to thank all the authors for devoting their precious time and providing their valuable contribution to this book. I would also like to express my gratitude to my fellow colleagues who encouraged me throughout the process.

Editor

1

Thauera aminoaromatica MZ1T Identified as a Polyhydroxyalkanoate-Producing Bacterium within a Mixed Microbial Consortium

Dana I. Colpa [1,†], Wen Zhou [1,†], Jan Pier Wempe [1], Jelmer Tamis [2], Marc C. A. Stuart [3], Janneke Krooneman [1] and Gert-Jan W. Euverink [1,*]

1 Products and Processes for Biotechnology Group, Engineering and Technology Institute Groningen, University of Groningen, Nijenborgh 4, 9747 AG Groningen, The Netherlands; d.i.colpa@rug.nl (D.I.C.); wen.zhou@rug.nl (W.Z.); j.p.wempe@student.rug.nl (J.P.W.); j.krooneman@rug.nl (J.K.)

2 Paques Technology B.V., Tjalke de Boerstrjitte 24, 8561 EL Balk, The Netherlands; j.tamis@tudelft.nl

3 Groningen Biomolecular Sciences and Biotechnology Institute, University of Groningen, Nijenborgh 7, 9747 AG Groningen, The Netherlands; m.c.a.stuart@rug.nl

* Correspondence: g.j.w.euverink@rug.nl

† These authors contributed equally to this work.

Abstract: Polyhydroxyalkanoates (PHAs) form a highly promising class of bioplastics for the transition from fossil fuel-based plastics to bio-renewable and biodegradable plastics. Mixed microbial consortia (MMC) are known to be able to produce PHAs from organic waste streams. Knowledge of key-microbes and their characteristics in PHA-producing consortia is necessary for further process optimization and direction towards synthesis of specific types of PHAs. In this study, a PHA-producing mixed microbial consortium (MMC) from an industrial pilot plant was characterized and further enriched on acetate in a laboratory-scale selector with a working volume of 5 L. 16S-rDNA microbiological population analysis of both the industrial pilot plant and the 5 L selector revealed that the most dominant species within the population is *Thauera aminoaromatica* MZ1T, a Gram-negative beta-proteobacterium belonging to the order of the Rhodocyclales. The relative abundance of this *Thauera* species increased from 24 to 40% after two months of enrichment in the selector-system, indicating a competitive advantage, possibly due to the storage of a reserve material such as PHA. First experiments with *T. aminoaromatica* MZ1T showed multiple intracellular granules when grown in pure culture on a growth medium with a C:N ratio of 10:1 and acetate as a carbon source. Nuclear magnetic resonance (NMR) analyses upon extraction of PHA from the pure culture confirmed polyhydroxybutyrate production by *T. aminoaromatica* MZ1T.

Keywords: *Thauera aminoaromatica* MZ1T; biopolymers; polyhydroxyalkanoate (PHA); polyhydroxybutyrate (PHB); 16S-rDNA analysis; biodiversity; mixed microbial consortium

---

1. Introduction

The production and use of fossil-based plastic materials became ubiquitous, especially after the second world war. Since then, plastics play an indispensable role in our daily life. However, in the past decades, it has become evident that these plastics persist in nature and have accumulated to incredibly high volumes and become severe ecological threats. The immense volume of the plastic soup in our oceanic ecosystems is one of the most eye-catching examples [1]. Currently, the global plastics production reached approximately 360 million tons and shows an increasing trend [1]. Obviously, the tide must be turned from an ecological point of view by (1) reducing the use of plastics, (2) recycling of existing plastics, (3) production of plastics that are not petrochemically derived and that are biodegradable once they end up in the environment.

Polyhydroxyalkanoates (PHAs) are examples of bio-based polymers, produced from natural organic resources and organic waste streams by a variety of microorganisms [2]. PHAs are known to be biodegradable and biocompatible with many materials and tissues, thereby having potential in diverse applications. For example, PHAs have been used in drug delivery, in tissue engineering [3], and in agriculture in mulch films and nets [4].

The PHA-producing microbes utilize diverse organic carbon substrates for growth and to synthesize intracellular storage materials in the form of PHAs. The accumulation of PHA occurs mainly under high carbon-substrate levels combined with nutritionally imbalanced environmental conditions. PHA accumulation is a way to store energy and carbon in order to survive starvation-conditions in scarcer times [5]. Once the micro-organisms experience starvation, thus lacking sufficient external carbon source for growth, the accumulated PHA will be used as growth substrate [6,7]. A feast–famine regime is a proven approach to specifically enrich for PHA-producing organisms [8–10]. During the feast-phase (excess carbon), PHA accumulation occurs and during the famine-phase (no carbon) the PHA can be used as an alternative carbon source. Organisms that stored PHA have a competitive advantage over those that did not store reserve materials.

Up until now, PHA-producing microorganisms were found in more than 70 archaeal and (cyano)bacterial genera [11,12]. Mixed microbial consortia are promising for accumulating PHAs because of their potential in using low-cost waste streams as substrates like wastewater, crude palm kernel oil, used cooking oil, cheese whey, and coffee grounds, as well as swine waste liquor [10,13–17]. After hydrolysis and acidogenic fermentation, most organic waste streams contain mixtures of (volatile) fatty acids, such as acetic acid, propionic acid, butyric acid, valeric acid, and hexanoic acid [18,19]. These fatty acids are perfect substrates for the production of PHAs, that consist of the polymerized hydroxylated fatty acid monomers. Mixtures of (volatile) fatty acids, as well as single fatty acids, could be used for PHA production.

To date, there are over 150 different types of PHAs [11]. Most of the PHAs are classified based on their chain length and monomer composition [2,11]. Examples of short-chain length PHA (*scl*-PHA, 3–5 carbon atoms per monomer) are polyhydroxybutyrate (PHB), polyhydroxyvalerate (PHV), and the copolymer hydroxybutyrate-*co*-hydroxyvalerate (PHBV). Medium-chain length PHAs (*mcl*-PHAs) are characterized by a monomer composition based on larger 3-hydroxy fatty acids (6–14 carbon atoms per monomer) like polyhydroxyhexanoate and polyhydroxyoctanoate and copolymers thereof. As described above, when acetic acid is used as a sole carbon source, PHB accumulates inside the cells [20,21]. The presence and concentration of acetic acid in mixed substrates is, to some extent, thought to regulate the ratio of 3-hydroxybutyrate (3HB) in the produced copolymers [22]. This holds for the production of PHAs by pure or defined mixed cultures growing on defined (mixtures of) substrates. However, substrate-directed PHA production is not fully understood when using undefined microbial consortia and complex mixtures of substrates derived from organic waste streams [23]. So, one can imagine that for successful conversion of undefined low-cost organic waste into PHA using undefined natural microbial consortia, many challenges exist. For example, what is the effect of the feed-composition on the PHA production? What is the efficiency of the cells to produce PHA? Which yields can be achieved? Which types of PHAs can be produced, and how can this be controlled? Which microbes within the consortia are involved in the PHA production? Therefore, it is relevant to gain more insight into the mixed microbial consortia and to understand their dynamics in relation to changes in environmental conditions and substrate compositions. Furthermore, it is important to identify the predominant species involved in the PHA production and to understand their eco-physiological characteristics in general, and towards PHA production specifically.

In this work, we aim to enrich, identify, isolate, and characterize the predominant bacterial species within a PHA-producing mixed microbial consortium. In our lab, we grew this microbial consortium to further enrich the predominant strain under a feast–famine regime. The substrate consumption and microbiological diversity was analyzed. Lastly, the predominant bacterial species was identified and characterized as a PHA-producing bacterial strain.

2. Materials and Methods

2.1. Strains and Chemicals

Media components, salts, and solvents were obtained from Sigma-(Aldrich) (St. Louis, MO, USA), Merck (Kenilworth, NJ, USA), Difco (Becton Dickinson Company, Franklin Lakes, NJ, USA), Biosolve Chimie (Dieuze, France), Becton Dickinson Company (Franklin Lakes, NJ, USA), and BOOM (Meppel, The Netherlands). The LCK 365 kit for organic acid detection was obtained from Hach Lange. The genomic DNA was extracted using the FastDNA Spin kit for soil from MP Biomedicals. *Thauera aminoaromatica* MZ1 (DSM No. 25461) was obtained from DSMZ (Braunschweig, Germany).

2.2. Operating Conditions Pilot Reactor

The pilot reactor (working volume 180 L) was operated as an aerobic SBR (sequence batch reactor). Air was supplied via a fine bubble diffuser at a rate of 100 L/min to provide oxygen and ensure mixing. The dissolved oxygen concentration was kept above 2 mg/L. The reactor was kept at 30 ± 3 °C. The pH in the system was not controlled but remained between 8.0 and 8.5 due to the buffering capacity of the substrate. The substrate was leachate from organic waste with a soluble COD (chemical oxygen demand) content of 5 ± 0.6 g/L (average ± standard deviation, n = 62). The substrates' COD content comprised 3 ± 1 g/L of volatile fatty acids: the fractions of individual n-VFA compared to the total n-VFA content were: acetic acid 25% ± 9%, propionic acid 15% ± 5%, butyric acid 29% ± 6% and valeric acid 32% ± 10% on COD basis. The remainder of the soluble COD (2 g/L) of COD could not be readily identified; nevertheless, it was observed that from the remaining fraction, about 1 ± 0.4 g/L was converted in the SBR cycle while the rest was inert. Total ammonium nitrogen concentrations in the substrate were around 300 mg/L. Additional nutrient solution (nutrimix) was added to prevent nitrogen and phosphorus limitation in the reactor. The substrate was fed pulse-wise in order to select for PHA-producing bacteria [24], with a cycle of 12 h, see Figure 1. Every cycle, the floc-forming PHA-producing bacteria were settled, and the upper half of the reactor volume (supernatant) was discarded. The remaining biomass was split in two: half of the volume was kept in the selector, and the other half was transferred to a fed-batch reactor for PHA accumulation. A part (5 L) of the broth of the pilot reactor was used to start the 5 L lab-scale bioreactor.

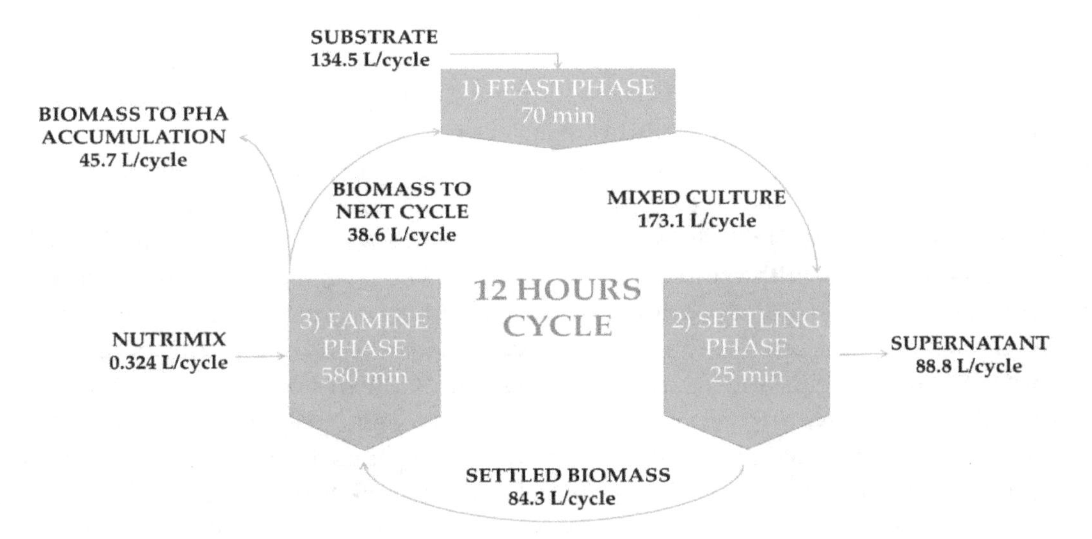

Figure 1. Operating conditions of the pilot selector reactor. The reactor was operated under a feast–famine regime with a cycle of 12 h. Leachate from organic waste was used as substrate.

2.3. Bacterial Growth Conditions—Lab-Scale Bioreactor

2.3.1. Growth Media

The culture medium for the enrichment bioreactor consisted of 36 mM sodium acetate, 16.6 mM NH_4Cl, 2.5 mM KH_2PO_4, 0.55 mM $MgSO_4$, and 0.72 mM KCl. This medium was supplemented with 1.5 mL/L trace metal solution, as described by Vishniac and Santer [25]. The pH of the growth medium was adjusted to 8.0 by a few drops of 5M KOH.

2.3.2. Growth Conditions: 5 L Enrichment Bioreactor under a Feast–Famine Regime

For the enrichment of PHA-accumulating bacteria on lab-scale, a 5 L glass autoclavable bioreactor (Applikon, Delft, The Netherlands) was operated under a feast–famine regime [9,24]. The operational set-up was as follows: The total cycle was 24 h with a feast-phase of approximately 4 h. Every 24 h (excluding the weekend), half of the culture medium was refreshed. At late famine-phase, the flock-forming PHA-producing bacteria were settled for 15 min (no stirring or aeration). Part of the upper liquid (2.5 L) was drained from the bioreactor, after which the feast-phase was initiated by adding 2.5 L of fresh medium (2× stock of the medium described above) to the bioreactor. After feeding with fresh medium, the bioreactor was stirred at 300 rpm, aerated by 0.5 L air per minute, and kept at a temperature of 30 °C. The pH and the dissolved oxygen level of the bioreactor were monitored continuously, but not adjusted. If necessary, a drop of Antifoam A (Sigma-Aldrich) was added. The acetic acid concentration was measured by the LCK 365 kit from Hach Lange for two cycles. The mixed microbial consortium from this reactor was used as inoculum for further batch experiments.

2.3.3. Growth Conditions: Polyhydroxyalkanoates (PHA) Production in 2 L Bioreactor

PHA accumulation experiments were performed in 2 L glass autoclavable bioreactors (Applikon), operating under batch mode, using 10% inoculum of the enriched culture taken after >40 cycles of selection. The reactor was operated at 30 ± 1 °C and stirred at 150 rpm. The growth medium was the same as for the enrichment bioreactor, with minor changes. An acetic acid concentration of 20 g/L, and a C:N ratio of 10:1 (mol/mol) were used. Antifoam A was added if necessary. Samples of 40 mL were taken daily and analyzed for total biomass production and PHA content. Cells were harvested by centrifugation at 12,000 rpm (Thermo Fisher, F15-6x 100y rotor) for 10 min and lyophilized. The PHA content within the samples was analyzed by GC, see materials and methods Section 2.8.2.

2.4. Bacterial Growth Conditions—Thauera aminoaromatica MZ1(T)

Thauera aminoaromatica MZ1(T) was ordered from DSMZ. The lyophilized cells were revived and aseptically grown in two different media. All media were sterilized by autoclaving for 20 min at 121 °C. First, the cells were grown on the rich medium recommended by DSMZ, medium 830 supplemented with 5 mL vitamin solution (of DSMZ medium 461). The cells were grown aerobically in Erlenmeyer flasks at 30 °C, 150 rpm for 4 days, or grown under the same conditions on solid agar (1.5%) plates. Secondly, *T. aminoaromatica* MZ1T was aerobically grown on a more defined medium for *Thauera aromatica*, DSMZ medium 586, with minor changes. The pH was changed to 8.0, as carbon source 10 g/L sodium acetate was added instead of sodium benzoate, KNO_3 was omitted, and 0.5 g/L casamino acids (Difco) were added. The C:N ratio of the medium was 24.6 (mol/mol), without taking the relatively small amounts of casamino acids and vitamins into account. Cultures on medium 586 were inoculated by 1:100 (vol/vol) of a culture grown on medium 830, and subsequently grown in 200 mL medium in 1 L non-baffled Erlenmeyer flasks that were sealed with cotton plugs for a few days to a week. Analysis of the optical density (OD) of the cultures was difficult due to flock formation. *T. aminoaromatica* MZ1T grows slowly under these conditions. The cells were harvested as described above and 0.08 g of lyophilized cell pellet (cell dry mass) was obtained from the 200 mL culture. When growing the culture for longer than a week, 10 g/L sodium acetate was added to the culture, to prevent PHA degradation.

2.5. *Chemical Extraction of PHA*

The chloroform extraction method was used to extract PHA from the lyophilized cells [26,27]. First, the cells were harvested by centrifugation at 12,000 rpm (Thermo Fisher, F15-6x 100y rotor) for 10 min and then washed with distilled water. After washing, the pellet was kept at −20 °C, and then lyophilized for 24 h. For polymer extraction, 1 g dried cells were dissolved in 20 mL chloroform in a 100-mL screw-capped glass bottle and incubated at 60 °C for 72 h. The remaining suspension was filtered over a Whatman GF/A glass microfiber filter to remove the cell debris. The remaining chloroform solution was transferred to a 250-mL flask. PHA within this solution was precipitated by adding cold absolute methanol (1:9), this mixture was incubated at 4 °C for 24 h (or −20 °C for 2h). The PHA precipitate was recovered by filtration over a Whatman GF/A glass microfiber filter and air-dried overnight.

2.6. *Microbial Characterization by Microscopy*

2.6.1. Phase-Contrast and Fluorescence Microscopy—Nile Blue A Staining

The PHA-producing mixed microbial consortium was regularly analyzed by phase-contrast microscopy. To analyze whether the cells contained PHA, Nile blue A staining was performed as described by Ostle and Holt, with minor changes [28]. The cells were stained on a slide by a 1% aqueous solution of Nile blue A and incubated at 55 °C for 10–15 min. After staining, the slides were washed in three steps; (1) with distilled water, (2) with an 8% acetic acid solution for 1 min, and (3) with distilled water. The slides were dried using scratch and dust-free paper and a glass cover slide was placed. The samples were analyzed with a Euromex Oxion fluorescence microscope using a green excitation wavelength of 560 nm and an emission wavelength of 595 nm.

2.6.2. Cryo-Transmission Electron Microscopy (Cryo-TEM)

Bacterial samples of 3 μL were applied on a holey-carbon coated grid (Quantifoil 3.5/1) and vitrified in ethane in a Vitrobot (FEI, The Netherlands). Samples in the frozen, hydrated grid were observed by a Tecnai T20 (FEI) electron microscope operating at 200 keV, equipped with a Gatan cryo-stage (Gatan, model 626). Images were recorded under low-dose conditions on a slow-scan CCD camera.

2.7. *Genomic Diversity Analysis*

The microbial diversity of the mixed microbial consortium was studied by 16S-rDNA analysis. Genomic DNA of 1–8 mL of the consortium was extracted using the FastDNA Spin kit for soil from MP Biomedicals. The extracted genomic DNA was send for 16S-rDNA analysis to MR DNA (Shallowater, TX, USA) and analyzed by bTEFAP PacBio Sequel sequencing technology using the primer set 27F and 1492R. Diversity profiles were obtained by plotting the obtained percentage from MR DNA in a Sunburst graph.

2.8. *Analytical Analysis of PHA*

2.8.1. Determination of the Acetic Acid Concentration by High-Performance Liquid Chromatography (HPLC) or Kit

Acetic acid (or acetate depending on the pH of the culture medium) was the sole carbon source added to the culture media used. The decrease in acetic acid concentration during growth was analyzed using the LCK 365 kit for organic acids from Hach Lange or by high-performance liquid chromatography (HPLC). For HPLC analysis, 2 mL samples of the culture medium were taken every 24 h [29]. Prior to injection, cell-free supernatant was obtained by centrifugation (13,500× *g*, Eppendorf Centrifuge 5424, 10 min, 4 °C), and subsequently filtered over a 0.2 μm cellulose acetate (CA) membrane. The filtered aqueous samples were analyzed with an Agilent 1200 HPLC equipped with an Agilent

1200 pump, a refractive index detector, and a standard ultraviolet detector. Samples were separated over a Bio-Rad organic acid column (Aminex HPX-87H) which was maintained at 60 °C. An eluent of 5 mM sulfuric acid was used, with a flow rate of 0.55 mL/min. The elution of acetic acid was followed at 210 nm. Calibration curves of acetic acid were prepared for accurate quantification and were based on a minimum of five data points within the range of 1–20 g/L acetic acid with an excellent linear fit ($R^2 > 0.99$).

2.8.2. Gas Chromatography–Mass Spectroscopy (GC–MS)

For characterization and quantitative analysis of the produced PHA, the extracted polymer was depolymerized by methanolysis and the corresponding methyl esters of the monomers were analyzed by GC–MS. Fifty mg lyophilized cells were resuspended in 2 mL chloroform, 1.7 mL methanol, and 0.3 mL 98% sulfuric acid. This mixture was incubated at 100 °C in a water bath shaker for 4 h. After cooling down, 1 mL of water was added to the reaction mixture for phase separation [30]. Benzoic acid was added as internal standard, and commercial PHB (Sigma-Aldrich) was analyzed together with the samples as an external standard. The lower organic phase, containing the methyl esters, was analyzed on a Hewlett-Packard 6890 gas chromatograph equipped with a Rxi-5Sil capillary column (30 m × 0.25 mm i.d. and 0.25 μm film thickness) and a Quadrupole Hewlett-Packard 5973 mass selective detector. Helium was used as carrier gas at a flow rate of 2 mL/min. Samples of 5 μL were injected at an inlet temperature of 60 °C. The oven temperature was kept at 60 °C for 3 min and then increased to 280 °C at a rate of 12 °C/min and held at 280 °C for 8 min [31]. All samples were diluted 10 times with chloroform and filtered over a 0.2 μm PTFE membrane prior to analysis.

2.8.3. Nuclear Magnetic Resonance (NMR)

For the characterization of the polymer, 10 mg of the extracted biopolymer was dissolved in 2 mL deuterated chloroform ($CDCl_3$). As a standard, commercial PHB (Sigma) was analyzed together with the samples. ^1H NMR spectra were recorded by an Agilent Technologies 400 MHz MR-DD2 supplemented with a 5 mm One nuclear magnetic resonance (NMR) probe, or by a Varian Mercury Plus 300 MHz supplemented with a 5 mm 4nuc probe. Settings: 45° duration of pulse, 1 s repetition delay, 8–32 scans, and an acquisition time of 2.923 s (400 MHz) or 1.953 s (300 MHz).

3. Results and Discussion

3.1. Description of the Mixed Microbial Consortium

To enhance the selection of the PHA-producing bacteria from the pilot reactor in our 5 L bioreactor, a feast–famine regime was combined with gravitational settling. The PHA-producing bacteria appeared to be floc forming and settling. The culture was allowed to settle daily before the medium was exchanged from the top of the bioreactor. If the amount of biomass became too much, the excess was removed from the reactor. The pH and the dissolved oxygen level were recorded continuously. Upon feeding with acetate, the dissolved oxygen level dropped to zero due to the increased metabolism of the culture. When acetate became limited, the dissolved oxygen level increased again. The oxygen limiting conditions lasted 5 h at most. The pH showed a repetitive cycle upon feeding too. The pH became first more alkaline, then more acidic, and subsequently recovered to a pH value of 8–8.5. PHB production by this mixed microbial consortium was confirmed. To further enrich the PHA-producing bacterial strain(s), the 5 L bioreactor was operated under these conditions (feast–famine regime with gravitational settling and oxygen limitation period) for several months, thereby obtaining a steady-state with a stable repetitive pattern, as shown in Figure 2.

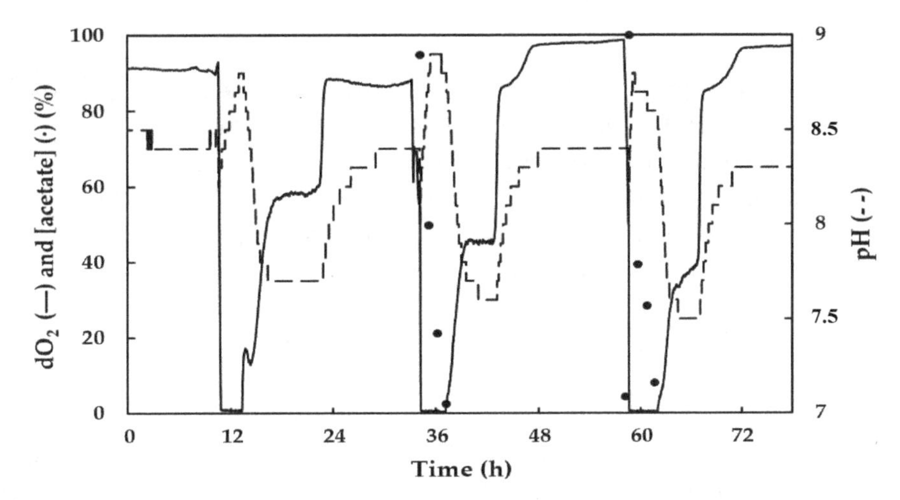

Figure 2. Daily pattern of the pH (dashed line) and dissolved oxygen concentration (solid line) of a polyhydroxyalkanoate (PHA)-producing mixed microbial consortium under a steady state feast–famine regime upon feeding with acetate (dots). An acetate concentration of 100% corresponds to 500 mg/L acetic acid.

3.2. Culture Growth and PHA Production by the Mixed Microbial Consortium

To study the production of PHA by the mixed microbial consortium, the consortium was grown in a 2 L batch bioreactor using 10 g/L acetic acid as the sole carbon source and a C:N ratio of 10:1 (mol/mol). Samples were taken every 12 h and analyzed for the substrate concentration, total cell dry mass (CDM), and for the PHA content within the CDM, see Figure 3. The PHA content within the dried biomass increased gradually during bacterial growth, reaching a maximum amount of PHA within the cell dry mass of 50.4% after 110–120 h. When acetate depleted, a decrease in the PHA content of the dried biomass was observed. This pattern was reproducible and observed before by other researchers [5]. Besides PHA synthases for biopolymer production, these bacteria can produce PHA depolymerases to initiate its metabolization. To prevent PHA degradation, the PHA depolymerase(s) could be knocked out, as was done for *Pseudomonas putida* KT2442 and *Rhodobacter sphaeroides* HJ [32,33].

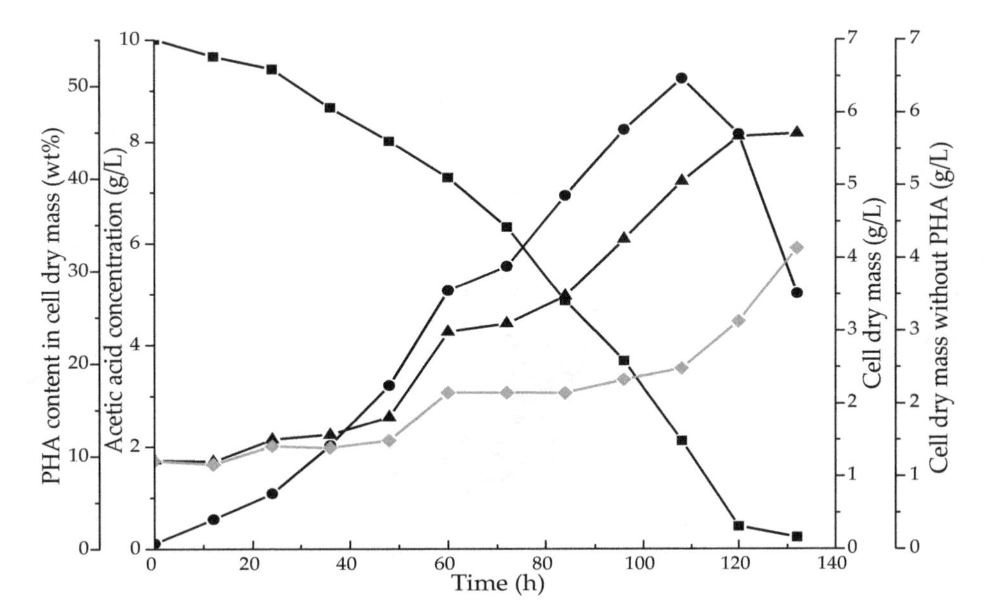

Figure 3. PHA accumulation by the mixed microbial consortium when grown in a 2 L batch bioreactor, with 10 g/L acetic acid as the carbon source and a C:N ratio of 10:1 (mol/mol). The acetate concentration is shown in black squares, the PHA content of the dried cells (wt%) in blue circles, and the cell dry mass including and excluding PHA in red triangles and orange diamonds, respectively.

3.3. Extraction and Identification of PHA

In order to study the PHA produced, it was extracted from the bacterial cells. Here, we used a chloroform extraction method, as described in the materials and methods section. A white product was obtained after chloroform extraction of the lyophilized red-brown colored cell pellets, see Figure S1 in the supplementary materials. Using both GC and ^{1}H NMR, the extracted product was confirmed to be pure PHB, see Figure S2 and Figure S3. The production of pure PHB was expected since the consortium was fed with acetate as the sole carbon source.

3.4. Genomic (Diversity) Analysis of the Mixed Microbial Consortium

In order to identify the predominant PHA-producing organism in the mixed microbial consortium, probably being a flock forming PHB-producing bacterium that settles by gravity, the genomic diversity of this mixed microbial consortium was studied by 16S-rDNA analysis. Two samples were analyzed, one of the original consortium retrieved from the industrial pilot reactor and one after two months of enrichment in the 5 L lab-scale selector. In the pilot reactor, several *Thauera spp.* were abundantly present, with one dominating with 24% of the total community. This predominant species became even more predominant after two months of growth in the feast–famine operated 5 L selector-reactor. An increase from 24 to approximately 40% was observed, see Figure 4 and Tables 1 and 2. However, the abundance of the other *Thauera* species reduced significantly over the two months. The predominant species was further identified, showing >99.9% homology with *Thauera aminoaromatica* MZ1T [34,35]. Additionally, a new group of organisms increased significantly after several months of enrichment: the planctomycetes, ubiquitous bacteria that live in all sorts of habitats [36]. The dominating species in this group is *Algisphaera* spp., belonging to the class *Phycisphaerae*, known to be associated with aquatic phototrophs [37]. Since these phototrophs were not present in detectable amounts in the original PHA-producing industrial pilot reactor, we did not focus on these microorganisms for their involvement in PHA production. The relative abundance of all species belonging to the Bacteroidetes-phylum did not change significantly during the enrichment-period, either. Therefore the focus of our study was on the proteobacteria, and on the *T. aminoaromatica sp.* specifically.

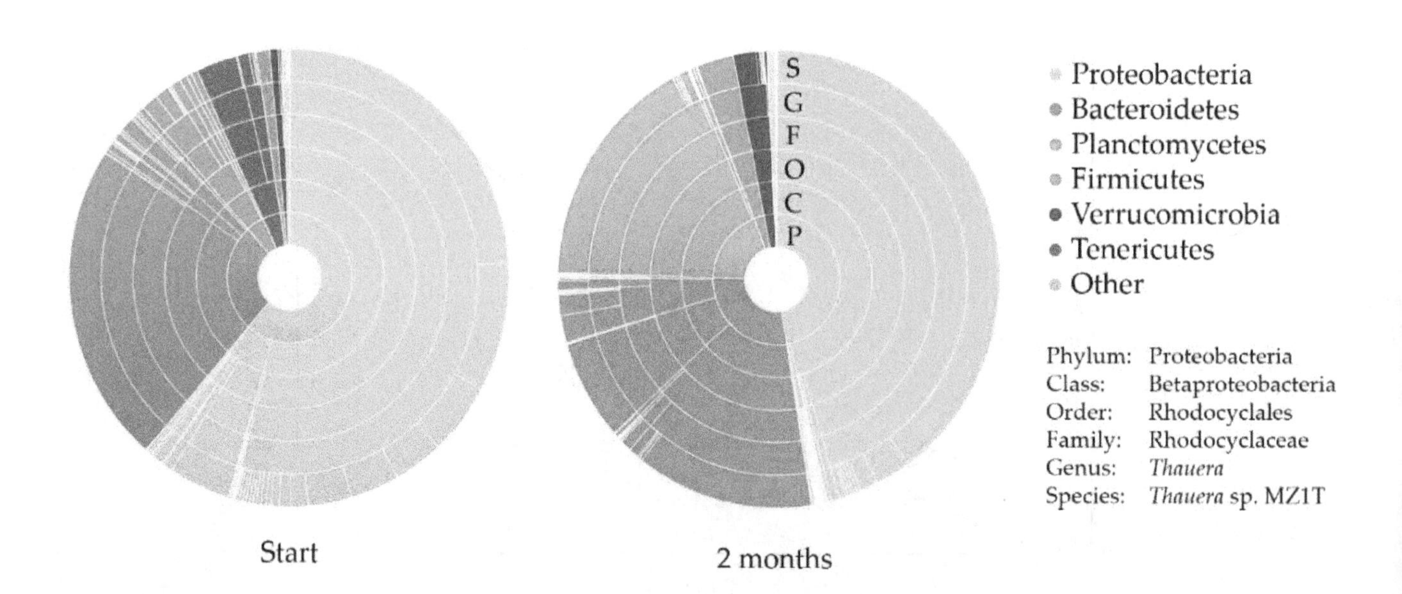

Figure 4. Sunburst plots of the genomic diversity of the PHA-producing mixed microbial consortium at the start (left) and after 2 months (right) of enrichment in the 5 L lab-scale selector-bioreactor.

Table 1. Top 10 bacterial species as determined by 16S-rDNA analysis of the mixed microbial consortium of the pilot plant used to start the 5 L selector-bioreactor in our lab.

	Species	Phylum	Relative Abundance (%)
1	*Thauera* sp. MZ1T *	Proteobacteria	23.9
2	Uncultured *Haliscomenobacter* sp.	Bacteroidetes	22.7
3	Uncultured *Thauera* sp.	Proteobacteria	9.2
4	*Thauera aminoaromatica*	Proteobacteria	5.1
5	*Micavibrio aeruginosavorus* arl_13	Proteobacteria	4.5
6	*Thauera aminoaromatica*	Proteobacteria	3.6
7	*Thauera phenylacetica*	Proteobacteria	3.6
8	Uncultured *Thauera*	Proteobacteria	3.3
9	Uncultured *Asteroleplasma* sp.	Tenericutes	3.1
10	Uncultured *Thauera* sp.	Proteobacteria	1.3

* The most abundant species showed >99.9% homology to *Thauera* sp. MZ1T (DSMZ strain 25461).

Table 2. Top 10 bacterial species as determined by 16S-rDNA analysis for the mixed microbial consortium after 2 months of growth in our 5 L lab-scale selector-bioreactor.

	Species	Phylum	Relative Abundance (%)
1	*Thauera* sp. MZ1T *	Proteobacteria	39.9
2	*Algisphaera agarilytica*	Planctomycetes	16.8
3	Uncultured *Ohtaekwangia* sp.	Bacteroidetes	13.0
4	Uncultured *Alkaliflexus* sp.	Bacteroidetes	7.1
5	Uncultured *Thermodesulfobium* sp.	Firmicutes	2.7
6	*Thauera aminoaromatica*	Proteobacteria	2.5
7	*Sediminibacterium salmoneum*	Bacteroidetes	1.9
8	*Prosthecobacter vanneervenii* str. dsm 12252	Verrucomicrobia	1.6
9	Uncultured *ferruginibacter* sp.	Bacteroidetes	1.4
10	*Thauera aminoaromatica*	Proteobacteria	1.3

* The most abundant species showed >99.9% homology to *Thauera* sp. MZ1T (DSMZ strain 25461).

T. aminoaromatica MZ1T is a Gram-negative beta-proteobacterium belonging to the family of the *Rhodocyclaceae*. Sayler deposited this species in the German collection of microorganisms and cell cultures (DSMZ) under no. 25461 [34]. *T. aminoaromatica* MZ1T was originally isolated from activated sludge mixed liquor from the Eastman Chemical company wastewater treatment facility (Kingsport, TN, USA) [34]. This species was shown to be a floc-forming facultative anaerobic strain capable of producing abundant amounts of exopolysaccharide (EPS) [35,38]. The ability to form flocs may be an important aspect in explaining why especially this *Thauera* species became more dominant in our selector-vessel: the selective pressure was towards the capability of storing reserve materials as PHA due to the feast–famine regime on the one hand, and on the capability of gravitational settling possibly due to floc-formation on the other hand (see above). A closely-related strain belonging to the same species, *T. aminoaromatica* S2, has been previously studied [35,39,40].

Kutralam-Muniasamy et al. analyzed the fully sequenced genomes of 66 beta-proteobacteria for the presence of a genetic pathway for PHA production [41], one of these being the genome of *T. aminoaromatica* MZ1T, that was fully sequenced by Jiang et al. [35]. They showed that *T. aminoaromatica* MZ1T contains a pathway for PHA production. When blasting the genes known to be involved in PHA production in *Cupriavidus necator* H16 to the genome of *T. aminoaromatica* MZ1T homologs of most genes were found, see Table 3. These genes cover the PHA synthesis genes (*pha*A, acetyl-CoA acetyltransferase; *pha*B, acetoacetyl-CoA reductase and *pha*C, PHA synthase), a gene for phasin expression (*pha*P, a surface protein of the PHA granules), a regulatory gene for phasin expression (*pha*R) and a PHA depolymerase (*pha*Z).

Table 3. Gene equivalents for PHA synthesis observed in the genome of *T. aminoaromatica* MZ1T. The genes known to be involved in PHA synthesis of *Cupriavidus necator* H16 [41] were blasted to the genome of *T. aminoaromatica* MZ1T.

Category	Genes	Description and Function	NCBI/GenBank Code *C. necator* H16 [41]	NCBI/GenBank Code *T. aminoaromatica* MZ1T
Synthesis	*pha*A	Acetyl-CoA acetyltransferase	CAJ92573.1	ACK53504.1 ACK53575.1 ACK53579.1 ACR01093.1
	*pha*B	Acetoacetyl-CoA reductase	CAJ92574.1	ACK53788.1 ACR01719.1
	*pha*C	PHA synthase	CAJ92572.1 CAJ93103.1	ACK53500.1 Class I ACK53786.1 Domain protein ACK53908.1 Domain protein
Surface proteins	*pha*P	Phasin	CAJ92517.1	ACR02450.1 ACR01124.1 ACK54768.1 ACK54704.1 ACK53642.1
Gene regulation	*pha*R	PHA repressor, regulates phasin expression	CAJ92575.1	ACR01718.1
Degradation	*pha*Z	PHA depolymerase	CAJ92291.1	ACK52971.1 PHA depolymerase ACK53308.1 Esterase

3.5. Does T. aminoaromatica MZ1T Produce PHA?

To study the ability of *T. aminoaromatica* MZ1T to produce PHA, the strain was obtained from DSMZ (no. 25461, called *T. aminoaromatica* MZ1, showing >99.9% resemblance with the dominant species in our selector-reactor). When grown in batch culture, it was observed that the organism initially grew as homogeneous suspension, but after several days floc formation occurred, see Supplementary materials Figure S4. This floc forming capacity was previously shown [35,38].

The production of PHB by *T. aminoaromatica* MZ1T was studied by exchanging the rich growth medium recommended by DSMZ (medium 830) for an adapted version of the defined medium recommended for *Thauera aromatica* (medium 586). This medium allowed us to grow the strain on a medium that was more comparable to the medium used for the mixed microbial consortium in the 5 L selector-bioreactor. The culture grew slowly but steadily on this medium, and after a few days, flocs appeared. The production of PHB was analyzed by cryo-TEM on the whole cells, and by ^{1}H NMR on the extracted biopolymer.

When *T. aminoaromatica* MZ1T was grown on the rich DSMZ medium 830 the cryo-TEM images showed nicely uniform rod-shaped bacterial cells with a size of approximately 1 μm × 2 μm, see Figure 5. Remarkably, when the cells were grown on the defined medium, multiple dense spheres of 0.2–0.5 μm where visible within the bacterial cells. This result suggests that *T. aminoaromatica* MZ1T stores some compounds within their cells under nitrogen limiting conditions, possibly being PHA. Applying chloroform-based extraction procedures for extraction of PHA and subsequent analysis of the extracted material by ^{1}H NMR confirmed this hypothesis: *T. aminoaromatica* MZ1T produced PHB when grown on acetate under nitrogen limiting conditions, see Supplementary materials Figure S5. This is the first evidence showing that *T. aminoaromatica* MZ1T is capable of PHA production as supposed by the presence of the PHA-producing genes in the genome as shown by Kutralam-Muniasamy et al. [41].

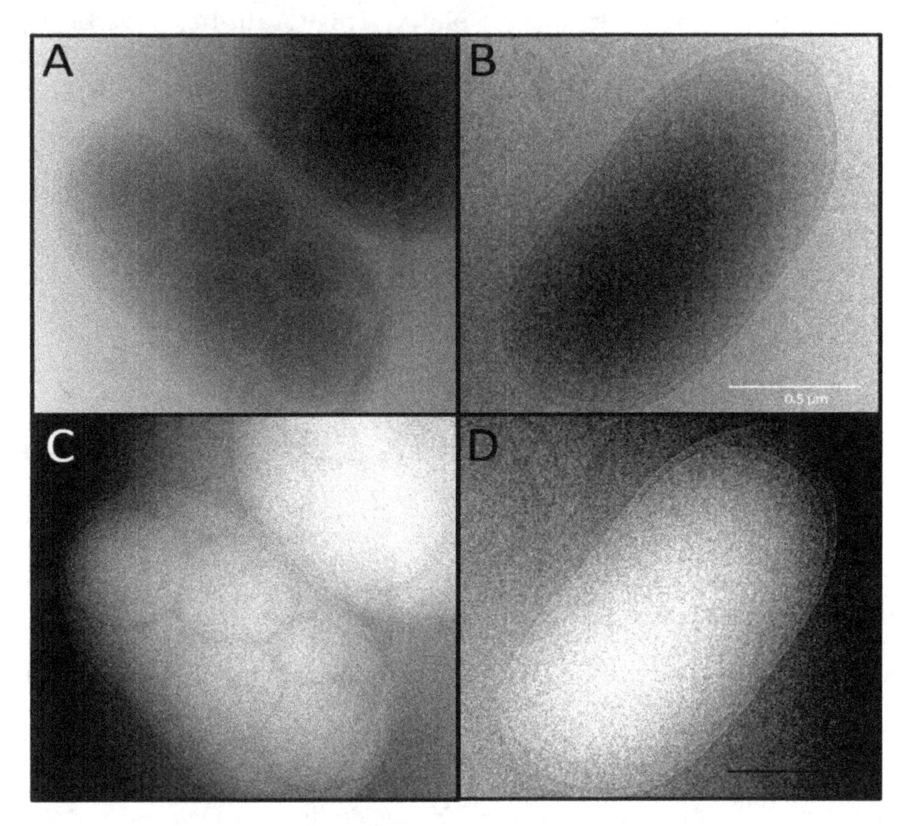

Figure 5. Cryo-TEM of *T. aminoaromatica* MZ1T cells with (panel (**A**)) and without polyhydroxybutyrate (PHB) (panel (**B**)). Panels (**C**) and (**D**) are negative images of panels (**A**) and (**B**) made in GIMP with a slightly adjusted contrast.

In this research, a PHA-producing mixed microbial consortium was studied and characterized. 16S-rDNA analysis showed that the predominant species within the consortium is the beta-proteobacterium *Thauera aminoaromatica* MZ1T. This bacterial species was isolated and characterized before [34,38], but not fully studied yet. Its genome was fully sequenced, and a pathway for PHA production had been identified previously [35,41]. We demonstrated that *T. aminoaromatica* MZ1T is indeed capable of PHA production under aerobic nitrogen limiting conditions. It produces PHB when grown on acetate as the sole carbon source, comparable to other PHA-producing bacteria [20]. This strain grows well, both as monoculture and within a natural mixed microbial consortium in which it could be enriched to at least 40%. A striking characteristic of this strain is its floc-forming ability and gravitational settling, allowing to separate the bacterial flocs from the majority of the medium just by gravity. This allows a relatively easy selective production of high volumes of PHA-producing biomass.

Besides PHA, *T. aminoaromatica* MZ1T is also reported to be able to produce large amounts of other biopolymers like exopolysaccharide (EPS) [38]. It would be interesting to study if *T. aminoaromatica* MZ1T is also capable of simultaneous production of both EPS and PHB, as previously shown for other bacteria [42], and how to direct production towards the most valuable polymer.

By identifying *Thauera aminoaromatica* MZ1T as a key-microbe in an industrial mixed microbial consortia (MMC) that produces PHA from organic waste, we will be able to further characterize this organism in relation to its PHA-producing capacities. For example, to optimize culturing conditions and substrate use to increase PHA production yields, and to direct the (co)-polymer composition. Knowledge of the cell-membrane composition and the structures of the PHA-granules will help us to develop more environmentally friendly extraction methods to keep the green label of this biopolymer.

4. Conclusions

In this work, we identified *Thauera aminoaromatica* as the dominant microorganism in a PHA-producing MMC from an industrial pilot plant. This Gram-negative beta-proteobacterium, belonging to the order of the Rhodocyclales, was further enriched by a feast–famine regime and gravitational settling from 24% to 40% in two months. Using cryo-electron microscopy analysis, intracellular dense granules, possibly consisting of PHB, were observed when grown on acetate as the sole carbon source and a C:N ratio of the medium of 10:1. NMR analysis confirmed PHB production by *T. aminoaromatica* MZ1T. Identification and characterization of the key-microbe(s), such as *T. aminoaromatica* MZ1T within PHA-producing MMCs, allows for a better understanding of PHA production in terms of yield and polymer composition.

Supplementary Materials:
Figure S1. PHA extracted from the mixed microbial consortium grown on acetate, Figure S2. Gas chromatogram of methanolized commercial PHB and PHB produced by the mixed microbial consortium, Figure S3. ^{1}H NMR (400 MHz, $CDCl_3$) spectrum of commercial PHB and PHB produced by the mixed microbial consortium, Figure S4. Flock formation of *T. aminoaromatica* MZ1T, Figure S5. ^{1}H NMR of extracted PHB produced by *T. aminoaromatica* MZ1T.

Author Contributions: D.I.C., J.K., and W.Z. designed the experiments; D.I.C., W.Z., J.P.W., J.T., and M.C.A.S. performed the experiments and analyzed the data; J.T. provided the mixed microbial consortium; D.I.C., J.K., and W.Z. wrote the paper; G.-J.W.E. reviewed and edited the paper. All authors have read and agreed to the published version of the manuscript.

Acknowledgments: We thank João Sousa and Bert Geurkink for meaningful discussions and Stratingh Institute for Chemistry of the University of Groningen for the use of the NMR facility.

References

1. Plastics—The Facts 2019: An Analysis of European Plastic Production, Demand and Waste Data. Available online: https://www.plasticseurope.org/en/resources/publications/1804-plastics-facts-2019 (accessed on 20 February 2020).
2. Kourmentza, C.; Plácido, J.; Venetsaneas, N.; Burniol-Figols, A.; Varrone, C.; Gavala, H.N.; Reis, M.A.M. Recent advances and challenges towards sustainable polyhydroxyalkanoate (PHA) production. *Bioengineering* **2017**, *4*, 55. [CrossRef] [PubMed]
3. Puppi, D.; Pecorini, G.; Chiellini, F. Biomedical processing of polyhydroxyalkanoates. *Bioengineering* **2019**, *6*, 108. [CrossRef] [PubMed]
4. Amelia, T.S.M.; Govindasamy, S.; Tamothran, A.M.; Vigneswari, S.; Bhubalan, K. Applications of PHA in agriculture. In *Biotechnological Applications of Polyhydroxyalkanoates*; Kalia, V.C., Ed.; Springer: Singapore, 2019; pp. 347–361.
5. Castro-Sowinski, S.; Burdman, S.; Matan, O.; Okon, Y. Natural functions of bacterial polyhydroxyalkanoates. In *Plastics from Bacteria*; Chen, G.Q., Ed.; Springer: Berlin/Heidelberg, Germany, 2010; pp. 39–61.
6. Beun, J.J.; Dircks, K.; van Loosdrecht, M.C.M.; Heijnen, J.J. Poly-β-hydroxybutyrate metabolism in dynamically fed mixed microbial cultures. *Water Res.* **2002**, *36*, 1167–1180. [CrossRef]
7. Tamis, J.; Marang, L.; Jiang, Y.; van Loosdrecht, M.C.M.; Kleerebezem, R. Modeling PHA-producing microbial enrichment cultures—Towards a generalized model with predictive power. *New Biotechnol.* **2014**, *31*, 324–334. [CrossRef] [PubMed]
8. Koller, M. Advances in polyhydroxyalkanoate (PHA) production. *Bioengineering* **2017**, *4*, 88. [CrossRef]
9. Huang, L.; Chen, Z.; Wen, Q.; Zhao, L.; Lee, D.-J.; Yang, L.; Wang, Y. Insights into feast-famine polyhydroxyalkanoate (PHA)-producer selection: Microbial community succession, relationships with system function and underlying driving forces. *Water Res.* **2018**, *131*, 167–176. [CrossRef]
10. Oliveira, C.S.S.; Silva, C.E.; Carvalho, G.; Reis, M.A. Strategies for efficiently selecting PHA producing mixed microbial cultures using complex feedstocks: Feast and famine regime and uncoupled carbon and nitrogen availabilities. *New Biotechnol.* **2017**, *37*, 69–79. [CrossRef]
11. Tan, G.-Y.A.; Chen, C.-L.; Li, L.; Ge, L.; Wang, L.; Razaad, I.M.N.; Li, Y.; Zhao, L.; Mo, Y.; Wang, J.-Y. Start a research on biopolymer polyhydroxyalkanoate (PHA): A review. *Polymers* **2014**, *6*, 706–754. [CrossRef]

12. Yan, Q.; Zhao, M.; Miao, H.; Ruan, W.; Song, R. Coupling of the hydrogen and polyhydroxyalkanoates (PHA) production through anaerobic digestion from Taihu blue algae. *Bioresour. Technol.* **2010**, *101*, 4508–4512. [CrossRef]
13. Obruca, S.; Benesova, P.; Kucera, D.; Petrik, S.; Marova, I. Biotechnological conversion of spent coffee grounds into polyhydroxyalkanoates and carotenoids. *New Biotechnol.* **2015**, *32*, 569–574. [CrossRef]
14. Kourmentza, C.; Costa, J.; Azevedo, Z.; Servin, C.; Grandfils, C.; De Freitas, V.; Reis, M.A.M. *Burkholderia thailandensis* as a microbial cell factory for the bioconversion of used cooking oil to polyhydroxyalkanoates and rhamnolipids. *Bioresour. Technol.* **2018**, *247*, 829–837. [CrossRef]
15. Altaee, N.; Fahdil, A.; Yousif, E.; Sudesh, K. Recovery and subsequent characterization of polyhydroxybutyrate from *Rhodococcus equi* cells grown on crude palm kernel oil. *J. Taibah Univ. Sci.* **2016**, *10*, 543–550. [CrossRef]
16. Valentino, F.; Morgan-Sagastume, F.; Campanari, S.; Villano, M.; Werker, A.; Majone, M. Carbon recovery from wastewater through bioconversion into biodegradable polymers. *New Biotechnol.* **2017**, *37*, 9–23. [CrossRef]
17. Cho, K.S.; Ryu, H.W.; Park, C.H.; Goodrich, P.R. Poly(hydroxybutyrate-co-hydroxyvalerate) from swine waste liquor by *Azotobacter vinelandii* UWD. *Biotechnol. Lett.* **1997**, *19*, 7–10. [CrossRef]
18. Moretto, G.; Russo, I.; Bolzonella, D.; Pavan, P.; Majone, M.; Valentino, F. An urban biorefinery for food waste and biological sludge conversion into polyhydroxyalkanoates and biogas. *Water Res.* **2020**, *170*, 1–12. [CrossRef]
19. Albuquerque, M.G.E.; Torres, C.A.V.; Reis, M.A.M. Polyhydroxyalkanoate (PHA) production by a mixed microbial culture using sugar molasses: Effect of the influent substrate concentration on culture selection. *Water Res.* **2010**, *44*, 3419–3433. [CrossRef]
20. Jiang, Y.; Hebly, M.; Kleerebezem, R.; Muyzer, G.; van Loosdrecht, M.C.M. Metabolic modeling of mixed substrate uptake for polyhydroxyalkanoate (PHA) production. *Water Res.* **2011**, *45*, 1309–1321. [CrossRef]
21. Kourmentza, C.; Kornaros, M. Biotransformation of volatile fatty acids to polyhydroxyalkanoates by employing mixed microbial consortia: The effect of pH and carbon source. *Bioresour. Technol.* **2016**, *222*, 388–398. [CrossRef]
22. Ling, C.; Qiao, G.-Q.; Shuai, B.-W.; Olavarria, K.; Yin, J.; Xiang, R.-J.; Song, K.-N.; Shen, Y.-H.; Guo, Y.; Chen, G.-Q. Engineering $NADH/NAD^+$ ratio in *Halomonas bluephagenesis* for enhanced production of polyhydroxyalkanoates (PHA). *Metab. Eng.* **2018**, *49*, 275–286. [CrossRef]
23. Hanson, A.J.; Guho, N.M.; Paszczynski, A.J.; Coats, E.R. Community proteomics provides functional insight into polyhydroxyalkanoate production by a mixed microbial culture cultivated on fermented dairy manure. *Appl. Microbiol. Biotechnol.* **2016**, *100*, 7957–7976. [CrossRef]
24. Johnson, K.; Jiang, Y.; Kleerebezem, R.; Muyzer, G.; van Loosdrecht, M.C.M. Enrichment of a mixed bacterial culture with a high polyhydroxyalkanoate storage capacity. *Biomacromolecules* **2009**, *10*, 670–676. [CrossRef] [PubMed]
25. Vishniac, W.; Santer, M. The Thiobacilli. *Bacteriol. Rev.* **1957**, *21*, 195–213. [PubMed]
26. Ramsay, J.A.; Berger, E.; Voyer, R.; Chavarie, C.; Ramsay, B.A. Extraction of poly-3-hydroxybutyrate using chlorinated solvents. *Biotechnol. Tech.* **1994**, *8*, 589–594. [CrossRef]
27. Kumar, M.; Ghosh, P.; Khosla, K.; Thakur, I.S. Recovery of polyhydroxyalkanoates from municipal secondary wastewater sludge. *Bioresour. Technol.* **2018**, *255*, 111–115. [CrossRef]
28. Ostle, A.G.; Holt, J. Nile blue A as a fluorescent stain for poly-β-hydroxybutyrate. *Appl. Environ. Microbiol.* **1982**, *44*, 238–241. [CrossRef]
29. Guo, W.; Heeres, H.J.; Yue, J. Continuous synthesis of 5-hydroxymethylfurfural from glucose using a combination of $AlCl_3$ and HCl as catalyst in a biphasic slug flow capillary microreactor. *Chem. Eng. J.* **2020**, *381*, 122754. [CrossRef]
30. Borrero-de Acuña, J.M.; Hidalgo-Dumont, C.; Pacheco, N.; Cabrera, A.; Poblete-Castro, I. A novel programmable lysozyme-based lysis system in *Pseudomonas putida* for biopolymer production. *Sci. Rep.* **2017**, *7*, 1–11. [CrossRef]
31. Kumar, M.; Gupta, A.; Thakur, I.S. Carbon dioxide sequestration by chemolithotrophic oleaginous bacteria for production and optimization of polyhydroxyalkanoate. *Bioresour. Technol.* **2015**, *213*, 249–256. [CrossRef]
32. Cai, L.; Yuan, M.-Q.; Liu, F.; Jian, J.; Chen, G.-Q. Enhanced production of medium-chain-length polyhydroxyalkanoates (PHA) by PHA depolymerase knockout mutant of *Pseudomonas putida* KT2442. *Bioresour. Technol.* **2009**, *100*, 2265–2270. [CrossRef]

33. Kobayashi, J.; Kondo, A. Disruption of poly (3-hydroxyalkanoate) depolymerase gene and overexpression of three poly (3-hydroxybutyrate) biosynthetic genes improve poly (3-hydroxybutyrate) production from nitrogen rich medium by *Rhodobacter sphaeroides*. *Microb. Cell Fact.* **2019**, *18*, 1–13. [CrossRef]
34. Lajoie, C.A.; Layton, A.C.; Gregory, I.R.; Sayler, G.S.; Taylor, D.E.; Meyers, A.J. Zoogleal clusters and sludge dewatering potential in an industrial activated-sludge wastewater treatment plant. *Water Environ. Res.* **2000**, *72*, 56–64. [CrossRef]
35. Jiang, K.; Sanseverino, J.; Chauhan, A.; Lucas, S.; Copeland, A.; Lapidus, A.; Del Rio, T.G.; Dalin, E.; Tice, H.; Bruce, D.; et al. Complete genome sequence of *Thauera aminoaromatica* strain MZ1T. *Stand. Genom. Sci.* **2012**, *6*, 325–335. [CrossRef] [PubMed]
36. Wiegand, S.; Jogler, M.; Jogler, C. On the maverick Planctomycetes. *FEMS Microbiol. Rev.* **2018**, *42*, 739–760. [CrossRef] [PubMed]
37. Fukunaga, Y.; Kurahashi, M.; Sakiyama, Y.; Ohuchi, M.; Yokota, A.; Harayama, S. *Phycisphaera mikurensis* gen. nov., sp. nov., isolated from a marine alga, and proposal of *Phycisphaeraceae* fam. nov., *Phycisphaerales* ord. nov. and *Phycisphaerae* classis nov. in the phylum *Planctomycetes*. *J. Gen. Appl. Microbiol.* **2009**, *55*, 267–275. [CrossRef] [PubMed]
38. Allen, M.S.; Welch, K.T.; Prebyl, B.S.; Baker, D.C.; Meyers, A.J.; Sayler, G.S. Analysis and glycosyl composition of the exopolysaccharide isolated from the floc-forming wastewater bacterium *Thauera* sp. MZ1T. *Environ. Microbiol.* **2004**, *6*, 780–790. [CrossRef] [PubMed]
39. Seyfried, B.; Tschech, A.; Fuchs, G. Anaerobic degradation of pheylacetate and 4-hydroxyphenylacetate by denitrifying bacteria. *Arch. Microbiol.* **1991**, *155*, 249–255. [CrossRef]
40. Mechichi, T.; Stackebrandt, E.; Gad'on, N.; Fuchs, G. Phylogenetic and metabolic diversity of bacteria degrading aromatic compounds under denitrifying conditions, and description of *Thauera phenylacetica* sp. nov., *Thauera aminoaromatica* sp. nov., and *Azoarcus buckelii* sp. nov. *Arch. Microbiol.* **2002**, *178*, 26–35. [CrossRef]
41. Kutralam-Muniasamy, G.; Marsch, R.; Pérez-Guevara, F. Investigation on the evolutionary relation of diverse polyhydroxyalkanoate gene clusters in Betaproteobacteria. *J. Mol. Evol.* **2018**, *86*, 470–483. [CrossRef]
42. Li, T.; Elhadi, D.; Chen, G.-Q. Co-production of microbial polyhydroxyalkanoates with other chemicals. *Metab. Eng.* **2017**, *43*, 29–36. [CrossRef]

2

Cyanobacterial Polyhydroxyalkanoates: A Sustainable Alternative in Circular Economy

Diana Gomes Gradíssimo [1,2,*], Luciana Pereira Xavier [2] and Agenor Valadares Santos [1,2,*]

1 Post Graduation Program in Biotechnology, Institute of Biological Sciences, Universidade Federal do Pará, Augusto Corrêa Street, Guamá, Belém, PA 66075-110, Brazil
2 Laboratory of Biotechnology of Enzymes and Biotransformations, Institute of Biological Sciences, Universidade Federal do Pará, Augusto Corrêa Street, Guamá, Belém, PA 66075-110, Brazil; lpxavier@ufpa.br
* Correspondence: dianagradissimo@gmail.com (D.G.G.); avsantos@ufpa.br (A.V.S.)

Abstract: Conventional petrochemical plastics have become a serious environmental problem. Its unbridled use, especially in non-durable goods, has generated an accumulation of waste that is difficult to measure, threatening aquatic and terrestrial ecosystems. The replacement of these plastics with cleaner alternatives, such as polyhydroxyalkanoates (PHA), can only be achieved by cost reductions in the production of microbial bioplastics, in order to compete with the very low costs of fossil fuel plastics. The biggest costs are carbon sources and nutrients, which can be appeased with the use of photosynthetic organisms, such as cyanobacteria, that have a minimum requirement for nutrients, and also using agro-industrial waste, such as the livestock industry, which in turn benefits from the by-products of PHA biotechnological production, for example pigments and nutrients. Circular economy can help solve the current problems in the search for a sustainable production of bioplastic: reducing production costs, reusing waste, mitigating CO_2, promoting bioremediation and making better use of cyanobacteria metabolites in different industries.

Keywords: biopolymer; biorefinery; cyanobacteria; circular economy; polyhydroxyalkanoate; waste

---

1. Introduction

An urgent demand for biotechnology is to find alternatives to conventional plastics, derived from hydrocarbons, which are harmful to the environment not only in its exploration and refining, but also in its disposal. In 2018, more than 359 million tons of plastic was produced worldwide [1]. Since traditional plastic is not biodegradable, it depends on human action for its degradation, however a very small portion of fossil plastic is actually recycled. About 35.4 million tons of plastic is discarded annually by the United States alone, and only an estimated 8.4% is sent for recycling [2]. In the last 50 years, we have primarily and almost exclusively depended on petrochemical plastics, due to its wide range of applications and cheap manufacture; for example, in 1995, a kilo of polypropylene cost less than US $1.00 to produce, which justifies the predilection for this type of polymer [3].

The consequences of its unbridled use are already visible and have been studied for a long time. Since the 1970s, researchers have been warning about the high prevalence of microplastics, the result of the wear of fossil fuel plastics with a size of less than 1 mm, in marine environments and its damage to this ecosystem [4]. Alternatives to conventional plastics are being studied and some are already on the market, these polymers can be classified as polynucleotides, polyamides, polysaccharides, polyoxoesters, polythioesters, polyphosphates, polyisoprenoides and polyphenols [5]. We focus here on polyhydroxyalkanoates (PHA), especially polyhydroxybutyrates (PHB), examples of polyesters, which have similar applications as polypropylene with physical-chemical characteristics comparable to this petrochemical plastic [6]. In addition to its favorable structural properties, this thermoplastic of natural origin is biodegradable, water resistant and liable to be manipulated by techniques that are already widespread in the industry, such as injection, being better absorbed by current industrial equipment [7].

An alternative has to be found, one that does not produce non-biodegradable waste such as petrochemical residues with its high molecular masses accumulating in the soil and water for a long period of time [8,9]. Despite the environmental advantages of PHA over conventional plastics, for its replacement to be a reality, it is necessary to reduce the costs associated with the microbiological production of these biopolymers. The main obstacle in the process is the carbon source used to maintain fermentation costs, the yield of the chosen entries, the productivity and the downstream processing, including purification [10,11].

The use of cyanobacteria as industrial PHA producers makes it possible to reduce the cost of nutritional inputs, since these photosynthetic organisms have fewer nutritional needs than heterotrophic bacteria [12,13]. The potential application of cyanobacteria by-products in industries with high added value [14,15] is interesting from an economic and environmental point of view, even more so if this system is implemented in light of the circular economy (Figure 1).

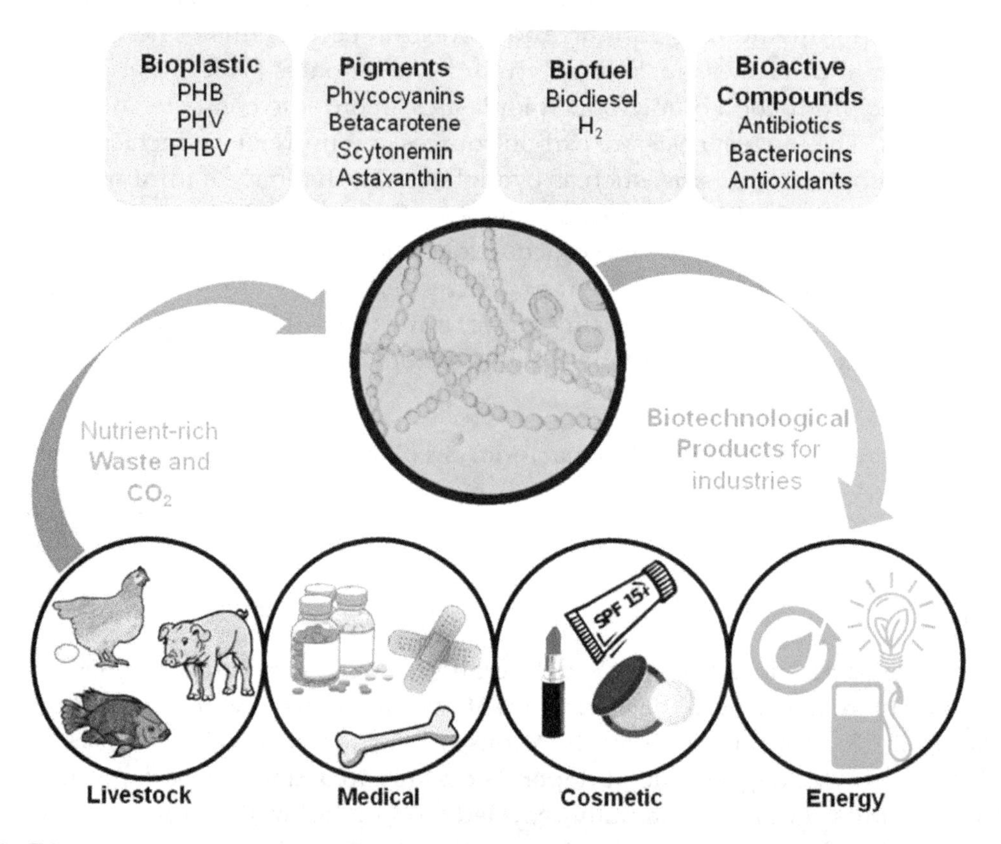

Figure 1. Diagrammatic representation showing cyanobacteria's role in a circular economy-based system for various industries, and its possible products and waste assimilation.

In this review, we seek to demonstrate the feasibility of applying the concept of circular economy in the production of PHA by cyanobacteria, a strategy that has been proposed for microalgae in general, including eukaryotic algae, and notably, for the production of biofuel [16–20]. We present an introduction to bioplastics, focusing on PHA, its biosynthesis, properties and applications. With respect to cyanobacteria, we bring PHB production by some species, and in the hopes of taking advantage of their potential to lower industrial PHB production costs, we also show the effectiveness of these organisms as bioremediators and waste nutrient removers. With the current work, we aim to add to the knowledge of this field, which is still somewhat deficient, in order to make the substitution, or at least reduction of petrochemical plastics, a more attainable goal.

2. Bioplastics

There is still controversy over the term "bioplastic" as there is still no standardized definition [21]. In this review, we embrace the two most broadly used definitions for this environment-friendly plastic: (1) bioplastics are polymers originated either entirely or in part from renewable natural sources, according to the Organization for Economic Co-operation and Development (OECD) [22], bioplastic can be viewed as synonymous with bio-based, and in addition to generating cleaner residues in its production, their decomposition is less harmful than that of petrochemical plastics, and its wear time is also considerably shorter [23–25]. The second definition (2) takes into account its biodegradability, as in German norm EN13432 [26], which refers not to the origin of the polymer, but to its ability to be degraded by organisms such as fungi, bacteria and algae [27,28]. The PHA and PHB addressed here are both bio-based and biodegradable.

Some of the most promising types of bioplastic are polysaccharides, such as starch and cellulose, and polyesters, including PHA [29]. When speaking of bioplastics used in packaging, worldwide, the two polysaccharides mentioned correspond to 30.7% of the global market for bio-based packaging—22.2% starch and 8.5% the representative portion of cellulose. As for polyesters—totaling 50.6%—the largest portion is of poly lactic acid plastic (PLA), 42.5%, followed by 6.7% of aliphatic and aromatic co-polyesters (AA) and 1.4% of PHA [30]. In terms of values, the global plastic packaging market was already estimated at 6.1 billion dollars in 2015, and the sector is expected to increase its value to more than 25 billion by 2022. The packaging segment is still the main use for plastics of natural origin, corresponding to the destination of 58% of all bioplastics produced in 2017. Next, we have the textile industry with 11% and the automobile and consumer goods industries with 7% each [31].

An important sector of application of bioplastics, especially PHA, is in medicine and the pharmaceutical industry, in the manufacture of prostheses, surgical material, as a scaffold in tissue engineering and used in drug carriers [23,32]. A characteristic that makes these biotechnological applications of PHA possible is the biocompatibility of these polymers, which can be implanted in the body without causing inflammation [33,34].

2.1. Polyhydroxyalkanoates

PHA are neutral lipids stored in the cells of cyanobacteria and other organisms as an energy reserve and carbon source. They are thermostable and elastomer bioplastics that have physical properties similar to plastics of fossil origin [23,35]. They are produced from the microbial fermentation of sugars, lipids, alkanes, alkenes and alkanoic acids and stored and accumulated in granules in the cytoplasm [36], with its occurrence extending to some archaea and bacteria, both Gram-negative and Gram-positive, and with no apparent prevalence in any specific phylum. This accumulation of polymers is not limited to taxonomic realms, nor to environmental niches, occurring in both terrestrial and aquatic organisms [37,38].

The main natural sources for obtaining PHA, especially different types of PHB, are heterotrophic bacteria, achieving good yields, with accumulation of PHB up to 85% dry cell weight (dcw) in *Cupriavidus necator* [39]. In addition to the good yields obtained, these species proved to be competent in assimilating alternative carbon sources for the production of PHA, using vinasse and molasses from sugar production and even waste frying oil [40,41]. Due to the relative ease in using genetic engineering with these bacteria, several studies deal with the production by recombinant organisms or even heterologous expression in *Escherichia coli*, with good results, such as the yield of 70–90% (dcw) [42–44]. Marine prokaryotes have a high production of PHA, with accumulation of up to 80% of its dry weight in bioplastic, and this good performance is observed in organisms that inhabit areas rich in nutrients [45], corroborating with what was proposed in the 1950s, regarding the role of excess carbon sources in bacterial biopolymer production [46].

The predilection for research and development of bioplastic production by heterotrophic bacteria is justified by the considerable production and accumulation of PHA in their cells, with some species such as *Cupriavidus metallidurans* and *E. coli*, wild or genetically modified, being used for the production of PHA on a larger scale [47,48]. Despite the good production of biopolymer by these strains, an operational problem hinders its use at an industrial level: the need for more elaborated carbon sources, which can compete with the food sector [49], and are expensive with the cost of nutritional inputs reaching up to 50% of production costs [50].

One way to mitigate these high costs in cultivation is using renewable and cheap carbon sources, such as domestic and industrial waste, which applies to heterotrophic bacteria [40,41,48] as well as photosynthetic organisms, that require fewer nutrients for their growth and production of biomass and biotechnological metabolites, such as bioplastic [51,52], thus, producing biopolymers basically from light and CO_2 [14]; in this context, we can understand the potential of the so-called blue algae, the cyanobacteria.

Perhaps the main advantage of these polymers, so versatile in their characteristics and applications, is the ability to be degraded naturally. Being of biological origin, these plastics are biodegradable, and they can be digested by PHA-depolymerases. The degradation of PHA biofilms at 25 °C in soil, sludge or seawater is in the range of 5 to 7 μm per week [23], thus being an attractive alternative to petrochemical plastics and an important research focus. The decomposition of PHA, either by the action of bacteria or natural elements, depends on some factors, such as the composition of the polymer, and the temperature and humidity of the environment, which also helps accelerate this process. In the case of microbial degradation, the decomposing microorganism also influences the degradation time, since different bacteria, for example, express different PHA-depolymerases, enzymes responsible for the degradation of the biopolymer [53]. The physical and chemical characteristics of PHA cause them to sink in aquatic environments, which also favor its conversion to carbon and water by decomposers [47].

PHA are gaining visibility as possibilities for replacing petrochemical plastics in an increasingly tangible way. Many stages of its production have not been improved yet, but much has been achieved in this area of research since the 1920s, when studies began on this biopolymer with the detection and extraction of poly(3-hydroxybutyrate) (P(3HB) or PHB) from *Bacillus* sp. [6,54,55]. The structural variations of a biopolymer, such as different monomers, will result in different physical-chemical characteristics and mechanical properties, which make different PHA more or less suitable for a given application.

2.1.1. PHA Structure

Polyhydroxyalkanoates are linear polyesters with a basic structure (Figure 2) formed by 3 to 6 hydroxy acids [56]. PHA polymers and copolymers can contain more than 150 monomers, reaching a molecular weight of up to 2 million Daltons [34]. The production of these polymers occurs in microorganisms through the use of substrates such as alcohols, sugars and alkanes, and the different chemical structures give polymers different physical properties, which may be more suitable for certain applications [23,56].

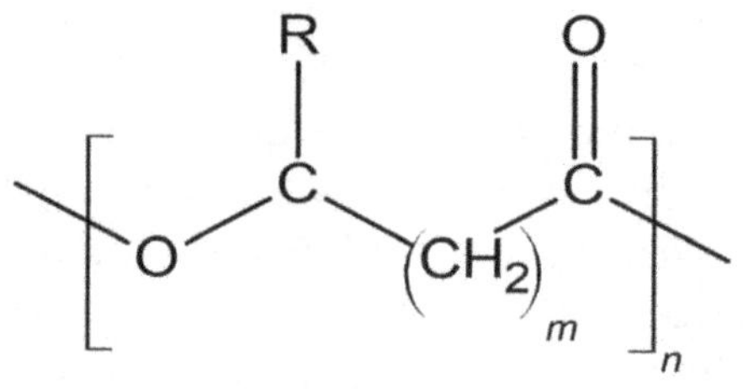

R O C C O (CH2) m n

Figure 2. Polyhydroxyalkanoates (PHA) general structure, where *m* ranges from 1 to 3, with 1 being most common, as in polyhydroxybutyrates (PHB), *n* is the degree of polymerization with values from 100 to 30,000, and the variable *R* is the alkyl group with different chain lengths and structures in PHB. *R* = methyl.

These polymers can be separated into three categories, according to the number of carbons, short chain length (SCL), with 3 to 5, medium chain length (MCL), formed by 6 to 14 carbons, and PHA with more than 14 carbons are called long chain length (LCL) [57]. Different organisms and bacteria genera produce different polyesters [47], and a determining factor in the production of PHA is the carbon source used by the producing organism, which can result in vastly different chemical structures. Some microbial products are said to be related to their carbon source, i.e., they are similar to the input used, and others are called unrelated, as they differ from the raw material consumed, presenting, for example, a different number of carbons [56]. PHA can also be presented as homo-polymers, such as PHB, or copolymers, like the poly(3-hydroxybutyrate-co-3-hydroxyvalerate) (PHVB), depending on the structural variation of its monomers [37].

2.1.2. PHA Biosynthetic Pathways

How microorganisms assimilate different forms of carbon into different polymers occurs through three biosynthetic pathways. The first, and most well-known pathway for PHA production, especially PHB, the one most found in cyanobacteria, is well described in archaea and heterotrophic bacteria as in the freshwater bacillus *Cupriavidus metallidurans*, and other species of the same genus. These granules are the result of a metabolic process that has two acetyl-CoA molecules as a precursor, derived from the tricarboxylic acid cycle (TCA) [13,58,59]. The reversible condensation of these two molecules is mediated by β-ketothiolase (encoded by gene *phaA*), the intermediate generated is reduced by the action of an NADPH-linked acetyl-CoA reductase (*phaB*), resulting in D(-)-3-hydroxybutyryl-CoA, which is then polymerized by the action of PHA polymerase (*phaC* and *phaE*), generating the poly(3-hydroxybutyrate) biopolymer (PHB) [17].

The second biosynthetic pathway uses lipid metabolism, and its medium-chain PHA product is based on the biotransformation of alkanes, alkenes and alkanoates, and the carbon source is directly related to the product's monomeric composition [60,61]. This production occurs through the β-oxidation pathway of fatty acids, in which different hydroxyalkanoate monomers are generated and then polymerized using PHA synthase enzymes [59–62]. An important step in this path is the conversion of trans-2-enoyl-CoA, a β-oxidation intermediate, to (R)-hydroxyacyl-CoA, an R-specific enoyl-CoA hydratase, encoded by *phaJ*. This enzyme acts in an (R)-specific manner and has already been reported in *Aeromonas caviae* and *Pseudomonas putida* [60,63].

The last biosynthesis pathway produces alkanoate monomers that also result in medium chain length PHA (MCL-PHA); however, the precursors of this pathway are simple carbon sources such as sucrose, glucose and fructose, which makes it a potentially less expensive industrial process [6,23]. In this pathway, we have the presence of both sugars and lipids, starting from a glycolic precursor and making use of fatty acid biosynthesis intermediates. The Entner–Doudoroff pathway is used by *Pseudomonas* sp., with the catalysis of the sugar source from glucose to pyruvic acid [64]. This feat is made possible by the action of PHA synthase which catalyzes the biosynthesis of PHA from fatty acids as well as from sugars [65]. As in the second pathway, a key enzyme in this reaction is an (R)-specific, acyl-ACP-CoA transacylase, encoded by the *phaG* gene [66].

In addition to the carbon source, other nutrients such as nitrogen, phosphate, sulfur, oxygen, or deprivation of these, can affect growth and also play an important role in PHA biosynthesis [67,68]. Changes in the C:N ratio have been used in culture optimization, including evaluation of waste as a nutritional source [46,69,70], with abundance of carbon being favorable to biomass production [71] and limiting phosphorus and/or nitrogen as most beneficial stress for PHA production [72,73]. Such strategies have shown good results: in the extremophilic archea *Haloferax mediterranei*, an accumulation of 47.22% (dcw) of PHB was found using a 1:35 ratio of C:N. This higher value found for PHB production differs from the best C:N condition for the production of extracellular polymeric substances (EPSs) in *H. mediterranei*, which was 1:5, this result is interesting because it allows to direct production to the biopolymer of interest according to the nutritional medium [74]. An increase in the ratio between the two nutrients resulted in an even greater accumulation, 1:125 of C:N, providing an accumulation of up to 59% (dcw) of PHB using activated sludge from food industry waste [68].

Nitrogen is a micronutrient present in proteins and nucleic acids and its limitation can affect these metabolisms [36,75], also influencing the concentration of NAD(P)H and the ratio NAD(P)H/NAD(P) within the cell [58,59]. Whereas in a balanced culture (Figure 3), without nutritional stress, the concentration of these co-factors remains constant since the flow in TCA is maintained, in a situation of nitrogen deprivation, the synthesis of amino acids is reduced, and the decrease in the conversion of a-ketoglutarate into glutamate, which assimilates ammonium ions into the cell causing an accumulation of NAD(P)H [76,77], can then be used to reduce acetoacetyl-CoA to R-3-hydroxybutyryl-CoA. Another interesting supplement for the accumulation of NADPH is citrate, since it reduces citrate synthase activity in the Krebs cycle, supplementation in *A. fertilissima* increased PHB production compared to control [78], and citrate synthase can also be inhibited by high concentrations of NAD and NADPH [58].

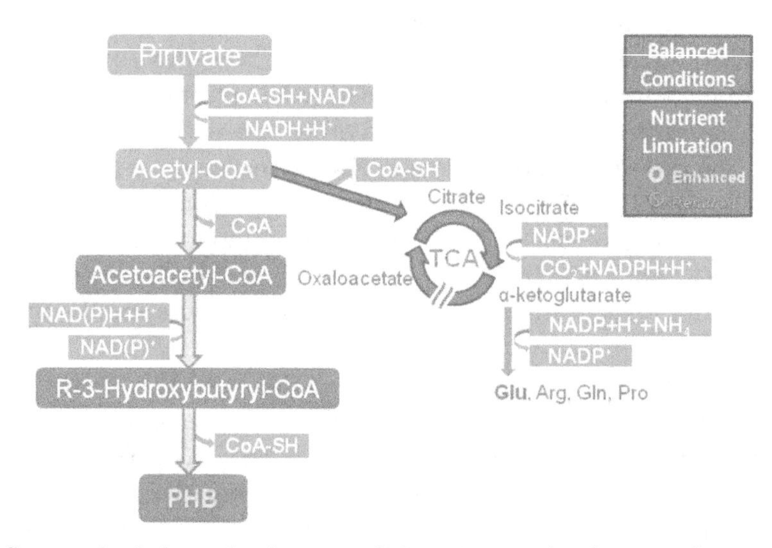

Figure 3. Carbon flow under balanced culture conditions, in purple, showing flux towards tricarboxylic acid cycle and under nutritional stress due to nitrogen and/or phosphorus limitation, enhanced flux or accumulation in blue and reduced activity in orange, with the carbon flux being directed to PHB biosynthesis.

Phosphorus, in its inorganic form as adenosine triphosphate, acts in protein and nucleic acid syntheses in addition to being important in the assimilation of lipids and carbohydrates and other roles in cell maintenance [36,39]. Despite the need for a minimum concentration of phosphorus in the cell to guarantee vital functions, the limitation of this nutrient proved to be a good strategy for inducing the production of PHA, and it is sometimes more significant as a limiting factor than nitrogen in cyanobacteria [77]. Nutritional stress due to phosphorus limitation also results in changes in TCA, restricting the Krebs cycle [73]. In balanced nutritional conditions, there is a greater amount of coenzyme-A (CoA-SH), which inhibits the synthesis of PHA. The withdrawal of nutrients like phosphorus and nitrogen from the metabolism promotes the accumulation of NADH, since citrate synthase and isocitrate dehydrogenase are inhibited with a consequent increase in acetyl-coA [36,59,79], the initial molecule for PHA biosynthesis [13,76].

2.1.3. Physical and Mechanical Properties

The chemical structure of PHA, such as number of monomers, directly influences the physical characteristics of the polymer as well as its mechanical properties. While shorter PHA have high crystallinity and are quite brittle, polymers with a greater number of monomers can be more elastic and flexible [39,80]. The composition of the biopolymer also determines its degradation: in medical applications, for example, this determines whether its rate of degradation is compatible with a patient's tissue regeneration [81,82].

In order for biopolymers to replace conventional plastics, they must play the same role. For this, they must have similar physical and mechanical properties, in order to be inserted in different industries without major losses and ideally being assimilated by the same machinery already installed, producing it through techniques such as injection molding and extrusion [83]. PHB has a range of Tm (melting temperature), which refers to the average melting temperature, similar to 190 °C of polypropylene and 176 °C of LDPE (Low-Density Polyethylene), that is, it supports similar temperatures. It is also comparable to polypropylene in its limit of tensile strength, which is the force necessary to stretch a material until it breaks [37,83]. However, the low value of Tg (glass transition temperature) of PHB makes for poor flexibility.

Another parameter in which PHB does not match petrochemicals is in fracture elongation, the mechanical property a material has to be stretched, in terms of percentage over the original size of the material. While the two conventional plastics mentioned above reach up to 620% of its size before rupture, PHB supports only up to 6%. This shows the importance of tailor-made PHA in order to adjust the properties of a polymer to the desired parameter, the most suitable for its purpose [84,85]. P(4HB), for example, which has two monomeric units instead of just one like PHB, manages to reach up to 1000% of its initial size before breaking [6], with the added advantage of a greater UV rays' resistance [86].

Mechanical properties can vary widely even among PHA, short chain length PHA (SCL-PHA), for example, are quite brittle thanks to their high crystallinity, and the addition of 3-hydroxyvalerate units enhances copolymer flexibility [87,88]. To further illustrate the relation between structure and proprieties, the addition of 20 mol% of PHV in the polymer P(3HB-co-20 mol% 3HV) make it more malleable but also more heat-sensible than pure PHB, enduring temperatures up to 150 °C, and no higher than 180 °C. Increasing the PHV content to 71 mol% in P(3HB-co-71 mol% 3HV) reduces its Tm even more, to only 83 °C [85]. Regarding the physical and mechanical properties of poly(3-hydroxybutyrate-co-3-hydroxyhexanoate) (P(3HB-co-3HHx)), which is already produced on an industrial scale in 20,000 L reactors by heterotrophic bacteria, namely *Aeromonas hydrophila* [89], its rapid degradation by PHA depolymerases can be attributed to its lower crystallinity [90]; hence, this copolymer can be well-used in disposable products. Its thermo-tolerance and overall mechanical characteristics, such as 150% enhancement in stiffness, were improved with the addition of alfalfa and hemp fibers forming a composite [91]. This approach is another way of benefiting natural polymers. For example, composite production using natural fibers with the incorporation of 30% *w/w* of pineapple leaf fibers with PHBV increased the polymer tensile strength by 100% [6,85]. Other mixtures show good results, like the greater impact resistance and heat tolerance achieved by blending PHB with polycaprolactone (PCL) [92].

The tailor-made modeling of PHA choosing each of its monomers or in blends with other polymers, including synthetic ones, is a way to improve the physical properties of a bioplastic and it can also reduce production costs [93,94]. This can be achieved with simple changes in cultivation, using block polymers as an example in the heterotrophic bacteria model *C. necator*, P(3HB) was obtained using fructose as a substrate, and it was possible to add 3-hydroxyvalerate units by applying supplementation with pentanoic acid pulses in a bioreactor, thus changing its physical properties [95,96]. Recombinant organisms can also be used for the production of block copolymer, with the copolymer consisting of poly-3-hydroxybutyrate (PHB) as one block, and random copolymer of 3-hydroxyvalerate (3HV) and 3-hydroxyheptanoate (3HHp) as another block, obtained with a *Pseudomonas putida* KTOY06ΔC (*phaPCJAc*), where the PHA synthesis gene *phaPCJAc* was cloned from *Aeromonas caviae* [97], resulting in improved mechanical properties.

2.1.4. Applications

The structural diversity of PHA makes it an attractive alternative in numerous industrial sectors, and different applications can benefit from different aspects of PHA, whether it be its biocompatibility, biodegradability or its natural origin, which make it cleaner compared to hydrocarbon plastics. The most obvious, and most explored, use of PHA is in the production of packaging and non-durable goods, taking advantage of its biodegradability. The first consumer good using PHA was launched to the public in the 1990s, with a line of shampoos and conditioners with a PHA packaging, made with BIOPOL, a copolymer of PHB and PHV, with good water resistance [45]. Since then, other companies have shown interest in the study and production of PHA, for example, the German company Biomer, which through research with *A. latus*, managed to produce and accumulate up to 90% (dcw) in PHB using commercial sucrose as feedstock [37].

North American Telles, formed by Metabolix Inc., among its PHB copolymer portfolio, has a food additive approved by the American Food and Drug Agency (FDA), the elastomer poly(3-hydroxyoctanate). The bacterium used for industrial production of this PHA is a genetically engineered *E. coli* K12, which can accumulate up to 90% (dcw) in just 24 h [98]. This high productivity in heterotrophic bacteria cheapens the industrial process, making this bioplastic accessible for large-scale production of disposables, such as combs, pens and other consumer goods.

PHB's piezoelectric characteristic also makes it suitable for manufacturing electronics, such as computer equipment, microphones, detectors and sensors. The ability of some copolymers, such as PHBV, to barrier gases is useful in the application of food packaging, delaying the action of microorganisms in industrialized foods and drinks [6]. This is interesting for usage of PHB as a biomaterial in medical applications, as its piezoelectric characteristic promotes osteogenesis, assisting in bone regeneration [23,99,100]. The application of PHA in the medical field has been gaining attention, together with fine chemistry, because it deals with high value-added products and processes, and manages to better afford the high costs of PHA production at an industrial level. In addition the biocompatibility of these biopolymers makes them an interesting target for research and application in regenerative medicine, as well as in the sustained release of drugs and hormones [86,101], the PHAs most used for this purpose are: P(3HB), P(3HB-3HV), P(4HB), P(3HO) and P(3HB-3HHx), all of which have already been tested in vivo, showing biocompatibility [102].

Due to structural specificities, some polymers are more suitable for certain uses. As biomaterial, these PHAs have been reported to be useful in tissue engineering, bone orthoses and surgical sutures. PHB with its aforementioned piezoelectric characteristic is interesting for use in bone plates because it has a high biodegradation rate, in addition to being resistant to hydrolysis in sterile tissues [103]. Its high biodegradability also makes it ideal for the manufacture of surgical devices and medical material in general, that have short-use life. In tissue regeneration, microfilaments formed by copolymer P (3HB-co-3HV) maintained their masses and other characteristics for up to 12 months after being implanted in rats, being suitable for use as scaffolds [91]. Studies with PHB biofilms have shown a loss of up to 80% of the initial mass after one year of implantation [104]. The different results obtained regarding PHB degradation can be attributed to a series of chemical and physical factors. The enzymes present in the patient's blood and tissues, for example, must be taken into account in the design of a biomaterial, as these can impair the rate of degradation [35,105].

The cardiovascular sector has benefited quite a lot from the application of PHA as a biomaterial: heart valves, pericardial adhesives, artery augers and implants in general are already available on the market, all from PHA, varying in their structures to better suit the final product [106]. In addition to contributions to the cardiovascular field, we mention P(4HB) for its importance as an exponent in medical biomaterials. Currently, this P(4HB) is the best option for use as a biomaterial compared to other thermoplastics available, being a strong biopolymer but with greater flexibility than synthetic absorbable polymers, such as polyglycolic acid (PGA) and polylactic acid (PLA), with a rate of bioabsorption of up to two years [107]. The sustained release of drugs is another possible application for PHA. In these systems, the drug or bioactive molecule is released gradually, without the need for new doses by the patient [86]. This approach is especially interesting for antibiotics, which maintain their therapeutic window, with a constant dosage of the drug, thus avoiding low dosage, where resistant strains of the pathogen could be selected.

Not only do PHA polymers have industrial applications, but also, the enzymes that produce them and their constituent monomers can be used in fine chemistry. PHA synthases are a good alternative for obtaining chiral compounds due to their enantioselectivity, they have specific action that can be used in the development of more specific chiral drugs, thus causing less adverse effects in patients, with the advantage of cleaner biosynthesis [108]. 3-hydroxybutyrate monomers can be used as precursors of new biopolymers, such as chiral polyesters—some with antibacterial, antiproliferative and hemolytic action—biocompatible dendrimers for carrying drugs and optically pure monomers, such as the ones that have already been used in the synthesis of sex hormones and fragrances [57,94,109].

3. Cyanobacteria

Known as blue algae, or cyanophytic algae, cyanobacteria are prokaryotic, Gram-negative, photosynthetic organisms of great biological importance, with fossils of cyanobacteria over 2.8 billion years [110–112]. Responsible for the oxygenation of the Earth's atmosphere, these bacteria still play an important role in the carbon and nitrogen cycle on Earth [113,114], and they also originated chloroplasts, which were transferred horizontally to other strains and are now present in eukaryotes, namely higher plants [115]. The metabolism of cyanobacteria is noteworthy because they are the only microorganisms with photosynthesis similar to that of plants. They possess chlorophyll A, in addition to the pigment phycocyanin (a photosynthetic phycobilin), which gives the phylum its name, and its photosynthetic metabolism is also divided into two complementary photosystems, as it occurs in plants [116,117].

Cyanobacteria are organisms with high adaptive capacity which survive in environments with extreme temperatures, salinity, pH and levels of solar radiation [116,118,119]. Since photosynthesis is the main source of nutrition, these organisms can survive in almost any environment that has light, from dimly lit caves to open spaces with a high incidence of solar radiation [112]. They are also present in deserts, assisting in their fertility through nitrogen fixation [120]. The fixation of nitrogen in the environment by cyanobacteria is possible due to nitrogenase, an enzyme present in some species which makes them capable of assimilating unstable nitrogen gas (N_2) as a direct source of nitrogen, in the form of more stable compounds such as nitrates, nitrites and ammonium salts. This makes them more independent in an environment with limited nitrogen [121]. A crucial factor for the functioning of the nitrogenase enzyme is its sensitivity to oxygen, which is a problem for cyanobacteria that produce oxygen via photosynthesis. As a solution, some cyanobacteria, such as those of the order Nostocales, developed specialized cells, the heterocysts, responsible for fixing nitrogen, spatially separating the two functions, while other orders simply operate photosynthesis and nitrogen fixation in different time periods [122].

Associating the photosynthetic process with nutrient metabolization, cyanobacteria are able to balance the electrons in their photosynthesis with their metabolism and thus neutralize the reduction caused by reactive oxygen species as a way of protecting against oxidative stress. Cyanobacteria UV-absorbing pigments also play a protective role [123,124]. The adaptive capacity of this phylum is explained in part by its production of secondary metabolites, which makes them competitive for survival in different environments. Such metabolites place cyanobacteria as highly promising microorganisms for biotechnological applications of commercial interest [125,126]. Compounds produced by cyanobacteria, which include fatty acids, amides, polyketides and lipopeptides, show several biological actions, including inhibition of glycosidases and of protein C kinases, tumor promoters, inhibitors of microtubulin aggregation [127] and immunosuppressive agents [128–130]. Some also have antibacterial, antialgal, antiplasmodial, antifungal and antiviral action, including anti-HIV, inhibiting human immunodeficiency virus synthesis [131].

In addition to application in segments with high added-value, such as the pharmaceutical market [132], the possibility of producing cyanobacterial biomass at a reduced cost allows its application in mass industries, which, as a rule, demand cheap production and on a large scale, such as the food and agriculture industry, with the use of cyanobacteria biomass and pigments as food and animal feed [133,134].

3.1. PHA Production in Cyanobacteria

The production of biopolymers is an example of the adaptation of cyanobacteria to environmental stimuli, especially nutritional deprivation, storing carbon in the form of glycogen and types of PHA produced by this phylum: PHB, PHV and their copolymers [12,135]. As PHB is the most prevalent polymer in cyanobacteria, we will focus on this short-chain bioplastic.

The accumulation of glycogen is a characteristic that is highly conserved, being found in a greater number of genera than the production of PHB or PHV [136]. Glycogen productivity is also higher than that of PHB, between 30% and 60% (dcw) [137], and glycogen is also produced quicker than PHB when the cell is deprived of nitrogen, being stored in several small granules in cells affected by stress [138]. Although the production of glycogen and PHB is activated by the same nutritional stress, the role of each polymer is different. In order to understand the role of each carbon storage, mutants of *Synechocystis* sp. PCC6803 with PHB and glycogen deficiency were compared [139]. In knockout mutants unable to produce glycogen, PHB accumulation increased between 8% and 13%, but most of the excess carbon was expelled from the cell in the form of pyruvate and α-ketoglutaric acid. These mutants were unable to turn on a dormant mode, that would save energy in the face of nutritional deprivation, and they also did not recover from nitrogen scarcity. PHB-deficient mutants, on the other hand, maintained the same rate of glycogen production as the wild-type, also maintaining recovery capacity once the nutrients were replenished. Only mutants that suffered knockout of genes related to both polymers showed deficiency in growth.

This competition among the production of both polymers in cyanobacteria can be attributed to 3-phosphoglycerate (3PG) usage by PHB biosynthetic pathways as well as glycogen, with this last one making greater use of the 3PG pool obtained by cyanobacterial CO_2 assimilation through Rubisco [7], hence for a robust PHB production, it is important to mitigate glycogen production [13]. Some authors compare PHB to triacylglycerol in green algae, since it also acts as an electron sink and consumes excess NADPH [138–140], while the storage of starch as an energy reserve in higher plants can be compared to the glycogen stock by cyanobacteria [139]. Although certainly important, the role of PHB in cyanobacteria has not been fully elucidated.

The first time PHB was reported in cyanobacteria was in 1966, in the single-celled *Chlorogloea fritschii*, which accumulated up to 10% (dcw) in PHB [141]. PHB and PHV have now been identified in different genera of cyanobacteria (Table 1).

Table 1. Examples of PHA-producing cyanobacteria, namely PHB and PHBV, with respective production in % (dcw) and nutritional conditions.

Cyanobacteria	Mode	Nutritional Deprivation	Nutritional Supplementation	PHA	Production % (dcw)	Reference
Synechocystis sp. PCC6803	Mixotrophic	P	Acetate	PHB	28.8	[77]
Synechocystis sp. PCC6803	Mixotrophic	N	Acetate	PHB	14.6	[77]
Synechocystis sp. PCC6803 (mutant)	Mixotrophic	-	Acetate	PHB	35	[142]
Synechocystis sp. PCC6803	Photoautotrophic	N, P	-	PHB	16.4	[143]
Synechococcus sp. MA19	Photoautotrophic	P	-	PHB	55	[144]
Nostoc muscorum Agardh	Mixotrophic	N	Glucose, acetate, valerate	PHBV	78	[145]
Chlorogloea fritschii	Mixotrophic	-	Acetate	PHB	10	[141]
Spirulina subsalsa	Photoautotrophic	N	-	PHB	14.7	[32]
Aulosira fertilissima	Mixotrophic	N, P	Acetate, citrate	PHB	85	[146]

P = phosphorus; N = nitrogen.

The biosynthetic pathway for PHB production in cyanobacteria is shared with archaea and heterotrophic bacteria, in which three enzymatic reactions occur mediated by enzymes encoded by genes *phaA*, *phaB*, *phaC* and *phaE*. In cyanobacteria, PHA synthase coding is mediated by the last two genes, similar to that of anoxygenic purple sulfur bacteria [147–149]. The role of Rubisco in CO_2

assimilation through the Calvin-Benson-Bassham cycle (CBB) [150], besides giving rise to the 3PG pool for glycogen and PHB production, is also a possible route for photosynthetic PHB production, with the 2-phosphoglycolate (2PG) resulting from CBB producing glycolate that in turn can be used as a starting blinding block for PHB synthesis. This was observed in a recombinant *Escherichia coli* JW2946 [151]. As for the production of PHV in cyanobacteria, it uses propionic acid as a starting point, and can occur in conjunction with PHB biosynthesis, resulting in the PHBV copolymer [12].

The production yield of these biopolymers in the literature in genus *Synechocystis* sp. ranges from 4% to 16% (dcw) under photo-autotrophy conditions [7,77,142,143]. Under heterotrophic conditions, higher yields are obtained, consistent with that obtained by other heterotrophic bacteria, although at lower levels, there is an accumulation of 28.8% (dcw) of PHB in the *Synechocystis* PCC6803 strain when cultivated with acetate supplementation and phosphorus deprivation [77]. Comparing wild and genetically modified organisms, production by *Synechocystis* sp. PCC6803 recombinant, grown only with acetate supplementation, showed a higher yield than that obtained previously, with an accumulation of 35% (dcw) of PHB [142]. The greatest yield in unicellular cyanobacteria was obtained with strain *Synechococcus* sp. MA19, reaching 55% (dcw), switching the phosphorus source for $Ca_3(PO_4)_2$ [144].

An even greater production is observed in the filamentous cyanobacteria *Nostoc muscorum* Agardh, with an accumulation of 78% (dcw) in heterotrophy with nitrogen limitation, and supplementation of acetate, valerate and glucose [108]. Another N_2-fixing cyanobacteria, *A. fertilissima*, under conditions of mixotrophic cultivation, with phosphorus and nitrogen deprivation and addition of citrate and acetate, showed an accumulation of 85% (dcw)—the highest obtained in cyanobacteria thus far [146].

3.2. *Waste Utilization and Bioremediation*

The use of agricultural residues as a carbon source or for supplementation of other nutrients has already been studied in wild and engineered organisms, such as heterotrophic bacteria and eukaryotic microalgae, in addition to cyanobacteria [48,152], being well assimilated by them in the production of metabolites for various industries, reducing input costs as well as agro-industrial waste. It is a logical path to use wastewater for the cultivation of cyanobacteria given their nutritional preferences, as the blooms of this phylum mostly occur in environments with an abundance of phosphorus and nitrogen, which are also widely found in aquaculture waste [153,154]. This strategy also makes it possible to save a key resource: water, which must have limited use in order to actually achieve a sustainable process [155].

Cyanobacteria and microalgae, in general, are ideal for removing nutrients and capturing CO_2 from the most varied residues, reaching nutrient removal rates between 50% and 100% [15,36]. The main sources of nitrogen in sewage are metabolic interconversions of extra derived compounds, while about half of the phosphorus comes from detergents [154]. The ability of cyanobacteria to assimilate different forms of nitrogen and phosphorus, such as NH_4+, NO_2^-, NO_3- and PO_4^{3-} [74], is well applied in wastewater treatment [156]. Sewage treatment is quite costly, in addition to being responsible for emissions to the environment [157,158], so bioremediation of these residues is an interesting alternative. Secondary sewage removal focuses on organic matter, and in order to reduce biological oxygen demand (BOD), this step benefits from heterotrophic bacteria that use said matter for growth and energy [154], but photosynthetic organisms also show potential for this role as they produce oxygen for other bacteria [159] and are also efficient in reducing oxygen demands in waste, as seen in the PHB-producing species *A. fertilissima* [78].

Secondary treated wastewater, although poorer in general as organic matter was mostly removed, is still loaded with micronutrients that can be further assimilated by cyanobacteria in a tertiary treatment [156,160]. *Phormidum* sp. is well applied in this sense because in addition to good nutrient removal rates, such as 48% of ammonium ions, 68% of total phosphate, 87% of nitrate and 100% removal of orthophosphate from swine wastewater [161], it has the capacity for self-aggregation, and this flocculation facilitates its later removal after bioremediation [162,163].

Pre-treated wastewater sewage was denitrified by a consortium of *Chlorella vulgaris*, *Botryococcus braunii* and *Spirulina platensis*, in a submerged membrane system in a photobioreactor, achieving good CO_2 capture results and 92% removal of the initial 7.5 mg of total nitrogen N/L [164]. Wastes such as secondary effluents and digestate treatment with higher nitrogen and phosphorus content can, and should be used, since the higher concentration of these micronutrients results in dominance of cyanobacteria in the medium [163,165]. This is advantageous for the proposed application of bioremediation by cyanobacteria allied with industrial production of metabolites, providing an environment more suitable for cyanobacteria than other microalgae or heterotrophic bacteria [11,166]. Water reused directly from fish farming tanks was the nutritional source for the production of PHB by *Synechocystis* sp. 6803, reaching an accumulation of up to 20% (dcw) [77]. The aquaculture waste was characterized according to pH, available phosphorus and assimilable forms of nitrogen, separating the residues into two groups, both containing a lower concentration of phosphorus and nitrogen than conventional BG-11 medium.

Another work, besides using water from the aquaculture tank directly, minimizes processing steps, and proposes the diazotrophic cyanobacteria *A. fertilissima*, not only as a producer of PHB, but as a bioremediator of the water for aquaculture in a recirculatory system [78]. The greatest accumulation of PHB obtained was around 80% (dcw), achieved in the summer period [78]; in addition, this good result was accompanied by better bioremediation efficiency with all parameters within the accepted range for fish farming [167], with the lowest nitrate concentrations, total organic carbon and biological oxygen demand (BOD), as well as chemical oxygen demand (COD), occurring in this season. Ammonia, nitrite and orthophosphate were not even detected in this condition, being initially present in concentrations of about 2.3, 3.5 and 3 mg/liter respectively, illustrating the bioremediation capacity of *A. fertilissima*. The increase in dissolved oxygen promoted by this treatment makes fish less susceptible to ammonia toxicity [78], an additional benefit of this symbiosis.

More recently in *Nostoc muscorum* Agardh, the production of PHB was verified after supplementation with poultry waste [168]. Using 10 g/L of this agro-industrial waste, an increase of about 11% in PHB production was observed in relation to the control culture, with a total accumulation of 65% (dcw). This culture also received supplementation of 10% of CO_2 in order to verify the capacity of this strain as a carbon dioxide mitigator, and therefore, adding yet another dimension in the search for sustainable production, reducing CO_2 by means of solar energy using the electrons of photosynthesis [169,170]. CO_2 fixation promoted a greater biomass production in *N. muscorum* and the same was verified in *S. platensis* [171], a species largely consumed as nutraceutical, with the goal of removing nutrients from pig farm waste, as a way to prevent eutrophication, reaching NH_4-N removal of up to 95% and phosphorus removal rates up to 87%, respectively. Sludge from this kind of waste is rich in nutrients, with about 6 kg of nitrogen per ton of manure and 3 kg/ton of phosphorus [171,172]. Experimental culture media composed of this sludge is about 4.4 times cheaper than the usual Zarrouk culture medium [173].

4. Cyanobacteria Potential Application in Circular Economy

Despite the advantages PHA has over conventional plastics in terms of sustainability, for fossil plastic to be viably replaced, it is necessary to reduce the costs associated with the microbiological plastic production. Research and investments in the area have been making production cheaper. In 2002, the cost for manufacturing conventional petroleum-based plastic was €1.00/kg, a fraction of the PHA cost of €9.00/kg [24]. Two decades later, microbiological production of PHA can be obtained at €2.49/kg, which is still expensive, even compared to other sustainable polymers, such as PLA, costing €1.72/kg [9].

The main obstacles in the process concerns the carbon source used [50], the costs of maintaining the fermentation, the yield of the chosen inputs, the productivity and the downstream processing, including the extraction and purification of the polymer [10,11]. There are different strategies to face these obstacles; here, we will address only a few that are related to circular economy and industrial ecosystems, an approach that has already been applied with microalgae and heterotrophic bacteria [16–20]. The use of cyanobacteria is interesting because of the possibility of integrated production of different metabolites—with more than one type of compound as a salable product—and application of a "cradle-to-cradle" system [113], using by-products or production residues as a substrate for another product. Like the use of carbon monoxide (CO) in synthesis gas (syngas) for the production of PHB by the proteobacterium *Rhodospirillum rubrumde* [174], this author evens refers to this process as "grave-to-cradle", turning a waste into a new product, bioplastic. Another example of waste being reapplied to the production process, now using microalgae, is the reuse of effluents from the refining of olive oil in the cultivation of microalgae for biodiesel and biopolymers [175]. This approach can benefit from the implementation and maintenance of an "inter-system ecology", associating different industries [15,176].

From an environmental point of view, cyanobacteria are well-used as bioremediators, feeding on nutrients from domestic and agro-industrial waste, promoting nutrient removal and detoxification, removing heavy metals [78,164,177]. The assimilation of atmospheric carbon dioxide for conversion into biotechnological products [140,169] is another positive environmental impact, making the implementation of a circular bioeconomy more tangible.

An alternative to make microbial PHB cheaper is to integrate the production of bioplastic with other desirable products, reusing by-products and residues of the microbiological production [14,178,179]. The production of acids for the cosmetic and pharmaceutical industry, such as eicosapentaenoic acid, by cyanobacteria of the genus *Nannochloropsis* sp., and γ-linoleic acid by cyanobacteria *Spirulina platensis,* is a viable alternative [180]. This species is also relevant for its expressive biomass production, with high protein content, suitable for application in nutraceuticals or animal feed [32,133]. The implementation of a biorefinery, integrating the PHB production of *Synechocystis salina*, with pigments of commercial interest, specifically phycocyanin and chlorophyll, commonly abundant in this phylum, and carotenoids, presented promising results [181]. The extraction of pigments without their degradation is not only possible, but essential, as the quality of the obtained polymer is directly affected by purification, which includes the removal of pigments, that can be used in production chains of higher value.

In addition to pigments, *S. salina* biomass has carbohydrates, lipids and proteins [181], which can be used for animal feed [134], provided that the necessary nutritional requirements and laws regarding the presence of contaminants such as heavy metals or mycotoxins are observed [182], and in this case, cyanotoxins [183], giving priority to non-toxin-producing cyanobacteria. The residual biomass of cyanobacteria would therefore be well-used in the nutrition of livestock and aquaculture, but it is possible to go further in the optimization of this production chain. Residues from these same livestock farming can be re-applied as supplementary nutrients to the growth of cyanobacteria in an integrated bio-factory [184]. The return of cyanobacterial by-products such as pigments and biomass to animal feed completes the proposed circular economy. Still, using *Spirulina* sp. as an example, its supplementation to animal feed has already been studied in shrimp, fish and chicken farming [134,185], valuing the production of this associated industry, improving the growth and coloring of tilapia [186] and egg yolks of chickens fed with *S. platensis* astaxanthin [187]. Animal health is also benefited by nutritional supplementation with cyanobacteria, with *Spirulina* sp. biomass improving the humoral and immunological response of chickens [129,130].

The dual advantage of production associated with bioremediation has already been described for cyanobacteria and microalgae in general, mainly aimed at the production of biodiesel [17,188–190]. The same concept can be applied to the production of biopolymers by cyanobacteria [15], naturally transformable organisms, which opens up possibilities for genetic engineering [143,191].

As a way to take advantage of Amazonian biodiversity in the search for microbial metabolites, in recent years, research with cyanobacteria from the Amazon has been developed with good results, and the sequencing of their genomes is an important tool in the search for compounds of biotechnological interest [192,193]. These organisms proved to be good producers of biodiesel, with yields higher than those in the literature and with parameters following international standards [194,195]. Biopolymer has also been detected in cyanobacteria in the region, with efforts being made to increase its production [196]. Subsequent work in this field would benefit from the approach proposed here of a circular economy, optimizing the resources employed, handling the waste and using its by-products and industrial "waste", which is, as seen here, a potential feedstock for new biotechnological processes.

5. Conclusions

PHB-producing cyanobacteria, due to its lower productivity, in comparison to heterotrophic bacteria, are commercially viable only if combined with exploration of their various metabolites. Feed costs must be reduced, and using non-conventional feedstock, preferably agro-industrial waste, this opens up the possibility of industrial implementation within a circular economy, with the production of more than one product from this bio-factory, or with the use of by-products in other sectors, returning, for example, as animal nutrition to the industry itself, whose residues supplement cyanobacterial growth. Another great advantage of cyanobacteria in relation to other microorganisms producing PHB is its photosynthetic capacity, acting as a solar cell, reducing the CO_2 of the environment, which is an assimilated carbon source and can be used as building blocks for the cyanobacterial production of bioplastic.

The critical reading of the studies already carried out in the area shows promising results for the manufacture and large-scale use of cyanobacterial biopolymers to be a reality in the near future, and more than that, demonstrates the need for cooperation between different knowledge and industries and through a circular economy to optimize the production process and reduce environmental impacts.

Author Contributions: Conceptualization, A.V.S. and D.G.G.; Investigation, D.G.G.; Writing—original draft preparation, D.G.G.; Writing—review and editing, A.V.S. and L.P.X.; Supervision, A.V.S. and L.P.X.; Project administration, A.V.S.; Funding acquisition, A.V.S. All authors have read and agreed to the published version of the manuscript.

Acknowledgments: The authors would like to thank Pró-Reitoria de Pesquisa e Pós-Graduação da Universidade Federal do Pará (PROPESP/UFPA).

References

1. Plastics—The Facts 2019. Available online: https://issuu.com/plasticseuropeebook/docs/final_web_version_plastics_the_facts2019_14102019 (accessed on 24 June 2020).
2. US EPA Plastics: Material-Specific Data. Available online: https://www.epa.gov/facts-and-figures-about-materials-waste-and-recycling/plastics-material-specific-data (accessed on 24 June 2020).
3. Koller, M.; Hesse, P.; Kutschera, C.; Bona, R.; Nascimento, J.; Ortega, S.; Agnelli, J.A.; Braunegg, G. Sustainable Embedding of the Bioplastic Poly-(3-Hydroxybutyrate) into the Sugarcane Industry: Principles of a Future-Oriented Technology in Brazil. In *Polymers-Opportunities and Risks II*; Eyerer, P., Weller, M., Hübner, C., Eds.; Springer: Berlin, Germany, 2009; Volume 12, pp. 81–96, ISBN 9783642027963.
4. Carpenter, E.J.; Smith, K.L. Plastics on the Sargasso Sea Surface. *Science* **1972**, *175*, 1240–1241. [CrossRef] [PubMed]
5. Dhaman, Y.; Ugwu, C.U. Poly[(R)-3-hydroxybutyrate]: The Green Biodegradable Bioplastics of the Future! *Ferment. Technol* **2012**, 01. [CrossRef]
6. Philip, S.; Keshavarz, T.; Roy, I. Polyhydroxyalkanoates: Biodegradable polymers with a range of applications. *J. Chem. Technol. Biotechnol.* **2007**, *82*, 233–247. [CrossRef]
7. Wu, G.F.; Wu, Q.Y.; Shen, Z.Y. Accumulation of poly-β-hydroxybutyrate in cyanobacterium *Synechocystis* sp. PCC6803. *Bioresour. Technol.* **2001**, *76*, 85–90. [CrossRef]
8. Akaraonye, E.; Keshavarz, T.; Roy, I. Production of polyhydroxyalkanoates: The future green materials of choice. *J. Chem. Technol. Biotechnol.* **2010**, *85*, 732–743. [CrossRef]
9. Karan, H.; Funk, C.; Grabert, M.; Oey, M.; Hankamer, B. Green Bioplastics as Part of a Circular Bioeconomy. *Trends Plant. Sci.* **2019**, *24*, 237–249. [CrossRef]
10. Choi, J.; Lee, S.Y. Factors affecting the economics of polyhydroxyalkanoate production by bacterial fermentation. *Appl. Microbiol. Biotechnol.* **1999**, *51*, 13–21. [CrossRef]

11. Lee, G.; Na, J. Future of microbial polyesters. *Microb. Cell Fact.* **2013**, *12*, 54. [CrossRef]
12. Balaji, S.; Gopi, K.; Muthuvelan, B. A review on production of poly β hydroxybutyrates from cyanobacteria for the production of bio plastics. *Algal Res.* **2013**, *2*, 278–285. [CrossRef]
13. Singh, A.K.; Mallick, N. Advances in cyanobacterial polyhydroxyalkanoates production. *FEMS Microbiol. Lett.* **2017**, 364. [CrossRef]
14. Drosg, B.; Fritz, I.; Gattermyer, F.; Silvestrini, L. Photo-autotrophic Production of Poly(hydroxyalkanoates) in Cyanobacteria. *Chem. Biochem. Eng. Q.* **2015**, *29*, 145–156. [CrossRef]
15. Arias, D.M.; García, J.; Uggetti, E. Production of polymers by cyanobacteria grown in wastewater: Current status, challenges and future perspectives. *New Biotechnol.* **2020**, *55*, 46–57. [CrossRef] [PubMed]
16. Pittman, J.K.; Dean, A.P.; Osundeko, O. The potential of sustainable algal biofuel production using wastewater resources. *Bioresour. Technol.* **2011**, *102*, 17–25. [CrossRef] [PubMed]
17. Rawat, I.; Ranjith Kumar, R.; Mutanda, T.; Bux, F. Dual role of microalgae: Phycoremediation of domestic wastewater and biomass production for sustainable biofuels production. *Appl. Energy* **2011**, *88*, 3411–3424. [CrossRef]
18. Stiles, W.A.V.; Styles, D.; Chapman, S.P.; Esteves, S.; Bywater, A.; Melville, L.; Silkina, A.; Lupatsch, I.; Fuentes Grünewald, C.; Lovitt, R.; et al. Using microalgae in the circular economy to valorise anaerobic digestate: Challenges and opportunities. *Bioresour. Technol.* **2018**, *267*, 732–742. [CrossRef]
19. Lai, Y.-C.; Chang, C.-H.; Chen, C.-Y.; Chang, J.-S.; Ng, I.-S. Towards protein production and application by using Chlorella species as circular economy. *Bioresour. Technol.* **2019**, *289*, 121625. [CrossRef]
20. Blank, L.M.; Narancic, T.; Mampel, J.; Tiso, T.; O'Connor, K. Biotechnological upcycling of plastic waste and other non-conventional feedstocks in a circular economy. *Curr. Opin. Biotechnol.* **2020**, *62*, 212–219. [CrossRef]
21. Markl, E.; Grünbichler, H.; Lackner, M. Cyanobacteria for PHB Bioplastics Production: A Review. In *Algae*; Keung Wong, Y., Ed.; IntechOpen: London, UK, 2018; ISBN 9781838805623.
22. Jim, P. OECD Policies for Bioplastics in the Context of a Bioeconomy. *Ind. Biotechnol.* **2014**, *10*, 19–21. [CrossRef]
23. Doi, Y. Microbial Synthesis and Properties of Polyhydroxy-alkanoates. *Mrs Bull.* **1992**, *17*, 39–42. [CrossRef]
24. Reis, M.A.M.; Serafim, L.S.; Lemos, P.C.; Ramos, A.M.; Aguiar, F.R.; Van Loosdrecht, M.C.M. Production of polyhydroxyalkanoates by mixed microbial cultures. *Bioprocess. Biosyst. Eng.* **2003**, *25*, 377–385. [CrossRef]
25. Schlebusch, M.; Forchhammer, K. Requirement of the Nitrogen Starvation-Induced Protein Sll0783 for Polyhydroxybutyrate Accumulation in *Synechocystis* sp. Strain PCC 6803. *AEM* **2010**, *76*, 6101–6107. [CrossRef] [PubMed]
26. EN13432. Available online: https://www.beuth.de/de/norm/din-en-13432/32115376 (accessed on 11 July 2020).
27. Rutkowska, M.; Heimowska, A.; Krasowska, K.; Janik, H. Biodegradability of Polyethylene Starch Blends in Sea Water. *Pol. J. Environ. Stud.* **2002**, *11*, 267–271.
28. European Bioplastics What Types of Bioplastics Do Exist and What Properties Do They Have? Available online: https://www.european-bioplastics.org/faq-items/what-types-of-bioplastics-do-exist-and-what-properties-do-they-have/ (accessed on 24 June 2020).
29. Storz, H.; Vorlop, K.-D. Bio-based plastics: Status, challenges and trends. *Landbauforsch. Appl. Agric. Res.* **2013**, 321–332. [CrossRef]
30. Bio-Plastic Packaging-Global Market Outlook (2016–2022). Available online: https://www.strategymrc.com/report/bio-plastic-packaging-market-2016 (accessed on 24 June 2020).
31. European Bioplastics Market. Available online: https://www.european-bioplastics.org/market/ (accessed on 24 June 2020).
32. Shrivastav, A.; Mishra, S.K.; Mishra, S. Polyhydroxyalkanoate (PHA) synthesis by Spirulina subsalsa from Gujarat coast of India. *Int. J. Biol. Macromol.* **2010**, *46*, 255–260. [CrossRef]
33. Zinn, M.; Witholt, B.; Egli, T. Occurrence, synthesis and medical application of bacterial polyhydroxyalkanoate. *Adv. Drug Deliv. Rev.* **2001**, *53*, 5–21. [CrossRef]
34. Chen, G.-Q.; Wu, Q. The application of polyhydroxyalkanoates as tissue engineering materials. *Biomaterials* **2005**, *26*, 6565–6578. [CrossRef]

35. Gadgil, B.S.T.; Killi, N.; Rathna, G.V.N. Polyhydroxyalkanoates as biomaterials. *Med. Chem. Commun.* **2017**, *8*, 1774–1787. [CrossRef]
36. Reddy, M.V.; Mohan, S.V. Polyhydroxy alkanoates Production by Newly Isolated Bacteria Serratia ureilytica Using Volatile Fatty Acids as Substrate: Bio-Electro Kinetic Analysis. *J. Microb. Biochem. Technol.* **2015**, 7. [CrossRef]
37. Chen, G.-Q. A microbial polyhydroxyalkanoates (PHA) based bio- and materials industry. *Chem. Soc. Rev.* **2009**, *38*, 2434. [CrossRef]
38. Poli, A.; Di Donato, P.; Abbamondi, G.R.; Nicolaus, B. Synthesis, Production, and Biotechnological Applications of Exopolysaccharides and Polyhydroxyalkanoates by Archaea. *Archaea* **2011**, *2011*, 693253. [CrossRef]
39. Anderson, A.J.; Dawes, E.A. Occurrence, metabolism, metabolic role, and industrial uses of bacterial polyhydroxyalkanoates. *Microbiol. Rev.* **1990**, *54*, 450–472. [CrossRef] [PubMed]
40. Dalsasso, R.R.; Pavan, F.A.; Bordignon, S.E.; de Aragão, G.M.F.; Poletto, P. Polyhydroxybutyrate (PHB) production by Cupriavidus necator from sugarcane vinasse and molasses as mixed substrate. *Process. Biochem.* **2019**, *85*, 12–18. [CrossRef]
41. Benesova, P.; Kucera, D.; Marova, I.; Obruca, S. Chicken feather hydrolysate as an inexpensive complex nitrogen source for PHA production by Cupriavidus necator on waste frying oils. *Lett. Appl. Microbiol.* **2017**, *65*, 182–188. [CrossRef] [PubMed]
42. Fidler, S.; Dennis, D. Polyhydroxyalkanoate production in recombinant *Escherichia coli*. *Fems. Microbiol. Lett.* **1992**, *103*, 231–235. [CrossRef]
43. Wang, F.; Lee, S.Y. Production of poly(3-hydroxybutyrate) by fed-batch culture of filamentation-suppressed recombinant Escherichia coli. *Appl. Environ. Microbiol.* **1997**, *63*, 4765–4769. [CrossRef] [PubMed]
44. Wang, F.; Lee, S.Y. High cell density culture of metabolically engineered Escherichia coli for the production of poly(3-hydroxybutyrate) in a defined medium. *Biotechnol. Bioeng.* **1998**, *58*, 325–328. [CrossRef]
45. Weiner, R. Biopolymers from marine prokaryotes. *Trends Biotechnol.* **1997**, *15*, 390–394. [CrossRef]
46. Macrae, R.M.; Wilkinson, J.F. Poly-hyroxybutyrate Metabolism in Washed Suspensions of Bacillus cereus and Bacillus megaterium. *J. Gen. Microbiol.* **1958**, *19*, 210–222. [CrossRef]
47. Raza, Z.A.; Abid, S.; Banat, I.M. Polyhydroxyalkanoates: Characteristics, production, recent developments and applications. *Int. Biodeterior. Biodegrad.* **2018**, *126*, 45–56. [CrossRef]
48. Westbrook, A.W.; Miscevic, D.; Kilpatrick, S.; Bruder, M.R.; Moo-Young, M.; Chou, C.P. Strain engineering for microbial production of value-added chemicals and fuels from glycerol. *Biotechnol. Adv.* **2019**, *37*, 538–568. [CrossRef]
49. Singh, A.; Nigam, P.S.; Murphy, J.D. Renewable fuels from algae: An answer to debatable land based fuels. *Bioresour. Technol.* **2011**, *102*, 10–16. [CrossRef] [PubMed]
50. Halami, P.M. Production of polyhydroxyalkanoate from starch by the native isolate Bacillus cereus CFR06. *World J. Microbiol. Biotechnol.* **2008**, *24*, 805–812. [CrossRef]
51. Markou, G.; Vandamme, D.; Muylaert, K. Microalgal and cyanobacterial cultivation: The supply of nutrients. *Water Res.* **2014**, *65*, 186–202. [CrossRef] [PubMed]
52. Troschl, C.; Meixner, K.; Drosg, B. Cyanobacterial PHA Production—Review of Recent Advances and a Summary of Three Years' Working Experience Running a Pilot Plant. *Bioengineering* **2017**, *4*, 26. [CrossRef] [PubMed]
53. Tokiwa, Y.; Calabia, B.P. Review Degradation of microbial polyesters. *Biotechnol. Lett.* **2004**, *26*, 1181–1189. [CrossRef]
54. Lemoigne, M. Products of dehydration and of polymerization of β-hydroxybutyric acid. *Bull. Soc. Chem. Biol.* **1926**, *8*, 770–782.
55. Griffin, G.J.L. (Ed.) *Chemistry and Technology of Biodegradable Polymers*, 1st ed.; Blackie Academic & Professional: London, UK, 1994; ISBN 9780751400038.
56. Lee, S.Y. Bacterial Polyb ydroxyalkanoates. *Biotechnol. Bioeng.* **1996**, *49*, 1–14. [CrossRef]
57. Kunasundari, B.; Sudesh, K. Isolation and recovery of microbial polyhydroxyalkanoates. *Express Polym. Lett.* **2011**, *5*, 620–634. [CrossRef]
58. Lee, S.Y.; Hong, S.H.; Park, S.J.; van Wegen, R.; Middelberg, A.P.J. *Metabolic Flux Analysis on the Production of Poly(3-hydroxybutyrate) Biopolymers*; Wiley: New York, NY, USA, 2005; pp. 249–257.

59. Lim, S.-J.; Jung, Y.-M.; Shin, H.-D.; Lee, Y.-H. Amplification of the NADPH-related genes zwf and gnd for the oddball biosynthesis of PHB in an E. coli transformant harboring a cloned phbCAB operon. *J. Biosci. Bioeng.* **2002**, *93*, 543–549. [CrossRef]
60. Fukui, T.; Doi, Y. Cloning and analysis of the poly(3-hydroxybutyrate-co-3-hydroxyhexanoate) biosynthesis genes of Aeromonas caviae. *J. Bact.* **1997**, *179*, 4821–4830. [CrossRef] [PubMed]
61. Tsuge, T.; Fukui, T.; Matsusaki, H.; Taguchi, S.; Kobayashi, G.; Ishizaki, A.; Doi, Y. Molecular cloning of two (R)-specific enoyl-CoA hydratase genes from Pseudomonas aeruginosa and their use for polyhydroxyalkanoate synthesis. *FEMS Microbiol. Lett.* **2000**, *184*, 193–198. [CrossRef] [PubMed]
62. Hein, J.; Paletta, A.; Steinbüchel, S. Cloning. Characterization and comparison of the Pseudomonas mendocina polyhydroxyalkanoate synthases PhaC1 and PhaC2. *Appl. Microbiol. Biotechnol.* **2002**, *58*, 229–236. [CrossRef] [PubMed]
63. Fukui, T.; Shiomi, N.; Doi, Y. Expression and characterization of (R)-specific enoyl coenzyme A hydratase involved in polyhydroxyalkanoate biosynthesis by Aeromonas caviae. *J. Bacteriol.* **1998**, *180*, 667–673. [CrossRef] [PubMed]
64. Kniewel, R.; Lopez, O.R.; Prieto, M.A. Biogenesis of medium-chain-length polyhydroxyalkanoates. In *Biogenesis of Fatty Acids, Lipids and Membranes*; Geiger, O., Ed.; Springer International Publishing: Cham, Switzerland, 2019; pp. 457–481, ISBN 9783319504292.
65. Huijberts, G.N.; Eggink, G.; de Waard, P.; Huisman, G.W.; Witholt, B. Pseudomonas putida KT2442 cultivated on glucose accumulates poly(3-hydroxyalkanoates) consisting of saturated and unsaturated monomers. *Appl. Environ. Microbiol.* **1992**, *58*, 536–544. [CrossRef]
66. Rehm, B.H.A.; Krüger, N.; Steinbüchel, A. A New Metabolic Link between Fatty Acid de Novo Synthesis and Polyhydroxyalkanoic Acid Synthesis: The phag gene from pseudomonas putida kt2440 encodes a 3-hydroxyacyl-acyl carrier protein-coenzyme a transferase. *J. Biol. Chem.* **1998**, *273*, 24044–24051. [CrossRef]
67. Salehizadeh, H.; Van Loosdrecht, M.C.M. Production of polyhydroxyalkanoates by mixed culture: Recent trends and biotechnological importance. *Biotechnol. Adv.* **2004**, *22*, 261–279. [CrossRef]
68. Wen, Q.; Chen, Z.; Tian, T.; Chen, W. Effects of phosphorus and nitrogen limitation on PHA production in activated sludge. *J. Environ. Sci.* **2010**, *22*, 1602–1607. [CrossRef]
69. Rhu, D.H.; Lee, W.H.; Kim, J.Y.; Choi, E. Polyhydroxyalkanoate (PHA) production from waste. *Water Sci. Technol.* **2003**, *48*, 221–228. [CrossRef]
70. Sureshkumar, M. Production of biodegradable plastics from activated sludge generated from a food processing industrial wastewater treatment plant. *Bioresour. Technol.* **2004**, *95*, 327–330. [CrossRef]
71. Morgan-Sagastume, F.; Karlsson, A.; Johansson, P.; Pratt, S.; Boon, N.; Lant, P.; Werker, A. Production of polyhydroxyalkanoates in open, mixed cultures from a waste sludge stream containing high levels of soluble organics, nitrogen and phosphorus. *Water Res.* **2010**, *44*, 5196–5211. [CrossRef]
72. Albuquerque, M.G.E.; Eiroa, M.; Torres, C.; Nunes, B.R.; Reis, M.A.M. Strategies for the development of a side stream process for polyhydroxyalkanoate (PHA) production from sugar cane molasses. *J. Biotechnol.* **2007**, *130*, 411–421. [CrossRef] [PubMed]
73. Montiel-Jarillo, G.; Carrera, J.; Suárez-Ojeda, M.E. Enrichment of a mixed microbial culture for polyhydroxyalkanoates production: Effect of pH and N and P concentrations. *Sci. Total Environ.* **2017**, *583*, 300–307. [CrossRef]
74. Cui, Y.-W.; Gong, X.-Y.; Shi, Y.-P.; Wang, Z. (Drew) Salinity effect on production of PHA and EPS by Haloferax mediterranei. *RSC Adv.* **2017**, *7*, 53587–53595. [CrossRef]
75. Albuquerque, M.G.E.; Torres, C.A.V.; Reis, M.A.M. Polyhydroxyalkanoate (PHA) production by a mixed microbial culture using sugar molasses: Effect of the influent substrate concentration on culture selection. *Water Res.* **2010**, *44*, 3419–3433. [CrossRef] [PubMed]
76. Liu, Z.; Wang, Y.; He, N.; Huang, J.; Zhu, K.; Shao, W.; Wang, H.; Yuan, W.; Li, Q. Optimization of polyhydroxybutyrate (PHB) production by excess activated sludge and microbial community analysis. *J. Hazard. Mater.* **2011**, *185*, 8–16. [CrossRef] [PubMed]
77. Panda, B.; Jain, P.; Sharma, L.; Mallick, N. Optimization of cultural and nutritional conditions for accumulation of poly-β-hydroxybutyrate in Synechocystis sp. PCC 6803. *Bioresour. Technol.* **2006**, *97*, 1296–1301. [CrossRef]
78. Samantaray, S.; Nayak, J.K.; Mallick, N. Wastewater Utilization for Poly-β-Hydroxybutyrate Production by the Cyanobacterium Aulosira fertilissima in a Recirculatory Aquaculture System. *Appl. Environ. Microbiol.* **2011**, *77*, 8735–8743. [CrossRef]

79. Dawes, E.A. Microbial energy reserve compounds. In *Microbial Energetics*; Blackie: Glasgow, UK, 1986; pp. 145–165.
80. Tan, G.-Y.; Chen, C.-L.; Li, L.; Ge, L.; Wang, L.; Razaad, I.; Li, Y.; Zhao, L.; Mo, Y.; Wang, J.-Y. Start a Research on Biopolymer Polyhydroxyalkanoate (PHA): A Review. *Polymers* **2014**, *6*, 706–754. [CrossRef]
81. Miller, N.D.; Williams, D.F. On the biodegradation of poly-β-hydroxybutyrate (PHB) homopolymer and poly-β-hydroxybutyrate-hydroxyvalerate copolymers. *Biomaterials* **1987**, *8*, 129–137. [CrossRef]
82. Biazar, E.; Heidari Keshel, S. A nanofibrous PHBV tube with Schwann cell as artificial nerve graft contributing to Rat sciatic nerve regeneration across a 30-mm defect bridge. *Cell Commun. Adhes.* **2013**, *20*, 41–49. [CrossRef]
83. Laycock, B.; Halley, P.; Pratt, S.; Werker, A.; Lant, P. The chemomechanical properties of microbial polyhydroxyalkanoates. *Prog. Polym. Sci.* **2013**, *38*, 536–583. [CrossRef]
84. Kusaka, S.; Iwata, T.; Doi†*, Y. Microbial Synthesis and Physical Properties of Ultra-High-Molecular-Weight Poly[(*R*)-3-Hydroxybutyrate]. *J. Macromol. Sci. Part A* **1998**, *35*, 319–335. [CrossRef]
85. Sudesh, K.; Abe, H.; Doi, Y. Synthesis, structure and properties of polyhydroxyalkanoates: Biological polyesters. *Prog. Polym. Sci.* **2000**, *25*, 1503–1555. [CrossRef]
86. Kundu, P.P.; Nandy, A.; Mukherjee, A.; Pramanik, N. Polyhydroxyalkanoates: Microbial synthesis and applications. In *Encyclopedia of Biomedical Polymers and Polymeric Biomaterials*; CRC Press: Boca Raton, FL, USA, 2015; Volume 11, p. 10444.
87. Bengtsson, S.; Pisco, A.R.; Reis, M.A.M.; Lemos, P.C. Production of polyhydroxyalkanoates from fermented sugar cane molasses by a mixed culture enriched in glycogen accumulating organisms. *J. Biotechnol.* **2010**, *145*, 253–263. [CrossRef]
88. Bengtsson, S. The utilization of glycogen accumulating organisms for mixed culture production of polyhydroxyalkanoates. *Biotechnol. Bioeng.* **2009**, 698–708. [CrossRef] [PubMed]
89. Chen, G.; Zhang, G.; Park, S.; Lee, S. Industrial scale production of poly(3-hydroxybutyrate-co-3-hydroxyhexanoate). *Appl. Microbiol. Biotechnol.* **2001**, *57*, 50–55. [CrossRef]
90. Shimamura, E.; Kasuya, K.; Kobayashi, G.; Shiotani, T.; Shima, Y.; Doi, Y. Physical Properties and Biodegradability of Microbial Poly(3-hydroxybutyrate-co-3-hydroxyhexanoate). *Macromolecules* **1994**, *27*, 878–880. [CrossRef]
91. Battegazzore, D.; Noori, A.; Frache, A. Hemp hurd and alfalfa as particle filler to improve the thermo-mechanical and fire retardant properties of poly(3-hydroxybutyrate-co-3-hydroxyhexanoate). *Polym. Compos.* **2019**, *40*, 3429–3437. [CrossRef]
92. Urakami, T.; Imagawa, S.; Harada, M.; Iwamoto, A.; Tokiwa, Y. Development of Biodegradable Plastic-Poly-.BETA.-hydroxybutyrate/polycaprolaetone Blend Polymer. *Kobunshi Ronbunshu* **2000**, *57*, 263–270. [CrossRef]
93. Savenkova, L.; Gercberga, Z.; Bibers, I.; Kalnin, M. Effect of 3-hydroxy valerate content on some physical and mechanical properties of polyhydroxyalkanoates produced by Azotobacter chroococcum. *Process. Biochem.* **2000**, *36*, 445–450. [CrossRef]
94. Park, S.H.; Lee, S.H.; Lee, S.Y. Preparation of optically active β-amino acids from microbial polyester polyhydroxyalkanoates. *J. Chem. Res. (S)* **2001**, 498–499. [CrossRef]
95. Pederson, E.N.; McChalicher, C.W.J.; Srienc, F. Bacterial Synthesis of PHA Block Copolymers. *Biomacromolecules* **2006**, *7*, 1904–1911. [CrossRef] [PubMed]
96. Han, D.; Tong, X.; Zhao, Y. Fast Photodegradable Block Copolymer Micelles for Burst Release. *Macromolecules* **2011**, *44*, 437–439. [CrossRef]
97. Li, S.Y.; Dong, C.L.; Wang, S.Y.; Ye, H.M.; Chen, G.-Q. Microbial production of polyhydroxyalkanoate block copolymer by recombinant Pseudomonas putida. *Appl. Microbiol. Biotechnol.* **2011**, *90*, 659–669. [CrossRef]
98. Clarinval, A.-M.; Halleux, J. 1-Classification of biodegradable polymers. In *Biodegradable Polymers for Industrial Applications*; Smith, R., Ed.; CRC Press: Boca Raton, FL, USA, 2005; pp. 3–56.
99. Holmes, P.A. Biologically Produced (R)-3-Hydroxy- Alkanoate Polymers and Copolymers. In *Developments in Crystalline Polymers*; Bassett, D.C., Ed.; Springer: Dordrecht, The Netherlands, 1988; pp. 1–65, ISBN 9789401070966.
100. Misra, S.K.; Valappil, S.P.; Roy, I.; Boccaccini, A.R. Polyhydroxyalkanoate (PHA)/Inorganic Phase Composites for Tissue Engineering Applications. *Biomacromolecules* **2006**, *7*, 2249–2258. [CrossRef] [PubMed]
101. Duan, J.; Zhang, Y.; Han, S.; Chen, Y.; Li, B.; Liao, M.; Chen, W.; Deng, X.; Zhao, J.; Huang, B. Synthesis and in vitro/in vivo anti-cancer evaluation of curcumin-loaded chitosan/poly(butyl cyanoacrylate) nanoparticles. *Int. J. Pharm.* **2010**, *400*, 211–220. [CrossRef]

102. Valappil, S.P.; Boccaccini, A.R.; Bucke, C.; Roy, I. Polyhydroxyalkanoates in Gram-positive bacteria: Insights from the genera Bacillus and Streptomyces. *Antonie Van Leeuwenhoek* **2006**, *91*, 1–17. [CrossRef]
103. Steinbüchel, A.; Füchtenbusch, B. Bacterial and other biological systems for polyester production. *Trends Biotechnol.* **1998**, *16*, 419–427. [CrossRef]
104. Duvernoy, O.; Malm, T.; Ramström, J.; Bowald, S. A Biodegradable Patch used as a Pericardial Substitute after Cardiac Surgery: 6- and 24-Month Evaluation with CT. *Thorac. Cardiovasc. Surg.* **1995**, *43*, 271–274. [CrossRef]
105. Atkins, T.W.; Peacock, S.J. The incorporation and release of bovine serum albumin from poly-hydroxybutyrate-hydroxyvalerate microcapsules. *J. Microencapsul.* **1996**, *13*, 709–717. [CrossRef]
106. Tepha Patents. Available online: https://www.tepha.com/news-events/patents/ (accessed on 25 June 2020).
107. Martin, D.P.; Williams, S.F. Medical applications of poly-4-hydroxybutyrate: A strong flexible absorbable biomaterial. *Biochem. Eng. J.* **2003**, *16*, 97–105. [CrossRef]
108. Steinbüchel, A.; Valentin, H.E. Diversity of bacterial polyhydroxyalkanoic acids. *FEMS Microbiol. Lett.* **1995**, *128*, 219–228. [CrossRef]
109. Seebach, D.; Herrmann, G.F.; Lengweiler, U.D.; Bachmann, B.M.; Amrein, W. Synthesis and Enzymatic Degradation of Dendrimers from(R)-3-Hydroxybutanoic Acid and Trimesic Acid. *Angew. Chem. Int. Ed. Engl.* **1996**, *35*, 2795–2797. [CrossRef]
110. Bekker, A.; Holland, H.D.; Wang, P.-L.; Rumble, D.; Stein, H.J.; Hannah, J.L.; Coetzee, L.L.; Beukes, N.J. Dating the rise of atmospheric oxygen. *Nature* **2004**, *427*, 117–120. [CrossRef] [PubMed]
111. Olson, J.M. Photosynthesis in the Archean Era. *Photosynth Res.* **2006**, *88*, 109–117. [CrossRef] [PubMed]
112. Oren, A. Cyanobacteria: Biology, ecology and evolution. In *Cyanobacteria*; Sharma, N.K., Rai, A.K., Stal, L.J., Eds.; John Wiley & Sons, Ltd.: Chichester, UK, 2013; pp. 1–20, ISBN 9781118402238.
113. Kasting, J.F. EARTH HISTORY: The Rise of Atmospheric Oxygen. *Science* **2001**, *293*, 819–820. [CrossRef] [PubMed]
114. Kasting, J.F. Life and the Evolution of Earth's Atmosphere. *Science* **2002**, *296*, 1066–1068. [CrossRef]
115. Cavalier-Smith, T. Chloroplast Evolution: Secondary Symbiogenesis and Multiple Losses. *Curr. Biol.* **2002**, *12*, R62–R64. [CrossRef]
116. Stal, L.J. Physiological ecology of cyanobacteria in microbial mats and other communities. *New Phytol.* **1995**, *131*, 1–32. [CrossRef]
117. Whitton, B.A.; Potts, M. *The Ecology of Cyanobacteria: Their Diversity in Time and Space*; Springer: Dordrecht, The Netherlands, 2002; ISBN 9780306468551.
118. Waterbury, J.B.; Watson, S.W.; Valois, F.W.; Franks, D.G. Biological and ecological characterization of the marine unicellular cyanobacterium Synechococcus. In *Photosynthetic Picoplankton*; Platt, T., Li, W.K.W., Eds.; Canadian Bulletin of Fisheries and Aquatic Sciences: Ottawa, ON, Canada, 1986; Volume 214, pp. 71–120, ISBN 066012243X.
119. Thajuddin, N.; Subramanian, G. Cyanobacterial biodiversity and potential applications in biotechnology. *Curr. Sci.* **2005**, *89*, 47–57.
120. Lindell, D.; Padan, E.; Post, A.F. Regulation of ntcA Expression and Nitrite Uptake in the Marine Synechococcus sp. Strain WH 7803. *J. Bacteriol.* **1998**, *180*, 1878–1886. [CrossRef] [PubMed]
121. Dodds, W.K.; Gudder, D.A.; Mollenhauer, D. The ecology of nostoc. *J. Phycol.* **1995**, *31*, 2–18. [CrossRef]
122. Tsinoremas, N.F.; Castets, A.M.; Harrison, M.A.; Allen, J.F.; Tandeau de Marsac, N. Photosynthetic electron transport controls nitrogen assimilation in cyanobacteria by means of posttranslational modification of the glnB gene product. *Proc. Natl. Acad. Sci. USA* **1991**, *88*, 4565–4569. [CrossRef] [PubMed]
123. Latifi, A.; Ruiz, M.; Zhang, C.-C. Oxidative stress in cyanobacteria. *FEMS Microbiol. Rev.* **2009**, *33*, 258–278. [CrossRef] [PubMed]
124. Xiong, Q.; Chen, Z.; Ge, F. Proteomic analysis of post translational modifications in cyanobacteria. *J. Proteom.* **2016**, *134*, 57–64. [CrossRef] [PubMed]
125. de Oliveira, D.T.; da Costa, A.A.F.; Costa, F.F.; da Rocha Filho, G.N.; do Nascimento, L.A.S. Advances in the Biotechnological Potential of Brazilian Marine Microalgae and Cyanobacteria. *Molecules* **2020**, *25*, 2908. [CrossRef]
126. Shimizu, Y. Microalgal metabolites. *Curr. Opin. Microbiol.* **2003**, *6*, 236–243. [CrossRef]

127. Koehn, F.E.; Longley, R.E.; Reed, J.K. Microcolins A and B, New Immunosuppressive Peptides from the Blue-Green Alga Lyngbya majuscula. *J. Nat. Prod.* **1992**, *55*, 613–619. [CrossRef]
128. Al-Batshan, H.A.; Al-Mufarrej, S.I.; Al-Homaidan, A.A.; Qureshi, M.A. Enhancement Of Chicken Macrophage Phagocytic Function And Nitrite Production By Dietary *Spirulina platensis*. *Immunopharm. Immunotoxicol.* **2001**, *23*, 281–289. [CrossRef]
129. Qureshi, M.A.; Garlich, J.D.; Kidd, M.T. Dietary *Spirulina Platensis* Enhances Humoral and Cell-Mediated Immune Functions in Chickens. *Immunopharmacol. Immunotoxicol.* **1996**, *18*, 465–476. [CrossRef]
130. Rajeev, K.J.; Xu, Z. Biomedical Compounds from Marine organisms. *Mar. Drugs* **2004**, *2*, 123–146. [CrossRef]
131. Proksch, P.; Edrada, R.; Ebel, R. Drugs from the seas-current status and microbiological implications. *Appl. Microbiol. Biotechnol.* **2002**, *59*, 125–134. [CrossRef] [PubMed]
132. Carmichael, W.W.; Gorham, P.R. The mosaic nature of toxic blooms of cyanobacteria. In *The Water Environment*; Carmichael, W.W., Ed.; Springer: Boston, MA, USA, 1981; pp. 161–172, ISBN 9781461332695.
133. Belay, A.; Kato, T.; Ota, Y. Spirulina (Arthrospira): Potential application as an animal feed supplement. *J. Appl. Phycol.* **1996**, *8*, 303–311. [CrossRef]
134. Ansari, S.; Fatma, T. Cyanobacterial Polyhydroxybutyrate (PHB): Screening, Optimization and Characterization. *PLoS ONE* **2016**, *11*, e0158168. [CrossRef] [PubMed]
135. Beck, C.; Knoop, H.; Axmann, I.M.; Steuer, R. The diversity of cyanobacterial metabolism: Genome analysis of multiple phototrophic microorganisms. *BMC Genom.* **2012**, *13*, 56. [CrossRef]
136. Aikawa, S.; Izumi, Y.; Matsuda, F.; Hasunuma, T.; Chang, J.-S.; Kondo, A. Synergistic enhancement of glycogen production in Arthrospira platensis by optimization of light intensity and nitrate supply. *Bioresour. Technol.* **2012**, *108*, 211–215. [CrossRef]
137. Hu, Q.; Sommerfeld, M.; Jarvis, E.; Ghirardi, M.; Posewitz, M.; Seibert, M.; Darzins, A. Microalgal triacylglycerols as feedstocks for biofuel production: Perspectives and advances. *Plant. J.* **2008**, *54*, 621–639. [CrossRef]
138. Damrow, R.; Maldener, I.; Zilliges, Y. The Multiple Functions of Common Microbial Carbon Polymers, Glycogen and PHB, during Stress Responses in the Non-Diazotrophic Cyanobacterium *Synechocystis* sp. PCC 6803. *Front. Microbiol.* **2016**, *7*. [CrossRef]
139. De Jaeger, L.; Verbeek, R.E.; Draaisma, R.B.; Martens, D.E.; Springer, J.; Eggink, G.; Wijffels, R.H. Superior triacylglycerol (TAG) accumulation in starchless mutants of Scenedesmus obliquus: (I) mutant generation and characterization. *Biotechnol. Biofuels* **2014**, *7*, 69. [CrossRef]
140. Bjornsson, W.J.; MacDougall, K.M.; Melanson, J.E.; O'Leary, S.J.B.; McGinn, P.J. Pilot-scale supercritical carbon dioxide extractions for the recovery of triacylglycerols from microalgae: A practical tool for algal biofuels research. *J. Appl. Phycol.* **2012**, *24*, 547–555. [CrossRef]
141. Carr, N.G. The occurrence of poly-β-hydroxybutyrate in the blue-green alga, *Chlorogloea fritschii*. *Biochim. Et Biophys. Acta (Bba) Biophys. Incl. Photosynth.* **1966**, *120*, 308–310. [CrossRef]
142. Khetkorn, W.; Incharoensakdi, A.; Lindblad, P.; Jantaro, S. Enhancement of poly-3-hydroxybutyrate production in Synechocystis sp. PCC 6803 by overexpression of its native biosynthetic genes. *Bioresour. Technol.* **2016**, *214*, 761–768. [CrossRef] [PubMed]
143. Kamravamanesh, D.; Pflügl, S.; Nischkauer, W.; Limbeck, A.; Lackner, M.; Herwig, C. Photosynthetic poly-β-hydroxybutyrate accumulation in unicellular cyanobacterium *Synechocystis* sp. PCC 6714. *AMB Expr.* **2017**, *7*, 143. [CrossRef] [PubMed]
144. Nishioka, M.; Nakai, K.; Miyake, M.; Asada, Y.; Taya, M. Production of poly-β-hydroxybutyrate by thermophilic cyanobacterium, Synechococcus sp. MA19, under phosphate-limited conditions. *Biotechnol. Lett.* **2001**, *23*, 1095–1099. [CrossRef]
145. Bhati, R.; Mallick, N. Poly(3-hydroxybutyrate-co-3-hydroxyvalerate) copolymer production by the diazotrophic cyanobacterium Nostoc muscorum Agardh: Process optimization and polymer characterization. *Algal Res.* **2015**, *7*, 78–85. [CrossRef]
146. Samantaray, S.; Mallick, N. Production and characterization of poly-β-hydroxybutyrate (PHB) polymer from Aulosira fertilissima. *J. Appl. Phycol.* **2012**, *24*, 803–814. [CrossRef]
147. Hein, S.; Tran, H.; Steinbüchel, A. *Synechocystis* sp. PCC6803 possesses a two-component polyhydroxyalkanoic acid synthase similar to that of anoxygenic purple sulfur bacteria. *Arch. Microbiol.* **1998**, *170*, 162–170. [CrossRef]

148. Matsusaki, H.; Manji, S.; Taguchi, K.; Kato, M.; Fukui, T.; Doi, Y. Cloning and molecular analysis of the Poly(3-hydroxybutyrate) and Poly(3-hydroxybutyrate-co-3-hydroxyalkanoate) biosynthesis genes in Pseudomonas sp. strain 61-3. *J. Bacteriol.* **1998**, *180*, 6459–6467. [CrossRef]
149. Lane, C.E.; Benton, M.G. Detection of the enzymatically-active polyhydroxyalkanoate synthase subunit gene, phaC, in cyanobacteria via colony PCR. *Mol. Cell. Probes* **2015**, *29*, 454–460. [CrossRef]
150. Erb, T.J.; Zarzycki, J. Biochemical and synthetic biology approaches to improve photosynthetic CO_2-fixation. *Curr. Opin. Chem. Biol.* **2016**, *34*, 72–79. [CrossRef]
151. Matsumoto, K.; Saito, J.; Yokoo, T.; Hori, C.; Nagata, A.; Kudoh, Y.; Ooi, T.; Taguchi, S. Ribulose-1,5-bisphosphate carboxylase/oxygenase (RuBisCO)-mediated de novo synthesis of glycolate-based polyhydroxyalkanoate in Escherichia coli. *J. Biosci. Bioeng.* **2019**, *128*, 302–306. [CrossRef] [PubMed]
152. Abo, B.O.; Odey, E.A.; Bakayoko, M.; Kalakodio, L. Microalgae to biofuels production: A review on cultivation, application and renewable energy. *Rev. Environ. Health* **2019**, *34*, 91–99. [CrossRef] [PubMed]
153. Marinho-Soriano, E.; Panucci, R.A.; Carneiro, M.A.A.; Pereira, D.C. Evaluation of Gracilaria caudata J. Agardh for bioremediation of nutrients from shrimp farming wastewater. *Bioresour. Technol.* **2009**, *100*, 6192–6198. [CrossRef] [PubMed]
154. Abdel-Raouf, N.; Al-Homaidan, A.A.; Ibraheem, I.B.M. Microalgae and wastewater treatment. *Saudi J. Biol. Sci.* **2012**, *19*, 257–275. [CrossRef] [PubMed]
155. Kumar, A.; Srivastava, J.K.; Mallick, N.; Singh, A.K. Commercialization of Bacterial Cell Factories for the Sustainable Production of Polyhydroxyalkanoate Thermoplastics: Progress and Prospects. *Recent Pat. Biotechnol.* **2015**, *9*, 4–21. [CrossRef] [PubMed]
156. Pouliot, Y.; Buelna, G.; Racine, C.; de la Noüe, J. Culture of cyanobacteria for tertiary wastewater treatment and biomass production. *Biol. Wastes* **1989**, *29*, 81–91. [CrossRef]
157. Oswald, W.J. Micro-algae and wastewater treatment. In *Micro-Algal Biotechnology*; Borowitzka, M.A., Borowitzka, L.J., Eds.; Cambridge University Press: Cambridge, UK, 1988; pp. 305–328.
158. Theregowda, R.B.; Vidic, R.; Landis, A.E.; Dzombak, D.A.; Matthews, H.S. Integrating external costs with life cycle costs of emissions from tertiary treatment of municipal wastewater for reuse in cooling systems. *J. Clean. Prod.* **2016**, *112*, 4733–4740. [CrossRef]
159. Sood, A.; Renuka, N.; Prasanna, R.; Ahluwalia, A.S. Cyanobacteria as Potential Options for Wastewater Treatment. In *Phytoremediation*; Ansari, A.A., Gill, S.S., Gill, R., Lanza, G.R., Newman, L., Eds.; Springer International Publishing: Cham, Switzerland, 2015; pp. 83–93, ISBN 9783319109688.
160. Su, Y.; Mennerich, A.; Urban, B. Comparison of nutrient removal capacity and biomass settleability of four high-potential microalgal species. *Bioresour. Technol.* **2012**, *124*, 157–162. [CrossRef]
161. Cañizares-Villanueva, R.O.; Ramos, A.; Corona, A.I.; Monroy, O.; de la Torre, M.; Gomez-Lojero, C.; Travieso, L. Phormidium treatment of anaerobically treated swine wastewater. *Water Res.* **1994**, *28*, 1891–1895. [CrossRef]
162. Chevalier, P.; Proulx, D.; Lessard, P.; Vincent, W.F.; de la Noüe, J. Nitrogen and phosphorus removal by high latitude mat-forming cyanobacteria for potential use in tertiary wastewater treatment. *J. Appl. Phycol.* **2000**, *12*, 105–112. [CrossRef]
163. Talbot, P.; de la Noüe, J. Tertiary treatment of wastewater with Phormidium bohneri (Schmidle) under various light and temperature conditions. *Water Res.* **1993**, *27*, 153–159. [CrossRef]
164. Honda, R.; Boonnorat, J.; Chiemchaisri, C.; Chiemchaisri, W.; Yamamoto, K. Carbon dioxide capture and nutrients removal utilizing treated sewage by concentrated microalgae cultivation in a membrane photobioreactor. *Bioresour. Technol.* **2012**, *125*, 59–64. [CrossRef] [PubMed]
165. Arias, D.M.; Uggetti, E.; García-Galán, M.J.; García, J. Cultivation and selection of cyanobacteria in a closed photobioreactor used for secondary effluent and digestate treatment. *Sci. Total Environ.* **2017**, *587–588*, 157–167. [CrossRef]
166. Md Din, M.F.; Ujang, Z.; van Loosdrecht, M.C.M.; Ahmad, A.; Sairan, M.F. Optimization of nitrogen and phosphorus limitation for better biodegradable plastic production and organic removal using single fed-batch mixed cultures and renewable resources. *Water Sci. Technol.* **2006**, *53*, 15–20. [CrossRef] [PubMed]
167. Srebotnjak, T.; Carr, G.; de Sherbinin, A.; Rickwood, C. A global Water Quality Index and hot-deck imputation of missing data. *Ecol. Indic.* **2012**, *17*, 108–119. [CrossRef]
168. Bhati, R.; Mallick, N. Carbon dioxide and poultry waste utilization for production of polyhydroxyalkanoate biopolymers by Nostoc muscorum Agardh: A sustainable approach. *J. Appl. Phycol.* **2016**, *28*, 161–168. [CrossRef]

169. Liebal, U.W.; Blank, L.M.; Ebert, B.E. CO_2 to succinic acid—Estimating the potential of biocatalytic routes. *Metab. Eng. Commun.* **2018**, *7*, e00075. [CrossRef]
170. Wang, B.; Li, Y.; Wu, N.; Lan, C.Q. CO_2 bio-mitigation using microalgae. *Appl. Microbiol. Biotechnol.* **2008**, *79*, 707–718. [CrossRef]
171. Olguín, E.J.; Galicia, S.; Mercado, G.; Pérez, T. Annual productivity of Spirulina (Arthrospira) and nutrient removal in a pig wastewater recycling process under tropical conditions. *J. Appl. Phycol.* **2003**, *15*, 249–257. [CrossRef]
172. Sweeten, J.M. Livestock and poultry waste management: A national overview. In *National Livestock, Poultry, And Aquaculture Waste Management*; American Society of Agricultural Engineers: San Jose, MI, USA, 1992; p. 414.
173. Chaiklahan, R.; Chirasuwan, N.; Siangdung, W.; Paithoonrangsarid, K.; Bunnag, B. Cultivation of Spirulina platensis Using Pig Wastewater in a Semi-Continuous Process. *J. Microbiol. Biotechnol.* **2010**, *20*, 609–614. [CrossRef]
174. Zinn, M.; Amstutz, V.; Hanik, N.; Pott, J.; Utsunomia, C. Grave-to-cradle: The potential of autotrophic bioprocesses in bioplastic production. *New Biotechnol.* **2018**, *44*, S64. [CrossRef]
175. Morillo, J.A.; Antizar-Ladislao, B.; Monteoliva-Sánchez, M.; Ramos-Cormenzana, A.; Russell, N.J. Bioremediation and biovalorisation of olive-mill wastes. *Appl. Microbiol. Biotechnol.* **2009**, *82*, 25–39. [CrossRef] [PubMed]
176. Korhonen, J.; Honkasalo, A.; Seppälä, J. Circular Economy: The Concept and its Limitations. *Ecol. Econ.* **2018**, *143*, 37–46. [CrossRef]
177. Wang, L.; Min, M.; Li, Y.; Chen, P.; Chen, Y.; Liu, Y.; Wang, Y.; Ruan, R. Cultivation of Green Algae Chlorella sp. in Different Wastewaters from Municipal Wastewater Treatment Plant. *Appl. Biochem. Biotechnol.* **2010**, *162*, 1174–1186. [CrossRef] [PubMed]
178. Chew, K.W.; Yap, J.Y.; Show, P.L.; Suan, N.H.; Juan, J.C.; Ling, T.C.; Lee, D.-J.; Chang, J.-S. Microalgae biorefinery: High value products perspectives. *Bioresour. Technol.* **2017**, *229*, 53–62. [CrossRef]
179. Haas, R.; Jin, B.; Zepf, F.T. Production of Poly(3-hydroxybutyrate) from Waste Potato Starch. *Biosci. Biotechnol. Biochem.* **2008**, *72*, 253–256. [CrossRef]
180. Cuellar-Bermudez, S.P.; Garcia-Perez, J.S.; Rittmann, B.E.; Parra-Saldivar, R. Photosynthetic bioenergy utilizing CO_2: An approach on flue gases utilization for third generation biofuels. *J. Clean. Prod.* **2015**, *98*, 53–65. [CrossRef]
181. Meixner, K.; Kovalcik, A.; Sykacek, E.; Gruber-Brunhumer, M.; Zeilinger, W.; Markl, K.; Haas, C.; Fritz, I.; Mundigler, N.; Stelzer, F.; et al. Cyanobacteria Biorefinery—Production of poly(3-hydroxybutyrate) with Synechocystis salina and utilisation of residual biomass. *J. Biotechnol.* **2018**, *265*, 46–53. [CrossRef]
182. Directive 2002/32/EC of the European Parliament and of the Council of 7 May 2002 on Undesirable Substances in Animal Feed-COUNCIL Statement. Available online: https://eur-lex.europa.eu/eli/dir/2002/32/2015-02-27 (accessed on 25 June 2020).
183. Gradíssimo, D.G.; Mourão, M.M.; Santos, A.V. Importância do Monitoramento de Cianobactérias e Suas Toxinas em Águas Para Consumo Humano. *J. Crim.* **2020**, *9*, 15–21. [CrossRef]
184. Price, S.; Kuzhiumparambil, U.; Pernice, M.; Ralph, P.J. Cyanobacterial polyhydroxybutyrate for sustainable bioplastic production: Critical review and perspectives. *J. Environ. Chem. Eng.* **2020**, *8*, 104007. [CrossRef]
185. Chaiklahan, R.; Chirasuwan, N.; Triratana, P.; Loha, V.; Tia, S.; Bunnag, B. Polysaccharide extraction from Spirulina sp. and its antioxidant capacity. *Int. J. Biol. Macromol.* **2013**, *58*, 73–78. [CrossRef] [PubMed]
186. Gomes, I.G.; Chaves, F.H.; Barros, R.N.; Moreira, R.L.; Teixeira, E.G.; Moreira, A.G.; Farias, W.R. Dietary supplementation with Spirulina platensis increases growth and color of red tilapia. *Rev. Colomb. Ciencias Pecu* **2012**, *25*, 462–471.
187. Zahroojian, N.; Moravej, H.; Shivazad, M. Effects of Dietary Marine Algae (*Spirulina platensis*) on Egg Quality and Production Performance of Laying Hens. *J. Agric. Sci. Technol.* **2013**, *15*, 1353–1360.
188. Briens, C.; Piskorz, J.; Berruti, F. Biomass Valorization for Fuel and Chemicals Production—A Review. *Int. J. Chem. React. Eng.* **2008**, 6. [CrossRef]
189. De Godos, I.; Blanco, S.; García-Encina, P.A.; Becares, E.; Muñoz, R. Long-term operation of high rate algal ponds for the bioremediation of piggery wastewaters at high loading rates. *Bioresour. Technol.* **2009**, *100*, 4332–4339. [CrossRef] [PubMed]

190. Balasubramanian, L.; Subramanian, G.; Nazeer, T.T.; Simpson, H.S.; Rahuman, S.T.; Raju, P. Cyanobacteria cultivation in industrial wastewaters and biodiesel production from their biomass: A review. *Biotechnol. Appl. Biochem.* **2011**, *58*, 220–225. [CrossRef]
191. Sarsekeyeva, F.; Zayadan, B.K.; Usserbaeva, A.; Bedbenov, V.S.; Sinetova, M.A.; Los, D.A. Cyanofuels: Biofuels from cyanobacteria. Reality and perspectives. *Photosynth. Res.* **2015**, *125*, 329–340. [CrossRef]
192. Lima, A.R.J.; Siqueira, A.S.; dos Santos, B.G.S.; da Silva, F.D.F.; Lima, C.P.; Cardoso, J.F.; Vianez Junior, J.L.d.S.G.; Dall'Agnol, L.T.; McCulloch, J.A.; Nunes, M.R.T.; et al. Draft Genome Sequence of the Brazilian *Cyanobium* sp. Strain CACIAM 14. *Genome Announc.* **2014**, *2*, e00669-14. [CrossRef]
193. Lima, A.R.J.; de Castro, W.O.; Moraes, P.H.G.; Siqueira, A.S.; Aguiar, D.C.F.; de Lima, C.P.S.; Vianez-Júnior, J.L.S.G.; Nunes, M.R.T.; Dall'Agnol, L.T.; Gonçalves, E.C. Draft Genome Sequence of Alkalinema sp. Strain CACIAM 70d, a Cyanobacterium Isolated from an Amazonian Freshwater Environment. *Genome Announc.* **2017**, *5*, e00635-17. [CrossRef]
194. Aboim, J.B.; Oliveira, D.; Ferreira, J.E.; Siqueira, A.S.; Dall'Agnol, L.T.; Rocha Filho, G.N.; Gonçalves, E.C.; Nascimento, L.A. Determination of biodiesel properties based on a fatty acid profile of eight Amazon cyanobacterial strains grown in two different culture media. *RSC Adv.* **2016**, *6*, 109751–109758. [CrossRef]
195. de Oliveira, D.T.; Turbay Vasconcelos, C.; Feitosa, A.M.T.; Aboim, J.B.; de Oliveira, A.N.; Xavier, L.P.; Santos, A.S.; Gonçalves, E.C.; da Rocha Filho, G.N.; do Nascimento, L.A.S. Lipid profile analysis of three new Amazonian cyanobacteria as potential sources of biodiesel. *Fuel* **2018**, *234*, 785–788. [CrossRef]
196. Gradíssimo, D.G.; Mourão, M.M.; do Amaral, S.C.; Lima, A.R.J.; Gonçalves, E.C.; Xavier, L.P.; Santos, A.V. Potencial produção de biomaterial pela cianobactéria amazônica Tolypothrix SP. CACIAM 22. In *A Produção do Conhecimento nas Ciências da Saúde*; Atena Editora: Belo Horizonte, Brazil, 2019; pp. 213–224, ISBN 9788572472982.

3

Production of Polyhydroxyalkanoates using Hydrolyzates of Spruce Sawdust: Comparison of Hydrolyzates Detoxification by Application of Overliming, Active Carbon and Lignite

Dan Kucera, Pavla Benesova, Peter Ladicky, Miloslav Pekar, Petr Sedlacek and Stanislav Obruca *

Faculty of Chemistry, Brno University of Technology, Purkynova 118, 612 00 Brno, Czech Republic; Dan.Kucera@vut.cz (D.K.); pavla.benesova@vut.cz (P.B.); peter.ladicky@vut.cz (P.L.); pekar@fch.vut.cz (M.P.); sedlacek-p@fch.vut.cz (P.S.)

* Correspondence: obruca@fch.vut.cz

Academic Editor: Martin Koller

Abstract: Polyhydroxyalkanoates (PHAs) are bacterial polyesters which are considered biodegradable alternatives to petrochemical plastics. PHAs have a wide range of potential applications, however, the production cost of this bioplastic is several times higher. A major percentage of the final cost is represented by the price of the carbon source used in the fermentation. *Burkholderia cepacia* and *Burkholderia sacchari* are generally considered promising candidates for PHA production from lignocellulosic hydrolyzates. The wood waste biomass has been subjected to hydrolysis. The resulting hydrolyzate contained a sufficient amount of fermentable sugars. Growth experiments indicated a strong inhibition by the wood hydrolyzate. Over-liming and activated carbon as an adsorbent of inhibitors were employed for detoxification. All methods of detoxification had a positive influence on the growth of biomass and PHB production. Furthermore, lignite was identified as a promising alternative sorbent which can be used for detoxification of lignocellulose hydrolyzates. Detoxification using lignite instead of activated carbon had lower inhibitor removal efficiency, but greater positive impact on growth of the bacterial culture and overall PHA productivity. Moreover, lignite is a significantly less expensive adsorbent in comparison with activated charcoal and; moreover, used lignite can be simply utilized as a fuel to, at least partially, cover heat and energetic demands of fermentation, which should improve the economic feasibility of the process.

Keywords: polyhydroxyalkanoates; detoxification; lignite; *Burkholderia*

---

1. Introduction

Polyhydroxyalkanoates (PHAs) are polyesters which are synthesized by numerous naturally occurring microorganisms as energy and carbon storage materials. Moreover, due to their mechanical and technological properties resembling those of some petrochemical plastics, PHAs are generally considered a biodegradable alternative to petrochemical-based synthetic polymers [1]. PHAs have a wide range of potential applications, however, the production cost of these bioplastics are several times higher which complicates their production at an industrial scale [2].

A substantial percentage of the final cost is represented by price of carbon substrate [3]. This is the motivation for seeking alternative sources for PHAs production. Among numerous inexpensive or even waste substrates, lignocellulose materials—with sn annual generation of 80 billion tons—represent one of the most promising resources for biotechnological production of (not only) PHAs [4]. Nevertheless, utilization of lignocellulosic materials is accompanied by numerous obstacles stemming from the complex nature of these materials. To access fermentable sugars from cellulose and hemicellulose, a hydrolytic step is required. The hydrolysis of complex lignocellulose biomass is usually performed in two steps. Diluted mineral acid is used in the first step to hydrolyze hemicellulose and to disrupt the complex structure of lignocellulose, which enables subsequent enzymatic hydrolysis of cellulose [5]. Nevertheless, apart from utilizable sugars, also numerous microbial inhibitors such as organic acids (e.g., acetic, formic or levulinic acid), furfurals, and polyphenols are generated by the hydrolysis process. These substances usually reduce fermentability of the hydrolyzates and decrease yields of the biotechnological processes. This problem can be solved by introduction of detoxification. Generally, the aim of detoxification is to selectively remove or eliminate microbial inhibitors from the hydrolyzate prior to biotechnological conversion of the hydrolyzate into desired products [6].

Numerous detoxification methods are based on more or less selective removal of inhibitors by their adsorption on various sorbents. The most commonly used sorbent for this purpose is active carbon, nevertheless, this detoxification strategy suffers from high cost of the sorbent [6]. On the contrary, lignite represents a very promising, low-cost, and effective sorbent which has already been used for the treatment of wastewater to remove various organic and inorganic contaminants [7].

The woodworking industry generates a variety of solid waste materials, such as sawdust, shavings, or bark. It is true that many of these waste materials are already used in various applications. Wood waste is very often burned and used for heat and electricity generation. On the contrary, it could be a potentially inexpensive and renewable feedstock for biotechnological production of PHAs. For instance, Pan et al. employed *Burkholderia cepacia* for biotechnological production of PHAs from detoxified maple hemicellulosic hydrolyzate [8]. Further, Bowers et al. studied PHA production wood chips of *Pinus radiata* which were subjected to high-temperature mechanical pre-treatment or steam explosion in the presence of sulphur dioxide before being enzymatically hydrolyzed. *Novosphingobium nitrogenifigens* and *Sphingobium scionense* were used for PHA production on these hydrolyzates [9]. *Brevundimonas vesicularis* and *Sphingopyxis macrogoltabida* were employed by Silva et al. [10] to produce ter-polymer consisting of 3-hydroxybutyrate, 3-hydroxyvalerate, and lactic acid (3-hydroxypropionate) from acid hydrolyzed sawdust. Despite the fact that PHA production capabilities are exhibited by many bacterial strains, *Burkholderia cepacia* and *Burkholderia sacchari* are the most commonly used for PHA production from hydrolyzates of lignocellulosic materials [8,11,12].

In this study, wood hydrolyzate was utilized as a carbon source for production of polyhydroxyalkanoates. Moreover, since hydrolyzates contain substantial concentrations microbial inhibitors, various detoxification methods including the novel application of lignite as a sorbent are used to improve the fermentability of wood hydrolyzate based media and thus, the PHA yields obtained on this promising substrate.

2. Materials and Methods

2.1. Wood Hydrolyzate (WH) Preparation

Spruce sawdust was supplied by a wood processing company. The waste material was firstly dried to constant weight (80 °C for 24 h). Sawdust was then pretreated with diluted acid and thereafter subjected to enzymatic hydrolysis. To hydrolyze the hemicelluloses of raw material, 20% (w/v) pre-dried sawdust was treated by 4% H_2SO_4 for 60 min at 121 °C. Enzymatic hydrolysis, as a following step, was used for digestion of cellulose structure to release further fermentable saccharides. It was performed by adjusting the pH of the suspension to 5.0 by NaOH and cellulose was treated by 0.5% of Viscozyme L (Sigma-Aldrich, Deisenhofen, Germany) at 37 °C under permanent shaking for 24 h. Subsequently, solids were removed by filtration and the permeate, called wood hydrolyzate (WH), was used in the preparation of the cultivation medium and for PHA production.

2.2. Microorganisms and Cultivation

Burkholderia cepacia (CCM 2656) was purchased from Czech Collection of Microorganisms, Brno, Czech Republic. *Burkholderia sacchari* (DSM 17165) was purchased from Leibnitz Institute

DSMZ-German Collection of Microorganism and Cell Cultures, Braunschweig, Germany. The mineral salt medium for *B. cepacia* and *B. sacchari* cultivation was composed of: 1 g L^{-1} $(NH_4)_2SO_4$, 1.5 g L^{-1} KH_2PO_4, 9.02 g L^{-1} $Na_2HPO_4{\cdot}12H_2O$, 0.1 g L^{-1} $CaCl_2{\cdot}2H_2O$, 0.2 g L^{-1} $MgSO_4{\cdot}7H_2O$, and 1 mL L^{-1} of microelement solution, the composition of which was as follows: 0.1 g L^{-1} $ZnSO_4{\cdot}7H_2O$, 0.03 g L^{-1} $MnCl_2{\cdot}4H_2O$, 0.3 g L^{-1} H_3BO_3, 0.2 g L^{-1} $CoCl_2$, 0.02 g L^{-1} $CuSO_4{\cdot}7H_2O$, 0.02 g L^{-1} $NiCl_2{\cdot}6H_2O$, 0.03 g L^{-1} $Na_2MoO_4{\cdot}2H_2O$. The cultivations were performed in Erlenmeyer flasks (volume 100 mL) containing 50 mL of the cultivation medium. The temperature was set to 30 °C and the agitation to 180 rpm. The cells were harvested after 72 h of cultivation.

2.3. Detoxification of Hydrolyzates

Overliming was carried out as described by Ranatunga et al. [13], whereupon pH of the hydrolyzate was adjusted to approx. pH 10.0 using solid calcium hydroxide. The samples were then kept at 50 °C for 30 min, the pH was adjusted back to 7, and the sample was subsequently filtered through filter paper.

Detoxification with activated charcoal was performed as described by Pan et al. [8]. Charcoal was added to hydrolyzate in the ratio 1:20 (w/v) and stirred for 1 h at 60 °C. Solid particles were removed by filtration. Furthermore, detoxification with lignite was performed similarly, finely milled lignite power (grain size of under 0.2 mm) from South Moravian Coalfield (the northern part of the Vienna basin in the Czech Republic) was used.

2.4. Analytical Methods

All analyses of hydrocarbons and furfural were performed with a Thermo Scientific UHPLC system–UltiMate 3000. REZEX-ROA column (150 × 4.6 mm, 5 μm; City, Phenomenex, Torrance, California, USA) was used for separation. The mobile phase was 5 mN H_2SO_4 at a flow rate of 0.5 mL per min. Xylose and other saccharides were detected using a refractive index detector (ERC RefractoMax 520). Acetate, levulinic acid and furfural were detected with a Diode Array Detector (DAD-3000) at 284 nm.

Total phenolics were determinated as described by Li et al. [14] with the Folin–Ciocalteu reagent (Sigma-Aldrich). Gallic acid was used for calibration and total phenolics were expressed as milligrams of gallic acid equivalents per liter of wood hydrolyzate.

2.5. PHA Extraction and Content Analysis

To determine biomass concentration and PHA content in cells, samples (10 mL) were centrifuged and the cells were washed with distilled water. The biomass concentration expressed as cell dry weight (CDW) was analyzed as reported previously [15]. PHA content of dried cells was analyzed by gas chromatography (Trace GC Ultra, Thermo Scientific, Waltham, Massachusetts, USA) as reported by Brandl et al. [16]. Commercially available P(3HB-co-3HV) (Sigma Aldrich) composed of 88 mol. % 3HB and 12 mol. % 3HV was used as a standard; benzoic acid (LachNer, Neratovice, Czech Republic) was used as an internal standard.

3. Results and Discussion

The composite formed by cellulose, hemicellulose, and lignin is responsible for the remarkable resistance against hydrolysis and enzymatic attack [17]. Generally, proper pre-treatment of lignocellulose prior to its enzymatic hydrolysis by cellulases significantly improves fermentable sugar yields [18]. The combination of diluted acid hydrolysis (1% H_2SO_4) and enzymatic digestion of cellulose was used for hydrolysis of spruce sawdust. This approach yielded liquid hydrolyzate of wood (WH) and its composition is shown in Table 1.

Table 1. Composition of wood hydrolyzate (50 g of sawdust per 1 L of 4% H_2SO_4).

	Concentration
Glucose	4.5 g/L
Xylose	10.4 g/L
Ash	52,6 g/L
Polyphenols	1205 mg/L
Furfural	52.0 mg/L
Acetic acid	0.53 g/L
Levulinic acid	9.9 mg/L
5-HMF	*not detected*

Glucose is formed by the cleavage of cellulose. Hemicelluloses can be hydrolyzed to yield molecules such as xylose, arabinose, mannose, galactose, and uronic acid [19]. The total concentration of sugars in WH was determined to be 14.9 g L^{-1} (by the hydrolysis of 50 g L^{-1} of spruce sawdust). The only identified saccharides in WH are xylose (10.4 g L^{-1}) and glucose (4.5 g L^{-1}). Unfortunately, WH contains high concentrations of inhibiting substances such as polyphenols (1205 mg L^{-1}), furfural (52.0 mg L^{-1}), acetic acid (0.53 mg L^{-1}), and levulinic acid (9.9 mg L^{-1}). Polyphenols are likely to be released from waste wood biomass during the partial degradation of lignin by acid hydrolysis. Furfural is formed by degradation of reducing sugars at high pressure and low pH. Levulinic acid, the degradation product of furfural or 5-hydroxymethylfurfural, is formed in the same manner. Acetic acid is probably formed by deesterification of acetylated wood components. Moreover, the amount of ash is significant, which is a consequence of the application of sulfuric acid and subsequent neutralization by NaOH. High concentrations of salts may theoretically cause the inhibition of bacterial growth due to the induction of osmotic stress. On the contrary, mild osmotic up-shock was reported to support PHB accumulation in *Cupriavdius necator* H16 [20,21].

The hydrolyzate of waste wood biomass was used as the sole carbon source for PHA production employing *B. cepacia* and *B. sacchari*. Figure 1 demonstrates the negative impact of the presence of inhibitors on the intended biotechnological processes. In both cases, the WH was twice diluted prior to culturing and supplemented by mineral medium.

Yields of biomass were relatively low, approximately 1.0–1.5 g L^{-1}, and PHB content in CDW was about 10%. Total yield of PHB was around 0.1 g L^{-1}, which is very low.

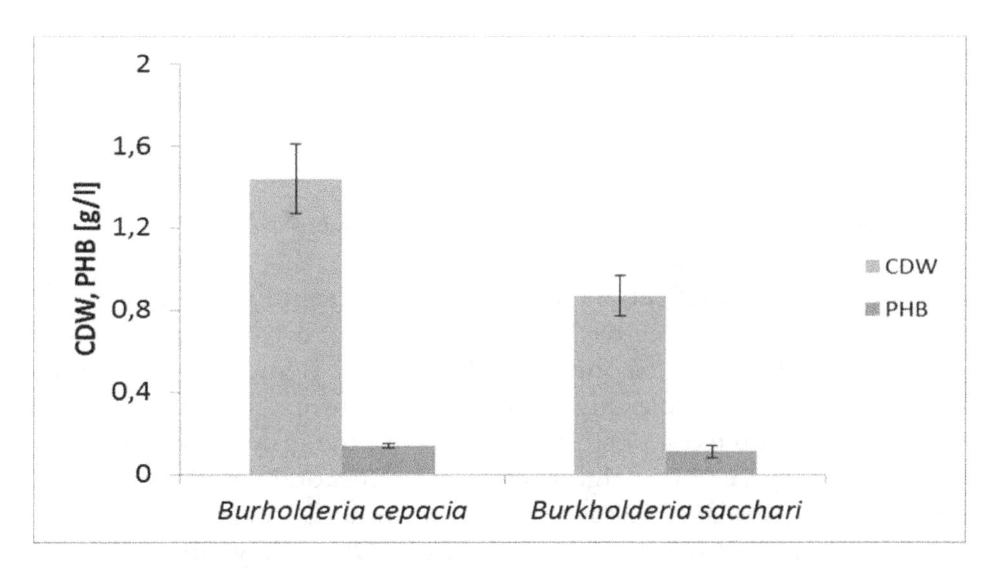

Figure 1. Cultivation of *B. cepacia* and *B. sacchari* on WH which composition is demonstrated in Table 1, WH was twice diluted and supplemented with mineral salts as described above. Cultivation conditions: 30 °C, 72 h, 180 rpm.

The effect of phenolic and other aromatic compounds, which may inhibit both microbial growth and product yield, are very variable, and can be related to specific functional groups. One possible mechanism is that phenolics interfere with the cell membrane by influencing its function and changing its protein-to-lipid ratio [22]. Undissociated acids enter the cell through diffusion over the cell membrane and then dissociate due to the neutral cytosolic pH. The dissociation of the acid leads to a decrease in the intracellular pH, which may cause cell death. This effect is promoted by furfural and 5-HMF which cause higher cell membrane permeation and disturb the proton gradient over the inner mitochondrial membrane which inhibits regeneration of ATP and eventually can lead to cellular death [23]. A different mechanism of action of growth inhibitors results in a stronger synergistic effect.

The presence of inhibitors, and especially polyphenols, in the wood hydrolyzate appears to be crucial for the intended biotechnological process. Therefore, we continued to focus on the elimination of microbial inhibitors. In the first phase, we compared two common detoxification procedures—separation by adsorption inhibitors on activated carbon and over-liming. Theoretically, overliming is effective due to precipitation or chemical destabilization of inhibitors [13] and activated charcoal could improve the fermentability of hydrolysate by absorbing phenolic compounds and other inhibitory substances [24].

The effect of various methods of detoxification on the concentration of the most important inhibitors present in hydrolysates and polyphenols is demonstrated in Figure 2.

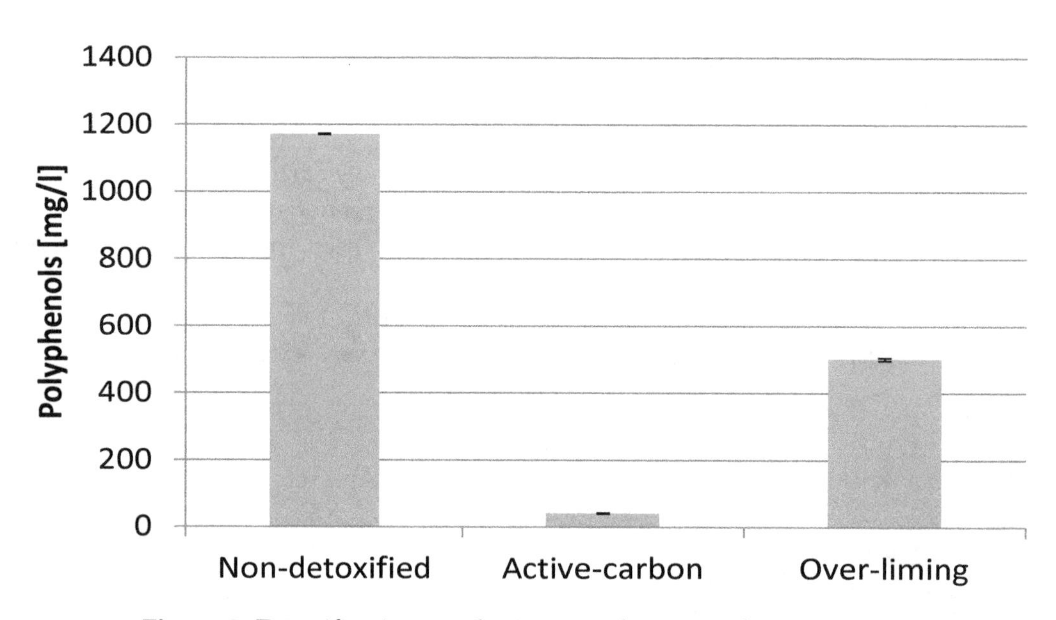

Figure 2. Detoxification employing over-liming and active carbon.

It is evident that both detoxification techniques significantly reduce the concentration of polyphenols in hydrolysates. More effective is the application of activated carbon, which can adsorb and thereby remove more than 90% of polyphenols. Table 2 demonstrates results of cultivation experiment with detoxified WH, employing the same PHB producers as in the previous test. Both methods of detoxification exhibited a positive influence on the growth of biomass, and this effect was more apparent with *B. cepacia*. More significantly, the effect of detoxification was reflected in the content of PHB in biomass. A positive effect on the biosynthesis of PHB occurred primarily in the strain of *B. sacchari*. PHB content reached nearly 90% of CDW. The yields were 8–12 times higher compared to the use of non-detoxified hydrolyzate. On the other hand, the detoxification process itself is time-consuming and particularly expensive, especially if activated carbon is used for detoxification of the hydrolyzates [25].

Table 2. Cultivation on detoxified hydrolyzates.

	Detoxification	Biomass (g/L)	PHB (%)	PHB (g/L)
Burkholderia sacchari	Non-detoxified	0.87	12.2	0.11
	Over-liming	1.57	88.7	1.39
	Activated carbon	1.01	87.6	0.89
Burkholderia cepacia	Non-detoxified	1.44	9.8	0.14
	Over-liming	2.86	30.0	0.86
	Activated carbon	1.40	74.7	1.05

A further aim of our experiments was to find an alternative sorbent, which would be comparable to activated carbon, but the cost of which would be significantly lower. After several pilot experiments, we focused on lignite. It is the youngest and the least carbonized brown coal, which consists of a macromolecular complex polyelectrolyte (e.g., humic acids), polysaccharides, polyaromatics, and carbon chains with sulfur, nitrogen, and oxygen-containing groups. Its cost is significantly lower than that of activated carbon. The price of activated carbon is currently around $1/kg [26] compared to a lignite price of $0.2/kg [26]. Moreover, the recovery of activated charcoal after its application as a sorbent in detoxification is practically impossible [27,28]. On the other side, lignite can be burned after absorbing the inhibitors and the energy released during the combustion process could provide energy which can at least partially cover energetic demands of the intended process of PHA production from waste wood biomass.

The adsorption capacity of lignite and its application as a sorbent is often a subject of interest. It is the price of conventional sorbents that leads to finding low-cost alternatives [29]. Over the last decade, there has been an increase in publications dealing with low-cost adsorbents for wastewater treatment [30]. For instance, lignite was used as a sorbent for removal of organic substances such as phenol [31] or inorganic components, especially heavy metals [32], from contaminated water solutions. Nevertheless, to our best of our knowledge despite its high sorption capacity and low cost, lignite has not been used as a sorbent for detoxification of complex lignocellulose hydrolyzates to increase their fermentability and yield of biotechnological products.

According to our results, lignite has a lower sorption capacity than activated charcoal. On the other hand, lignite is also able to eliminate a substantial amount of inhibitors, and thus potentially increase the fermentability of WH. Comparison of lignite and activated charcoal as a sorbent for microbial inhibitors is displayed in Table 3.

Table 3. Detoxification using active carbon and lignite

	Glucose (g/L)	Xylose (g/L)	Polyphenols (mg/L)	Furfural (mg/L)	Levulinic Acid (mg/L)	Acetic Acid (g/L)
Non-detoxified	4.4	10.0	998.8	41.4	10.0	0.5
Lignite	4.6	10.3	772.4	35.7	7.9	0.4
Activated carbon	4.5	10.1	23.8	3.8	3.6	0.4

The sorption properties of lignite depend on the number of sorption sites or functionalities [33]. Understandably, WH detoxified with lignite were also tested for cultivation. Detoxified hydrolyzates using lignite and activated carbon were used for the biotechnological production of PHB employing *B. cepacia* and *B. sacchari*. Figure 3 shows the results.

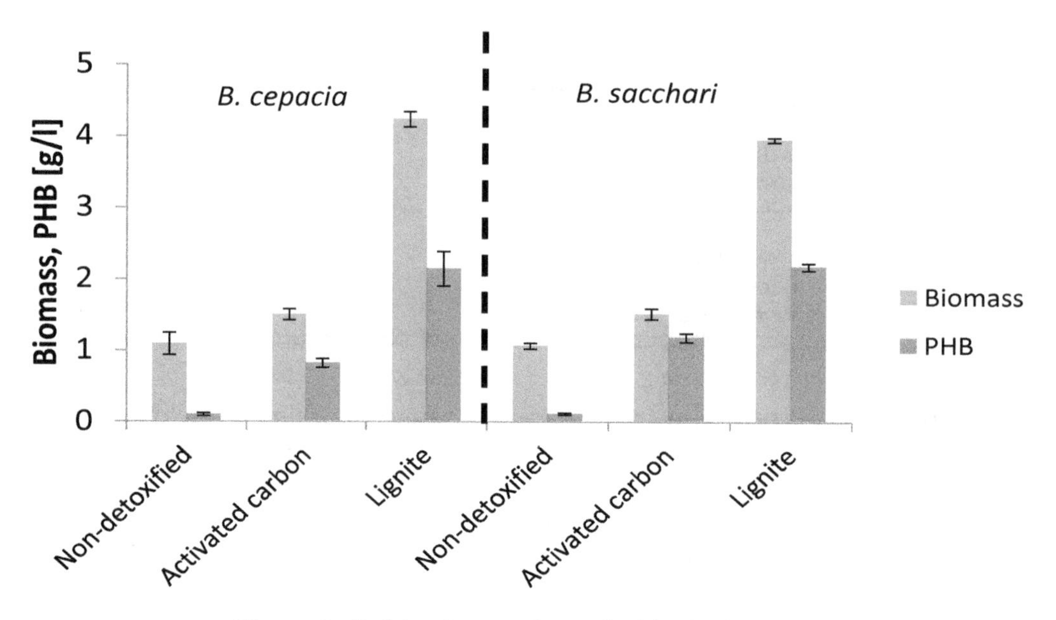

Figure 3. Cultivation on detoxified hydrolyzates.

It is interesting that the sorption capacity of lignite as detoxification strategy does not provide as good a removal of monitored inhibitors as in the case of activated carbon. However, lignite is comparable with activated charcoal, considering the overall yields of PHB. *B. Sacchari* reached markedly higher yields. These surprisingly higher yields obtained by replacing active carbon with lignite should be explained and our further experiments will be focused in this direction. We assume that, during the detoxification, lignite released substances that had a positive effect on the growth of soil bacteria. Lignite is a complex material compared to activated carbon which contains only carbon. Therefore, use of lignite could have an enriching effect on the composition of the production medium. This interesting and surprising feature could also be used in other biotechnological processes in which soil originating microorganisms are employed.

4. Conclusions

This article demonstrates the possibilities of utilization of lignite in biotechnology. Lignite can be used as a sorbent to detoxify wood hydrolyzate and its efficiency was comparable with commonly used activated carbon. This detoxification method was evaluated directly using the hydrolyzates to produce PHAs employing *Burkholderia cepacia* and *Burkholderia sacchari*. The results showed that the use of lignite considerably improved fermentability of wood hydrolyzates and enhanced PHA yields. Therefore, lignite can cope with significantly more expensive activated carbon.

Acknowledgments: This work was supported by the project Materials Research Centre at FCH BUT–Sustainability and Development No. LO1211 and by national COST project LD15031 of the Ministry of Education, Youth, and Sports of the Czech Republic and by the project GA15-20645S of the Czech Science Foundation (GACR).

Author Contributions: Dan Kucera, Peter Ladicky and Stanislav Obruca conceived and designed the experiments; Dan Kucera, Peter Ladicky, Stanislav Obruca, Petr Sedlacek, Miloslav Pekar and Pavla Benesova performed the experiments; Dan Kucera, Peter Ladicky, Pavla Benesova and Stanislav Obruca analyzed the data; Miloslav Pekar, Petr Sedlacek and Stanislav Obruca contributed reagents/materials/analysis tools; Dan Kucera and Stanislav Obruca wrote the paper.

References

1. Steinbüchel, A. Perspectives for Biotechnological Production and Utilization of Biopolymers: Metabolic Engineering of Polyhydroxyalkanoate Biosynthesis Pathways as a Successful Example. *Macromol. Biosci.* **2001**, *1*, 1–24. [CrossRef]
2. Ivanov, V.; Stabnikov, V.; Ahmed, Z.; Dobrenko, S.; Saliuk, A. Production and applications of crude polyhydroxyalkanoate-containing bioplastic from the organic fraction of municipal solid waste. *Int. J.*

Environ. Sci. Technol. **2015**, *12*, 725–738. [CrossRef]

3. Choi, J.; Lee, S.Y. Factors affecting the economics of polyhydroxyalkanoate production by bacterial fermentation. *Appl. Microbiol. Biotechnol.* **1999**, *51*, 13–21. [CrossRef]
4. Obruca, S.; Benesova, P.; Kucera, D.; Petrik, S.; Marova, I. Biotechnological conversion of spent coffee grounds into polyhydroxyalkanoates and carotenoids. *New Biotechnol.* **2015**, *32*, 569–574. [CrossRef] [PubMed]
5. Amidon, T.E.; Wood, C.D.; Shupe, A.M.; Wang, Y.; Graves, M.; Liu, S. Biorefinery: Conversion of Woody Biomass to Chemicals, Energy and Materials. *J. Biobased Mater. Bioenergy* **2008**, *2*, 100–120. [CrossRef]
6. Canilha, L.; Kumar Chandel, A.; dos Santos Milessi, T.S.; Fernandes Antunes, F.A.; da Costa Freitas, W.L.; das Graças Almeida Fellipe, M.; da Silva, S.S. Bioconversion of Sugarcane Biomass into Ethanol: An Overview about Composition, Pretreatment Methods, Detoxification of Hydrolysates, Enzymatic Saccharification, and Ethanol Fermentation. *J. Biomed. Biotechnol.* **2012**, *2012*, 1–15. [CrossRef] [PubMed]
7. Doskočil, L.; Grasset, L.; Enev, V.; Kalina, L.; Pekař, M. Study of water-extractable fractions from South Moravian lignite. *Environ. Earth Sci.* **2015**, *73*, 3873–3885. [CrossRef]
8. Pan, W.; Perrotta, J.A.; Stipanovic, A.J.; Nomura, C.T.; Nakas, J.P. Production of polyhydroxyalkanoates by *Burkholderia cepacia* ATCC 17759 using a detoxified sugar maple hemicellulosic hydrolysate. *J. Ind. Microbiol. Biotechnol.* **2012**, *39*, 459–469. [CrossRef] [PubMed]
9. Bowers, T.; Vaidya, A.; Smith, D.A.; Lloyd-Jones, G. Softwood hydrolysate as a carbon source for polyhydroxyalkanoate production. *J. Chem. Technol. Biotechnol.* **2014**, *89*, 1030–1037. [CrossRef]
10. Silva, J.A.; Tobella, L.M.; Becerra, J.; Godoy, F.; Martínez, M.A. Biosynthesis of poly-β-hydroxyalkanoate by *Brevundimonas vesicularis* LMG P-23615 and *Sphingopyxis macrogoltabida* LMG 17324 using acid-hydrolyzed sawdust as carbon source. *J. Biosci. Bioeng.* **2007**, *103*, 542–546. [CrossRef] [PubMed]
11. Keenan, T.M.; Tanenbaum, S.W.; Stipanovic, A.J.; Nakas, J.P. Production and Characterization of Poly-β-hydroxyalkanoate Copolymers from *Burkholderia cepacia* Utilizing Xylose and Levulinic Acid. *Biotechnol. Prog.* **2004**, *20*, 1697–1704. [CrossRef] [PubMed]
12. Wang, Y.; Liu, S. Production of (R)-3-hydroxybutyric acid by *Burkholderia cepacia* from wood extract hydrolysates. *AMB Express* **2014**, *4*. [CrossRef] [PubMed]
13. Ranatunga, T.D.; Jervis, J.; Helm, R.F.; McMillan, J.D.; Wooley, R.J. The effect of overliming on the toxicity of dilute acid pretreated lignocellulosics: The role of inorganics, uronic acids and ether-soluble organics. *Enzyme Microb. Technol.* **2000**, *27*, 240–247. [CrossRef]
14. Li, H.-B.; Wong, C.-C.; Cheng, K.-W.; Chen, F. Antioxidant properties in vitro and total phenolic contents in methanol extracts from medicinal plants. *Food Sci. Technol.* **2008**, *41*, 385–390. [CrossRef]
15. Obruca, S.; Marova, I.; Melusova, S.; Mravcova, L. Production of polyhydroxyalkanoates from cheese whey employing *Bacillus megaterium* CCM 2037. *Ann. Microbiol.* **2011**, *61*, 947–953. [CrossRef]
16. Brandl, H.; Gross, R.A.; Lenz, R.W.; Fuller, R.C. *Pseudomonas oleovorans* as a source of poly(beta-hydroxyalkanoates) for potential application as a biodegradable polyester. *Appl. Environ. Microbiol.* **1988**, *54*, 1977–1982. [PubMed]
17. Peters, D. Raw Materials. In *White Biotechnology*; Ulber, R., Sell, D., Eds.; Springer: Berlin, Germany, 2007; pp. 1–30.
18. Obruca, S.; Benešová, P.; Maršálek, L.; Márová, I. Use of Lignocellulosic Materials for PHA Production. *Chem. Biochem. Eng. Q.* **2015**, *29*, 135–144. [CrossRef]
19. Wyman, C.E.; Decker, S.R.; Himmel, M.E.; Brady, J.W.; Skopec, C.E.; Viikari, L. *Hydrolysis of Cellulose and Hemicellulose*; Dumitriu, S., Ed.; CRC Press: New York, NY, USA, 2004; pp. 995–1034.
20. Obruca, S.; Marova, I.; Svoboda, Z.; Mikulikova, R. Use of controlled exogenous stress for improvement of poly (3-hydroxybutyrate) production in *Cupriavidus necator*. *Folia Microbiol.* **2010**, *55*, 17–22. [CrossRef] [PubMed]
21. Passanha, P.; Kedia, G.; Dinsdale, R.M.; Guwy, A.J.; Esteves, S.R. The use of NaCl addition for the improvement of polyhydroxyalkanoate production by *Cupriavidus necator*. *Bioresour. Technol.* **2014**, *163*, 287–294. [CrossRef] [PubMed]
22. Jönsson, L.J.; Alriksson, B.; Nilvebrant, N.-O. Bioconversion of lignocellulose: Inhibitors and detoxification. *Biotechnol. Biofuels* **2013**, *6*. [CrossRef] [PubMed]
23. Chandel, A.K.; da Silva, S.S.; Singh, O.V. Detoxification of Lignocellulose Hydrolysates: Biochemical and Metabolic Engineering Toward White Biotechnology. *Bioenergy Res.* **2013**, *6*, 388–401. [CrossRef]

24. Mussatto, S.I.; Roberto, I.C. Evaluation of nutrient supplementation to charcoal-treated and untreated rice straw hydrolysate for xylitol production by *Candida guilliermondii*. *Braz. Arch. Biol. Technol.* **2005**, *48*, 497–502. [CrossRef]
25. Babel, S.; Kurniawan, T.A. Low-cost adsorbents for heavy metals uptake from contaminated water: A review. *J. Hazard. Mater.* **2003**, *97*, 219–243. [CrossRef]
26. Lignite. Available online: https://www.alibaba.com (accessed on 26 April 2017).
27. Obruca, S.; Benesova, P.; Kucera, D.; Marova, I. Novel Inexpensive Feedstocks from Agriculture and Industry for Microbial Polyester Production. In *Recent Advances in Biotechnology; Microbial Biopolyester Production, Performance and Processing Microbiology, Feedstocks, and Metabolism*; Koller, M., Ed.; Bentham Science: Berlin, Germany, 2016; pp. 3–99.
28. Parawira, W.; Tekere, M. Biotechnological strategies to overcome inhibitors in lignocellulose hydrolysates for ethanol production: Review. *Crit. Rev. Biotechnol.* **2011**, *31*, 20–31. [CrossRef] [PubMed]
29. Gautam, R.K.; Mudhoo, A.; Lofrano, G.; Chattopadhyaya, M.C. Biomass-derived biosorbents for metal ions sequestration: Adsorbent modification and activation methods and adsorbent regeneration. *J. Environ. Chem. Eng.* **2014**, *2*, 239–259. [CrossRef]
30. De Gisi, S.; Lofrano, G.; Grassi, M.; Notarnicola, M. Characteristics and adsorption capacities of low-cost sorbents for wastewater treatment: A review. *Sustain. Mater. Technol.* **2016**, *9*, 10–40. [CrossRef]
31. Polat, H.; Molva, M.; Polat, M. Capacity and mechanism of phenol adsorption on lignite. *Int. J. Miner. Process.* **2006**, *79*, 264–273. [CrossRef]
32. Klucakova, M.; Pavlikova, M. Lignitic Humic Acids as Environmentally-Friendly Adsorbent for Heavy Metals. *J. Chem.* **2017**, *2017*, 1–5. [CrossRef]
33. Robles, I.; Bustos, E.; Lakatos, J. Adsorption study of mercury on lignite in the presence of different anions. *Sustain. Environ. Res.* **2016**, *26*, 136–141. [CrossRef]

4

Polyhydroxybutyrate Production from Natural Gas in a Bubble Column Bioreactor: Simulation using COMSOL

Mohsen Moradi [1], Hamid Rashedi [2], Soheil Rezazadeh Mofradnia [1], Kianoush Khosravi-Darani [3,*], Reihaneh Ashouri [4] and Fatemeh Yazdian [5,*]

1 Department of Chemical Engineering, Faculty of Engineering, Islamic Azad University North Tehran Branch, Tehran 1651153311, Iran; khakeraheyar110@yahoo.com (M.M.); srezazadeh69@gmail.com (S.R.M.)
2 Biotechnology Group, School of Chemical Engineering, College of Engineering, University of Tehran, Tehran 11155-4563, Iran; hrashedi@ut.ac.ir
3 Department of Food Technology Research, Faculty of Nutrition Sciences and Food Technology/National Nutrition and Food Technology Research Institute, Shahid Beheshti University of Medical Sciences, Tehran 19395–4741, Iran
4 Department of Environment, Faculty of Environment and Energy, Science and Research Branch, Islamic Azad University, Tehran 1477893855, Iran; reihaneh.ashouri@yahoo.com
5 Department of Life Science Engineering, Faculty of New Science & Technology, University of Tehran, Tehran 1417466191, Iran
* Correspondence: k.khosravi@sbmu.ac.ir (K.K.-D.); yazdian@ut.ac.ir (F.Y.)

Abstract: In this study, the simulation of microorganism ability for the production of poly-β-hydroxybutyrate (PHB) from natural gas (as a carbon source) was carried out. Based on the Taguchi algorithm, the optimum situations for PHB production from natural gas in the columnar bubble reactor with 30 cm length and 1.5 cm diameter at a temperature of 32 °C was evaluated. So, the volume ratio of air to methane of 50:50 was calculated. The simulation was carried out by COMSOL software with two-dimensional symmetric mode. Mass transfer, momentum, density-time, and density-place were investigated. The maximum production of biomass concentration reached was 1.63 g/L, which shows a 10% difference in contrast to the number of experimental results. Furthermore, the consequence of inlet gas rate on concentration and gas hold up was investigated Andres the simulation results were confirmed to experimental results with less than 20% error.

Keywords: bubble column bioreactor; COMSOL; microorganism; PHB; simulation

1. Introduction

The facility of use and desirability of plastic materials properties led to their growing utilization in packaging and food industries [1]. At present, 30% of urban waste includes plastic waste. The extremely long durability of plastic (about three hundred years) has led to many environmental problems and destroyed the attractiveness of cities and nature. Burning polymeric lesions causes air pollution and according to the type of material used in their preparation, gases, such as hydrogen cyanide and other hazardous gases are released into the environment [1,2]. On the other hand, recycling has a lot of economic problems and costs. Since 1970, with the decline of the landfill problem worldwide, the issue of the use of biodegradable polymers was raised [3]. Polymers can be generally classified into two major biodegradable and non-degradable groups [4]. Non-degradable polymers, such as polyethylene, polypropylene, and polystyrene are produced from monomeric sources of oil and are resistant to environmental issues. Biodegradable polymers are separated according to their constituent components, their preparation process or their application. Biodegradable plastics or bioplastics with properties, such as biodegradability, environmental compatibility, the ability to produce renewable sources, less energy consumption in production, the production of water and carbon dioxide when degrading, compete with conventional plastics obtained from oil, especially with declining demand in the global market [4–7].

Amongst the biodegradable plastics, polyhydroxyalkanoates (PHAs) were considered due to the similarity of these materials with conventional plastics. The flexibility and expandable strength of PHA are similar to that of polypropylene and polystyrene polymers [2,8]. The agglomeration of PHAs in the cell can be enhanced by creating inhomogeneous growth conditions by limiting some nutrients like the source of nitrogen, phosphorus or sulfate, by reducing the concentration of oxygen, or by increasing the fraction of carbon to nitrogen in the feed [9,10].

One of the most important PHAs is polyhydroxy butyrate (PHB), which forms in the form of intracellular granules in different microorganisms. The high price of these biopolymers has led to limitations in their use of petrochemical polymers [11]. The main factor influencing the final cost of these polymers is the carbon origin, the carbon substrate yield, the method of fermentation, and its extraction method [12,13].

PHB can be synthesized by microorganisms with a special application in food packing [14]. However, their use is presently limited owing to the high production cost. PHB production cost is related to several key factors including the substrate, chosen strain, cultivation strategy, and downstream processing. The utilization of smart and cheap [14–22] modeling [23], proper bioreactors, experimental design [24,25], and the development of a new recovery method [24–26], as well as chances for their competition in the global market have recently been addressed. Efforts were made to optimize the growth of *Ralstonia eutropha* NRRL B14690 in the existence of nutrients, which would not only decrease the production cost of PHB but also help in increasing productivity [9].

Methane is the most suitable substrate for the production of PHB, both natural gas and biogas, between the most widely used carbon sources. Natural gas includes 85–90% of methane and is also produced by methane-producing bacteria in biological degradation of organic matter [27–30]. In many European countries, methane produced in low-cost biotechnology is accessible while in the United States, methane production units of urban waste are increasing rapidly; therefore, switching from PHB to natural gas to replace biodegradable plastics is necessary [7,31,32].

One way to enhance the production of PHB from natural gas is to use new bacterial types in a hydrodynamic biosystem. In a study, the PHB from natural gas was studied. After selecting the appropriate method, the effect of two key parameters of methane to air ratio and nitrogen on the production of PHB in a bubble pillar reactor revealed that both factors had a significant effect on the PHB density. The production of PHB by *Methylocystis* (*M.*) *hirsuta* was obtained in contrast to other methanotrophic bacteria to increase metabolism. After sampling from southern oilfields, suitable microorganisms were isolated for the production of PHB from natural gas (as a carbon source). Then, according to the Taguchi model, optimal conditions for the production of PHB from natural gas were appraised in the bubble column reactor [33]. The results showed the growth of microorganisms and the production of PHB in the existence of methane in the liquid phase. The variables affecting the production of PHB include temperature, air volume to methane gas, nitrogen source pH, phosphate source, age of inoculation, and culture medium [7,34].

In this study, first, the studies conducted in line with PHB production are conducted for familiarity and knowledge of the operation of the bioreactor activity. Then, the process is simulated using COMSOL software to find optimal circumstances and to reach a precise view for the production of PHB in a bubbler bioreactor. After this stage, the process of production of PHB in the bioreactor is investigated. Since no study was published in, the field of simulating the performance of the Methylocystis bacteria in the production of PHB in a bubble column bioreactor, the function of microorganisms in the production of PHB is simulated, which precipitates the production process and reduces costs using the results of the experiment.

2. Materials and Methods

In order to examine the simulation of the process by using COMSOL software (Version 5.2: downloady.ir) and the production review in the bioreactor, first, a number of effective simplifying assumptions were selected:

- The temperature of the system is always constant at 32 °C.
- The physical properties of the solute with time are discarded.
- The velocity of all compounds in the same phase is equal.
- The culture medium and the microorganism mixture are considered as a single phase.

2.1. *Reaction Kinetics*

The production of PHB involves a type of methanotrophic bacterium, such as *M. hirsuta*, which is an aerobic and gram-negative bacteria that can produce PHB from the serum pathway. This methylotrophic bacterium of type II is used to investigate the production of PHB in the bioreactor, and the mathematical model equations provided by Equation (1) [35]:

$$r_{PHB} = \frac{dC_{PHB}}{dt} = \mu C_{PHB} - K_d C_{PHB} \tag{1}$$

while C_{PHB} is the dry weight of cell (g/L), K_d is the cell death rate (s/1), and μ is specific growth rates for cell mass (1/s) (Appendix A). The modular kinetic model for producing PHB by Equation (2):

$$\mu = \mu_{max}\left(\frac{s}{K_s + s}\right), \tag{2}$$

In Equation (2), μ_{max} is the maximum specific growth rate of microorganisms (1/s), S is the concentration of limited substrate for growth (g/L), and K_S is the Monod constant (g/L). The kinetic of the Monod causes the substrate outlet concentration to be low in the CSTR because the K_S is usually small. In order to create a delay phase, it is suggested that K_d changes with time in Equation (3):

$$K_d = K_d(\infty)(1 - \exp(\alpha t)) \tag{3}$$

α is the reversal of fixed cell death time (s^{-1}), K_d (∞) is the infinite cell death rate (s^{-1}), and t is the time in seconds. It should be noted that for the implementation of the simulation reaction, the following simple reaction to the metabolic reactions is used for simulation:

$$6(CH_4) + NH_3 + 7(O_2) \rightarrow C_5H_7NO_2 + CO_2 + 10\,(H_2O)$$

2.2. *The Equation of the Governing Model*

The bubble flow model with two phases (diffused air in the liquid phase) was used in this study. In this model, the equations of continuity, momentum, and energy are solved for each step. The equation of motion is calculated by Equation (4) [34,35]:

$$\phi l \rho l \partial u_l/\partial t + \phi l \rho l(ul \cdot \nabla)ul = \nabla \cdot [-pl + \phi l \mu l(\nabla ul + (\nabla ul)T - 2/3(\nabla \cdot ul)l)] + \phi l \rho l g + F \tag{4}$$

In this equation, u_l indicates the velocity value in (m/s), p is the pressure in (Pa), and ϕ is the volume fraction indicated with m^3/m^3. ρ is density value with kg/m^3, g is the gravity unit with m/s^2, F is (N/m^3), μl is the dynamic velocity of the liquid replaced Pa.s in the equation. The values I and g, respectively, show the values of the liquid phase and the gas phase. The right-hand side of Equation (4) shows all forces that involve gradient, pressure, stress, adhesion, gravity, and force between the two phases, such as pulling, lifting, and virtual collective forces. In this study, a tension force is involved in the model. Based on the explanation, the continuity equation is written as Equation (5):

$$\frac{\partial}{\partial t}\left(\phi_l\rho_l + \phi_g\rho_g\right) + \nabla.\left(\phi_l\rho_l\mu_l + \phi_g\rho_g\mu_g\right) \tag{5}$$

and the gas phase transfer equation is calculated as in Equation (6):

$$\frac{\partial\phi_g\rho_g}{\partial t} + +\nabla.\left(\phi_g\rho_g\mu_g\right) = -m_{gl} \tag{6}$$

m_{gl} is the mass transfer rate from gas to liquid (kg/m^3). The gas velocity ug is equal to the sum of the velocity of Equation (7):

$$U_g = U_l + U_{slip} + U_{drift} \tag{7}$$

U_{slip} is the relative velocity among phases and U_{drift} is the drift velocity. The physics relation calculates the density of gas from the ideal gas law by Equation (8):

$$\rho_g = \frac{\left(p + p_{ref}\right)M}{RT} \tag{8}$$

M is the molecular weight of the gas (kg/mol), R is the ideal gas constant (J/(mol·K) 3/3141472), and T is the temperature (K). p_{ref} is a scalar variable being 1 at (1 at or 101.325 Pa) as a default. While a drift velocity is calculated by Equation (9)

$$U_{drif} = \frac{\mu\nabla\phi_g}{\rho\phi_g} \tag{9}$$

μ is the effective viscosity, which causes it to fall. By putting Equations (9) and (7) in (6), we will have Equation (10):

$$\frac{\partial\phi_g\rho_g}{\partial t} = \nabla\cdot\left(\phi_g\rho_g\left(U_l + U_{slip}\right)\right) = \nabla\cdot\left(\frac{\mu\nabla\phi_g}{\rho\phi_g}\right) - m_{gl} \tag{10}$$

The drift velocity of the transfer equation is introduced in the gas transfer equation. This means that the gas transport equation is actually implemented in the physics interface. The equation of the bubble flow equation is relatively simple but it can indicate non-physical behavior. An artificial accumulation of bubbles, for example, is at the base of the walls in which the pressure gradients raise the bubbles while the bubbles have no place to go and there is no model for modifying the amount of gas fraction to grow. In order to prevent this, μ is set to μl for laminar. The only clear effect in most cases when the bubble flow equations are applicable is that the non-physical accumulation of bubbles reduces. The small effective viscosity in the transfer equation for φ_g has beneficial effects on the numerical properties of the equation system.

2.3. Simulation Operations

In order to define the system, the properties and repercussions contained in COMSOL software were given to the simulation system. COMSOL simulation software has a complete set of information, properties, and constancy of materials, and if there is no specific information for a system, it can be found with the help Perry's handbook and enter into the physics of the problem. The geometry of the system was cylindrical with a radius of 0.015 m and a stanchion at the end of the cylinder with a radius of 0.1 mm. In order to prevent computing and convergence, and in view of the symmetry in the given geometry, the problem was solved in the symmetric two-dimensional mode, which did not have an effect on the overall solution due to the application of boundary conditions. All three components of water, methane, and ammonium were regarded as a phase. In this way, the liquid environment indicates the environment of the wastewater in which the ammonium and methane are soluble, and the properties of the bacteria are applied to the environment. The reaction occurred to produce biomass of an irreversible first-degree reaction with a reaction constant $k_1 = 10^{-5}$ in form of $Va \rightarrow W$ with the reaction rate, $r = k_1 C_{va}$. The system temperature was set to 305.15 K. Since the environment is water, the incompressible flow with density and dynamic viscosity, respectively, were regarded as 1000 kg/m^3 and 10^{-3} Pa.s. The initial pressure level (P_0) was considered zero and the permeation coefficient of the environment 10^{-9} m^2/s applied. The system under the boundary conditions of sleep with the equation $u = 0$ means that the system boundary was fixed and the simulation conducted in the symmetric two-dimensional mode. Figure 1 shows a schematic view of system geometry in the two-dimensional mode.

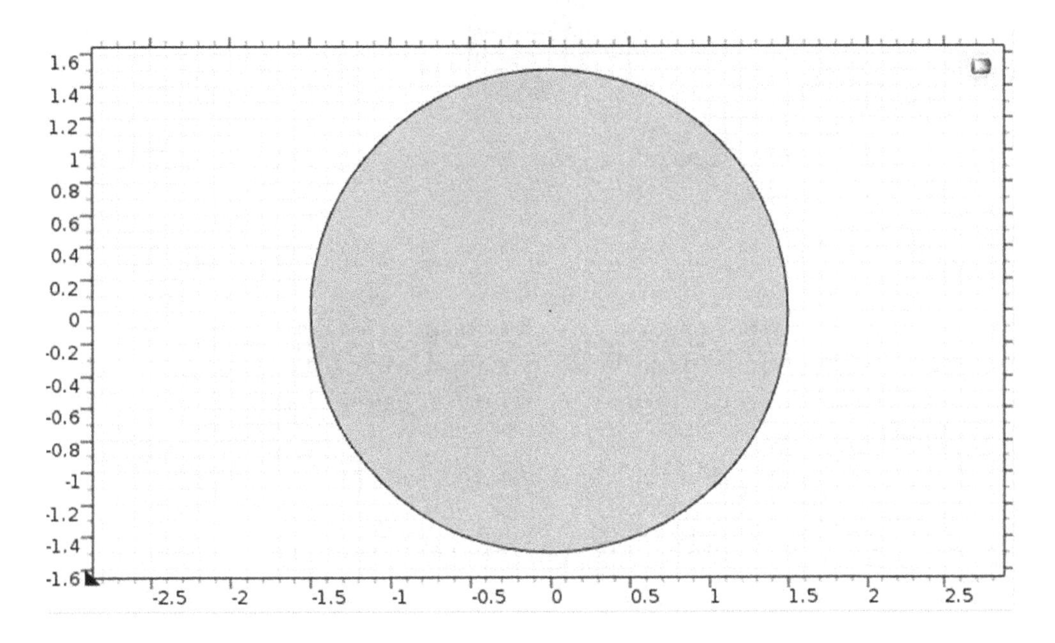

Figure 1. A schematic view of the system geometry.

2.4. *Resolution Independence Analysis of Numerical Grid*

The number of model components is designed to meet the conditions of numerical resolution independence from the computational grid size and gradually increases the number of components. The resolution independence test of the computational grid to a point in which the difference in response is less than 5%. The number of components in resolution independence tests of the numerical grid is given in Table 1.

Table 1. Constants and values of the parameters inputted.

Constants	Value
Temperature	32 °C
Physical properties of the solute	discard
Velocity	Equal in all part
Situation of flow	Single-phase
Columnar bubble reactor	30 cm and 1.5 cm

2.5. *Meshing*

At this stage, the geometry of the system was considered under meshing. In a two-dimensional model, the symmetrical mesh arrangement was selected based on the free triangular model and fine mesh size. The geometry of this system was split into 34,279 components. In Figure 2, a schematic view of the system meshing is presented in a two-dimensional mode.

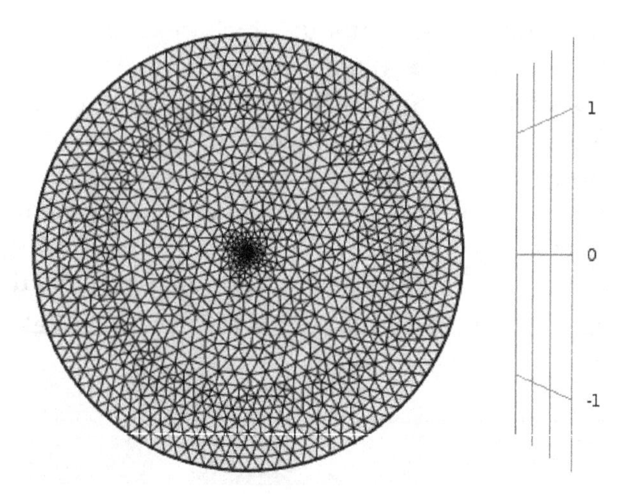

Figure 2. A schematic view of the geometry meshing with 34,269 components.

3. Results

In the production of PHB from natural gas, the objective of optimizing the composition of the culture medium is to provide important and sensitive food, increase the yield of the product, prevent product degradation, and decrease the formation of harmful side effects.

3.1. *The Results of Resolution Independence Analysis of Numerical Grid*

As indicated in Table 2, three large, medium, and fine grid sizes were applied separately in the model and the final concentration of the biomass in each mesh size was studied to examine the sensitivity of the computational grid. Table 3 indicates the final concentration of biomass in the last step of solving equations in three computational grids.

Table 2. The number of elements in different groups to check the resolution independence of the numerical grid.

Mesh Size	Fine	Medium	Coarse
Number of elements	70,563	34,269	16,396

Table 3. Final concentration of biomass in the last step of solving equations in three computational grids.

Mesh Size	Fine	Medium	Coarse
Concentration (g/L)	1.63474	1.63338	1.63233

3.2. *Concentration of Biomass*

3.2.1. Concentration Contour

Biomass ($C_5H_7NO_2$) is generated under the applied conditions according to the reaction from Va $\rightarrow$ W. After the simulation, the concentration contour was determined at 14 min. Figure 3 displays the concentration contour of biomass. Figure 3 shows the concentration gradient is from the center toward the wall. As indicated in Figure 4, the concentration contour at time t = 2 min and t =10 min can be seen for the circulation and mixing of PHB [36].

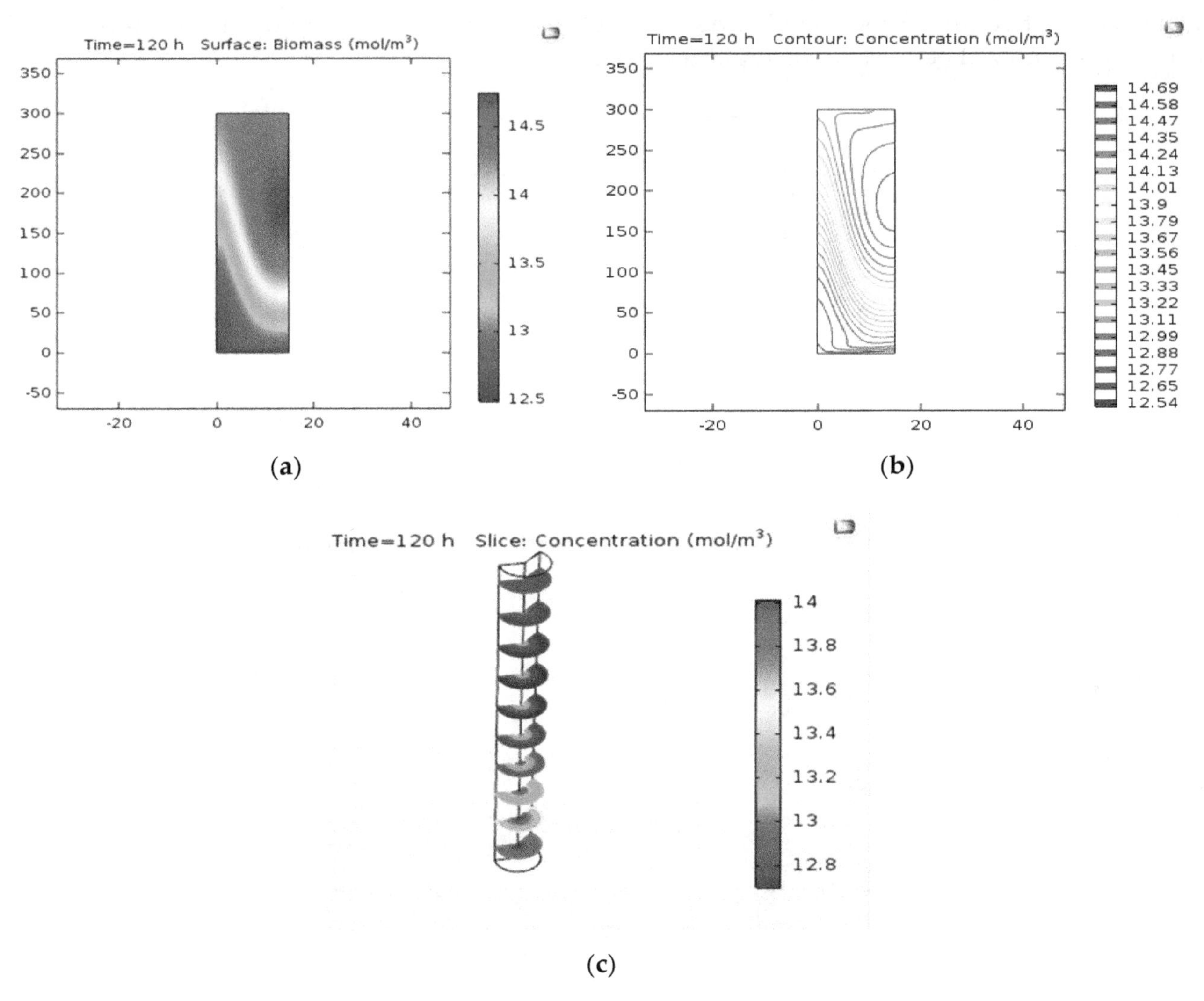

Figure 3. (**a**) The concentration surface contour of biomass. (**b**) The concentration contour of biomass. (**c**) The slice concentration contour of biomass.

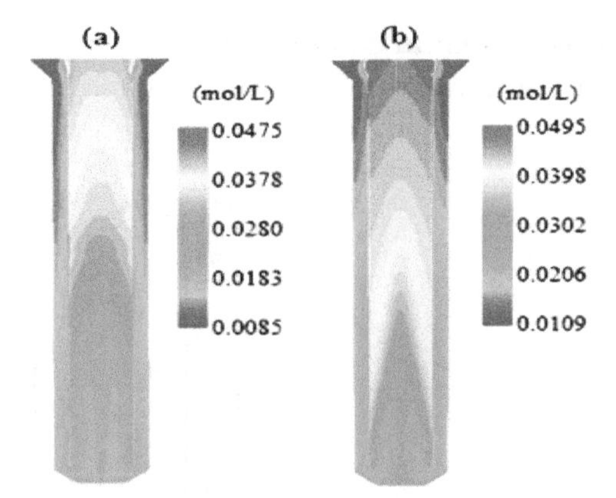

Figure 4. The concentration contour of polyhydroxy butyrate (PHB) at time (**a**) t = 2min and (**b**) t = 10 min [36].

3.2.2. Concentration Variations versus Time

The changes in the concentration of biomass were studied with time and the results are shown in Figure 5. As can be seen in this curve, the final concentration of biomass is almost 14 mol/m^3, which is less than 5% in comparison to laboratory values (14.5-mol/m^3). Figure 6 indicates concentration-location changes in the two-dimensional model. As can be observed, the concentration at the beginning of the reactor is maximized because of the rapid reaction.

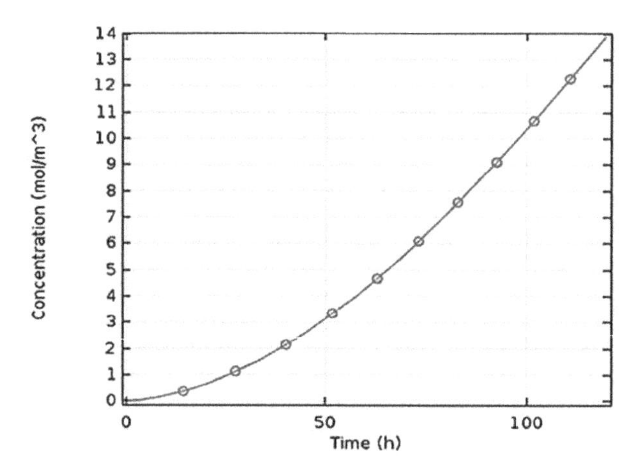

Figure 5. Changes in the concentration of biomass with time.

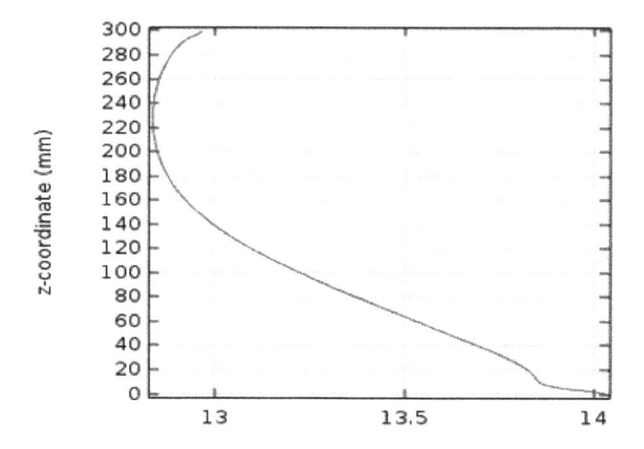

Figure 6. Linear changes of the reactor location concentration.

3.3. *Velocity Contour*

Figure 7a,b indicate the results of spatial changes of fluid velocity in the bioreactor. On the right side of Figure 7, the velocity assigned to each color is represented.

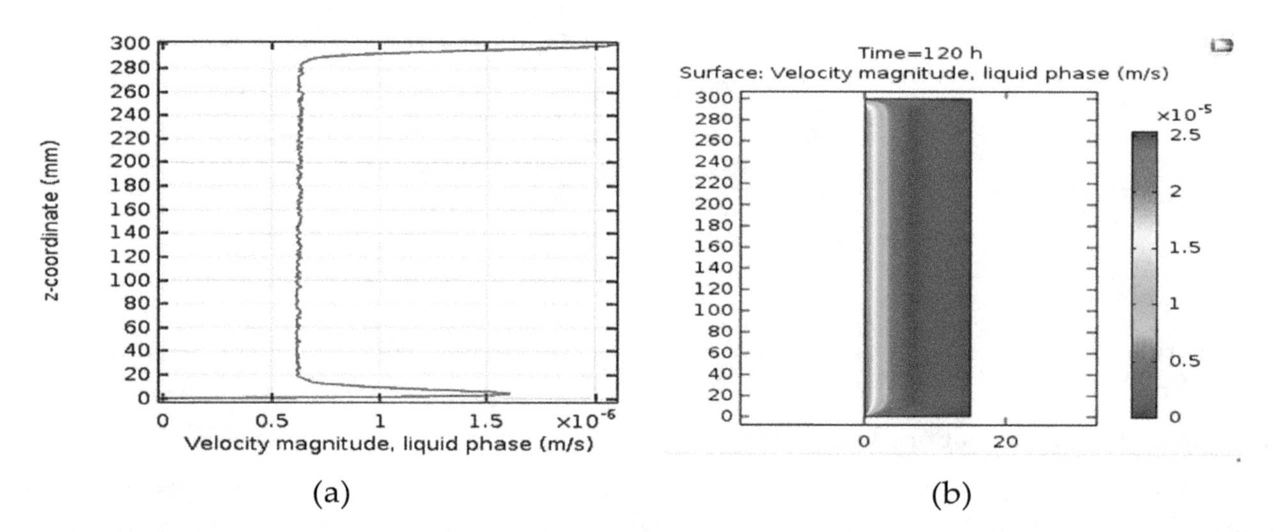

Figure 7. (**a**) Spatial changes of velocity in the liquid phase along the reactor. (**b**) Surface spatial changes of velocity in the liquid phase along the reactor.

3.4. *Analysis of Variations in the Input Gas Velocity*

The gas flow rate inside the reactor will influence the number of bubbles and the mass transfer rate. Thus, the simulation was studied at four different velocities of 0.0015, 0.065, and 0.15 m/s. The results are displayed in Figure 8a–c.

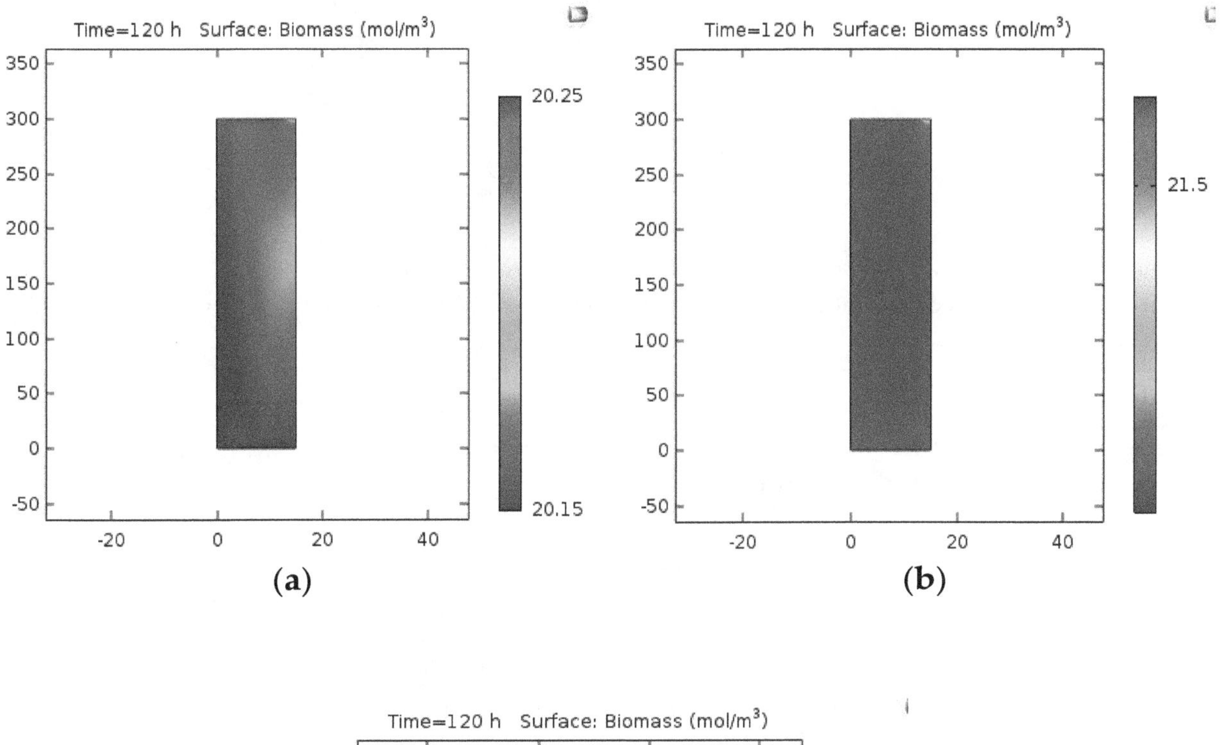

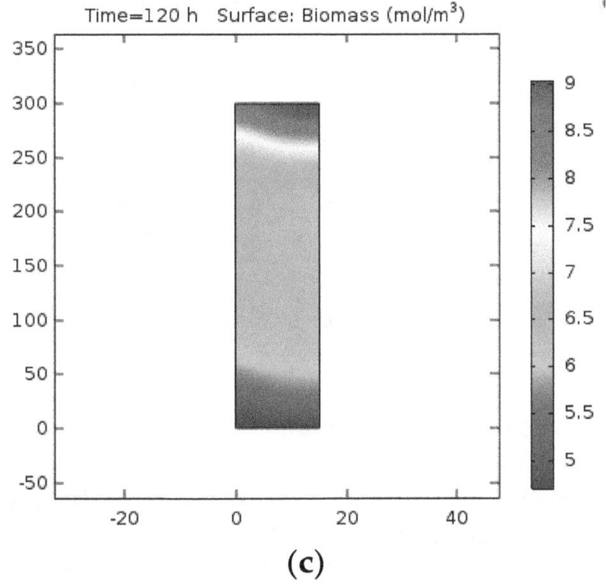

Figure 8. (**a**) Biomass concentration at the gas velocity of 0.0015 m/s, (**b**) biomass concentration at the gas velocity of 0.065 m/s, (**c**) biomass concentration at the gas velocity of 0.15 m/s.

3.5. *Effect of Changing the Bubble Diameter on the Concentration*

In Figure 9a,b, the change of concentration of the biomass can be observed with a bubble diameter of 3.5 and 1.5 mm.

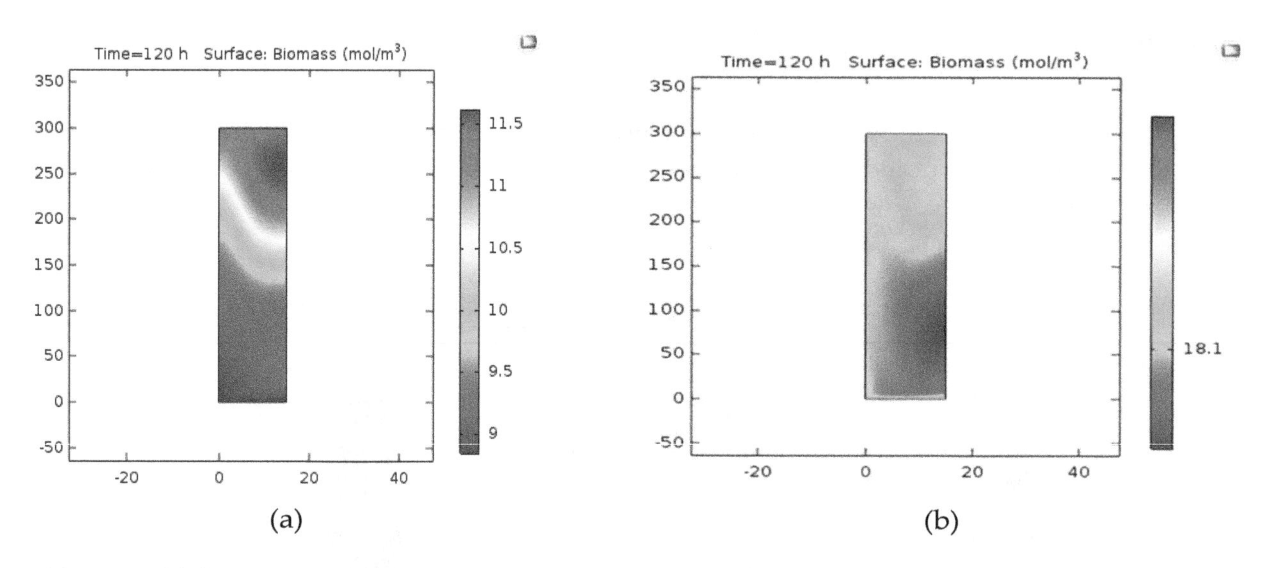

Figure 9. (**a**) Biomass concentration with a bubble diameter of 3.5 mm. (**b**) Biomass concentration with a bubble diameter of 1.5 mm.

3.6. Pressure Analysis

The relation between the amount of gravity and the bubbles are directed which are effected on pressure changes as presented in Figure 10.

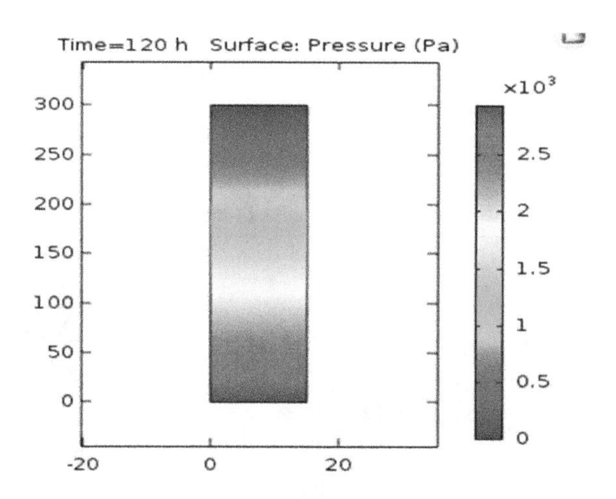

Figure 10. Pressure contour in two-dimensional mode.

3.7. Gas Accumulation

Gas accumulation is a dimensionless key dimensional parameter for design purposes identifying the phenomenon of transition in bubble column systems (Equation (11)). It is basically defined as the volume fraction of gas-phase occupied by gas bubbles. Similarly, liquid and solid phases can be determined as a liquid and solid phase coefficient. All studies examined gas accumulation because it plays an essential role in the design and analysis of bubble columns. The volume of gas accumulation with the mathematical relation was studied. In similar, the results achieved in the simulation show the very low gas accumulation in the lower part of the reactor and around the central axis, which can be observed in Figure 11a,b. Along the reaction, gas accumulation increases throughout the reactor but it is false since the accumulated gas is eliminated from the upper end of the reactor and ultimately the amount of accumulated gas at the bottom of the reactor is computed. The graph of gas volume coefficient throughout the reactor can be seen in Figure 11c, such as the value of gas volume coefficient and their increase and decrease tendency with time and position.

$$\varepsilon\_g = u\_g/(0.3+2u\_g) \quad (11)$$

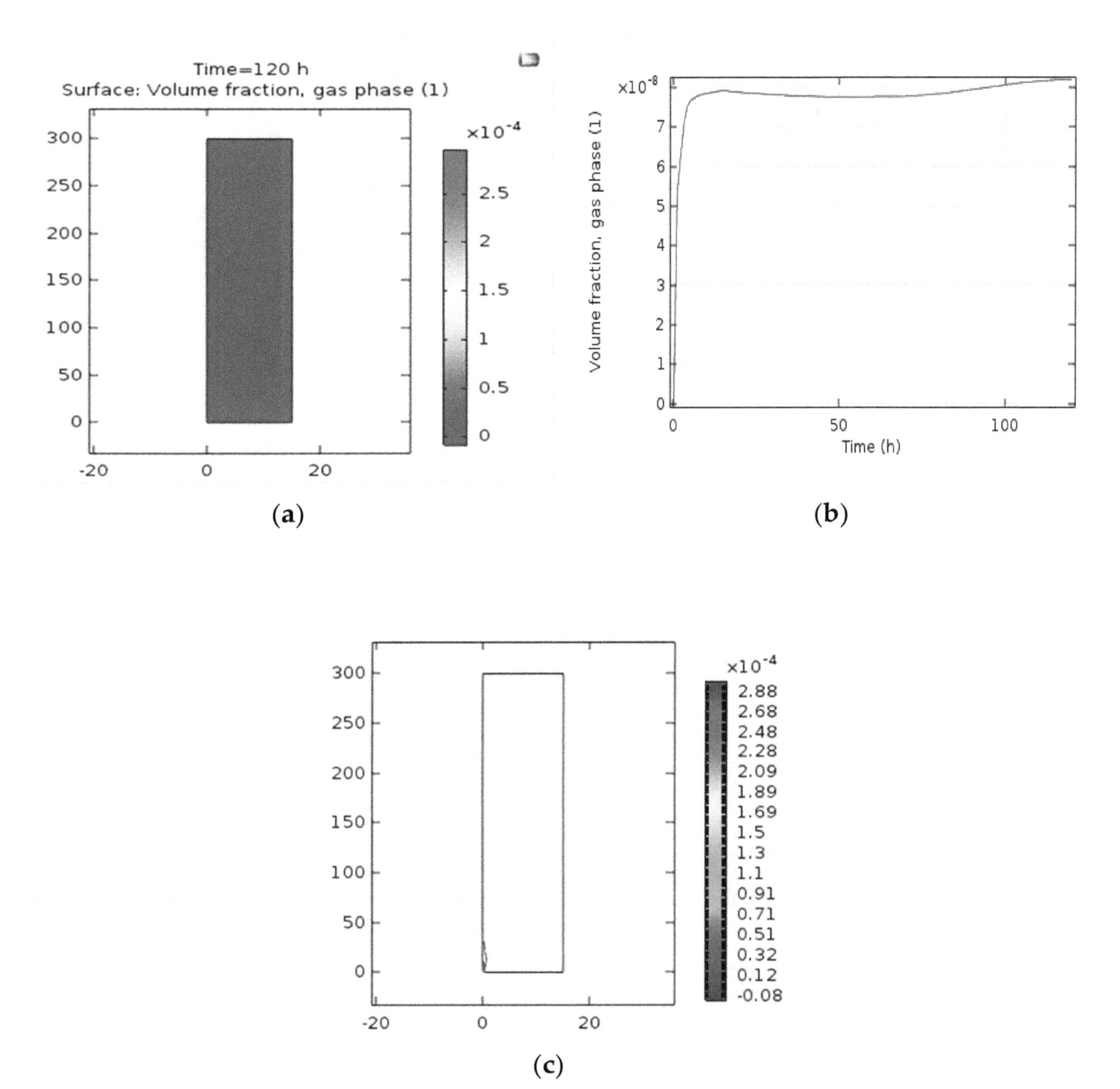

Figure 11. (**a**) The surface rate of gas accumulation along the reactor. (**b**) The rate of gas accumulation along the reactor. (**c**) Gas volume coefficient versus time.

3.8. Shear Stress

In the present project, the shear stress in different input flows was studied indicating (20) and (21) the stress rate, which is based on the scale available on the right of these shapes, the results achieved in line with the expectation of the sign, which contributes to an increase in the stress rate at the vicinity of the oven in line with the increase in the rate of intake gas and discharge. Figure 12 indicates the stress rate over time.

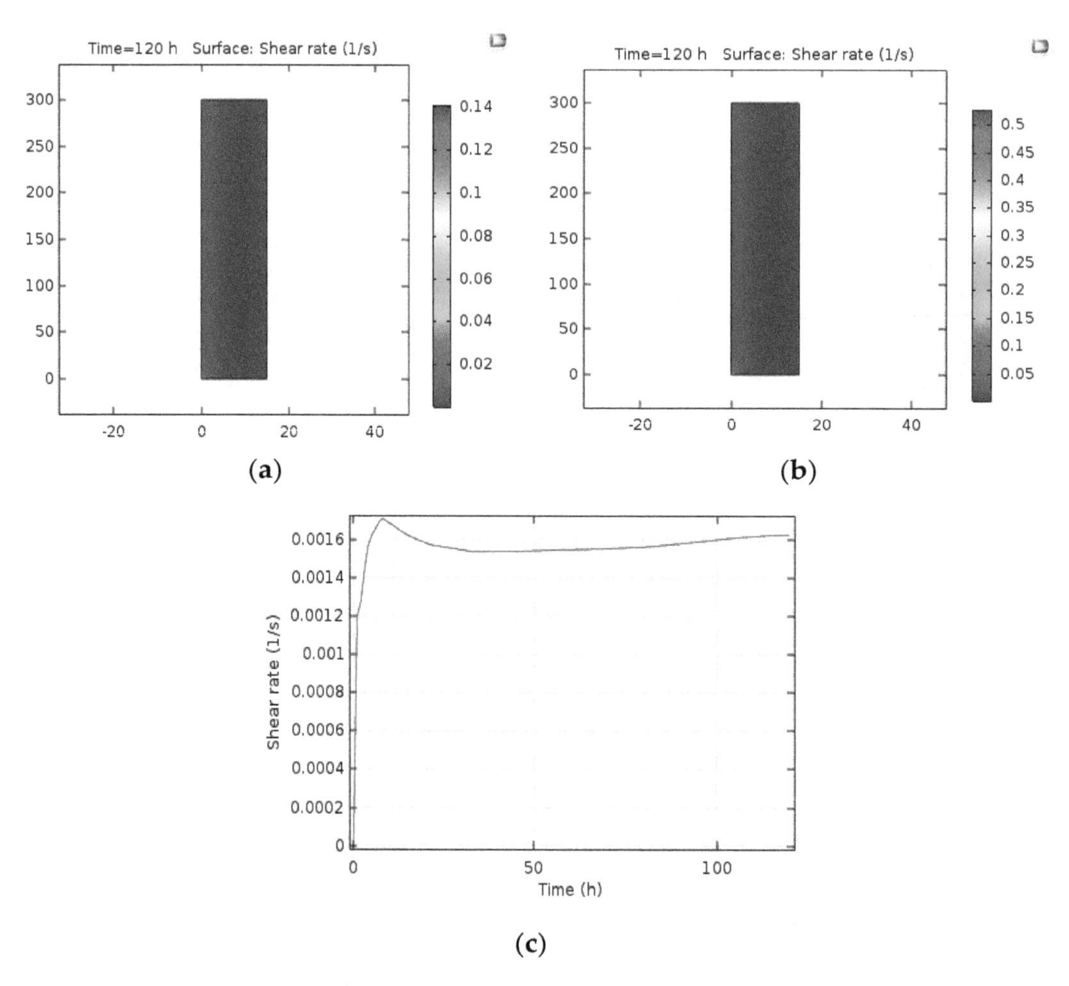

Figure 12. (**a**) The stress rate at the gas input velocity of 0.015 m/s. (**b**) The stress rate at the gas input velocity of 0.055 m/s. (**c**) The stress rate over time.

4. Discussion

4.1. The Results of Resolution Independence Analysis of Numerical Grid

As Table 2 shows, the variance in the three large, medium, and fine grid sizes is less than 1% indicating the independence of the obtained responses on the type and size of the grid.

4.2. Concentration of Biomass

4.2.1. Concentration Contour

Figure 3 shows that from the center toward the wall, concentrations are reducing being observed in Figure 3b. Thus, the largest value of production is located near the central axis. By enhancing the velocity of the areas around the center of the bioreactor, the maximum concentration, and with a distance less than the center of the percentage of production means that the increase in velocity would lead to a shorter spatial dimension to the final production.

Mousavi et al. examined the simulation of PHB production in an aircraft bioreactor with Fluent software. The average molecular concentration of PHB in the bioreactor was almost 0.331 and 0.3337 mol/L [36].

4.2.2. Concentration Variations versus Time

Figure 5 shows the changes in the concentration of biomass were investigated vs. time. The final concentration of biomass is 14 mol/m^3. The researchers showed the production of PHB was 8 g/L in a bubble column bioreactor at optimized conditions [17]. In addition, Shah Hosseini et al. applied a dynamic optimization program written in MATLAB software to specify the optimal amount of carbon and nitrogen sources on PHB production. The proposed model presented was confirmed on the experimental data. Using the preferred feed strategy, PHB was enhanced by 100% [37].

4.3. Velocity Contour

Figure 7 shows the results of spatial changes in fluid velocity in the bioreactor. Dean et al. (2013) evaluated the use of large-spin simulations (LES) in mathematical simulations of microwave in bubble column reactors. The Euler–Euler approach is used to clarify two-phase motion equations that correspond slightly to the experimental data for both mid-velocity and motor velocities. LES provides better matching with simulation data through the k-ε model [38].

Mousavi et al. studied the fluid velocity fluctuations of the aircraft bioreactor in which the maximum fluid velocity can be seen in the input aperture at the bottom of the reactor. Similarly, the velocity of the fluid in the Down Comer is much lower than the Riser section [36].

Liquid velocity can directly effect production and methane gas enhances the growth structure of the biomass because these two parameters change in parallel. The flow rate can also change the behavior of the bioreactor and can change the production speed according to the size of the meshes designed [39,40].

4.4. Analysis of Variations in the Input Gas Velocity

The result of change of the gas flow rate inside the reactor vs. the number of bubbles and mass transfer rate (at four different velocities of 0.055, 0.015, 0.065, and 0.15 m/s) are displayed in Figure 8. As can be observed, as the gas flow rate increases, the content of biomass increases because of the increase in the number of bubbles and the number of encounters, which finally leads to an increase in the mass transfer rate and biomass production. The effects of gas and liquid velocity in a bubble pillar biomass with COMSOL software varying in the gas-liquid bubble column system, the velocity of gas and liquid with time and space in the column. Weather velocity vectors were achieved after a semi-stable mode at an input velocity of 0.001 m/s [41]. Such vectors indicated the velocity of the path along the path, which is useful in specifying the patterns of flow in the bubble column system. That contour indicated a gas concentration of 0.001 m/s.

4.5. Effect of Changing the Bubble Diameter on the Concentration

Figure 9 shows that the concentration of the biomass is reduced by increasing the diameter of the bubble to reduce the flow rate. In addition, it enhances the final concentration of biomass by decreasing the diameter of the bubble.

4.6. Pressure Analysis

As it can be observed in Figure 10, the pressure across the reactor from bottom to top reduces because of the presence of gravity on the bubbles.

4.7. Gas Accumulation

The graph of changes in gas volume throughout the reactor can be seen in Figure 11. In 2010, Mousavi et al. studied gas cumulating in aircraft bioreactors at numerous aeration rates [36]. The results obtained from the simulation were compared to the results of different mathematical relations related to the bubble columns and aircraft reactors. The simulation results are relatively suitable in comparison to the relationships and it approves the reliability of the simulation. For the aeration rate of 30 L/g. min, the mean diameter of the bubble was 5.1 mm [36].

Gas accumulation occurs during the five different phases of aeration. As expected, the amount of gas accumulation increased with aeration. The highest volume is always increased initially and then gas accumulation decreases because of the gas outflow from the bioreactor. However, in cases with higher aeration rates (40 and 50 L/min), the final amount of gas accumulation is higher than the peak value. In higher aeration conditions (40 and 50 L/min), the amount of gas accumulation is higher than the initial rate of gas accumulation in comparison to the lower aeration rate due to the accumulation of gas in the liquid phase.

4.8. Shear Stress

Wall shear stress is a critical parameter for determining energy transfer and movement in a two-phase flow system. The shear stress properties play an essential role in understanding the internal state of the two-phase flow because the liquid velocity and the highest grade of turbulence fluctuations are specified with the highest gradient. In order to clarify the mechanism of the two-phase bubble flow, it is significant to know the performance of the bubbles in the fluid flow. The movement of isolated bubbles seems to be related to the bubble distortion, the location of the injection point, and the average fluid flow rate. They found the only bubbles with diameters less than 5 mm along the wall while all the spherical bubbles and elliptical bubbles with a diameter greater than 5 milligram meters in the core of the stream [42,43]. Mousavi et al. in a study on shear stress in aircraft bioreactors specified the average shear stress in the bioreactor by applying the equations in FLUENT software for five different conditions. As expected, aeration enhances the shear stress in the bioreactor. Since airborne bioreactors and bubble columns fail at having a propellant, their tensions are lower than those of cohesive bioreactors, and the shear stress in the bioreactor is not likely to be significant, while the ability of the software to calculate shear stress is critical. The greatest shear stress relates to areas where gas is flowing from the opening. Vortices are located at the vicinity of upstream and downstream sections as well as the walls with higher shear stresses compared to other points [36].

5. Conclusions

In this study, a simulation of microorganisms capability for the production of PHB from natural gas (source of carbon) was carried out. The error interval is normally up to 20% for the difference between laboratory data and simulation because the resulting error can be because of simple simulation assumptions; thus, the simulation results were in good accordance with empirical results. Regarding the concentration contour, concentration is the lowest in the center of maximum concentration near the wall. The concentration gradient from the center is toward the wall and produces the highest percentage of production near the central axis of the bioreactor. Substrates and microorganisms with the environment were considered homogeneous, thus we would not have much difference in production compared to the place. Enhancing the gas flow rate has a direct relation to the production rate. Based on the three-velocity concentration-time graph, enhancing the intake rate of the agitator stirring increases and the percent of production enhances. The flow pattern in a bioreactor is dependent on the diameter of the bioreactor and the flow rate of the inlet gas. In this study, the bubble flow studied it placed due to the low velocity of the inlet gas (0.015 m/s) and the low diameter of the bioreactor (1.5 cm). Because of the concentration contour with enhancing the flux, the number of bubbles increases and results in more mass transfer and an increase in the value of biomass production. With increasing fluid velocity, a decrease can be seen in the amount of gas accumulation because liquid bubbles are rapidly displaced by liquid at higher velocities. Moreover, increasing fluid velocity decreases the duration of bubble stays.

Furthermore, based on the results in volume concentration, 50/50 from air and methane, the highest rate of microorganism growth and PHB production was achieved. According to presented graphs, such as time concentration and location concentration, the final amount of biomass production extracted was 1.6338 g/L under optimal conditions. The reason for selecting the volume of 50/50 was the coordination with the Taguchi algorithm that is based on research in the same conditions is the ideal conditions for evaluation of chosen sizes of the system. In addition, the amount of mesh size in the system was not significantly affected by the final concentration of production, so the ideal condition was regularly 1.6338 g/L. Thus, in different conditions of meshes, less than 1% difference was observed. In addition, the amount of gas accumulation decreases with increasing gas velocity.

Author Contributions: Conceptualization, M.M., F.Y., and S.R.M.; methodology, F.Y.; software, M.M.; validation, R.A. and F.Y.; formal analysis, R.A. and K.K.-D.; investigation, M.M.; resources, F.Y.; data curation, K.K.-D.; guidance and validation: H.R.; writing—original draft preparation, F.Y. and K.K.-D.; writing—review and editing, F.Y. and K.K.-D.; supervision, F.Y.

Acknowledgments: Authors wish to acknowledge all administrative and technical support supports given by Tehran University.

Appendix A

List of symbols in equations:

Parameters	Description
C_{PHB}	Dry weight of cell (g/L)
K_d	Cell death rate (s/1)
μ	Growth rates for cell mass (1/s)
μ_{max}	The maximum specific growth rate of microorganisms (1/s)
S	The concentration of limited substrate for growth (g/L)
Ks	Monod constant (g/L)
α	Fixed cell death time (s^{-1})
K_d (∞)	Infinite cell death rate (s^{-1})
u_l	Velocity value in (m/s)
p	Pressure in (Pa)
ϕ	Volume fraction indicated with m^3/m^3
ρ	Density value with kg/m^3
g	Gravity unit with m/s^2
F	N/m^3
μL	Dynamic velocity Pa.s
m_{gl}	The mass transfer rate from gas to liquid (kg/m^3)
U_{slip}	Relative velocity
U_{drift}	Drift velocity
M	The molecular weight of the gas (kg/mol)
R	Ideal gas constant (J/(mol·K) 3/3141472)
T	Temperature (K)
P_{ref}	Scalar variable at (1 at or 101.325 Pa)
μ	Effective viscosity

References

1. Brandl, H.; Gross, R.A.; Lenz, R.W.; Fuller, R.C. Plastics from bacteria and for bacteria: Poly (β-hydroxyalkanoates) as natural, biocompatible, and biodegradable polyesters. In *Microbial Bioproducts;* Springer: Berlin, Germany, 1990; pp. 77–93.
2. Reddy, C.S.K.; Ghai, R.; Kalia, V. Polyhydroxyalkanoates: An overview. *Bioresour. Technol.* **2003**, *87*, 137–146. [CrossRef]
3. Shimao, M. Biodegradation of plastics. *Curr. Opin. Biotechnol.* **2001**, *12*, 242–247. [CrossRef]
4. Chen, C.S.; Liu, T.G.; Lin, L.W.; Xie, X.D.; Chen, X.H.; Liu, Q.C.; Liang, B.; Yu, W.W.; Qiu, C.Y. Multi-walled carbon nanotube-supported metal-doped ZnO nanoparticles and their photocatalytic property. *J. Nanopart. Res.* **2013**, *15*, 1295. [CrossRef] [PubMed]

5. Braunegg, G.; Lefebvre, G.; Genser, K.F. Polyhydroxyalkanoates, biopolyesters from renewable resources: Physiological and engineering aspects. *J. Biotechnol.* **1998**, *65*, 127–161. [CrossRef]
6. Mofradnia, S.R.; Tavakoli, Z.; Yazdian, F.; Rashedi, H.; Rasekh, B. Fe/starch nanoparticle-Pseudomonas aeruginosa: Bio-physiochemical and MD studies. *Int. J. Biol. Macromol.* **2018**, *117*, 51–61. [CrossRef] [PubMed]
7. Mofradnia, S.R.; Ashouri, R.; Tavakoli, Z.; Shahmoradi, F.; Rashedi, H.; Yazdian, F. Effect of zero-valent iron/starch nanoparticle on nitrate removal using MD simulation. *Int. J. Biol. Macromol.* **2018**, *121*, 727–733. [CrossRef] [PubMed]
8. Ashouri, R.; Ghasemipoor, P.; Rasekh, B.; Yazdian, F.; Mofradnia, S.R.M.; Ghasemipoor, R.A.P.; Yazdian, B.R.F.; Mofradnia, S.R.M. The effect of ZnO-based carbonaceous materials for degradation of benzoic pollutants: A review. *Int. J. Environ. Sci. Technol.* **2018**, *16*, 1–12. [CrossRef]
9. Khanna, S.; Srivastava, A.K. Statistical media optimization studies for growth and PHB production by Ralstonia eutropha. *Process Biochem.* **2005**, *40*, 2173–2182. [CrossRef]
10. Young, F.K.; Kastner, J.R.; May, S.W. Microbial production of poly-β-hydroxybutyric acid from D-xylose and lactose by Pseudomonas cepacia. *Appl. Environ. Microbiol.* **1994**, *60*, 4195–4198. [PubMed]
11. Lee, S.Y. Bacterial polyhydroxyalkanoates. *Biotechnol. Bioeng.* **1996**, *49*, 1–14. [CrossRef]
12. Yamane, T. Yield of poly-D (−)-3-hydroxybutyrate from various carbon sources: A theoretical study. *Biotechnol. Bioeng.* **1993**, *41*, 165–170. [CrossRef] [PubMed]
13. Saratale, R.G.; Saratale, G.D.; Cho, S.K.; Kim, D.S.; Ghodake, G.S.; Kadam, A.; Kumar, G.; Bharagava, R.N.; Banu, R.; Shin, H.S. Pretreatment of kenaf (*Hibiscus cannabinus* L.) biomass feedstock for polyhydroxybutyrate (PHB) production and characterization. *Bioresour. Technol.* **2019**, *282*, 75–80. [CrossRef] [PubMed]
14. Khosravi-Darani, K.; Bucci, D.Z. Application of poly (hydroxyalkanoate) in food packaging: Improvements by nanotechnology. *Chem. Biochem. Eng. Q.* **2015**, *29*, 275–285. [CrossRef]
15. Darani, K.K.; Vasheghani-Farahani, E.; Tanaka, K. Hydrogen oxidizing bacteria as poly (hydroxybutyrate) producers. *Iran. J. Biotechnol.* **2006**, *4*, 193–196.
16. Khosravi-Darani, K.; Yazdian, F.; Babapour, F.; Amirsadeghi, A.R. Poly (3-hydroxybutyrate) Production from Natural Gas by a Methanotroph Native Bacterium in a Bubble Column Bioreactor. *Chem. Biochem. Eng. Q.* **2019**, *33*, 69–77. [CrossRef]
17. Ghoddosi, F.; Golzar, H.; Yazdian, F.; Khosravi-Darani, K.; Vasheghani-Farahani, E. Effect of carbon sources for PHB production in bubble column bioreactor: Emphasis on improvement of methane uptake. *J. Environ. Chem. Eng.* **2019**, *7*, 102978. [CrossRef]
18. Mokhtari-Hosseini, Z.B.; Vasheghani-Farahani, E.; Heidarzadeh-Vazifekhoran, A.; Shojaosadati, S.A.; Karimzadeh, R.; Darani, K.K. Statistical media optimization for growth and PHB production from methanol by a methylotrophic bacterium. *Bioresour. Technol.* **2009**, *100*, 2436–2443. [CrossRef] [PubMed]
19. Mokhtari-Hosseini, Z.B.; Vasheghani-Farahani, E.; Shojaosadati, S.A.; Karimzadeh, R.; Heidarzadeh-Vazifekhoran, A. Effect of feed composition on PHB production from methanol by HCDC of Methylobacterium extorquens (DSMZ 1340). *J. Chem. Technol. Biotechnol. Int. Res. Process. Environ. Clean Technol.* **2009**, *84*, 1136–1139.
20. Bozorg, A.; Vossoughi, M.; Kazemi, A.; Alemzadeh, I. Optimal medium composition to enhance poly-β-hydroxybutyrate production by Ralstonia eutropha using cane molasses as sole carbon source. *Appl. Food Biotechnol.* **2015**, *2*, 39–47.
21. Koller, M.; Hesse, P.; Fasl, H.; Stelzer, F.; Braunegg, G. Study on the Effect of Levulinic Acid on Whey-Based Biosynthesis of Poly (3-hydroxybutyrate-co-3-hydroxyvalerate) by Hydrogenophaga pseudoflava. *Appl. Food Biotechnol.* **2017**, *4*, 65–78.
22. Khosravi-Darani, K.; Mokhtari, Z.B.; Amai, T.; Tanaka, K. Microbial production of poly (hydroxybutyrate) from C 1 carbon sources. *Appl. Microbiol. Biotechnol.* **2013**, *97*, 1407–1424. [CrossRef] [PubMed]
23. Shahhosseini, S.; Sadeghi, M.T.; Khosravi Darani, K. Simulation and Model Validation of Batch PHB production process using Ralstonia eutropha. *Iran. J. Chem. Chem. Eng.* **2003**, *22*, 35–42.
24. Khosravi Darani, K.; Vasheghani Farahani, E.; Shojaosadati, S.A. Application of the Plackett-Burman Statistical Design to Optimize Poly (Î²-hydroxybutyrate) Production by Ralstonia eutropha in Batch Culture. *Iran. J. Biotechnol.* **2003**, *1*, 155–161.
25. Vasheghani Farahani, E.; Khosravi Darani, K.; Shojaosadati, S.A. Application of the Taguchi Design for Production of Poly (β-hydroxybutyrate) by Ralstonia eutropha. *Iran. J. Chem. Chem. Eng.* **2004**, *23*, 131–136.

26. Khosravi-Darani, K.; Vasheghani-Farahani, E. Application of supercritical fluid extraction in biotechnology. *Crit. Rev. Biotechnol.* **2005**, *25*, 231–242. [CrossRef] [PubMed]
27. Asenjo, J.A.; Suk, J.S. Microbial conversion of methane into poly-β-hydroxybutyrate (PHB): Growth and intracellular product accumulation in a type II methanotroph. *J. Ferment. Technol.* **1986**, *64*, 271–278. [CrossRef]
28. Yazdian, F.; Shojaosadati, S.A.; Nosrati, M.; Pesaran Hajiabbas, M.; Vasheghani-Farahani, E. Investigation of gas properties, design, and operational parameters on hydrodynamic characteristics, mass transfer, and biomass production from natural gas in an external airlift loop bioreactor. *Chem. Eng. Sci.* **2009**, *64*, 2455–2465. [CrossRef]
29. Bozorg, M.; Mofradnia, R. Simulation of Bioreactors for Poly (3-hydroxybutyrate) Production from Natural Gas. *IJCCE* **2018**. (In press)
30. Pérez, R.; Casal, J.; Muñoz, R.; Lebrero, R. Polyhydroxyalkanoates production from methane emissions in Sphagnum mosses: Assessing the effect of temperature and phosphorus limitation. *Sci. Total Environ.* **2019**, *688*, 684–690. [CrossRef]
31. Zinn, M.; Witholt, B.; Egli, T. Occurrence, synthesis and medical application of bacterial polyhydroxyalkanoate. *Adv. Drug Deliv. Rev.* **2001**, *53*, 5–21. [CrossRef]
32. Rezapour, N.; Rasekh, B.; Mofradnia, S.R.; Yazdian, F.; Rashedi, H.; Tavakoli, Z. Molecular dynamics studies of polysaccharide carrier based on starch in dental cavities. *Int. J. Biol. Macromol.* **2018**, *121*, 616–624. [CrossRef] [PubMed]
33. Bordel, S.; Rodríguez, E.; Muñoz, R. Genome sequence of Methylocystis hirsuta CSC1, a polyhydroxyalkanoate producing methanotroph. *Microbiologyopen* **2019**, *8*, e00771. [CrossRef] [PubMed]
34. Rahnama, F.; Vasheghani-Farahani, E.; Yazdian, F.; Shojaosadati, S.A. PHB production by Methylocystis hirsuta from natural gas in a bubble column and a vertical loop bioreactor. *Biochem. Eng. J.* **2012**, *65*, 51–56. [CrossRef]
35. Yang, H.; Ye, H.; Zhai, S.; Wang, G.; Mavaddat, P.; Mousavi, S.M.; Amini, E.; Azargoshasb, H.; Shojaosadati, S.A. Modeling and CFD-PBE simulation of an airlift bioreactor for PHB production. *Asia Pac. J. Chem. Eng.* **2014**, *9*, 562–573.
36. Mousavi, S.M.; Shojaosadati, S.A.; Golestani, J.; Yazdian, F. CFD simulation and optimization of effective parameters for biomass production in a horizontal tubular loop bioreactor. *Chem. Eng. Process. Process Intensif.* **2010**, *49*, 1249–1258. [CrossRef]
37. Shah, A.A.; Hasan, F.; Hameed, A.; Ahmed, S. Biological degradation of plastics: A comprehensive review. *Biotechnol. Adv.* **2008**, *26*, 246–265. [CrossRef]
38. Lau, Y.M.; Deen, N.G.; Kuipers, J.A.M. Development of an image measurement technique for size distribution in dense bubbly flows. *Chem. Eng. Sci.* **2013**, *94*, 20–29. [CrossRef]
39. Yazdian, F.; Shojaosadati, S.A.; Nosrati, M.; Pesaran Hajiabbas, M.; Malek Khosravi, K. On-Line Measurement of Dissolved Methane Concentration During Methane Fermentation in a Loop Bioreactor. *Iran. J. Chem. Chem. Eng.* **2009**, *28*, 85–93.
40. Yazdian, F.; Shojaosadati, S.A.; Fatemi, S. Study of Growth Kinetic Models of a Methanotroph Bacterium Growing on Natural Gas. *J. Chem. Pet. Eng.* **2009**, *43*. [CrossRef]
41. Šimčík, M.; Mota, A.; Ruzicka, M.C.; Vicente, A.; Teixeira, J. CFD simulation and experimental measurement of gas holdup and liquid interstitial velocity in internal loop airlift reactor. *Chem. Eng. Sci.* **2011**, *66*, 3268–3279. [CrossRef]
42. Wallen, L.L.; Rohwedder, W.K. Poly-beta-hydroxyalkanoate from activated sludge. *Environ. Sci. Technol.* **1974**, *8*, 576–579. [CrossRef]
43. Koller, M. Switching from petro-plastics to microbial polyhydroxyalkanoates (PHA): The biotechnological escape route of choice out of the plastic predicament? *EuroBiotech J.* **2019**, *3*, 32–44. [CrossRef]

5

Cyanobacterial PHA Production—Review of Recent Advances and a Summary of Three Years' Working Experience Running a Pilot Plant

Clemens Troschl [1], Katharina Meixner [2,*] and Bernhard Drosg [2]

[1] Institute of Environmental Biotechnology, Department of Agrobiotechnology, IFA-Tulln, University of Natural Resources and Life Sciences, Vienna, Tulln 3430, Austria; clemens.troschl@boku.ac.at
[2] Bioenergy2020+ GmbH, Tulln 3430, Austria; bernhard.drosg@bioenergy2020.eu
* Correspondence: katharina.meixner@bioenergy2020.eu

Academic Editor: Martin Koller

Abstract: Cyanobacteria, as photoautotrophic organisms, provide the opportunity to convert CO_2 to biomass with light as the sole energy source. Like many other prokaryotes, especially under nutrient deprivation, most cyanobacteria are able to produce polyhydroxyalkanoates (PHAs) as intracellular energy and carbon storage compounds. In contrast to heterotrophic PHA producers, photoautotrophic cyanobacteria do not consume sugars and, therefore, do not depend on agricultural crops, which makes them a green alternative production system. This review summarizes the recent advances in cyanobacterial PHA production. Furthermore, this study reports the working experience with different strains and cultivating conditions in a 200 L pilot plant. The tubular photobioreactor was built at the coal power plant in Dürnrohr, Austria in 2013 for direct utilization of flue gases. The main challenges were the selection of robust production strains, process optimization, and automation, as well as the CO_2 availability.

Keywords: cyanobacteria; polyhydroxyalkanoates; CO_2 mitigation; flue gas utilization; photobioreactor

---

1. Introduction

Polyhydroxyalkanoates (PHAs) are considered as one of the most promising bioplastics. Their mechanical properties are similar to polypropylene and they can be processed in a similar way, including extrusion, injection molding, or fiber spinning [1]. One of the major advantages of PHAs are their biodegradability. They are degraded relatively rapidly by soil organisms, allowing easy composting of PHA waste material [2].

Currently, PHA is produced in large fermenters by heterotrophic bacteria, like *Cupriavidus necator* or recombinant *Escherichia coli* [3]. For these fermentation processes large amounts of organic carbon sources like glucose are necessary, accounting for approximately 50% of the total production costs [4]. An alternative way of producing PHA is the use of prokaryotic algae, better known as cyanobacteria. As part of the phytoplankton, they are global primary biomass producers using light as the sole energy source to bind atmospheric CO_2 [5]. Burning of fossil fuels has increased the atmospheric CO_2 concentration from approximately 300 ppm in 1900 to over 400 ppm today. The latest report of the intergovernmental panel on climate change (IPCC) clearly indicates anthropogenic CO_2 emissions as the main driver for climate change [6]. Given these facts, cultivation of cyanobacteria for PHA production could be a more sustainable way of producing bioplastics.

2. Cyanobacteria and Cyanobacterial Energy and Carbon Storage Compounds

Cyanobacteria are Gram-negative prokaryotes that perform oxygenic photosynthesis. They are abundant in illuminated aquatic ecosystems and contribute significantly to the world carbon and oxygen cycle [7]. According to current evidence, oxygen was nearly absent in the Earth's early atmosphere until 2.4 billion years ago [8]. Due to oxygenic photosynthesis of early cyanobacteria the CO_2-rich atmosphere gradually turned into an oxygen-rich atmosphere, providing the conditions for multicellular life [9,10]. Today there are an estimated 6000 species of cyanobacteria with great diversity, for example ranging in size from the 1 μm small unicellular *Synechocystis* sp. to the several millimeter-long multicellular filaments of *Oscillatoria* sp. [11]. The common feature of cyanobacteria is the presence of the pigment phycocyanin, which gives them their typical blue-green color. Figure 1 shows photographs of four different cyanobacterial species.

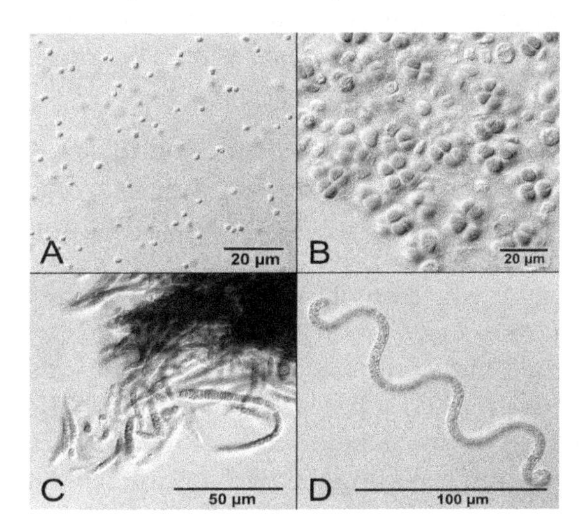

Figure 1. Microscopic photographs of different cyanobacterial species made in DIC (differential interference contrast). (**A**) *Synechocystis* sp.; (**B**) *Cyanosarcina* sp.; (**C**) *Calothrix* sp.; and (**D**) *Arthrospira* sp.

2.1. Cyanobacteria–Microalgae or Not?

For more than a century, cyanobacteria were considered as an algal group under the general name "blue-green algae". They were classified under the International Code of Botanical Nomenclature, nowadays called the International Code of Nomenclature for Algae, Fungi, and Plants (ICN). In 1980 the International Code of Nomenclature of Bacteria, nowadays called the International Code of Nomenclature of Prokaryotes (ICNP), was established. Stanier, one of the leading cyanobacteria researchers at that time, proposed the inclusion of cyanobacteria in the ICNP [12]. Nevertheless, the ICNP was not consistently applied for cyanobacteria and cyanobacteria are still covered by the ICN as well. The latest preamble of the ICN clarifies, that this code applies to all organisms traditionally treated as algae, fungi, or plants, including cyanobacteria [13]. Today cyanobacteria continue to be covered by both the Botanical Code (ICN) and Prokaryotic Code (ICNP). An effort to reconcile the status of this group of bacteria has been underway for several decades. Although some progress has been made, a final decision has not yet been reached [14]. From a phylogenetic point of view, there is a clear distinction between prokaryotic cyanobacteria and eukaryotic green algae. However, phycologists regard any organism with chlorophyll *a* and a thallus not differentiated into roots, stem, and leaves to be an alga. Therefore, in phycology, the term microalgae refers to both eukaryotic green algae and cyanobacteria, microscopic in size [15].

2.2. Cyanobacterial PHA

Polyhydroxyalkanoates (PHAs) can be classified into three groups: short-chain-length-PHA (scl-PHA), medium-chain-length-PHA (mcl-PHA), and long-chain-length-PHA (lcl-PHA). They differ in mechanical and thermal properties [16]. Among the different PHAs, polyhydroxybutyrate (PHB) is by far the most common and the only PHA produced under photoautotrophic conditions reported so far. Other scl-PHAs, like P[3HB-co-3HV], are only produced when adding organic carbon precursors, like valerate, to the medium. No mcl-PHA or lcl-PHA have been reported in cyanobacteria. Therefore, the term PHB is used in this study, if no other specific PHA is described.

PHB is frequently found in cyanobacteria as an energy and carbon storage compound. In the biosphere they often have to cope with unfavourable environmental conditions. One of the most important growth limiting factors is the absence of nutrients. Nitrogen limitation is the most important and best studied trigger for PHB production in cyanobacteria [17–19]. Non-diazotrophic strains are not able to bind molecular nitrogen and depend on nitrogen in the form of nitrate or ammonium. Nitrogen-depleted cells cannot synthesize the necessary proteins for reproduction and, therefore, start to accumulate storage compounds like PHB. Another important function of PHB synthesis is to compensate imbalanced metabolic situations, as it acts as an electron sink and delivers new reduction equivalents in the form of $NADP^+$ [18–20].

The model organism *Synechocystis* PCC6803 is the best-studied cyanobacterium, and its genome was fully sequenced in 1997 [21]. Most of the understanding of cyanobacterial PHB formation was gained by research done with *Synechocystis* PCC6803. Biosynthesis of PHB from the precursor acetyl-CoA takes place in three steps. Acetoacetyl-CoA is produced from two molecules of acetyl-CoA in a Claisen type condensation by β-ketothiolase. Next step is the reduction of acetoacetyl-CoA by the acetoacetyl-CoA reductase to form D-3-hydroxybutyryl-CoA. Ultimately, PHB is formed in a polymerization reaction by the PHA-synthase. The necessary three enzymes are encoded by the four genes phaA (slr1993), phaB (slr1994), phaC (slr1830), and phaE (slr1829). phaA and phaB are organized in one operon encoding for the β-ketothiolase and acetoacetyl-CoA reductase. phaC and phaE are also organized in one operon encoding the two subunits of the type III PHA synthase [22,23].

2.3. *Cyanobacterial Glycogen*

Regarding PHB synthesis, it should be kept in mind that cyanobacteria also produce glycogen as a second carbon and energy storage compound under nitrogen depletion. In fact, the glycogen content is most often higher than the PHB content and varies between 20% and 60% [24–28]. While PHB is produced in 3–8 larger granules, glycogen is stored in many small granules [18,29–33]. Glycogen is synthesized instantly after nitrogen depletion while PHB synthesis is slower [34]. Glycogen is also produced in non-depleted cells with lower content, aiding the cell to cope with short term energy deficits like the day-night cycle. Glycogen deficient mutants were shown to be highly sensitive to day-night cycles [35]. Glycogen synthesis is a highly-conserved feature abundant in all cyanobacterial genomes reported so far [36]. PHB synthesis on the other hand is common in many, but not all, cyanobacteria [37,38]. Glycogen shows similarities to starch in green algae, while PHB synthesis shows some similarities to triacylglycerol (TAG) synthesis in green algae, where TAG synthesis also serves as an electron sink and consumes excess NADPH [39,40].

In a recent study Damrow and colleagues compared PHB-deficient mutants to glycogen-deficient mutants of *Synechocystis* PCC6803. Glycogen-deficient mutants could not switch to a dormant metabolic state and could not recover from nitrogen depletion. Excess carbon was mostly secreted into the medium in the form of 2-oxoglutaric acid and pyruvate, although the PHB content also increased from 8% to 13%. PHB-deficient mutants, on the other hand, behaved very much like the wild-type with the same amount of glycogen accumulation and the same recovery capability. Only double-knockout mutants (glycogen and PHB deficient) were most sensitive and showed a reduced growth rate, signs for a very specific role of PHB in cyanobacteria, which is still not totally clear [41]. The reported studies show that inhibiting glycogen synthesis increases the PHB production, although cells suffer as glycogen plays an important role.

2.4. Nitrogen Chlorosis and Photosynthetic Activity

During nitrogen starvation the cells gradually change from a vegetative state to a dormant state. The most obvious feature of this is the change in colour from blue-green to brownish-yellow. This phenomenon is called "nitrogen chlorosis" and was described already at the begin of the 20th century [42]. It is caused by the degradation of the pigments phycocyanin and chlorophyll. When transferring *Synechococcus* PCC7942 to a nitrogen depleted medium, 95% phycocyanin was degraded within 24 h, and after 10 days 95% of the chlorophyll was also degraded [43]. Concomitantly, the activities of the photosystems (PS) I and II decrease strongly and are only about 0.1% compared to vegetative cells [44]. A recent and very interesting study examined the awakening of a dormant *Synechocystis* PCC6803 cell. After the addition of nitrate the yellow culture turned green again within 36 hours. Transmission electron microscopy revealed the rapid degradation of glycogen and PHB. During the first 24 h of this process the cells consumed oxygen. Transcriptome analysis showed the induction of RuBisCO and carboxysom associated RNAs, as well as the photosystem-related RNAs to prepare the cells for vegetative photoautotrophic growth [34]. The results indicate the decrease in photosynthetic activity during nitrogen starvation, which can be considered a significant challenge to photoautotrophic PHB production.

3. Different Cyanobacteria as PHA Producers

3.1. Synechocystis and Synechococcus

Synechocystis and *Synechococcus* are very small (0.5–2 μm) unicellular cyanobacteria abundant in almost all illuminated saltwater and freshwater ecosystems. One of the first detailed descriptions of PHB accumulation in *Synechocystis* PCC6803 was provided by Wu and colleagues. Nitrogen starved cells produced 4.1% PHB of cdw while under-balanced culturing conditions PHB content were under the detection limit [45]. The same strain was examined for PHB production some years later, where 9.5% PHB of cdw were produced under nitrogen limitation. Phosphorous-depleted cells showed 11.2% PHA of cdw. Interestingly, balanced cultivated control cultures already contained 4.5% PHB of cdw. Supplementation of acetate and fructose lead to a PHB content of 38% of cdw [46]. Recently, recombinant *Synechocystis* PCC6803 with overexpression of the native PHA genes were constructed. They showed a PHB content of 26% of cdw under nitrogen-depleted culturing conditions compared to 9.5% of cdw of the wild-type [47]. However, it must be considered that there are legal issues in most countries when cultivating recombinant strains outdoors. In another study the thermophilic strain *Synechococcus* MA19 showed a PHB content of 55% under phosphate-limited culturing conditions. This study was published in 2001 and still reports the highest PHB content under photoautotrophic conditions [48]. Table 1 shows reported PHA values of *Synechocystis* and *Synechococcus*.

Table 1. *Synechocystis* and *Synechococcus* as PHA producers. (cdw = cell dry weight, n.r. = not reported).

Carbon Source	Cyanobacterium	Culture Condition	%PHA of cdw	PHA Composition	Total cdw	Reference
Photoautotrophic	*Synechocystis* PCC6803	Photoautotrophic, nitrogen lim.	4.1%	PHB	0.65 g/L	[45]
	Synechocystis PCC6803	Photoautotrophic, nitrogen lim.	9.5%	PHB	n.r.	[46]
	Synechocystis PCC6803	Photoautotrophic, phosphate lim.	11.2%	PHB	n.r.	[46]
	Synechocystis PCC6803 (recombinant)	Photoautotrophic, nitrogen lim.	26%	PHB	n.r.	[47]
	Synechococcus MA19	Photoautotrophic, phosphate lim., 50 °C	55%	PHB	4.4 g/L	[48]
Heterotrophic	*Synechocystis* PCC6803	Acetate + Fructose supplementation	38%	PHB	n.r.	[46]
	Synechocystis PCC6803 (recombinant)	Acetate supplementation	35%	PHB	n.r.	[47]

3.2. *Arthrospira (Spirulina)*

Arthrospira (formally *Spirulina*) is a species of filamentous cyanobacteria that grows naturally in alkaline salt lakes. It has a high protein and vitamin content and is mainly grown as a food supplement. Recent studies have shown its antioxidant, immunomodulatory, and anti-inflammatory activities [49]. From all cyanobacterial species known, only *Arthrospira* sp. is produced at an industrial scale. The main reason for that is the possibility of cultivation in a highly alkaline environment that prevents contamination and enables the maintenance of a stable culture in open ponds. No exact data are available; however, we estimate the world annual production of around 5000–15,000 tons *Arthrospira* sp. dry weight per year [50–53].

The first description of PHB accumulation in *Arthrospira* was reported by Campbell and colleagues, who described a PHB content of 6% of cdw in a non-optimized mineral medium. Interestingly, the highest PHB content was measured at the end of exponential growth and decreased during stationary phase [54]. In a screening of 23 cyanobacterial strains, *Arthrospira platensis* had the lowest PHB concentration of only 0.5% in a non-optimized medium [37]. In a screening of several *Arthrospira* species the PHB amount never exceeded 1% of cdw in photoautotrophic growth. Addition of sodium acetate led to a PHB amount of 2.5% of cdw [55]. In another experiment *Arthrospira platensis* was grown under phosphate limitation and reached 3.5% PHB of cdw [56]. *Arthrospira subsalsa*, a strain isolated from the Gujarat coast, India, produced 14.7% PHB of cdw under increased salinity [57]. A detailed ultrastructural analysis of *Arthrospira* strain PCC8005 was conducted by Deschoenmaker and colleagues. Under nitrogen depleted conditions PHB granules were more abundant and larger. The nitrogen-starved cells showed an estimated four times higher PHB concentration [27]. Nitrogen starvation was performed in *Arthrospira maxima* and glycogen and PHB content was measured. While the glycogen content increased from around 10% to 60%–70% of cdw, PHB amount remained low at 0.7% of cdw. The addition of sodium acetate increased the PHB amount to 3% of cdw [26].

The performed studies support the idea, that PHB production in *Arthrospira* is highly strain-dependent. Most *Arthrospira* species produce PHB only in amounts of lower than 5%, even with the addition of sodium acetate. *Arthrospira* produces glycogen as storage compound, what has been shown in ultrastructural research, too [27]. Nevertheless, it must be emphasized that *Arthrospira*, at an industrial scale, is still one of the most promising candidates for PHB production with cyanobacteria. Indeed, PHB nanofibers were produced recently from *Arthrospira* PHB and showed highly favourable properties [58,59]. The biggest challenge for further research is to increase the relatively low PHB content of *Arthrospira*. Table 2 shows reported PHA values of *Arthrospira*.

Table 2. *Arthrospira* as a PHA producer. (cdw = cell dry weight, n.r. = not reported).

Carbon Source	Cyanobacterium	Culture Condition	%PHA of cdw	PHA Composition	Total cdw	Reference
Photoautotrophic	*Arthrospira platensis*	Photoautotrophic	6%	PHB	n.r.	[54]
	Arthrospira sp.	Photoautotrophic	<1%	PHB	n.r.	[55]
	Arthrospira platensis	Photoautotrophic, phosphate lim.	3.5%	PHB	0.3 g/L	[56]
	Arthrospira subsalsa	Photoautotrophic, nitrogen lim.	14.7%	PHB	1.97 g/L	[57]
	Arthrospira platensis	n.r.	22%	PHB	n.r.	[59]
Heterotrophic	*Arthrospira maxima*	Acetate + CO_2	5%	PHB	1.4 g/L	[26]
	Arthrospira sp.	Acetate + CO_2	2.5%	PHB	n.r.	[55]

3.3. *Nostoc*

Nostoc is a group of filamentous cyanobacteria very common in terrestrial and aquatic habitats. They are capable of fixing atmospheric nitrogen with specialized heterocysts and are suspected to maintain soil fertility in rice fields due to nitrogen fixation [60]. Ge-Xian-Mi, an edible *Nostoc* species, forms spherical colonies that have been collected in China for centuries [61]. The first reports found for PHB production in *Nosctoc muscorum* are from 2005, when Sharma and Mallick showed that *Nostoc muscroum* produced 8.6% PHB of cdw under phosphate and nitrogen limitation during the stationary phase. PHB content could be boosted to 35% of cdw with 0.2% acetate and seven days dark incubation [62]. Limited gas exchange and supply with 0.4% acetate increased the PHB content to 40% [63]. *Nostoc muscorum* was grown photoautotrophically without combined nitrogen sources and four days of phosphate deficiency increased PHB content from 4% to 22% [56]. The co-polymer P[3HB-co-3HV] could be produced by *Nostoc* in a propionate- and valerate-supplied medium. The 3HV fraction ranged from 10–40 mol% and showed desirable properties in terms of flexibility and toughness. Nitrogen and phosphate depletion led to a PHA content of 58%–60% of cdw, however, the total cdw did not exceed 1 g/L [64]. Further process optimization led to a PHA productivity of 110 mg/L/d and a P[3HB-co-3HV] content of 78% of cdw, the highest yield in heterotrophic grown cyanobacteria reported so far [65]. Recently, poultry litter was used for cultivation of *Nostoc muscorum agardh*. The poultry litter contained phosphate, ammonium, nitrate, and nitrite as nutrients for cyanobacterial growth. Optimized conditions, which included the addition of acetate, glucose, valerate, and CO_2-enriched air, led to a P[3HB-co-3HV] content of 70% cdw. However, total cdw remained relatively low at 0.68 g/L [66].

The reported studies show that PHB content in *Nostoc* can be significantly increased with organic carbon sources, especially in the form of acetate. However, those organic carbon sources lead to heterotrophic growth and may suppress CO_2 uptake by the cells, which is the most important argument for using cyanobacteria as PHA producers. All of the reported experiments of *Nostoc* were performed in shaking flasks or small reactors under sterile conditions. In mass cultivation *Nostoc* would have to be cultivated under non-sterile conditions and organic carbon sources could cause problems maintaining stable cultures. Although optimized conditions of several experiments lead to PHA contents of more than 50% of cdw, the total cdw remained mostly under 1 g/L and the overall productivity and growth rate of *Nostoc* is relatively low. Table 3 shows reported PHA values of *Nostoc*.

Table 3. *Nostoc* as a PHA producer. (cdw = cell dry weight, n.r. = not reported).

Carbon Source	Cyanobacterium	Culture Condition	%PHA of cdw	PHA Composition	Total cdw	Reference
Photoautotrophic	*Nostoc muscorum*	Photoautotrophic, nitrogen and phosphorous lim.	8.7%	PHB	n.r.	[62]
	Nostoc muscorum agardh	Photoautotrophic, 10% CO_2	22%	PHB	1.1 g/L	[66]
	Nostoc muscorum	Photoautotrophic, nitrogen and phosphorous lim.	22%	PHB	0.13 g/L	[56]
Heterotrophic	*Nostoc muscorum agardh*	Acetate, valerate, nitrogen lim.	58%	P[3HB-co-3HV]	0.29 g/L	[64]
	Nostoc muscorum	Acetate, limited gas exchange	40%	PHB	n.r.	[63]
	Nostoc muscorum agardh	Acetate, glucose, valerate, 10% CO_2	70%	P[3HB-co-3HV]	0.98 g/L	[66]
	Nostoc muscorum agardh	Acetate, glucose, valerate, nitrogen lim.	78%	P[3HB-co-3HV]	0.56 g/L	[65]
	Nostoc muscorum	Acetate, dark incubation, nitrogen and phosphorous lim.	35%	PHB	n.r.	[62]

3.4. Other Cyanobacteria

Recently, the PHB content of 137 cyanobacterial strains representing 88 species in 26 genera was determined under photoautotrophic conditions. High PHB content was highly strain-specific and was not associated with the genera. From the 137 tested strains, 134 produced PHB and the highest content was measured in *Calothrix scytonemicola* TISTR 8095 (Thailand Institute of Scientific and Technological Research). This strain produced 356.6 mg/L PHB in 44 days and reached a PHB content of 25% of cdw and a total biomass of 1.4 g/L. The PHB content of 25% was reached under nitrogen depletion, while cells with nitrogen supply reached a PHB content of only 0.4%. From the 19 tested *Calothrix* strains, only six produced more than 5% PHB of cdw. One of the greatest advantages of *Calothrix* is the relative ease of harvesting the dense flocs of algae, but cultivation of *Calothrix* is still at a very early stage [38].

The filamentous diazotroph cyanobacterium *Aulosira fertilissima* produced 10% PHB of cdw under photoautotrophic conditions and phosphate deficiency. The PHB content was boosted to 77% under phosphate deficiency with 0.5% acetate supplementation. This study also shows the positive effect of other carbon sources like citrate, glucose, fructose and maltose on PHB production [67]. Anabaena cylindrica, a filamentous cyanobacterium, was examined for PHB and P[3HB-co-3HV] production. Under nitrogen depletion with acetate supply, *Anabeana cylindrica* produced 2% PHB of cdw and a total biomass of 0.6 g/L. This organism was also able to produce the co-polymer P[3HB-co-3HV] when supplemented with valerate and propionate [68]. Table 4 shows reported PHA values of different cyanobacterial species.

Table 4. Different cyanobacterial species as PHA producers. (cdw = cell dry weight, n.r. = not reported).

Carbon Source	Cyanobacterium	Culture Condition	%PHA of cdw	PHA Composition	Total cdw	Reference
Photoautotrophic	*Phormidium* sp. TISTR 8462	Photoautotrophic, nitrogen lim.	14.8%	PHB	n.r.	[38]
	Oscillatoria jasorvensis TISTR 8980	Photoautotrophic, nitrogen lim.	15.7%	PHB	n.r.	[38]
	Calothrix scytonemicola TISTR 8095	Photoautotrophic, nitrogen lim.	25.2%	PHB	n.r.	[38]
	Anabaena sp.	Photoautotrophic	2.3%	PHB	n.r.	[69]
	Aulosira fertilissima	Photoautotrophic, phosphorous lim.	10%	PHB	n.r.	[67]
Heterotrophic	*Aulosira fertilissima*	Acetate, phosphorous lim.	77%	PHB	n.r.	[67]
	Aulosira fertilissima	Maltose, balanced	15.9%	PHB	2.3 g/L	[67]

4. CO_2 and Nutrient Supply for Mass Cultivation of Cyanobacteria

4.1. CO_2 Supply

Today, commercial microalgae production is still mainly taking place in open ponds. Here, the C source is normally sodium bicarbonate or atmospheric CO_2. In order to boost productivities in open systems, or if photobioreactor systems are employed, the use of commercial CO_2 from gas cylinders is common [70].

However, current production systems are used for the production of high value products (food, feed additives), where CO_2 price is not critical. If PHA is to be produced, which has a lower economic value, cheap CO_2 sources are of interest. Although there is considerable literature on various CO_2-sources (e.g., flue gases) and microalgae growth, there is very limited literature available on cyanobacteria and alternative CO_2-sources. Table 5 summarizes the literature on cyanobacterial growth on flue gases or fermentation gases.

Table 5. Growing cyanobacteria with alternative CO_2-sources.

Type of Gas	Cyanobacterium	CO_2 Source	Reference
Flue gases	*Phormidium valderianum*	Coal combustion flue gas	[71]
	Atrhrospira platensis	Coal combustion flue gas	[72]
	Arthrospira sp.	Synthetic flue gas	[73]
	Synechocystis sp.	Flue gas from natural gas combustion	[74]
CO_2 rich fermentation gases	*Arthrospira platensis*	CO_2-offgas from ethanol fermentation	[75]
	Arthrospira platensis	Biogas	[76]

4.2. Nutrient Supply

The cultivation of microalgae and cyanobacteria consume high amounts of nutrients, mainly nitrogen and phosphorous [77]. For research, and even cultivation, mainly synthetic nutrient sources are used [78]. By using alternative nutrient sources, like agro-industrial effluents, waste waters, or anaerobic digestate, questions concerning sustainability of cyanobacteria cultivation, which arise by using fertilizer as a synthetic nutrient source, can be answered [78]. The biomass concentration achieved in open, as well as in closed, cultivation systems are 0.5–1 g/L and 2–9 g/L, respectively [79]. Therefore, large amounts of water are needed. Recycling of process water is another important approach for a more sustainable microalgae cultivation.

In addition to their low costs, the advantages of using alternative nitrogen and phosphorous sources include the production of valuable biomass while removing nutrients from wastewaters, as well as the prevention of competition with food and feed production [78]. On the other hand, new challenges arise, including microbial contaminations, heavy metals and growth inhibitors, suspended solids, or dissolved organic compounds contained in wastewaters, as well as the seasonal composition and fluctuation in amounts [80]. To cope with these challenges recent research focused on cultivating cyanobacteria in anaerobic digestate and agro-industrial effluents or wastewaters for removing nutrients [81–86] (see Table 6) and on integrating cultivation processes into biorefinery systems [83].

Additionally, process water and nutrients after harvesting cyanobacterial biomass [79] and product extraction can be directly recycled. Biomass can also be anaerobically digested [87,88] or hydrothermally liquefied via HTL (mineralization of organic nutrients) [89,90] and then recycled. Recycling process water directly can increase the concentration of inhibitory substances and dissolved organic matter from the previous batch produced by cyanobacteria [91], which decrease the productivity of cyanobacteria. Furthermore, nutrient competition may arise by enhanced bacterial growth [79].

Although many publications deal with alternative nutrient sources for cultivating cyanobacteria, hardly any of them focus on cyanobacterial PHA production [66,92]. Reasons for that may be that PHA production requires nutrient limitation [93] and the balance between nutrient limitation, decreased growth and production rates is difficult. The colouring of the nutrient source must be respected as well [94].

Table 6. Overview of agro-industrial effluents and wastewaters and anaerobic digestates used as nutrient sources for cultivating cyanobacteria.

Nutrient Source		Cyanobacterium	Total cdw/Growth Rate	Product/Purpose	Reference
Agro-industrial effluents and waste waters	Raw cow manure	*Arthrospira maxima*	3.15 g/L	Biomass production	[80]
	Molasses	*Arthrospira platensis*	2.9 g/L	Biomass production	[95]
	Olive-oil mill wastewater	*Arthrospira platensis*	1.69 g/L	Nutrient removal	[84]
	Poultry litter	*Nostoc muscorum agardh*	0.62 g/L	PHA production	[66]
Anaerobic digestate	Waste from pig farm	*Arthrospira platensis*	20 g/m^2/d	Nutrient removal	[81]
	Digested sago effluent	*Arthrospira platensis*	0.52–0.61 g/L	Nutrient removal	[96]
	Digestate from municipal solid waste	*Arthrospira platensis*	Growth rate 0.04 d^{-1}	Nutrient removal	[97]
	Digestate from vegetable waste	*Arthrospira platensis*	Growth rate 0.20 d^{-1}	Nutrient removal	[97]
	Waste from pig farm	*Arthrospira* sp.	15 g/m^2/d	Nutrient removal	[85]
	Algal digestate	*Chroococcus* sp.	0.79 g/L	Nutrient removal	[86]
	Digestate sludges	*Lyngbya aestuarii*	0.28 g/L	Biomass production	[83]
	Digestates of *Scenedesmus* spp.	*Lyngbya aestuarii*	0.11 g/L	Biomass production	[83]
	Thin stillage digestate	*Synechocystis* cf. *salina Wislouch*	1.6 g/L	PHB production	[92]
	Anaerobic digester effluent	*Synechocystis* sp.	0.15 g/L	Lipid production	[98]

5. Three Years' Working Experience Running a Pilot Plant for Photoautotrophic PHB Production

5.1. Location and Reactor Description

The photobioreactor is situated in a glass house at the coal power station in Dürnrohr, Austria. It is a tubular system built from glass elements of Schott AG with an inner diameter of 60 mm, a total length of 80 m and a volume of 200 L (Figure 2). The main design of the photobioreactor is described elsewhere [99,100]. A central degassing unit serves to remove the oxygen as well as to compensate filling level. The medium is circulated with a 400 W centrifugal pump. pH value can be controlled through injection of pure CO_2 via a mass flow controller. Additional artificial light is provided by six 250 W gas-discharge lamps. Temperature is controlled with an air conditioner.

Figure 2. Two-hundred litre tubular photobioreactor with *Synechocystis salina* CCALA192. The central tower serves as a degasser. The centrifugal pump is situated at the lowest point of the reactor on the left side.

5.2. CO_2 Supply of the Reactor

The flue gases of the power plant at Dürnrohr usually contain between 11%–13% CO_2. Next to the chimney is a CO_2 separation plant (acronym SEPPL), providing the possibility to concentrate the CO_2 and fill it into gas bottles [101] though, for a more economic approach, the CO_2 should be used directly without prior compression. The SEPPL provides this option, as well as the possibility to wash the flue gases after the flue gas cleaning of the plant itself to remove residual NO_x and SO_x. Unfortunately for our research project, due to the current situation on the energy market, the power station is no longer run in full operation and only runs occasionally for balancing peak demands of the electrical grid. Therefore, a continuous cultivation with direct utilization of flue gas is not possible. Aspects like this must be respected when planning large industrial cultivation plants.

5.3. Automation and pH Control

The pH value is one of the most crucial parameters and needs to be controlled carefully. Due to CO_2 consumption, the pH value rises during photoautotrophic growth. This can be observed when turning on illumination. The tubular photobioreactor is equipped with a PI (Proportional–Integral) controller for pH setting which adjusts the mass-flow controller for CO_2 inlet. This allows an online control of the currently consumed CO_2, which is a suitable parameter for growth monitoring. Figure 3 shows a 24-h course of the pH value and the CO_2 mass flow. Lamps turned on at 02:00 and off at 22:00, causing a rise and decrease of the pH value, respectively. The setpoint of 8.5 is reached after first overshooting and held during the day. The decrease of CO_2 consumption at noon is caused by the shadow of the power plant's chimney that casts upon the greenhouse at this time.

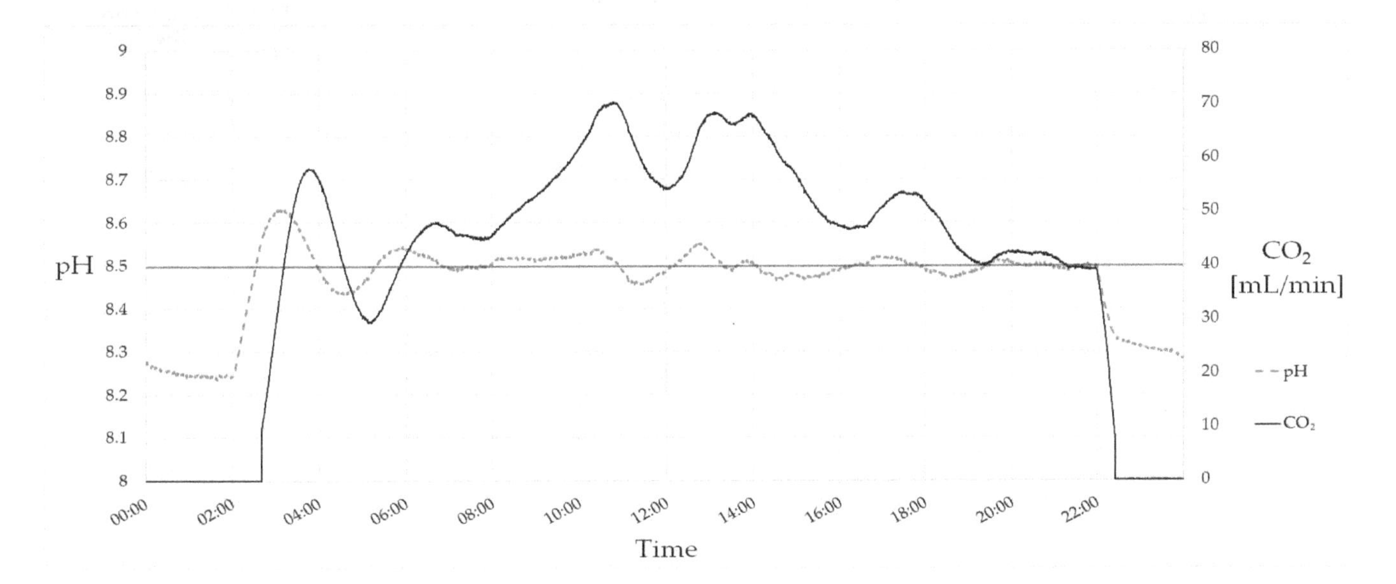

Figure 3. PI-controlled pH value. The setpoint of the pH is 8.5. Lamps turn on at 02:00 and turn off at 22:00, causing a rise and decrease of the pH value due to CO_2 consumption. In total, 59 L (118 g) of CO_2 were consumed on this day.

5.4. *Overview of PHB Production Trials*

Most of the trials (overview shown in Table 7) were performed using a modified BG 11 medium [102]. Modification in terms of PHB production is necessary, as normal BG 11 medium contains high amounts of nitrogen (1.5 g/L $NaNO_3$) and would not lead to nitrogen limitation. The modified BG 11 contains 0.45 g/L $NaNO_3$ and leads to a self-limitation of the culture. After 8–12 days nitrogen is consumed, PHB production starts and the color of the culture gradually turns yellow. This approach is necessary, as it is not possible to transfer large-scale cultures into a nitrogen-free medium [103].

Synechocystis salina CCALA192 was found to be a very suitable cyanobacterium. It is easy to handle and grows with small inoculation volumes of 1:50. Final biomass and PHB concentrations were in the range of 0.9–2.1 g/L and 4.8% to 9% of cdw, respectively. *Synechocystis salina* CCALA192 also grew with the addition of acetate, but no significant increase of biomass and PHB concentration was observed compared to photoautotrophic growth. When using acetate, contaminations with fungi were likely to occur and trials had to be stopped. Therefore, this approach was finally abandoned.

Digestate from a biogas reactor was successfully tested as an alternative nutrient source. The supernatant was produced by centrifugation with prior addition of precipitating agents. Before usage the supernatant was autoclaved and diluted 1:3 with water [92]. Figure 4 shows biomass and PHB production using digestate as nutrient source.

After one and a half years a new degassing system was installed, as the oxygen concentration was mostly above 250% saturation during the day. For an ideal cultivation of cyanobacteria the oxygen saturation should not exceed 200%. The new degasser led to a rise in biomass production with a maximum production rate of 0.25 g/L/d. Efficient degassing affected the cyanobacteria positively. However, during installation of the degasser dirt from the surrounding soil was brought into the reactor and from that time on culture crashes occurred due to ciliate contaminations (see Section 5.6).

The other tested cyanobacteria *Chlorogloeopsis fritschii* and *Arthrospira* sp. could not be successfully cultivated in the photobioreactor. It is assumed that these strains were sensitive to shear stress caused by the centrifugal pump [104].

Table 7. Overview of selected trials conducted in a tubular photobioreactor at pilot scale.

Trial	Strain	Nutrient Solution	Cultivation Time	Final Biomass Concentration	Final PHB-Concentration of cdw
1. Mineral medium	*Synechocystis salina* CCALA192	Optimized BG11	June 21 days	2.0 ± 0.12 g/L	6.6% ± 0.5%
2. Acetate addition	*Synechocystis salina* CCALA192	Optimized BG11, 20 mM acetate	July 26 days	1.9 ± 0.02 g/L	6.0% ± 0.1%
3. Acetate addition	*Synechocystis salina* CCALA192	Optimized BG11, 60 mM acetate	September 24 days	Trial cancelled, due to contaminations with fungi	
4. 24 h illumination	*Synechocystis salina* CCALA192	Optimized BG11	October 27 days	1.8 ± 0.02 g/L	4.8% ± 0.0%
5. Alternative nutrient source	*Synechocystis salina* CCALA192	Digestate supernatant	November–December 40 days	1.6 ± 0.02 g/L	5.5% ± 0.3%
6. Mineral medium	*Synechocystis salina* CCALA192	Optimized BG11	December–January 30 days	2.1 ± 0.03 g/L	6.0% ± 0.02%
7. Optimal degassing	*Synechocystis salina* CCALA192	Optimized BG11	May 7 days	0.9 ± 0.03 g/L (Trial prematurely cancelled due to ciliates)	9% ± 0.1% (Trial prematurely cancelled due to ciliates)
8. Chlorogloeopsis fritschii CCALA39	*Chlorogloeopsis fritschii* CCALA39	Optimized BG11	February 11 days	Trial cancelled, due to lack of growth	
9. Arthrospira	*Arthrospira* sp.	Spirulina Medium	October 7 days	Trial cancelled, due to lack of growth	

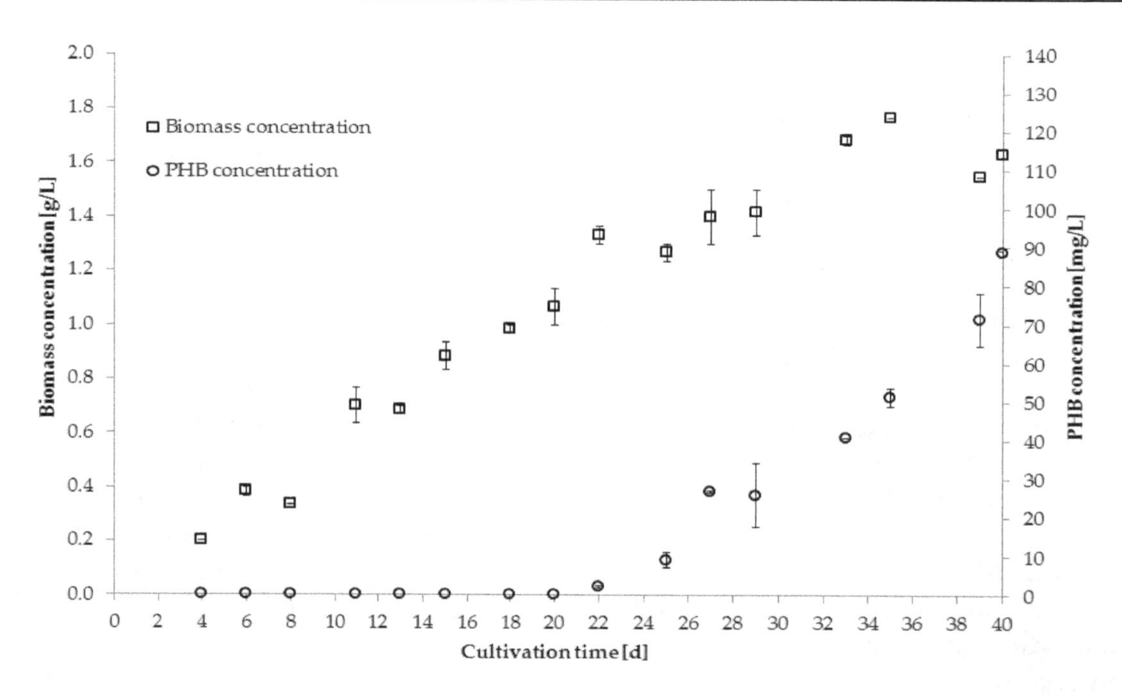

Figure 4. Biomass [g/L] and PHB [g/L] concentration of *Synechocystis salina* using digestate supernatant as nutrient source (Trial 5).

5.5. Downstream Processing of Cyanobacterial Biomass

Downstreaming of cyanobacterial cultures is particularly difficult, as cell densities are much lower compared to heterotrophic bacteria. Typical harvesting methods are sedimentation, filtration, or centrifugation [105]. The cyanobacterial biomass was harvested with a nozzle separator and stored at −20 °C. The biomass was then used to evaluate processing steps necessary to gain clean

PHB-samples for quality analysis. These downstream trials include (i) different cell disruption methods (milling, ultrasound, French press); (ii) different pigment extraction methods (with acetone and ethanol/methanol before or after extracting PHB); and (iii) different PHB extraction methods (soxhlet extraction with chloroform, biomass digestion with sodium hypochlorite) [106].

These trials showed that cell disruption with French press worked quite well but is very time consuming. Milling is assumed to decrease the molecular weight (polymer chain length). Pigment removal turned out to be necessary prior to PHB extraction, as pigments influenced the PHB properties negatively. This process step can be of advantage due to the generation of phycocyanin as a valuable side product [107]. A mixture of acetone and ethanol (70:30) was most suitable for this purpose. PHB extraction was performed with hot chloroform via a soxhlet extractor. Figure 5 compares the necessary processing steps of heterotrophs and cyanobacteria.

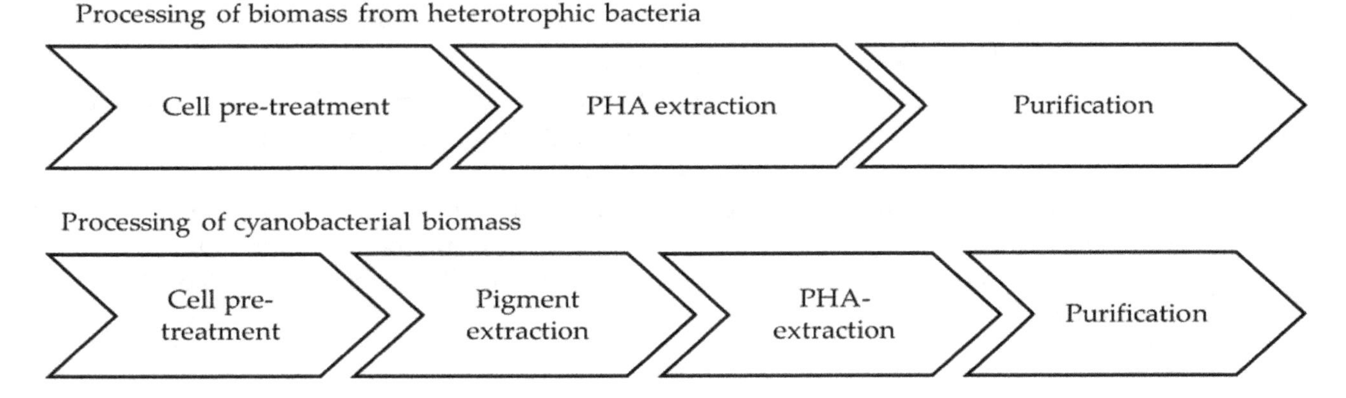

Figure 5. Comparison of processing steps needed to extract PHA from heterotrophic bacteria and cyanobacteria.

The PHB analysis showed that the polymers extracted from cyanobacterial biomass are comparable to commercially available PHB. Furthermore, it was shown that not only did the nutrient source, but also biomass pre-treatment and the method of polymer extraction influence the PHB properties. Pigment extraction and sample pre-treatment increased the average molecular weight (M_w) from 0.3 to 1.4 MD, but decreased degradation temperatures and crystallinity from 282 °C to 275 °C (T_{onset}) and from 296 °C to 287 °C (T_{max}), respectively. The M_w ranged from 5.8 to 8.0 MD, by using mineral medium and digestate, respectively. The thermal properties (T_{onset}: 283–282 °C; T_{max}: 293–292 °C), which are important for processing the polymer, are only slightly influenced by the nutrient source and are lower than, but comparable to, commercially available PHB. The crystallinity, responsible for higher final brittleness of the products, is about 17% lower than commercially available PHA.

5.6. Contaminations

Contaminations in non-sterile mass cultivation of microalgae are inevitable. It is only a matter of time before first contaminations appear, whether cultivation is done in open ponds or closed photobioreactors [108]. We observed certain bacterial and fungal contaminations with minor effects on *Synechocystis salina* CCALA192, when using CO_2 as the sole carbon source. Though, when adding acetate to the medium fungal contaminations were prevalent and difficult to control. After one and a half years the pilot plant was revised and a new degassing system was installed. From this moment on rapid culture crashes occurred. The microscopic image revealed a ciliated protozoa ingesting *Synechocystis* rapidly.

This ciliate forms highly resistant cysts and it is assumed that cysts from the surrounding soil were brought into the system during the revision work. Facts about contaminations in mass cultivation of *Synechocystis* are scarcely reported. Touloupakis and colleagues reported the grazing of *Synechocystis* PCC6803 by golden algae *Poterioochromonas* sp. [109]. High pH values of 10 and above helped to control the contaminant and maintain a stable culture. Unfortunately, the ciliate in our cultures survives those high pH values. Thoroughly cleaning and sanitizing the photobioreactor brought some success, but the ciliate is still occurring and leading to culture crashes. Due to the ciliate's capability to form cysts, it is very difficult to completely eliminate it from the reactor. Heat sterilization is not possible in tubular photobioreactors. Addressing further research, there is need for special cultivation methods for robustly growing *Synechocystis* in non-sterile environment.

6. Conclusions

Although not economical today, the idea of a sustainable PHB production with cyanobacteria, CO_2, and sunlight is still attractive and, more and more, researchers are working in this field. The main challenges today are similar to biofuel production with green algae: (i) realization of efficient low-cost cultivation systems at large scale; (ii) maintaining stable cultures under non-sterile conditions; (iii) increasing the total productivity and yield; and (iv) economic downstream processing and utilization of the residual biomass.

Looking for suitable production strains it must be considered that PHB production is a very common feature of many, but not all, cyanobacteria. The PHB content of cyanobacteria is highly strain specific, as strains of the same genus were reported with highly varying PHB contents. In addition to the PHB content the growth rate and robustness of a strain is particularly important. The only cyanobacterium cultivated in mass cultivation today is *Arthrospira* sp. and, therefore, one of the most promising candidates for photoautotrophic PHB production, although most *Arthrospira* sp. strains still show little PHB content.

Heterotrophic cultivation with acetate boosts the PHB content remarkably, as most reported values over 30% were achieved this way. However, it needs to be considered that using an organic carbon source impairs the most attractive feature of cyanobacteria, converting CO_2 to PHB. Using organic sources will also complicate non-sterile mass cultivation and could easily lead to contaminations and culture crashes. PHB production with organic carbon sources should be performed with heterotrophic bacteria, as their PHB productivity, as well as their cell density, are 10–100 times higher.

Nitrogen and phosphorous depletion are the most important factors to increase the PHB content and are often necessary to produce any PHB at all. Therefore, a two-stage cultivation with a self-limiting medium is necessary for large-scale photoautotrophic PHB production. With this strategy PHB was successfully produced in our 200 L photobioreactor. In tubular systems small unicellular organisms, like *Synechocystis* sp., are preferred over filamentous organisms, mainly because of the shear stress of the pump. Considering all of the difficulties to overcome, establishing a stable cyanobacterial culture is most important and most difficult to achieve.

Acknowledgments: The research project CO2USE was thankfully financed by the Austrian climate and energy fund and FFG (Austrian research promotion agency).

Author Contributions: Clemens Troschl designed and performed experiments, analyzed data, and wrote about 75% of the article. Katharina Meixner designed and performed experiments, analyzed data and wrote about 15% of the article. Bernhard Drosg supervised the work, analyzed data, and wrote about 10% of the article.

References

1. Khosravi-Darani, K.; Mokhtari, Z.-B.B.; Amai, T.; Tanaka, K. Microbial production of poly(hydroxybutyrate) from C1 carbon sources. *Appl. Microbiol. Biotechnol.* **2013**, *97*, 1407–1424. [CrossRef] [PubMed]
2. Dawes, E. Polyhydroxybutyrate: An intriguing biopolymer. *Biosci. Rep.* **1988**, *8*, 537–547. [CrossRef] [PubMed]
3. Chen, G.-Q. A microbial polyhydroxyalkanoates (PHA) based bio- and materials industry. *Chem. Soc. Rev.* **2009**, *38*, 2434–2446. [CrossRef] [PubMed]
4. Halami, P.M. Production of polyhydroxyalkanoate from starch by the native isolate Bacillus cereus CFR06. *World J. Microbiol. Biotechnol.* **2008**, *24*, 805–812. [CrossRef]

5. Ting, C.S.; Rocap, G.; King, J.; Chisholm, S.W. Cyanobacterial photosynthesis in the oceans: The origins and significance of divergent light-harvesting strategies. *Trends Microbiol.* **2002**, *10*, 134–142. [CrossRef]
6. IPCC. *2014: Climate Change 2014: Synthesis Report*; Core Writing Team, Pachauri, R.K., Meyer, L.A., Eds.; Contribution of Working Groups I, II and III to the Fifth Assessment Report of the Intergovernmental Panel on Climate Change; IPCC: Geneva, Switzerland, 2014; p. 151.
7. Stanier, R.J.; Cohen-Bazire, G. Phototrophic prokaryotes: The cyanobacteria. *Ann. Rev. Microbiol.* **1977**, *31*, 225–274. [CrossRef] [PubMed]
8. Bekker, A.; Holland, H.D.; Wang, P.-L.; Rumble, D.; Stein, H.J.; Hannah, J.L.; Coetzee, L.L.; Beukes, N.J. Dating the rise of atmospheric oxygen. *Nature* **2004**, *427*, 117–120. [CrossRef] [PubMed]
9. Schirrmeister, B.E.; Sanchez-Baracaldo, P.; Wacey, D. Cyanobacterial evolution during the Precambrian. *Int. J. Astrobiol.* **2016**, *15*, 1–18. [CrossRef]
10. Fischer, W.W.; Hemp, J.; Johnson, J.E. Evolution of Oxygenic Photosynthesis. *Annu. Rev. Earth Planet. Sci.* **2016**, *44*, 647–683. [CrossRef]
11. Nabout, J.C.; da Silva Rocha, B.; Carneiro, F.M.; Sant'Anna, C.L. How many species of Cyanobacteria are there? Using a discovery curve to predict the species number. *Biodivers. Conserv.* **2013**, *22*, 2907–2918. [CrossRef]
12. Stanier, R.Y.; Sistrom, W.R.; Hansen, T.A.; Whitton, B.A.; Castenholz, R.W.; Pfennig, N.; Gorlenko, V.N.; Kondratieva, E.N.; Eimhjellen, K.E.; Whittenbury, R.; et al. Proposal to Place the Nomenclature of the Cyanobacteria (Blue-Green Algae) Under the Rules of the International Code of Nomenclature of Bacteria. *Int. J. Syst. Bacteriol.* **1978**, *28*, 335–336. [CrossRef]
13. McNeill, J.; Barrie, F.R. *International Code of Nomenclature for Algae, Fungi, and Plants (Melbourne Code)*; Koeltz Scientific Books: Oberreifenberg, Germany, 2012.
14. Parker, C.T.; Tindall, B.J.; Garrity, G.M. International Code of Nomenclature of Prokaryotes. *Int. J. Syst. Evol. Microbiol.* **2015**. [CrossRef] [PubMed]
15. Richmond, A. *Handbook of Microalgal Culture*, 1st ed.; Blackwell Science Ltd.: Oxford, UK, 2004.
16. Tan, G.-Y.; Chen, C.-L.; Li, L.; Ge, L.; Wang, L.; Razaad, I.; Li, Y.; Zhao, L.; Mo, Y.; Wang, J.-Y. Start a Research on Biopolymer Polyhydroxyalkanoate (PHA): A Review. *Polymers (Basel.)* **2014**, *6*, 706–754. [CrossRef]
17. Nakaya, Y.; Iijima, H.; Takanobu, J.; Watanabe, A.; Hirai, M.Y.; Osanai, T. One day of nitrogen starvation reveals the effect of sigE and rre37 overexpression on the expression of genes related to carbon and nitrogen metabolism in *Synechocystis* sp. PCC 6803. *J. Biosci. Bioeng.* **2015**, 1–7. [CrossRef] [PubMed]
18. Schlebusch, M.; Forchhammer, K. Requirement of the nitrogen starvation-induced protein s110783 for polyhydroxybutyrate accumulation in *Synechocystis* sp. strain PCC 6803. *Appl. Environ. Microbiol.* **2010**, *76*, 6101–6107. [CrossRef] [PubMed]
19. Hauf, W.; Schlebusch, M.; Hüge, J.; Kopka, J.; Hagemann, M.; Forchhammer, K. Metabolic Changes in Synechocystis PCC6803 upon Nitrogen-Starvation: Excess NADPH Sustains Polyhydroxybutyrate Accumulation. *Metabolites* **2013**, *3*, 101–118. [CrossRef] [PubMed]
20. Stal, L.J. Poly(hydroxyalkanoate) in cyanobacteria: An overview. *FEMS Microbiol. Lett.* **1992**, *103*, 169–180. [CrossRef]
21. Kaneko, T.; Sato, S.; Kotani, H.; Tanaka, A.; Asamizu, E.; Nakamura, Y.; Miyajima, N.; Hirosawa, M.; Sugiura, M.; Sasamoto, S.; et al. Sequence analysis of the genome of the unicellular cyanobacterium *Synechocystis* sp. strain PCC6803. II. Sequence determination of the entire genome and assignment of potential protein-coding regions. *DNA Res.* **1996**, *3*, 109–136. [CrossRef] [PubMed]
22. Hein, S.; Tran, H.; Steinbüchel, A. *Synechocystis* sp. PCC6803 possesses a two-component polyhydroxyalkanoic acid synthase similar to that of anoxygenic purple sulfur bacteria. *Arch. Microbiol.* **1998**, *170*, 162–170. [CrossRef] [PubMed]
23. Taroncher-Oldenburg, G.; Nishina, K.; Stephanopoulos, G. Identification and analysis of the polyhydroxyalkanoate-specific beta-ketothiolase and acetoacetyl coenzyme A reductase genes in the cyanobacterium *Synechocystis* sp. strain PCC6803. *Appl. Environ. Microbiol.* **2000**, *66*, 4440–4448. [CrossRef] [PubMed]
24. Aikawa, S.; Izumi, Y.; Matsuda, F.; Hasunuma, T.; Chang, J.S.; Kondo, A. Synergistic enhancement of glycogen production in *Arthrospira platensis* by optimization of light intensity and nitrate supply. *Bioresour. Technol.* **2012**, *108*, 211–215. [CrossRef] [PubMed]
25. Monshupanee, T.; Incharoensakdi, A. Enhanced accumulation of glycogen, lipids and polyhydroxybutyrate under optimal nutrients and light intensities in the cyanobacterium *Synechocystis* sp. PCC 6803. *J. Appl. Microbiol.* **2014**, *116*, 830–838. [CrossRef] [PubMed]

26. De Philippis, R.; Sili, C.; Vincenzini, M. Glycogen and poly-β-hydroxybutyrate synthesis in Spirulina maxima. *J. Gen. Microbiol.* **1992**, *138*, 1623–1628. [CrossRef]
27. Deschoenmaeker, F.; Facchini, R.; Carlos, J.; Pino, C.; Bayon-Vicente, G.; Sachdeva, N.; Flammang, P.; Wattiez, R. Nitrogen depletion in *Arthrospira* sp. PCC 8005, an ultrastructural point of view. *J. Struct. Biol.* **2016**, *196*, 385–393. [CrossRef] [PubMed]
28. Aikawa, S.; Nishida, A.; Ho, S.-H.; Chang, J.-S.; Hasunuma, T.; Kondo, A. Glycogen production for biofuels by the euryhaline cyanobacteria *Synechococcus* sp. strain PCC 7002 from an oceanic environment. *Biotechnol. Biofuels* **2014**, *7*, 88. [CrossRef] [PubMed]
29. Yoo, S.H.; Keppel, C.; Spalding, M.; Jane, J.L. Effects of growth condition on the structure of glycogen produced in cyanobacterium *Synechocystis* sp. PCC6803. *Int. J. Biol. Macromol.* **2007**, *40*, 498–504. [CrossRef] [PubMed]
30. Ostle, A.G.; Holt, J.G. Fluorescent Stain for Poly-3-Hydroxybutyrate. *Appl. Environ. Microbiol.* **1982**, *44*, 238–241. [PubMed]
31. Gorenflo, V.; Steinbüchel, A.; Marose, S.; Rieseberg, M.; Scheper, T. Quantification of bacterial polyhydroxyalkanoic acids by Nile red staining. *Appl. Microbiol. Biotechnol.* **1999**, *51*, 765–772. [CrossRef] [PubMed]
32. Tsang, T.K.; Roberson, R.W.; Vermaas, W.F.J. Polyhydroxybutyrate particles in *Synechocystis* sp. PCC 6803: Facts and fiction. *Photosynth. Res.* **2013**, *118*, 37–49. [CrossRef] [PubMed]
33. Hauf, W.; Watzer, B.; Roos, N.; Klotz, A.; Forchhammer, K. Photoautotrophic Polyhydroxybutyrate Granule Formation Is Regulated by Cyanobacterial Phasin PhaP in *Synechocystis* sp. Strain PCC 6803. *Appl. Environ. Microbiol.* **2015**, *81*, 4411–4422. [CrossRef] [PubMed]
34. Klotz, A.; Georg, J.; Bučínská, L.; Watanabe, S.; Reimann, V.; Januszewski, W.; Sobotka, R.; Jendrossek, D.; Hess, W.R.; Forchhammer, K. Awakening of a dormant cyanobacterium from nitrogen chlorosis reveals a genetically determined program. *Curr. Biol.* **2016**, *26*, 2862–2872. [CrossRef] [PubMed]
35. Gründel, M.; Scheunemann, R.; Lockau, W.; Zilliges, Y. Impaired glycogen synthesis causes metabolic overflow reactions and affects stress responses in the cyanobacterium *Synechocystis* sp. PCC 6803. *Microbiol. (UK)* **2012**, *158*, 3032–3043. [CrossRef] [PubMed]
36. Beck, C.; Knoop, H.; Axmann, I.M.; Steuer, R. The diversity of cyanobacterial metabolism: Genome analysis of multiple phototrophic microorganisms. *BMC Genom.* **2012**, *13*, 56. [CrossRef] [PubMed]
37. Ansari, S.; Fatma, T. Cyanobacterial polyhydroxybutyrate (PHB): Screening, optimization and characterization. *PLoS ONE* **2016**, *11*, 1–20. [CrossRef] [PubMed]
38. Kaewbai-ngam, A.; Incharoensakdi, A.; Monshupanee, T. Increased accumulation of polyhydroxybutyrate in divergent cyanobacteria under nutrient-deprived photoautotrophy: An efficient conversion of solar energy and carbon dioxide to polyhydroxybutyrate by Calothrix scytonemicola TISTR 8095. *Bioresour. Technol.* **2016**, *212*, 342–347. [CrossRef] [PubMed]
39. Hu, Q.; Sommerfeld, M.; Jarvis, E.; Ghirardi, M.; Posewitz, M.; Seibert, M.; Darzins, A. Microalgal triacylglycerols as feedstocks for biofuel production: Perspectives and advances. *Plant J.* **2008**, *54*, 621–639. [CrossRef] [PubMed]
40. de Jaeger, L.; Verbeek, R.E.; Draaisma, R.B.; Martens, D.E.; Springer, J.; Eggink, G.; Wijffels, R.H. Superior triacylglycerol (TAG) accumulation in starchless mutants of Scenedesmus obliquus: (I) Mutant generation and characterization. *Biotechnol. Biofuels* **2014**, *7*, 69. [CrossRef] [PubMed]
41. Damrow, R.; Maldener, I.; Zilliges, Y. The multiple functions of common microbial carbon polymers, glycogen and PHB, during stress responses in the non-diazotrophic cyanobacterium *Synechocystis* sp. PCC 6803. *Front. Microbiol.* **2016**, *7*, 1–10. [CrossRef] [PubMed]
42. Allen, M.M.; Smith, A.J. Nitrogen chlorosis in blue-green algae. *Arch. für Mikrobiol.* **1969**, *69*, 114–120. [CrossRef]
43. Gorl, M.; Sauer, J.; Baier, T.; Forchhammer, K. Nitrogen-starvation-induced chlorosis in Synechococcus PCC 7942: Adaptation to long-term survival. *Microbiology* **1998**, *144*, 2449–2458. [CrossRef] [PubMed]
44. Sauer, J.; Schreiber, U.; Schmid, R.; Völker, U.; Forchhammer, K. Nitrogen starvation-induced chlorosis in Synechococcus PCC 7942. Low-level photosynthesis as a mechanism of long-term survival. *Plant Physiol.* **2001**, *126*, 233–243. [CrossRef] [PubMed]
45. Wu, G.F.; Wu, Q.Y.; Shen, Z.Y. Accumulation of poly-beta-hydroxybutyrate in cyanobacterium *Synechocystis* sp. PCC6803. *Bioresour. Technol.* **2001**, *76*, 85–90. [CrossRef]

46. Panda, B.; Mallick, N. Enhanced poly-β-hydroxybutyrate accumulation in a unicellular cyanobacterium, *Synechocystis* sp. PCC 6803. *Lett. Appl. Microbiol.* **2007**, *44*, 194–198. [CrossRef] [PubMed]
47. Khetkorn, W.; Incharoensakdi, A.; Lindblad, P.; Jantaro, S. Enhancement of poly-3-hydroxybutyrate production in *Synechocystis* sp. PCC 6803 by overexpression of its native biosynthetic genes. *Bioresour. Technol.* **2016**, *214*, 761–768. [CrossRef] [PubMed]
48. Nishioka, M.; Nakai, K.; Miyake, M.; Asada, Y.; Taya, M. Production of poly-β-hydroxybutyrate by thermophilic cyanobacterium, *Synechococcus* sp. MA19, under phosphate-limited conditions. *Biotechnol. Lett.* **2001**, *23*, 1095–1099. [CrossRef]
49. Wu, Q.; Liu, L.; Miron, A.; Klimova, B.; Wan, D.; Kuca, K. The antioxidant, immunomodulatory, and anti-inflammatory activities of Spirulina: An overview. *Arch. Toxicol.* **2016**, *90*, 1817–1840. [CrossRef] [PubMed]
50. Spolaore, P.; Joannis-Cassan, C.; Duran, E.; Isambert, A. Commercial applications of microalgae. *J. Biosci. Bioeng.* **2006**, *101*, 87–96. [CrossRef] [PubMed]
51. Shimamatsu, H. Mass production of *Spirulina*, an edible microalga. *Hydrobiologia* **2004**, *512*, 39–44. [CrossRef]
52. Choi, S.-L.; Suh, I.S.; Lee, C.-G. Lumostatic operation of bubble column photobioreactors for Haematococcus pluvialis cultures using a specific light uptake rate as a control parameter. *Enzyme Microb. Technol.* **2003**, *33*, 403–409. [CrossRef]
53. Lu, Y.M.; Xiang, W.Z.; Wen, Y.H. Spirulina (Arthrospira) industry in Inner Mongolia of China: Current status and prospects. *J. Appl. Phycol.* **2011**, *23*, 265–269. [CrossRef] [PubMed]
54. Campbell, J.; Stevens, S.E.; Balkwill, D.L. Accumulation of poly-beta-hydroxybutyrate in Spirulina platensis. *J. Bacteriol.* **1982**, *149*, 361–363. [PubMed]
55. Vincenzini, M.; Sili, C.; de Philippis, R.; Ena, A.; Materassi, R. Occurrence of poly-beta-hydroxybutyrate in Spirulina species. *J. Bacteriol.* **1990**, *172*, 2791–2792. [CrossRef] [PubMed]
56. Panda, B.; Sharma, L.; Mallick, N. Poly-β-hydroxybutyrate accumulation in Nostoc muscorum and Spirulina platensis under phosphate limitation. *J. Plant Physiol.* **2005**, *162*, 1376–1379. [CrossRef] [PubMed]
57. Shrivastav, A.; Mishra, S.K.; Mishra, S. Polyhydroxyalkanoate (PHA) synthesis by Spirulina subsalsa from Gujarat coast of India. *Int. J. Biol. Macromol.* **2010**, *46*, 255–260. [CrossRef] [PubMed]
58. De Morais, M.G.; Stillings, C.; Roland, D.; Rudisile, M.; Pranke, P.; Costa, J.A.V.; Wendorff, J. Extraction of poly(3-hydroxybutyrate) from Spirulina LEB 18 for developing nanofibers. *Polímeros* **2015**, *25*, 161–167. [CrossRef]
59. De Morais, M.G.; Stillings, C.; Dersch, R.; Rudisile, M.; Pranke, P.; Costa, J.A.V.; Wendorff, J. Biofunctionalized nanofibers using *Arthrospira* (*spirulina*) biomass and biopolymer. *Biomed Res. Int.* **2015**, *2015*. [CrossRef] [PubMed]
60. Dodds, W.K.; Gudder, D.A.; Mollenhauer, D. Review the Ecology of Nostoc. *J. Phycol.* **1995**, *18*, 2–18. [CrossRef]
61. Qiu, B.; Liu, J.; Liu, Z.; Liu, S. Distribution and ecology of the edible cyanobacterium Ge-Xian-Mi (Nostoc) in rice fields of Hefeng County in China. *J. Appl. Phycol.* **2002**, *14*, 423–429. [CrossRef]
62. Sharma, L.; Mallick, N. Accumulation of poly-β-hydroxybutyrate in Nostoc muscorum: Regulation by pH, light-dark cycles, N and P status and carbon sources. *Bioresour. Technol.* **2005**, *96*, 1304–1310. [CrossRef] [PubMed]
63. Sharma, L.; Mallick, N. Enhancement of poly-β-hydroxybutyrate accumulation in Nostoc muscorum under mixotrophy, chemoheterotrophy and limitations of gas-exchange. *Biotechnol. Lett.* **2005**, *27*, 59–62. [CrossRef] [PubMed]
64. Bhati, R.; Mallick, N. Production and characterization of poly(3-hydroxybutyrate-co-3-hydroxyvalerate) co-polymer by a N 2-fixing cyanobacterium, Nostoc muscorum Agardh. *J. Chem. Technol. Biotechnol.* **2012**, *87*, 505–512. [CrossRef]
65. Bhati, R.; Mallick, N. Poly(3-hydroxybutyrate-co-3-hydroxyvalerate) copolymer production by the diazotrophic cyanobacterium Nostoc muscorum Agardh: Process optimization and polymer characterization. *Algal Res.* **2015**, *7*, 78–85. [CrossRef]
66. Bhati, R.; Mallick, N. Carbon dioxide and poultry waste utilization for production of polyhydroxyalkanoate biopolymers by Nostoc muscorum Agardh: A sustainable approach. *J. Appl. Phycol.* **2016**, *28*, 161–168. [CrossRef]
67. Samantaray, S.; Mallick, N. Production and characterization of poly-β-hydroxybutyrate (PHB) polymer from Aulosira fertilissima. *J. Appl. Phycol.* **2012**, *24*, 803–814. [CrossRef]

68. Lama, L.; Nicolaus, B.; Calandrelli, V.; Manca, M.C.; Romana, I.; Gambacorta, A. Effect of growth conditions on endo- and exopolymer biosynthesis in Anabaena cylindrical 10 C. *Phytochemistry* **1996**, *42*, 655–659. [CrossRef]
69. Gopi, K.; Balaji, S.; Muthuvelan, B. Isolation Purification and Screening of Biodegradable Polymer PHB Producing Cyanobacteria from Marine and Fresh Water Resources. *Iran. J. Energy Environ.* **2014**, *5*, 94–100. [CrossRef]
70. Benemann, J. Microalgae for biofuels and animal feeds. *Energies* **2013**, *6*, 5869–5886. [CrossRef]
71. Dineshbabu, G.; Uma, V.S.; Mathimani, T.; Deviram, G.; Arul Ananth, D.; Prabaharan, D.; Uma, L. On-site concurrent carbon dioxide sequestration from flue gas and calcite formation in ossein effluent by a marine cyanobacterium Phormidium valderianum BDU 20041. *Energy Convers. Manag.* **2016**, in press. [CrossRef]
72. Chen, H.W.; Yang, T.S.; Chen, M.J.; Chang, Y.C.; Lin, C.Y.; Wang, E.I.C.; Ho, C.L.; Huang, K.M.; Yu, C.C.; Yang, F.L.; et al. Application of power plant flue gas in a photobioreactor to grow Spirulina algae, and a bioactivity analysis of the algal water-soluble polysaccharides. *Bioresour. Technol.* **2012**, *120*, 256–263. [CrossRef] [PubMed]
73. Kumari, A.; Kumar, A.; Pathak, A.K.; Guria, C. Carbon dioxide assisted Spirulina platensis cultivation using NPK-10:26:26 complex fertilizer in sintered disk chromatographic glass bubble column. *J. CO_2 Util.* **2014**, *8*, 49–59. [CrossRef]
74. He, L.; Subramanian, V.R.; Tang, Y.J. Experimental analysis and model-based optimization of microalgae growth in photo-bioreactors using flue gas. *Biomass Bioenergy* **2012**, *41*, 131–138. [CrossRef]
75. Ferreira, L.S.; Rodrigues, M.S.; Converti, A.; Sato, S.; Carvalho, J.C.M. *Arthrospira (spirulina)* platensis cultivation in tubular photobioreactor: Use of no-cost CO_2 from ethanol fermentation. *Appl. Energy* **2012**, *92*, 379–385. [CrossRef]
76. Sumardiono, S.; Syaichurrozi, I.; Budi Sasongko, S. Utilization of Biogas as Carbon Dioxide Provider for Spirulina platensis Culture. *Curr. Res. J. Biol. Sci.* **2014**, *6*, 53–59.
77. Cuellar-Bermudez, S.P.; Aleman-Nava, G.S.; Chandra, R.; Garcia-Perez, J.S.; Contreras-Angulo, J.R.; Markou, G.; Muylaert, K.; Rittmann, B.E.; Parra-Saldivar, R. Nutrients utilization and contaminants removal. A review of two approaches of algae and cyanobacteria in wastewater. *Algal Res.* **2016**. [CrossRef]
78. Markou, G.; Vandamme, D.; Muylaert, K. Microalgal and cyanobacterial cultivation: The supply of nutrients. *Water Res.* **2014**, *65*, 186–202. [CrossRef] [PubMed]
79. Kumar, K.; Mishra, S.K.; Shrivastav, A.; Park, M.S.; Yang, J.W. Recent trends in the mass cultivation of algae in raceway ponds. *Renew. Sustain. Energy Rev.* **2015**, *51*, 875–885. [CrossRef]
80. Markou, G.; Georgakakis, D. Cultivation of filamentous cyanobacteria (blue-green algae) in agro-industrial wastes and wastewaters: A review. *Appl. Energy* **2011**, *88*, 3389–3401. [CrossRef]
81. Chaiklahan, R.; Chirasuwan, N.; Siangdung, W.; Paithoonrangsarid, K.; Bunnag, B. Cultivation of spirulina platensis using pig wastewater in a semi-continuous process. *J. Microbiol. Biotechnol.* **2010**, *20*, 609–614. [CrossRef] [PubMed]
82. Cicci, A.; Bravi, M. Production of the freshwater microalgae scenedesmus dimorphus and arthrospira platensis by using cattle digestate. *Chem. Eng. Trans.* **2014**, *38*, 85–90. [CrossRef]
83. Fouilland, E.; Vasseur, C.; Leboulanger, C.; Le Floc'h, E.; Carré, C.; Marty, B.; Steyer, J.-P.P.; Sialve, B. Coupling algal biomass production and anaerobic digestion: Production assessment of some native temperate and tropical microalgae. *Biomass Bioenergy* **2014**, *70*, 564–569. [CrossRef]
84. Markou, G.; Chatzipavlidis, I.; Georgakakis, D. Cultivation of *Arthrospira* (*Spirulina*) platensis in olive-oil mill wastewater treated with sodium hypochlorite. *Bioresour. Technol.* **2012**, *112*, 234–241. [CrossRef] [PubMed]
85. Olguín, E.J.; Galicia, S.; Mercado, G.; Pérez, T. Annual productivity of *Spirulina* (*Arthrospira*) and nutrient removal in a pig wastewater recycling process under tropical conditions. *J. Appl. Phycol.* **2003**, *15*, 249–257. [CrossRef]
86. Prajapati, S.K.; Kumar, P.; Malik, A.; Vijay, V.K. Bioconversion of algae to methane and subsequent utilization of digestate for algae cultivation: A closed loop bioenergy generation process. *Bioresour. Technol.* **2014**, *158*, 174–180. [CrossRef] [PubMed]
87. Daelman, M.R.J.; Sorokin, D.; Kruse, O.; van Loosdrecht, M.C.M.; Strous, M. Haloalkaline Bioconversions for Methane Production from Microalgae Grown on Sunlight. *Trends Biotechnol.* **2016**, *34*, 450–457. [CrossRef] [PubMed]

88. Nolla-Ardevol, V.; Strous, M.; Tegetmeyer, H.E.E. Anaerobic digestion of the microalga Spirulina at extreme alkaline conditions: Biogas production, metagenome and metatranscriptome. *Front. Microbiol.* **2015**, *6*, 1–21. [CrossRef] [PubMed]
89. Zhou, Y.; Schideman, L.; Yu, G.; Zhang, Y. A synergistic combination of algal wastewater treatment and hydrothermal biofuel production maximized by nutrient and carbon recycling. *Energy Environ. Sci.* **2013**, *6*, 3765. [CrossRef]
90. Zheng, M.; Schideman, L.C.; Tommaso, G.; Chen, W.-T.; Zhou, Y.; Nair, K.; Qian, W.; Zhang, Y.; Wang, K. Anaerobic digestion of wastewater generated from the hydrothermal liquefaction of Spirulina: Toxicity assessment and minimization. *Energy Convers. Manag.* **2016**, in press. [CrossRef]
91. Depraetere, O.; Pierre, G.; Noppe, W.; Vandamme, D.; Foubert, I.; Michaud, P.; Muylaert, K. Influence of culture medium recycling on the performance of Arthrospira platensis cultures. *Algal Res.* **2015**, *10*, 48–54. [CrossRef]
92. Meixner, K.; Fritz, I.; Daffert, C.; Markl, K.; Fuchs, W.; Drosg, B. Processing recommendations for using low-solids digestate as nutrient solution for production with *Synechocystis salina*. *J. Biotechnol.* **2016**, *240*, 1–22. [CrossRef] [PubMed]
93. Balaji, S.; Gopi, K.; Muthuvelan, B. A review on production of poly-b-hydroxybutyrates from cyanobacteria for the production of bio plastics. *Algal Res.* **2013**, *2*, 278–285. [CrossRef]
94. Marcilhac, C.; Sialve, B.; Pourcher, A.M.; Ziebal, C.; Bernet, N.; Béline, F. Digestate color and light intensity affect nutrient removal and competition phenomena in a microalgal-bacterial ecosystem. *Water Res.* **2014**, *64*, 278–287. [CrossRef] [PubMed]
95. Andrade, M.R.; Costa, J.A.V. Mixotrophic cultivation of microalga Spirulina platensis using molasses as organic substrate. *Aquaculture* **2007**, *264*, 130–134. [CrossRef]
96. Phang, S.M.; Miah, M.S.; Yeoh, B.G.; Hashim, M.A. Spirulina cultivation in digested sago starch factory wastewater. *J. Appl. Phycol.* **2000**, *12*, 395–400. [CrossRef]
97. Massa, M.; Buono, S.; Langellotti, A.L.; Castaldo, L.; Martello, A.; Paduano, A.; Sacchi, R.; Fogliano, V. Evaluation of anaerobic digestates from different feedstocks as growth media for Tetradesmus obliquus, Botryococcus braunii, Phaeodactylum tricornutum and Arthrospira maxima. *New Biotechnol.* **2017**, *36*, 8–16. [CrossRef] [PubMed]
98. Cai, T.; Ge, X.; Park, S.Y.; Li, Y. Comparison of *Synechocystis* sp. PCC6803 and Nannochloropsis salina for lipid production using artificial seawater and nutrients from anaerobic digestion effluent. *Bioresour. Technol.* **2013**, *144*, 255–260. [CrossRef] [PubMed]
99. Molina, E.; Fernández, J.; Acién, F.G.; Chisti, Y. Tubular photobioreactor design for algal cultures. *J. Biotechnol.* **2001**, *92*, 113–131. [CrossRef]
100. Acién Fernández, F.G.; Fernández Sevilla, J.M.; Sánchez Pérez, J.A.; Molina Grima, E.; Chisti, Y. Airlift-driven external-loop tubular photobioreactors for outdoor production of microalgae: Assessment of design and performance. *Chem. Eng. Sci.* **2001**, *56*, 2721–2732. [CrossRef]
101. Rabensteiner, M.; Kinger, G.; Koller, M.; Gronald, G.; Hochenauer, C. Pilot plant study of ethylenediamine as a solvent for post combustion carbon dioxide capture and comparison to monoethanolamine. *Int. J. Greenh. Gas Control* **2014**, *27*, 1–14. [CrossRef]
102. Rippka, R.; Deruelles, J.; Waterbury, J.B.; Herdman, M.; Stanier, R.Y. Generic Assignments, Strain Histories and Properties of Pure Cultures of Cyanobacteria. *J. Gen. Microbiol.* **1979**, *111*, 1–61. [CrossRef]
103. Drosg, B.; Fritz, I.; Gattermayr, F.; Silvestrini, L. Photo-autotrophic Production of Poly(hydroxyalkanoates) in Cyanobacteria. *Chem. Biochem. Eng. Q.* **2015**, *29*, 145–156. [CrossRef]
104. Michels, M.H.A.; van der Goot, A.J.; Vermuë, M.H.; Wijffels, R.H. Cultivation of shear stress sensitive and tolerant microalgal species in a tubular photobioreactor equipped with a centrifugal pump. *J. Appl. Phycol.* **2016**, *28*, 53–62. [CrossRef] [PubMed]
105. Milledge, J.J.; Heaven, S. A review of the harvesting of micro-algae for biofuel production. *Rev. Environ. Sci. Biotechnol.* **2013**, *12*, 165–178. [CrossRef]
106. Heinrich, D.; Madkour, M.H.; Al-Ghamdi, M.; Shabbaj, I.I.; Steinbüchel, A. Large scale extraction of poly(3-hydroxybutyrate) from Ralstonia eutropha H16 using sodium hypochlorite. *AMB Express* **2012**, *2*, 59. [CrossRef] [PubMed]

107. Ramos, A.; Acién, F.G.; Fernández-Sevilla, J.M.; González, C.V.; Bermejo, R. Development of a process for large-scale purification of C-phycocyanin from *Synechocystis aquatilis* using expanded bed adsorption chromatography. *J. Chromatogr. B Anal. Technol. Biomed. Life Sci.* **2011**, *879*, 511–519. [CrossRef] [PubMed]
108. Wang, H.; Zhang, W.; Chen, L.; Wang, J.; Liu, T. The contamination and control of biological pollutants in mass cultivation of microalgae. *Bioresour. Technol.* **2013**, *128*, 745–750. [CrossRef] [PubMed]
109. Touloupakis, E.; Cicchi, B.; Benavides, A.M.S.; Torzillo, G. Effect of high pH on growth of *Synechocystis* sp. PCC 6803 cultures and their contamination by golden algae (*Poterioochromonas* sp.). *Appl. Microbiol. Biotechnol.* **2015**, *100*, 1333–1341. [CrossRef] [PubMed]

6

Production of Polyhydroxyalkanoates and Extracellular Products using *Pseudomonas Corrugata* and *P. Mediterranea*

Grazia Licciardello [1,*], Antonino F. Catara [2] and Vittoria Catara [3]

[1] Consiglio per la Ricerca in agricoltura e l'analisi dell'Economia Agraria-Centro di ricerca Olivicoltura, Frutticoltura e Agrumicoltura (CREA), Corso Savoia 190, 95024 Acireale, Italy
[2] Formerly, Science and Technologies Park of Sicily, ZI Blocco Palma I, Via V. Lancia 57, 95121 Catania, Italy; antoninocatara@virgilio.it
[3] Dipartimento di Agricoltura, Alimentazione e Ambiente, Università degli studi di Catania, Via Santa Sofia 100, 95130 Catania, Italy; vcatara@unict.it
* Correspondence: grazia.licciardello@crea.gov.it

Abstract: Some strains of *Pseudomonas corrugata* (*Pco*) and *P. mediterranea* (*Pme*) efficiently synthesize medium-chain-length polyhydroxyalkanoates elastomers (mcl-PHA) and extracellular products on related and unrelated carbon sources. Yield and composition are dependent on the strain, carbon source, fermentation process, and any additives. Selected *Pco* strains produce amorphous and sticky mcl-PHA, whereas strains of *Pme* produce, on high grade and partially refined biodiesel glycerol, a distinctive filmable PHA, very different from the conventional microbial mcl-PHA, suitable for making blends with polylactide acid. However, the yields still need to be improved and production costs lowered. An integrated process has been developed to recover intracellular mcl-PHA and extracellular bioactive molecules. Transcriptional regulation studies during PHA production contribute to understanding the metabolic potential of *Pco* and *Pme* strains. Data available suggest that *pha* biosynthesis genes and their regulations will be helpful to develop new, integrated strategies for cost-effective production.

Keywords: medium-chain-length polyhydroxyalkanoate (mcl-PHA); alginate; biosurfactants; biopolymer; *Pseudomonas*; blends; film

---

1. Introduction

Polyhydroxyalkanoates (PHAs) are microbial polyesters synthesized by both Gram-negative and Gram-positive eubacteria, and an increasing number of archaea isolated from environmentally extreme habitats, to increase their survival and competition in environments where carbon and energy sources are limited, such as soil and rhizosphere [1–3].

Based on their repeat unit composition, the up to 150 different PHA structures identified so far [4] are classified mainly in two distinct groups: (i) short chain length (scl) PHAs where the repeat units are hydroxy fatty acids (HFAs) of 3–5 carbon chain length (C3–C5); and (ii) medium chain length (mcl) PHAs with repeat units of C6–14. In general, scl-PHAs are crystalline polymers with a fragile, rigid structure, whereas mcl-PHAs are amorphous thermoplastics, which have various degrees of crystallinity as well as elastomeric and adhesive properties [5]. Less common and least studied are long chain length (lcl) PHAs, constituted of monomers with more than 14 carbon atoms.

Thanks to two metabolic pathways based on the degradation of aliphatic carbon sources or *de novo* synthesis of fatty acids from unrelated carbon sources, *Pseudomonas* species included in the rRNA homology group I are among the most important producers of PHA [6–9]. Historically, fatty acids have been the preferred substrate for the microbial synthesis of mcl-PHA. Glucose, gluconate or ethanol, as well soy molasses [10], biodiesel co-product stream [11] and glycerol [12,13], have been successfully used. The fatty acyl composition of the substrate reflects the repeat unit composition of biopolymers [14]. Biodiesel glycerol has been recognized as a suitable and cost attractive substrate for PHA production, and therefore constitutes the main focus of this review [15,16].

Medium chain length-PHAs in *Pseudomonas* bacteria were first detected in *P. oleovorans* [17] and later in a variety of *Pseudomonas* [10]. *Pseudomonas*-PHAs are biodegradable, non-toxic and biocompatible and can be produced using a wide range of carbon sources. In fact, there has been considerable research exploring their potential in medical devices, foods, agriculture and consumer products [18,19]. Their elastic and flexibility properties improve the processability and mechanical properties of blends with other biodegradable polymers [20–22]. Of the various species tested worldwide, *P. aeruginosa, P. putida, P. resinovorans, P. mendocina,* and *P. chlororaphis* are the most extensively studied to clarify the metabolic processes of the production of PHA and to enhance the bioconversion efficiency [3,23].

This review focuses on the two taxonomically related Gram-negative rod ubiquitous bacteria, *P. corrugata* and the strictly related *P. mediterranea*, which cause disease on several crop species [24] and can produce an arsenal of secondary metabolites [25]. Among them, are biosurfactants (BSs) [26] as well as poly-mannuronic acid alginate [27], bioactive cyclic lipopeptides (CLPs), such as cormycin A and corpeptins [28–30], and a lipopeptide siderophore, corrugatin [31].

These bacteria produce different cellular mcl-PHAs and extracellular products, on waste fried edible oils, biodiesel glycerol and high-grade glycerol [13,32,33]. Selected *P. corrugata* strains produce intracellular mcl-PHA with a molecular weight of 120–150 kDa on waste edible oils, whereas strains of *P. mediterranea* generate a distinctive filmable PHA around 55–65 kDa on high-grade and partially refined biodiesel glycerol. Extracellular products, such as biosurfactants, exopolysaccharides (EPS, mostly alginate) and bioactive molecules, accumulate in the supernatant during the bioconversion process. Genome analysis of nine *P. corrugata* and *P. mediterranea* strains has helped to develop molecular and genetic investigations to enhance productivity [25].

2. Production of mcl-PHA and Extracellular Products

The first strain of *P. corrugata* investigated for its capacity to convert triacylglycerols to produce mcl-PHA was strain 388 [7,14,34]. The positive results led to the screening of different carbon sources of 56 strains of *P. corrugata* and 21 strains of its closely related *P. mediterranea* [9]. Flask-scale tests, carried out on related and unrelated carbon sources, have been reported [9,13,32–34].

Subsequently, some strains of *P. corrugata* producing lipase have been reported as being able to bioconvert waste exhausted fried edible oils, from a licensed collector, in mcl-PHAs [9,32]. One of these strains, namely *P. corrugata A1* (DSM 18227) (hereafter *Pco A1*), obtained through culturing *P. corrugata* CFBP5454 in E* medium with triolein, helped to patent a fermentation process validated on a 5000 L fermenter [35]. This increased the productivity of the process from 2.90 g/L up to 26 g/L of dry cell weight with 38% of PHA [32,35].

To overcome the variable composition of licensed exhausted edible oils and the difficulty in collecting adequate stocks for industrial production, several sources of glycerol have been extensively tested to screen many strains of *P. corrugata* and *P. mediterranea*. Three of them, *Pco 388, Pco A1* and *P. mediterranea 9.1* (deposited as CFBP5447, hereafter *Pme 9.1*), have been selected to study the bioconversion processes exploiting commercial high-grade glycerol (≥99%, pH 7) and biodiesel glycerol obtained from the transesterification of rapeseed oils (*Brassica carenata* and *B. napus*). Crude biodiesel (15% glycerol), oil free (40%) and partially refined glycerol (87.5%) performed differently from commercial high grade glycerol in terms of yield, composition and properties of the mcl-PHA and extracellular products. It has also been highlighted that some apparently small differences in the carbon sources may have a large impact, and that the genetic and metabolic system of the strain are key to the bioconversion process [11–13,33].

P. mediterranea 9.1 reached a production of 2.93 g/L of mcl-PHA on 2% crude biodiesel glycerol with a PHA/cell dry weight ratio >60%, whereas on high-grade glycerol it yielded 0.81 g/L (Table 1) [33]. In parallel tests, *P. corrugata* A.1 produced 1.8 g/L of PHA with 51.5% in cultivation for 72 h [33]. The productivity of batch fermentations on E* medium with the addition of 1% or 2% of glycerol showed only minor differences, but decreased yields and mass molecular weight (Mw) were observed when 5% of glycerol was added. The same results were reported by Ashby et al. [11], in the case of Mw of PHB produced by *P. oleovorans* and mcl-PHAs accumulated by *P. corrugata 388*.

Another flask-scale experiment, carried out with *Pme 9.1* growing on a medium containing 2% refined glycerol, yielded 3.3 g/L in cell dry weight (CDW) after cultivation for 48 h, with a PHA/CDW ratio close to 18% (Table 1) [13]. No significant changes were observed after 60 and 72 h of cultivation. Soxhlet extraction of biomass with acetone produced 0.75 g/L of a thin opalescent film of crude mcl-PHA. Parallel fermentation carried out with partially refined glycerol (87.5%) obtained from the esterification of *B. napus* oil produced 3.1 g/L of biomass and 0.5 g/L of raw PHA (PHA/CDW = 16.5%) [13].

Besides the different conversion efficiencies, other chemical and technological properties of the PHA were even more relevant. Regardless of the carbon source, both strains of *P. corrugata* (*388* and *A1*) produced very similar mcl-PHA elastomers, whereas PHA obtained from *P. mediterranea* 9.1 grown on refined biodiesel glycerol, generated a transparent filmable polymer with a low molecular weight (56,000 Da) and very distinctive characteristics (Figure 1) [13].

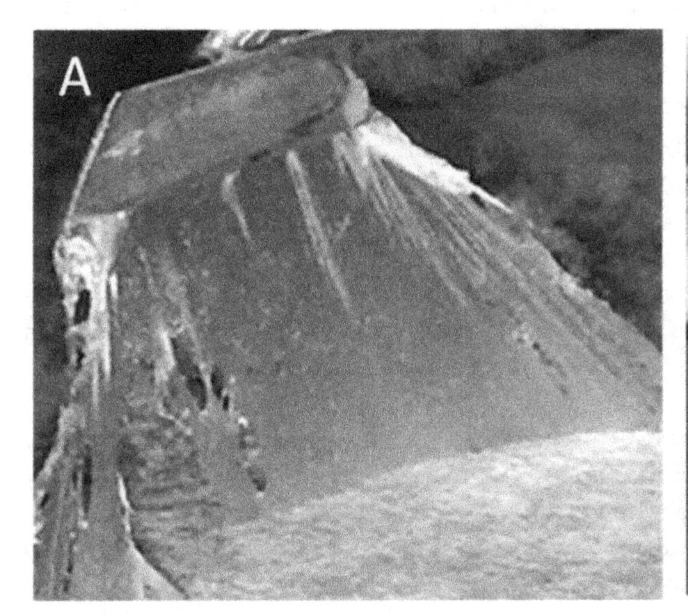

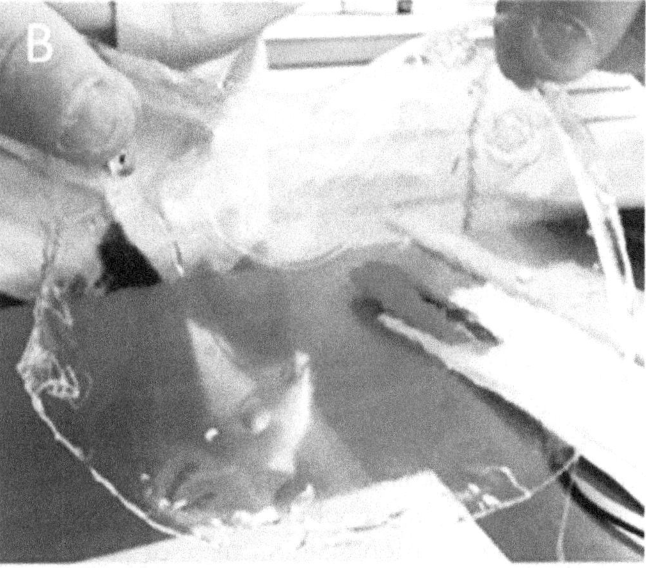

Figure 1. Crude PHA film (**A**) and transparent PHA film obtained after floating a toluene solution on a water surface (**B**) achieved from *Pseudomonas mediterranea* 9.1 using refined glycerol as carbon source (Figure 1B courtesy of Copyright Elsevier from [13]).

The extracellular biosurfactants released by these strains during the bioconversion process showed their dependence on the carbon sources, and the highest yields were reached much later than the PHA. *Pme 9.1* grown on crude glycerol (15%) obtained from *Brassica* spp. seed oil, was able to recover up to 14 g/L of surfactants with E24 (emulsification index) 54%, via chloroform:methanol (2:1) [36]. The highest accumulation occurred after 96–144 h. At the early stationary phase (48 h) *P. mediterranea 9.1* yielded 6.9 g/L of partially purified EPS, 17-fold higher than in *Pco A1* (0.39 g/L). PHA production was slightly higher in *Pco A1* than in *Pme 9.1* (respectively 0.92 g/L and 0.52 g/L) [37].

Table 1. Cell dry weight and raw PHA percentage obtained through bioconversion of different carbon sources by selected strains of *Pseudomonas mediterranea* and *P. corrugata*.

Carbon Source	Grade	% V:V	Time (h) [1]	*P. mediterranea* 9.1		*P. corrugata* A1		*P. corrugata* 388		References
				CDW (g/L)	Raw PHA (%)	CDW (g/L)	Raw PHA (%)	CDW (g/L)	Raw PHA (%)	
Glycerol	15%	1	72	3.4	50.2	4.7	50	4	28.5	[33]
	≥99%			3	25.3	3.5	29.4	4.2	18.7	
	15%	2	72	4.8	61.6	3.5	51.5	3.8	33.6	
	≥99%			3.2	26.1	3.4	30.2	3.6	15.7	
	15%	5	72	4.2	38	4.1	48.5	3.2	32.1	
	≥99%			3.3	21.5	4.1	22.1	2.8	14.3	
Glycerol	87.5%	2	48	3.1	16.5					[13]
	≥99%			3.3	18					
Glycerol	≥99%	2	66	2.9	17.9	3.1	29.4			[37]
Glycerol	≥99%	2	66	3.6 [2]	38.8 [2]					[38]
Glucose	≥99%	0.5	72					1.5	31.3	[34]
Oleic acid	≥99%							1.6	61.8	
Oleic acid	≥99%	2	72					3.1	24	[39]
Glucose	≥99%		48					1.3	2	

[1] time of cultivation; [2] this specific test was carried out with a modified strain of *P.mediterranea 9.1 VVC1GI.*

3. Conversion Process and Recovery

In order to establish standard and suitable protocols to scale up the production of PHAs, many strategies have been investigated using batch and fed-batch processes in flasks and low-medium volume fermenters (3–30 L). Fed-batch fermentation has always been shown to be more productive than the batch mode, as reported for *Cupriavidus* sp. [40].

A fed-batch cultivation has also been used in a process of glycerol conversion by growing *P. mediterranea 9.1* in a substrate with 2% glycerol. The cultivation was conducted in a 30 L bioreactor, 30 °C, pH 7.0, and dissolved oxygen maintained at 20% saturation, using E* medium (pH 7.0) containing 5.8 g/L K_2HPO_4, 3.7 g/L KH_2PO_4, 10 mL/L $MgSO_4$ 0.1 M, supplemented with 1 mL/L of a microelement solution, with the addition of 2% glycerol (1% at the start, and 1% after 24 h) [13,33].

The biomass obtained has been routinely harvested by centrifugation, washed with saline solution and lyophilized. The extraction of the PHA using acetone [41] in an automatic Soxhlet was found to be more effective than chloroform extraction and less impactful for the environment. Treatments with mild alkaline solution [42] or maceration, attempted considering the potential use of mcl-PHA in the biomedical field, have yielded a lower recovery of products.

The analysis of PHA composition was carried out by gas chromatography/mass spectrometry (GC/MS) of the 3-hydroxymethylesters, after the removal of all the residual free glycerol [13]. Overall, different approaches have been evaluated to reduce the very high production costs by increasing the yield or by recovering both the PHA and extracellular products simultaneously from the fermentation process. The addition of either meat or yeast extracts at 0.1% to crude glycerol or glucose eliminated the prolonged lag-phase (5–12 h) [43].

Rizzo et al. [44] showed that adding 5 mM glutamine as a co-feeder significantly increased the biomass and PHA production, inducing the early expression of *phaC1* and *phaC2* genes. This was due to the improvement in the specific growth rate and cell metabolic activity, and to the enhanced uptake of the unrelated (glycerol and glucose) and related (sodium octanoate) carbon sources.

An integrated process for the bioconversion of crude biodiesel glycerol to simultaneously produce biosurfactants and PHAs by *Pme 9.1*, has also been established by applying a mathematical mechanistic model to define nutritional requirements, as well as pH and temperature, which mutually influenced PHA and BSs production within a narrow range of variation [43]. Surface response methodology analysis showed that, after 72 h, up to 1.1 g/L of crude PHA and 0.72 g/L of biosurfactants were recovered. On the other hand, the respective best single yields were obtained after 48 h for PHA (60% of CDW) and 96 h for BSs (0.8 g/L) [43].

4. Composition and Technological Properties of mcl-PHA

GC/MS profiles of mcl-PHA were largely affected by the carbon source and bacteria species. PHAs obtained on waste food oils have been found to be very different from those obtained on glycerol, and different types of glycerol produced different mcl-PHAs [13,32].

GC/MS chromatograms of mcl-PHAs obtained by *Pco A1* and *Pme 9.1* on crude glycerol revealed similar profiles and technological properties, whereas substantial differences were observed with respect to those obtained on partially refined biodiesel glycerol and high-grade glycerol. They showed monomeric units of side chains from C12 to C19 in length on crude glycerol (15% glycerol), and from C5 to C16 on refined glycerol (≥99%) (Table 2) [33,45]. Interestingly, mcl-PHA produced by *Pme 9.1* on high-grade glycerol was less sticky and produced a thin film (Figure 1A) [33]. These properties have been shown to be associated with differences at transcriptomic level [37], and in the genetic organization of *pha* gene locus which affects *pha* polymerase gene expression, PHA composition, and granule morphology [39].

Other experiments on *Pme 9.1* have been conducted in Erlenmeyer flasks containing 500 mL volumes of E* medium (pH 7.0) with 2% high grade glycerol or a partially refined glycerol (87.5%) obtained from a biodiesel process of *Brassica napus* [13]. The polyesters obtained on high grade commercial glycerol highlighted a structure composed of six monomers, indicative of elastic and flexibility properties: 3-hydroxyhexanoate (C6), 3-hydroxyoctanoate (C8), 3-hydroxydeca-noate (C10), 3-hydroxydodecanoate (C12), cis 3-hydroxydodec-5-enoate ($C12{:}1\Delta^5$), and cis 3-hydroxydodec-6-enoate ($C12{:}1\Delta^6$). The molecular weight (Mw) was 55,480 Da and polydispersity index (PDI = Mw/Mn) was 1.34 (Table 2). On the other hand, PHA obtained from glycerol 87.5% had a small variation in monomeric composition, a Mw of 63,200 Da, and a PDI of 1.38. Tsuge et al. [46] also observed that a higher glycerol concentration induced a considerable reduction in the molecular mass of PHA, caused by a termination of the PhaC polymerization activity. The NMR spectra and MALDI-TOF data were almost identical regardless of the glycerol grade, but different in intensity. The degradation temperature started at 230 °C, higher than the melting temperatures, with a volatilization rate of about −40%/min.

Table 2. Molecular weight and monomer composition of PHAs obtained in different bioconversion processes of different carbon sources by *Pseudomonas corrugata* and *P. mediterranea*.

Strain	Carbon Source	Grade	% V:V	Time (h) [1]	Mw (kDa)	PDI	Molar Composition (mol %)								Reference
							C6	C8	C10	C12:0	C12:1	C12:0	C14	C14:1	
Pme 9.1	Waste fried oil						2	34	44	14			5		Pappalardo et al., unpublished
	Glycerol	80%	2				1	7	71	8	13		1		
	Glycerol	40%	2				1	15	43	11	7		24		
	Glycerol	≥99%	2	48	55.5	1.34	4.2	17.0	60.8	1.1	11.2	5.7	-	-	[13]
		87.5%			63.2	1.38	0.1	9.3	66.6	1.5	14.8	7.7	-	-	
	Glycerol	≥99%	2	66			4	17	60	7	12		0.4		[37]
Pme 9.1 VVC1GI	Glycerol	≥99%						0.9	13.5	57.5	12.8	11.8		3.7	[38]
Pco A1	Glucose		0.5	72	125.8	2.4	2	14	52	11	17		0.4	3.6	[47]
	Oleic acid				159.0	1.5	10	48	28	8				6	
	Na octanoate				183.2	2.1	11	82	7						
	Glycerol	≥99%	2	66			2	12	53	14	17		5		[37]
Pco 388	Oleic acid		0.5	72	735	4.1		47	24.5					16.5	[8]
	Glucose		0.5	72	nd		2	19	56	11			2	9	[34]
	Oleic acid				nd		5	37	33	12			2	12	
	Na octanoate		0.5	168	114	1.8	7	82	11						[47]
	Oleic acid		2	-			5	54	20	5			15		[39]
	Glucose			-			2	28	35	9	14		9		

[1] time of cultivation.

Drop casting a toluene solution of polymers in Petri dishes resulted in quite different films, depending on the carbon source used to produce the PHA (Figure 1B). The PHA obtained on high-grade glycerol produced an optically transparent film with a UV–vis absorption spectrum that was above the 800–350 nm range, comparable to the polyester film used for laser printer transparency.

The mechanical proprieties (tensile strength, Young's modulus and elongation at break of both PHAs) were not substantially affected by the different purities of the glycerol grade (87.5% and ≥99%). All these characteristics make the PHA obtained from *P. mediterranea 9.1* on glycerol quite different to most mcl-PHAs produced from bacteria of the same phylogenetic group.

5. Evaluation of Mixed Blends and Coatings

The distinctive characteristics of a mcl-PHA obtained from *Pme 9.1* grown on glycerol led to the investigation of the processability of blends with polylactide acid (PLA) to improve the mechanical and gas/vapors barrier properties of PLA [48]. Rheological tests indicated a significant increase in the elongation at break, while the elastic modulus was significantly lower only at higher contents of PHA. This suggests that the PHA macromolecules exert both a plasticization and lubricant action, which enable the PLA macromolecules subjected to solid deformation to slide more efficiently [48].

Preliminary investigation of blends of polyhydroxybutyrate (PHB) and a glycerol mcl-PHA obtained from *B. napus* oil showed an increase in crystallization temperature and a small increase in elongation at break, but at low concentrations of PHA (5%) the blend revealed some spaces between the two polymers.

Blends of mcl-PHA obtained from *Pco A1* on exhausted edible oils with Mater-Bi ZI01U/C polymers have poorly improved the processability of blends prepared by compression [22]. Soil mulching tests of paper sheets coated with blends based on PHA suggested some positive effects of coating. However, the expensive costs, as well the difficulty to obtain standardized exhausted edible oils as a carbon source, have discouraged further research [49,50].

6. *P. corrugata* and *P. mediterranea* PHA Locus

Genomic studies which investigated potential correlations between the phenotype and genotype of *Pco 388*, *Pco A1* and *Pme 9.1* have shown that, similarly to other *Pseudomonas*, the three strains have a class II PHA genetic system consisting of two synthase genes (*phaC1*, *phaC2*), separated by a gene coding for the depolymerization of PHA (*phaZ*) [8,9,23]. This genetic system allows *Pseudomonas* strains to utilize medium-chain-length (mcl) monomers (C6–C14), whereas class I, III and IV systems polymerize short-chain-length (scl) monomers (C3–C5) [6].

Sequence analysis of the *pha* locus revealed that the strains *Pco A1* (AY910767), *Pmed 9.1* (AY910768) and *Pco 388* (EF067339) share a high homology at nucleotide (93–95%) and amino acid levels (96–98%) [39, 51,52]. An additional 121 bp in the *phaC1–phaZ* intergenic region containing a predicted strong hairpin structure were present in both strains *388* and *A1* of *P. corrugata*, but not in *P. mediterranea* [39]. According to the authors, in *Pco A1* and *388* strains this additional sequence likely acts as a rho-independent terminator for the transcriptional terminator of *phaC1*, which would appear to be responsible for the slight variation in the PHA composition and granule organization [39].

Subsequently, genome analysis of strains *Pco A1* (ATKI01000000) and *Pme 9.1* (AUPB01000000) enabled the six genes of the entire *pha* locus to be studied (*phaC1*, *phaZ*, *phaC2*, *phaD*, *PhaF*, *PhaI*) [53,54]. Genome mining also identified gene coding for enzymes involved in β-oxidation (*fad*), fatty acid *de novo* synthesis (*fab*), and mcl-PHA precursor availability (*phaG* and *phaJ*) [53,54]. Pfam search domain and Blastp analysis on the *glp* operon, responsible for glycerol catabolism, revealed that both *P. corrugata* and *P. mediterranea* lack the *glpF* gene, coding for the glycerol uptake facilitator protein [53]. This condition had already been verified in all the *Pco* and *Pme* strains sequenced, which explains the prolonged lag growth phase observed during *P. mediterranea 9.1* growth with glycerol as the sole carbon source [25,55].

7. Transcriptional Regulation during PHA Production

To improve the knowledge about PHA biosynthesis genes and their regulations, helpful to increase mcl-PHAs production and also to obtain new, tailor-made polymers [6], regulatory mechanisms during

PHA accumulation have been investigated in *Pco A1*, *Pco 388* and *Pme 9.1* through different gene expression studies and full transcriptome analysis.

7.1. Expression of phaC1 and phaC2 under Different Carbon Sources

Preliminary studies on the transcriptional levels of *phaC1* and *phaC2* genes in *Pco 388* and *Pco A1* showed an up-regulation of *phaC1* in cultures with oleic acid as the sole carbon source (Table 3) [47]. On the other hand, both *phaC1* and *phaC2* were induced in cultures with glucose or sodium octanoate [47]. The significant correlation between PHA production and *phaC1/phaC2* expression suggested at least two distinct networks for the regulation of the two PHA polymerases genes, and that a putative promoter(s) is likely present upstream of *phaC2*. In addition, the lack of polycistronic transcripts under any culture conditions indicated that *phaC1* and *phaC2* were not co-transcribed (Table 3) [47].

Table 3. Gene expression detected in *P. corrugata* and *P. mediterranea* strains during mcl-PHA biosynthesis on different carbon sources.

Bacterial Strain	Carbon Source	% V/V	Time (h)	Detection Method	*PhaC1*	*PhaC2*	*PhaI*	*Alg* Genes	Operon	Reference
Pco 388	Oleic acid	2	48	Real-time PCR [1]	1.2	1.4	nt	nt	nt	[39]
Pco 388 clone XI 32-1					6.6	4.7	nt	nt	nt	
Pco 388 clone XI 32-4					7.0	5.4	nt	nt	nt	
Pco 388	Glucose	2			5.6	No change	nt	nt	nt	
Pco 388 clone XI 32-1					6.3	No change	nt	nt	nt	
Pco 388 clone XI 32-4					8.2	No change	nt	nt	nt	
Pco A1	Oleic acid	0.5	48 72	Real-time PCR [1]	6.8 10.5	No change	nt	nt	NO	[47]
Pco 388			48 72		2.7 2	No change	nt	nt	NO	
Pco A1	Glucose	2	72		6.2	3.5	nt	nt	NO	
Pco 388			72		3.8	3	nt	nt	NO	
Pme 9.1	Glycerol	2	24 48	β-gal [2]	420 U 300 U	340 U 400 U	2200 U 7000 U	nt	*PhaC1ZC2D* *PhaIF*	[56]
Pme 9.1 VVD (*phaD*-)			24 48	β-gal	140 U 300 U	350 U 400 U	45 U 45 U	nt	*PhaC1ZC2D* *PhaIF*	
Pme 9.1	Glycerol	2	48	RNA-Seq [3]	No change	No change	No change	5.53–2.32	nt	[37]
Pco A1			48	RNA-Seq	No change	No change	No change		nt	

[1] The relative quantification was performed by comparing ΔCt (i.e., Ct of the 16S rRNA housekeeping gene subtracted to the Ct of the target gene). The ΔCt value of the control sample (time 0) was used as the calibrator and fold-activation was calculated by the expression: $2^{-\Delta\Delta Ct}$. [2] β–galactosidase activities detected by transcriptional fusion plasmids for *phaC1*, *phaC2*, and *phaI* promoter regions based on the pMP220 promoter probe vector and expressed as Miller units. [3] Pairwise comparison of mRNA levels analysis, using the *Pme 9.1* sample as a reference (log2 fold change ≥ 2 and *p*-value ≤ 0.05).

Parallel studies showed that, in *Pco 388*, the *phaC1–phaZ* intergenic region plays an important but unclear role in the regulation of the carbon source-dependent expression of *phaC1* and *phaC2* genes [39]. Derivative mutants XI 32-1 and XI 32-4 of this strain (obtained by replacing the *phaC1–phaZ* intergenic region with a kanamycin resistance gene), showed a significant increase in *phaC1* and *phaC2* expression when grown for 48 h with oleic acid, but not with glucose. In addition, the wild type strain produced only a few large PHA inclusion bodies when grown with oleic acid, whereas the mutants showed numerous smaller PHA granules that line the periphery of the cells, as result of phasin activities [3,39]. A high content of the monounsaturated 3-hydroxydodecanoate as a repeat unit monomer was observed in the PHA of the mutant strains [39].

Diversely, the study of the promoter activity of *pha* genes in *Pme 9.1* grown on high-grade glycerol, revealed that the upstream regions of *phaC1* (PC1) and *phaI* genes (PI) are the most active [56]. PC1 is responsible for the *phaC1ZC2D* polycistronic unit transcription (Table 3). On the other hand, PI regulates the *phaIF* operons, as confirmed by the presence of three and two putative rho independent terminators, respectively located downstream of *phaD* and *phaF* [56]. In turn, PI and PC1 are controlled by PhaD, which acts as a transcriptional activator, as shown by the reduced promoter activities in the *phaD*- mutant [55]. Similar results were observed in *P. putida* KT2442 [23,57].

7.2. Transcriptome Analysis on Glycerol-Grown Strains

The transcriptional profiles of *Pco A1* and *Pme 9.1* growing on a substrate with 2% of high-grade glycerol under inorganic nutrient-limited conditions were investigated at the early stationary phase of the bioconversion into mcl-PHAs [37]. RNA-seq analysis revealed that in *P. mediterranea*, 175 genes were significantly upregulated and 217 downregulated, compared to *P. corrugata*. The genes responsible for stress response, central and peripheral metabolic routes and transcription factors involved in mcl-PHA biosynthesis, made up 39% of the genes differently transcribed by the two bacteria. Nonetheless, among the genes directly involved in PHA biosynthesis, slight differences were observed only in *phaZ* depolymerase and *phaG* transacylase genes (Table 3). Weak differences occurred in the expression levels of genes that are crucial for glycerol catabolism and pyruvate metabolism, transcriptionally downregulated, and fatty acid *de novo* biosynthesis pathways.

Interestingly, a significantly increased expression of 21 genes involved in alginate exopolysaccharide production was observed in *Pme 9.1* compared to *Pco A1* (Table 3), related to a 17-fold higher production of EPS (6.9 g/L compared to 0.39 g/L). A simultaneous production of PHA and alginate has been reported in some *P. mendocina* strains [58,59]. The increased EPS production, associated with the different transcriptome profiling between the two bacteria, suggests competition for the acetyl-CoA precursor amongst PHA and alginate metabolic pathways. Further studies on *P. corrugata A1* showed that regulation of alginate production is controlled by quorum sensing and the RfiA regulator [60].

8. Genetically Modified Bacteria to Improve the Production of mcl-PHAs

Two different approaches have been made to evaluate the feasibility of improving the efficiency of the conversion by using genetically modified-*Escherichia coli* and *P. mediterranea*.

Cloning of *pha* synthases genes of *Pco A.1* and *Pme 9.1* in *Escherichia coli,* a well-known organism in the research on PHA biosynthesis, yielded 2%–4% of PHA/CDW on sodium decanoate [61].

In a second approach, additional copies of *phaC1*, *phaG* and *phaI* genes, cloned in two plasmids under the control of strong promoters, were transferred into *Pme 9.1* [38]. When grown on high-grade glycerol, the modified *Pme VVC1GI* showed a higher cell fluorescence than WT, due to the presence of larger granules (Figure 2). It also showed a 40% increase in the cellular accumulation of crude PHA and a better PHA/CDW ratio (1.4 g/L, 38.8%), whereas the WT strain produced 1 g/L of PHA (23%) (Table 1). Biomass yield was 3.6 g/L in *VVC1GI* and 4.2 g/L in WT [38]. GC/MS analysis showed that the PHA structure was composed of six monomers, like PHA produced by the WT, although there were some differences in the ratio between C12 and C12:1 (12.8:11.8 in strain *VVC1GI* and 6:12 in Wt) (Table 2).

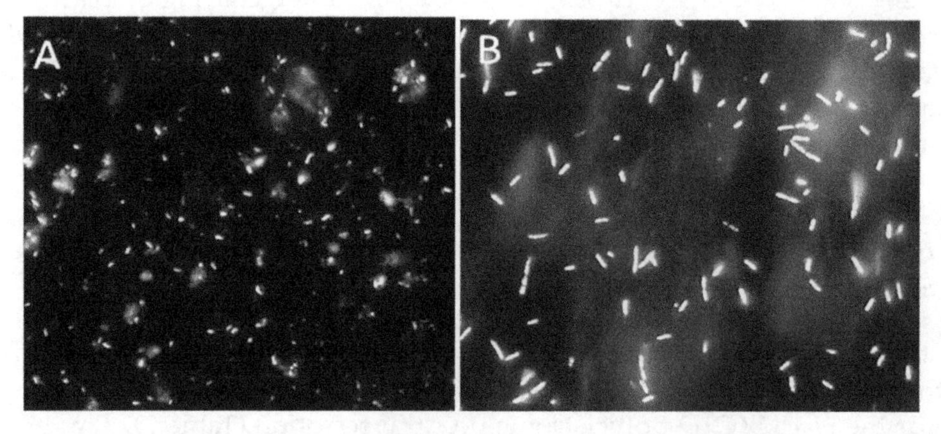

Figure 2. Fluorescent granules of PHA after Nile red-staining of *Pseudomonas mediterranea 9.1* wild-type strain (**A**) and *VVC1GI* recombinant strain (**B**) grown on high-grade glycerol (≥99%) as carbon source and limited nitrogen condition after 66 h of incubation.

9. Conclusions

Despite the fact that the PHAs market is still very small, the worldwide focus on the development of bio-polymers highlights that one day polyhydroxyalkanoates will replace some petroleum-based plastics. *Pseudomonas* species are a potential cell factory for their production [3]. Some strains of *P. corrugata* and its related *P. mediterranea* are able to convert different carbon sources and provide interesting mcl-PHA and co-products. They are naturally present in the soil and produce valuable extracellular co-products [37,43].

A manageable mcl-PHA film, unlike other mcl-PHAs reported to date, can be obtained by *P. mediterranea 9.1*. Although the yields are currently not particularly viable, their distinctive characteristics suggest a potential application as a softener in (bio) polymeric blends, for food packaging or medical devices. The unsaturated double bonds in the side chains could be used to enhance its properties and/or to help extend its applications to other biomaterials for food packaging or biomedicine [13].

P. mediterranea 9.1 also produces high-quality extracellular products (above all alginate) on a proper medium, which is very promising for high-level applications and which may orient further investigation towards an efficient co-production of cellular mcl-PHA and extracellular biosurfactants, EPS and other bioactive molecules [43]. These results and those available for other systems highlight the potential of such integrated microbial conversion processes [62]. Further top strategies are required to find solutions for the industrial production of such compounds and new ones.

Pioneering work on other PHA producers *Pseudomonas*, such as on expanding the number of inexpensive carbon sources [63], increasing the productivity [64,65] or making the PHA deposition extracellular [66], highlight the potential for successful future investments in this sector. In the meantime, given that robust strains are needed to reduce the high production costs, using genetic engineering and metabolic studies on these two bacteria should focus on developing over-producer strains of mcl-PHA, as well as the co-production of other valuable products, such as EPS and biosurfactants.

Author Contributions: The authors have several years of research experience on *Pseudomonas corrugata* and *P. mediterranea* and their bioconversion of waste oils and biodiesel co-products stream, mostly based on the long term projects "Biopolimeri" and "Polybioplast" (acknowledged below). All authors have contributed the writing and editing of the manuscript and approved the manuscript.

Acknowledgments: The results presented were partly generated by the projects MIUR PON 2000–2006 N.12842 "Biopolimeri" and MIUR PON 2007-2013N. 01 01377 "Polybioplast", both co-funded by EU.

References

1. Wu, Q.; Sun, S.Q.; Yu, P.H.F.; Chen, A.X.Z.; Chen, G.Q. Environmental dependence of microbial synthesis of polyhydroxyalkanoates. *Acta Polym. Sin.* **2000**, *6*, 751–756.
2. Kadouri, D.; Jurkevitch, E.; Okon, Y.; Castro-Sowinski, S. Ecological and agricultural significance of bacterial polyhydroxyalkanoates. *Crit. Rev. Microbiol.* **2005**, *31*, 55–56. [CrossRef]
3. Mozejko-Ciesielska, J.; Szacherska, K.; Marciniak, P. *Pseudomonas* species as producers of eco-friendly polyhydroxyalkanoates. *J. Polym. Environ.* **2019**, *27*, 1151–1166. [CrossRef]
4. Chen, G.Q. A microbial polyhydroxyalkanoates (PHA) based bio- and materials industry. *Chem. Soc. Rev.* **2009**, *38*, 2434–2446. [CrossRef]
5. Zinn, M.; Witholt, B.; Egli, T. Occurrence, synthesis and medical application of bacterial polyhydroxyalkanoate. *Adv. Drug Del. Rev.* **2001**, *53*, 5–21. [CrossRef]
6. Możejko-Ciesielska, J.; Kiewisz, R. Bacterial polyhydroxyalkanoates: Still fabulous? *Microbiol. Res.* **2016**, *192*, 271–282. [CrossRef] [PubMed]
7. Kessler, B.; Palleroni, N. Taxonomic implications of synthesis of poly-beta-hydroxybutyrate and other poly-beta-hydroxyalkanoates by aerobic pseudomonads. *Int. J. Syst. Evol. Microbiol.* **2000**, *50*, 711–713. [CrossRef]

8. Solaiman, D.K.Y.; Ashby, R.D.; Foglia, T.A. Rapid and specific identification of medium-chain-length polyhydroxyalkanoate synthase gene by polymerase chain reaction. *Appl. Microbiol. Biotechnol.* **2000**, *53*, 690–694. [CrossRef]
9. Solaiman, D.K.Y.; Catara, V.; Greco, S. Poly(hydroxyalkanoate) synthase genotype and PHA production of *Pseudomonas corrugata* and *P. mediterranea*. *J. Ind. Microbiol. Biotechnol.* **2005**, *32*, 75–82. [CrossRef]
10. Solaiman, D.K.Y.; Ashby, R.D.; Foglia, T.A.; Marmer, W.N. Conversion of agricultural feedstockand coproducts into poly(hydroxyalkanoates). *Appl. Microbiol. Biotechnol.* **2006**, *71*, 783–789. [CrossRef]
11. Ashby, R.D.; Solaiman, D.K.Y.; Foglia, T.A. Bacterial poly(hydroxyalcanoate) polymer production from the biodiesel co-product stream. *J. Polym. Environ.* **2004**, *12*, 105–112. [CrossRef]
12. Ashby, R.D.; Solaiman, D.K.Y.; Foglia, T.A. Synthesis of short-/medium-chain-length poly(hydroxyalkanoate) blends by mixed culture fermentation of glycerol. *Biomacromolecules* **2005**, *6*, 2106–2112. [CrossRef] [PubMed]
13. Pappalardo, F.; Fragalà, M.; Mineo, P.G.; Damigella, A.; Catara, A.F.; Palmeri, R.; Rescifina, A. Production of filmable medium-chain-length polyhydroxyalkanoates produced from glycerol by *P. mediterranea*. *Int. J. Biol. Macromol.* **2014**, *65*, 89–96. [CrossRef] [PubMed]
14. Ashby, R.D.; Foglia, T.A. Poly(hydroxyalkanoates) biosynthesis from triglyceride substrates. *Appl. Microbiol. Biotechnol.* **1998**, *49*, 431–437. [CrossRef]
15. Da Silva, G.P.; Mack, M.; Contiero, J. Glycerol: A promising and abundant carbon source for industrial microbiology. *Biotechnol. Adv.* **2009**, *27*, 30–39. [CrossRef]
16. Koller, M.; Marsalek, L. Principles of Glycerol-Based Polyhydroxyalkanoate Production. *Appl. Food Biotechnol.* **2015**, *2*, 3–10.
17. De Smet, M.J.; Eggink, G.; Witholt, B.; Kingma, J.; Wynberg, H. Characterization of intracellular inclusions formed by *Pseudomonas oleovorans* during growth on octane. *J. Bacteriol.* **1983**, *154*, 870–878.
18. Panith, N.; Assavanig, A.; Lertsiri, S.; Bergkvist, M.; Surarit, R.; Niamsiri, N. Development of tunable biodegradable polyhydroxyalkanoates microspheres for controlled delivery of tetracycline for treating periodontal disease. *J. Appl. Polym. Sci.* **2016**, *133*, 44128–44140. [CrossRef]
19. Zhang, J.; Shishatskaya, E.I.; Volova, T.G.; da Silva, L.F.; Chen, G.Q. Polyhydroxyalkanoates (PHA) for therapeutic applications. *Mater. Sci. Eng. C* **2018**, *86*, 144–150. [CrossRef]
20. Basnett, P.; Ching, K.Y.; Stolz, M.; Knowles, J.C.; Boccaccini, A.R.; Smith, C.; Locke, I.C.; Keshavarz, T.; Roy, I. Novel poly(3-hydroxyoctanoate)/poly(3-hydroxybutyrate) blends for medical applications. *React. Funct. Polym.* **2013**, *73*, 1340–1348. [CrossRef]
21. Takagi, Y.; Yasuda, R.; Yamaoka, M.; Yamane, T. Morphologies and mechanical properties of polylactide blends with medium chain length poly(3-hydroxyalkanoate) and chemically modified poly(3-hydroxyalkanoate). *J. Appl. Polym. Sci.* **2004**, *93*, 2363–2369. [CrossRef]
22. Scaffaro, R.; Dintcheva, N.T.; Marino, R.; La Mantia, F.P. Processing and properties of biopolymer/polyhydroxyalkanoates blends. *J. Polym. Environ.* **2012**, *20*, 267–272. [CrossRef]
23. Prieto, A.; Escapa, I.F.; Martínez, V.; Dinjaski, N.; Herencias, C.; de la Peña, F.; Tarazona, N.; Revelles, O. A holistic view of polyhydroxyalkanoate metabolism in *Pseudomonas putida*. *Environ. Microbiol.* **2016**, *18*, 341–357. [CrossRef] [PubMed]
24. Catara, V. *Pseudomonas corrugata*: Plant pathogen and/or biological resource? *Mol. Plant. Pathol.* **2007**, *8*, 233–244. [CrossRef]
25. Trantas, E.A.; Licciardello, G.; Almeida, N.F.; Witek, K.; Strano, C.P.; Duxbury, Z.; Ververidis, F.; Goumas, D.E.; Jones, J.D.G.; Guttman, D.S.; et al. Comparative genomic analysis of multiple strains of two unusual plant pathogens: *Pseudomonas corrugata* and *Pseudomonas mediterranea*. *Front. Microbiol.* **2015**, *6*, 811. [CrossRef] [PubMed]
26. Corsaro, N.M.; Piaz, F.D.; Lanzetta, R.; Naldi, T.; Parrilli, M. Structure of Lipid A from *Pseudomonas corrugata* by electrospray ionisation quadrupole time-of-flight tandem mass spectrometry. *Rapid Commun. Mass Spectrom.* **2004**, *18*, 853–858. [CrossRef]
27. Fett, W.F.; Cescutti, P.; Wijey, C. Exopolysaccharides of the plant pathogens *Pseudomonas corrugata* and *Ps. flavescens* and the saprophyte *Ps. chlororaphis*. *J. Appl. Bacteriol.* **1996**, *81*, 181–187. [CrossRef]
28. Scaloni, A.; Dalla Serra, M.; Amodeo, P.; Mannina, L.; Vitale, R.M.; Segre, A.L.; Cruciani, O.; Lodovichetti, F.; Greco, M.L.; Fiore, A.; et al. Structure, conformation and biological activity of a novel lipodepsipeptide from *Pseudomonas corrugata*: Cormycin A. *Biochem. J.* **2004**, *384*, 25–36. [CrossRef]

29. Emanuele, M.C.; Scaloni, A.; Lavermicocca, P.; Jacobellis, N.S.; Camoni, L.; Di Giorgio, D.; Pucci, P.; Paci, M.; Segre, A.; Ballio, A. Corpeptins, new bioactive lipodepsipeptides from cultures of *Pseudomonas corrugata*. *FEBS Lett.* **1998**, *433*, 317–320. [CrossRef]
30. Strano, C.P.; Bella, P.; Licciardello, G.; Fiore, A.; Lo Piero, A.R.; Fogliano, V.; Venturi, V.; Catara, V. *Pseudomonas corrugata crpCDE* is part of the cyclic lipopeptide corpeptin biosynthetic gene cluster and is involved in bacterial virulence in tomato and in hypersensitive response in *Nicotiana benthamiana*. *Mol. Plant Pathol.* **2014**, *9*. [CrossRef]
31. Risse, D.; Beiderbeck, H.; Taraz, K.; Budzikiewicz, H.; Gustine, D. Bacterial constituents part LXXVII. Corrugatin, a lipopeptide siderophore from *Pseudomonas corrugata*. *Z. Naturforsch. C* **1998**, *53*, 295–304.
32. Alicata, R.; Ballistreri, A.; Catara, V.; Conte, E.; Di Silvestro, S.; Ferreri, A.; Greco, S.; Guglielmino, S.; Impallomeni, G.; La Porta, S.; et al. Used cooking oils as renewable source for polyhydroxyalkanoates production. In Proceedings of the 3th European Symposium on Biopolymers CIB-CSIC, Madrid, Spain, 24–25 November 2005; p. 35.
33. Palmeri, R.; Pappalardo, F.; Fragalà, M.; Tomasello, M.; Damigella, A.; Catara, A.F. Polyhydroxyalkanoates (PHAs) production through conversion of glycerol by selected strains of *Pseudomonas mediterranea* and *Pseudomonas corrugata*. *Chem. Eng. Trans.* **2012**, *27*, 121–126.
34. Solaiman, D.K.Y.; Ashby, R.D.; Foglia, T.A. Physiological characterization and genetic engineering of *Pseudomonas corrugata* for medium-chain-length polyhydroxyalkanoates synthesis from triacylglycerols. *Curr. Microbiol.* **2002**, *44*, 189–195. [CrossRef]
35. Bella, P.; Catara, A.; Catara, V.; Conte, E.; Di Silvestro, S.; Ferreri, A.; Greco, S.; Guglielmino, S.; Immirzi, B.; Licciardello, G.; et al. Fermentation Process for the Production of Polyhydroxyalkanoates and the Digestion of Cooked Oils by Strains of *Pseudomonas* Producing Lipase. Italian Industrial Patent No. RM2005A000190, 20 April 2005.
36. Samadi, N.; Abadian, N.; Akhavan, A.; Fazeli, M.R.; Tahzibi, A.; Jamalifar, H. Biosurfactant production by the strain isolated from contaminated soil. *J. Biol. Sci.* **2007**, *7*, 1266–1269.
37. Licciardello, G.; Ferraro, R.; Russo, M.; Strozzi, F.; Catara, A.F.; Bella, P.; Catara, V. Transcriptome analysis of *Pseudomonas mediterranea* and *P. corrugata* plant pathogens during accumulation of medium-chain-length PHAs by glycerol bioconversion. *New Biotechnol.* **2017**, *37*, 39–47. [CrossRef] [PubMed]
38. Licciardello, G.; Russo, M.; Pappalardo, F.; Fragalà, M.; Catara, A. Genetically Modified Bacteria Producing Mcl-PHAs. Italian Industrial Patent No.15MG37I, 12 May 2015.
39. Solaiman, D.K.Y.; Ashby, R.D.; Licciardello, G.; Catara, V. Genetic organization of pha gene locus affects phaC expression, poly(hydroxyalkanoate) composition and granule morphology in *Pseudomonas corrugata*. *J. Ind. Microb. Biotech.* **2008**, *35*, 111–120. [CrossRef] [PubMed]
40. Shantini, K.; Yahya, A.R.; Amirul, A.A. Influence of feeding and controlled dissolved oxygen level on the production of poly(3-hydroxybutyrate-co-3-hydroxyvalerate) copolymer by *Cupriavidus* sp. USMAA2-4 and its characterization. *Appl. Biochem. Biotechnol.* **2015**, *176*, 1315–1334. [CrossRef]
41. Jiang, X.; Ramsay, J.A.; Ramsay, B.A. Acetone extraction of mcl-PHA from *Pseudomonas putida* KT2440. *J. Microb. Methods* **2006**, *67*, 212–219. [CrossRef]
42. Koller, M.; Niebelschütz, H.; Braunegg, G. Strategies for recovery and purification of poly[(R)-3-hydroxyalkanoates] (PHA) biopolyesters from surrounding biomass. *Eng. Life Sci.* **2013**, *13*, 549–562. [CrossRef]
43. Nicolò, M.S.; Franco, D.; Camarda, V.; Gullace, R.; Rizzo, M.G.; Fragalà, M.; Licciardello, G.; Catara, A.F.; Guglielmino, S.P.P. Integrated microbial process for bioconversion of crude glycerol from biodiesel into biosurfactants and PHAs. *Chem. Eng. Trans.* **2014**, *38*, 187–192.
44. Rizzo, M.G.; Chines, V.; Franco, D.; Nicolò, M.S.; Guglielmino, S.P.P. The role of glutamine in *Pseudomonas mediterranea* in biotechnological processes. *New Biotechnol.* **2017**, *37*, 144–151. [CrossRef] [PubMed]
45. Fragalà, M.; Palmeri, R.; Ferro, G.; Damigella, A.; Pappalardo, F.; Catara, A.F. Production of mcl-PHAs by *Pseudomonas mediterranea* conversion of biodiesel-glycerol. In Proceedings of the European Symposium on Biopolymers, Lisbon, Portugal, 7–9 October 2013.
46. Tsuge, T. Fundamental factors determining the molecular weight of polyhydroxyalkanoate during biosynthesis. *Polym. J.* **2016**, *48*, 1051–1057. [CrossRef]

47. Conte, E.; Catara, V.; Greco, S.; Russo, M.; Alicata, R.; Strano, L.; Lombardo, A.; Di Silvestro, S.; Catara, A. Regulation of polyhydroxyalcanoate synthases (*phaC1* and *phaC2*) gene expression in *Pseudomonas corrugata*. *Appl. Microbiol. Biotechnol.* **2005**, *72*, 1054–1062. [CrossRef] [PubMed]
48. Botta, L.; Mistretta, M.C.; Palermo, S.; Fragalà, M.; Pappalardo, F. Characterization and processability of blends of polylactide acid with a new biodegradable medium-chain-length polyhydroxyalkanoate. *J. Polym. Environ.* **2015**, *23*, 478–486. [CrossRef]
49. Cascone, G.; D'Emilio, A.; Buccellato, E.; Mazzarella, R. New biodegradable materials for greenhouse soil mulching. *Acta Hortic.* **2008**, *801*, 283–290. [CrossRef]
50. Salemi, F.; Lamagna, G.; Coco, V.; Barone, L.G. Preparation and characterization of biodegradable paper coated with blends based on PHA. *Acta Hortic.* **2008**, *801*, 203–210. [CrossRef]
51. Catara, V.; Bella, P.; Greco, S.; Licciardello, G.; Pitman, A.; Arnold, D.L. Cloning and sequencing of *Pseudomonas corrugata* polyhydroxyalkanoates biosynthesis genes. In Proceedings of the National Biotechnology Congress (CNB7), Catania, Italy, 8–10 September 2004; p. 199.
52. Bella, P.; Licciardello, G.; Lombardo, A.; Pitman, A.; Arnold, D.L.; Solaiman, D.K.Y.; Catara, V. Cloning and sequencing of *Pseudomonas mediterranea* PHA locus. In Proceedings of the 4th European Symposium on Biopolymers, Kuşadası, Turkey, 2–4 October 2007; p. 131.
53. Licciardello, G.; Bella, P.; Devescovi, G.; Strano, C.P.; Catara, V. Draft genome sequence of *Pseudomonas mediterranea* strain CFBP 5447T, a producer of filmable medium-chain-length polyhydroxyalkanoates. *Genome Announc.* **2014**, *2*, e01260-14. [CrossRef]
54. Licciardello, G.; Jackson, R.W.; Bella, P.; Strano, C.P.; Catara, A.F.; Arnold, D.L.; Venturi, V.; Silby, M.W.; Catara, V. Draft genome sequence of *Pseudomonas corrugata*, a phytopathogenic bacterium with potential industrial applications. *J. Biotechnol.* **2014**, *17*, 65–66. [CrossRef]
55. Zachow, C.; Müller, H.; Laireiter, C.M.; Tilcher, R.; Berg, G. Complete genome sequence of *Pseudomonas corrugata* strain RM1-1-4, a stress protecting agent from the rhizosphere of an oilseed rape bait plant. *Stand. Genomic Sci.* **2017**, *12*, 66. [CrossRef]
56. Licciardello, G.; Devescovi, G.; Bella, P.; De Gregorio, C.; Catara, A.F.; Gugliemino, S.P.P.; Venturi, V.; Catara, V. Transcriptional analysis of pha genes in *Pseudomonas mediterranea* CFBP 5447 grown on glycerol. *Chem. Eng. Trans.* **2014**, *38*, 289–294.
57. De Eugenio, L.I.; Escapa, I.F.; Morales, V.; Dinjaski, N.; Galán, B.; García, J.L.; Prieto, M.A. The turnover of medium-chain-length polyhydroxyalkanoates in *Pseudomonas putida* KT2442 and the fundamental role of PhaZ depolymerase for the metabolic balance. *Environ. Microbiol.* **2010**, *12*, 207–221. [CrossRef] [PubMed]
58. Guo, W.; Song, C.; Kong, M.; Geng, W.; Wang, Y.; Wang, S. Simultaneous production and characterization of medium-chain-length polyhydroxyalkanoates and alginate oligosaccharides by *Pseudomonas mendocina* NK-01. *Appl. Microbiol. Biotechnol.* **2011**, *92*, 791–801. [CrossRef] [PubMed]
59. Chanasit, W.; Hodgson, B.; Sudesh, K.; Umsakul, K. Efficient production of polyhydroxyalkanoates (PHAs) from *Pseudomonas mendocina* PSU using a biodiesel liquid waste (BLW) as the sole carbon source. *Biosci. Biotechnol. Biochem.* **2016**, *80*, 1440–1450. [CrossRef] [PubMed]
60. Licciardello, G.; Caruso, A.; Bella, P.; Gheleri, R.; Strano, C.P.; Anzalone, A.; Trantas, E.A.; Sarris, P.F.; Almeida, N.F.; Catara, V. The LuxR regulators PcoR and RfiA co-regulate antimicrobial peptide and alginate production in Pseudomonas corrugata. *Front. Microbiol.* **2018**, *9*, 521. [CrossRef]
61. Lombardo, A.; Bella, P.; Licciardello, G.; Palmeri, R.; Catara, V.; Catara, A. Poly(hydroxyalkanoate) synthase genes in *Pseudomonads* strains, isolation and heterologous expression. *J. Biotechnol.* **2010**, *150*, 420–421. [CrossRef]
62. Hori, K.; Marsudi, S.; Unno, H. Simultaneous production of polyhydroxyalkanoates and rhamnolipids by *Pseudomonas aeruginosa*. *Biotechnol. Bioeng.* **2002**, *78*, 699–707. [CrossRef]
63. Wang, Q.; Tappei, R.C.; Zhu, C.; Nomura, T.C. Development of a new strategy for production of medium-chain-length polyhydroxyalkanoates by recombinant *Escherichia coli* via inexpensive non fatty acid feedstocks. *Appl. Environ. Microbiol.* **2012**, *78*, 519–527. [CrossRef]
64. Lopez-Cortes, A.; Lanz-Landazuri, A.; Garcia-Maldonado, J.Q. Screening and isolation of PHB-producing bacteria in a polluted marine microbial mat. *Microb. Ecol.* **2008**, *56*, 112–120. [CrossRef]
65. Luengo, J.M.; García, B.; Sandoval, A.; Naharro, G.; Olivera, E.R. Bioplastics from microorganisms. *Curr. Opin. Microbiol.* **2003**, *6*, 251–260. [CrossRef]

7

Recent Advances and Challenges towards Sustainable Polyhydroxyalkanoate (PHA) Production

Constantina Kourmentza [1,*], Jersson Plácido [2], Nikolaos Venetsaneas [3,4], Anna Burniol-Figols [5], Cristiano Varrone [5], Hariklia N. Gavala [5] and Maria A. M. Reis [1]

[1] UCIBIO-REQUIMTE, Department of Chemistry, Faculdade de Ciências e Tecnologia/Universidade Nova de Lisboa, 2829-516 Caparica, Portugal; amr@fct.unl.pt
[2] Centre for Cytochrome P450 Biodiversity, Institute of Life Science, Swansea University Medical School, Singleton Park, Swansea SA2 8PP, UK; j.e.placidoescobar@swansea.ac.uk
[3] Faculty of Engineering and the Environment, University of Southampton, Highfield, Southampton SO17 1BJ, UK; n.venetsaneas@aston.ac.uk
[4] European Bioenergy Research Institute (EBRI), Aston University, Aston Triangle, Birmingham B4 7ET, UK
[5] Department of Chemical and Biochemical Engineering, Center for Bioprocess Engineering, Søltofts Plads, Technical University of Denmark, Building 229, 2800 Kgs. Lyngby, Denmark; afig@kt.dtu.dk (A.B.-F.); cvar@kt.dtu.dk (C.V.); hnga@kt.dtu.dk (H.N.G.)
* Correspondence: ckourmentza@gmail.com or c.kourmentza@fct.unl.pt

Academic Editor: Martin Koller

Abstract: Sustainable biofuels, biomaterials, and fine chemicals production is a critical matter that research teams around the globe are focusing on nowadays. Polyhydroxyalkanoates represent one of the biomaterials of the future due to their physicochemical properties, biodegradability, and biocompatibility. Designing efficient and economic bioprocesses, combined with the respective social and environmental benefits, has brought together scientists from different backgrounds highlighting the multidisciplinary character of such a venture. In the current review, challenges and opportunities regarding polyhydroxyalkanoate production are presented and discussed, covering key steps of their overall production process by applying pure and mixed culture biotechnology, from raw bioprocess development to downstream processing.

Keywords: polyhydroxyalkanoates; biopolymers; renewable feedstock; mixed microbial consortia; enrichment strategy; pure cultures; synthetic biology; downstream processing

---

1. Introduction

Polyhydroxyalkanoates (PHAs) are a class of renewable, biodegradable, and bio-based polymers, in the form of polyesters. Together with polylactic acid (PLA) and polybutylene succinate (PBS), they are considered the green polymers of the future since they are expected to gradually substitute conventional plastics with similar physicochemical, thermal, and mechanical properties such as polypropylene (PP) and low-density polyethylene (LDPE) [1,2]. While PLA and PBS are produced upon polymerization of lactic and succinic acid respectively, PHA polymerization is performed naturally by bacteria.

A wide variety of bacteria are able to accumulate PHAs in the form of intracellular granules, as carbon and energy reserves. PHA accumulation is usually promoted when an essential nutrient for growth is present in limited amount in the cultivation medium, whereas carbon is in excess. Although, several bacteria are able to produce PHAs during growth and do not require growth-limiting conditions. This carbon storage is used by bacteria as an alternate source of fatty acids, metabolized under stress conditions, and is the key mechanism for their survival [3]. Up to 150 different PHA structures have been identified so far [4]. In general, PHAs are classified into two groups according to the carbon atoms that comprise their monomeric unit. Short-chain-length PHAs (scl-PHAs) consist of 3–5 carbon atoms, whereas medium-chain-length PHAs (mcl-PHAs) consist of 6–14 carbon atoms. PHB, the most well-known scl-PHA member, is characterized as a stiff and brittle material and is difficult to be processed due to its crystalline nature. The incorporation of 3-hydroxyvalerate (HV) units in PHB, results in the production of the copolymer poly(3-hydroxybutyrate-co-3-hydroxyvalerate), or else PHBV. PHBV is a material that becomes tougher, more flexible, and broader to thermal processing when its molar fraction in the copolymer increases [5]. scl-PHAs are mostly used for the production of disposable items and food packaging materials. On the other hand, mcl-PHAs are characterized as elastomers and they are suitable for high value-added application, such as surgical sutures, implants, biodegradable matrices for drug delivery, etc. [4].

Taking into account the recalcitrance of conventional plastics in the environment, replacement of synthetic plastics with PHAs would have huge benefits for the society and the environment [6]. Wide commercialization and industrialization of PHAs is still struggling due to their high production cost, resulting in higher prices compared to conventional polymers. While the price of polymers such as PP and PE is around US$0.60–0.87/lb, PHA biopolymer cost is estimated to be 3–4 times higher, ranging between US$2.25–2.75/lb [7,8]. Although several companies have initiated and industrialized the production of PHAs, as presented in Table 1, there are still major issues that need to be addressed in an effort to reduce the overall production cost. The main reasons for their high cost is the high price of high purity substrates, such as glucose, production in discontinuous batch and fed-batch cultivation modes, and large amount of solvents and/or labor regarding their downstream processing. The increasing availability of raw renewable materials and increasing demand and use of biodegradable polymers for bio-medical, packaging, and food applications along with favorable green procurement policies are expected to benefit PHA market growth. According to a recent report, published in 2017, the global PHA market is expected to reach US$93.5 million by 2021, from an estimated US$73.6 million within 2016, characterized by a compound annual growth rate (CAGR) of 4.88% [9].

Table 1. Pilot and industrial scale PHA manufacturers currently active worldwide.

Name of Company	Product (Trademark)	Substrate	Biocatalyst	Production Capacity
Biomatera, Canada	PHA resins (Biomatera)	Renewable raw materials	Non-pathogenic, non-transgenic bacteria isolated from soil	
Biomer, Germany	PHB pellets (Biomer®)	Sugar (sucrose)		
Bio-On Srl., Italy	PHB, PHBV spheres (minerv®-PHA)	Sugar beets	*Cupriavidus necator*	10,000 t/a
BluePHA, China	Customized PHBVHHx, PHV, P3HP3HB, P3HP4HB, P3HP, P4HB synthesis		Development of microbial strains via synthetic biology	
Danimer Scientific, USA	mcl-PHA (Nodax® PHA)	Cold pressed canola oil		
Kaneka Corporation, Japan	PHB-PHHx (AONILEX®)	Plant oils		3500 t/a
Newlight Technologies LLC, USA	PHA resins (AirCarbon™)	Oxygen from air and carbon from captured methane emissions	Newlight's 9X biocatalyst	
PHB Industrial S.A., Brazil	PHB, PHBV (BIOCYCLE®)	Saccharose	*Alcaligenes* sp.	3000 t/a
PolyFerm, Canada	mcl-PHA (VersaMer™ PHA)	Sugars, vegetable oils	Naturally selected microorganisms	
Shenzhen Ecomann Biotechnology Co. Ltd., China	PHA pellets, resins, microbeads (AmBio®)	Sugar or glucose		5000 t/a
SIRIM Bioplastics Pilot Plant, Malaysia	Various types of PHA	Palm oil mill effluent (POME), crude palm kernel oil		2000 t/a
TianAn Biologic Materials Co. Ltd., China	PHB, PHBV (ENMAT™)	Dextrose deriving from corn of cassava grown in China	*Ralstonia eutropha*	10,000 t/a, 50,000 t/a by 2020
Tianjin GreenBio Material Co., China	P (3, 4HB) films, pellets/foam pellets (Sogreen®)	Sugar		10,000 t/a

PHB, P3HB: poly(3-hydroxybutyrate); PHBV: poly(3-hydroxybutyrate-co-3-hydroxyvalerate); PHBVHHx: poly(3-hydroxybutyrate-co-3-hydroxyvalerate-co-3-hydroxyhexanoate); PHV: poly-3-hydroxyvalerate; P3HP3HB: poly(3-hydroxypropionate-co-3-hydroxybutyrate); P3HP4HB: poly(3-hydroxypropionate-co-4-hydroxybutyrate); P3HP: poly(3-hydroxypropionate); P4HB: poly(4-hydroxybutyrate); mcl-PHA: medium-chain length PHA; P(3,4HB): poly(3-hydroxybutyrate-co-4-hydroxybutyrate).

In the following sections, the advantages and drawbacks of PHA production employing both pure and mixed culture biotechnology are presented and discussed, as well as several approaches regarding their downstream processing in order to identify bottlenecks and opportunities to leverage PHA production.

2. PHA Production by Pure Bacterial Cultures

Pure culture biotechnology is implemented on an industrial scale, since a wide variety of food, pharmaceutical, and cosmetic agents derive as metabolic compounds from certain bacterial strains. Within the last few decades, research has been focused on finding ways to decrease the high production cost of PHAs. One of the main contributors to their high cost is the use of high purity substrates, which can account for 45% of the total production cost [10]. Instead, renewable feedstocks are being explored and researchers have been developing bioprocesses for the valorization of waste streams and by-products. In addition, current legislation and policies promote biodegradable waste management solutions other than disposal in landfills. Since every type of waste stream or by-product has different composition, selecting the appropriate biocatalyst is of great importance. In cases where the raw material is rich in carbon and nutrients, a growth-associated PHA producer would be selected, such as *Alcaligenes latus* or *Paracoccus denitrificans*. Conversely, in cases where the feedstock lacks an essential nutrient for growth such as nitrogen, phosphorous, etc., PHA accumulation using non-growth-associated bacteria would be ideal, i.e., *Cupriavidus necator*.

Apart from well-known species involved in industrial PHA production such as *Alcaligenes latus*, *Cupriavidus necator*, and *Pseudomonas putida*, bacteria need to combine several features in order be selected and regarded as promising PHA producers. Such features include their performance utilizing renewable feedstocks and/or environmental pollutants, seawater instead of fresh water, possibility of PHA production under open, non-sterile conditions, and their potential to develop contamination-free continuous bioprocesses. The use of agricultural byproducts and forest residues as an abundant and renewable source of lignocellulosic material for PHA production, is mainly considered after its physicochemical or biological hydrolysis. However, a few microorganisms possess the ability to saccharify cellulose and simultaneously produce PHAs. Moreover, PHA producers isolated from contaminated sites may be regarded for combined PHA production and bioremediation of toxic pollutants and post-consumer plastics. In addition, microorganisms isolated from hypersaline environments are considered the most promising ones, since they combine several benefits with a huge potential for reducing PHA production cost, namely in the downstream step. Last but not least, within recent years synthetic biology tools are continuously being developed in order to provide solutions to industrial challenges such as maximizing cellular capacity to 'make more space' for PHA accumulation, manipulating PHA composition to design polymers for high value-added applications and enhancing PHA efficiency.

2.1. Lignocellulose Degraders

2.1.1. Saccharophagus Degradans

Saccharophagus degradans, formerly known as *Microbulbifer degradans*, refers to a species of marine bacteria capable of degrading complex marine polysaccharides, such as cellulose of algal origin and higher plant material [11–13]. So far, the only strain reported is *S. degradans* 2-40, that was isolated from decaying marsh cord grass *Spartina alterniflora*, in the Chesapeake Bay watershed [14]. It is a Gram-negative, aerobic, rod-shaped, and motile γ-proteobacterium and is able to use a variety of different complex polysaccharides as its sole carbon and energy source, including agar, alginate, cellulose, chitin, β-glucan, laminarin, pectin, pullulan, starch, and xylan [11,13,15–17].

The key enzymes involved in PHA biosynthesis—β-ketothiolase, acetoacetyl-CoA reductase, and PHA synthase—have been identified in the genome of *S. degradans* 2-40 [12,18]. Preliminary studies have been performed in order to evaluate the feasibility of *S. degradans* to produce PHAs from

D-glucose and D-cellobiose as the sole carbon source in minimal media comprised of sea salts [19]. In addition, the authors evaluated the capability of the strain to degrade lignocellulosic material in the form of tequila bagasse (*A. tequilana*). According to the results obtained, it was shown that *S. degradans* successfully degraded and utilized cellulose as the primary carbon source to grow and produce PHB. However, PHB yields were not reported, so as to evaluate the efficiency of the process, but it became evident that prior hydrolysis of the lignocellulosic material is not required. This is considered positive since it can contribute to up-stream processing cost reduction and thus encourage further research employing the certain strain. In another study, Gonzalez-Garcia et al. [20], investigated PHA production using glucose as the sole carbon source and a culture medium designed according to bacterial biomass and seawater composition. Experiments were performed using a two-step batch strategy, where in the first step bacterial growth was performed under balanced conditions for 24 h, whereas in the second step cells were aseptically transferred to a fresh nitrogen deficient medium and incubated for 48 h. Under these conditions PHA content reached up to 17.2 $\pm$ 2.7% of the cell dry weight (CDW).

PHB biosynthesis from raw starch in fed-batch mode was also investigated and the results were compared to the ones obtained using glucose as the carbon source under the same conditions [21]. When starch was used PHA yield, content, and productivity reached up to 0.14 $\pm$ 0.02 g/g, 17.5 $\pm$ 2.7% of CDW and 0.06 $\pm$ 0.01 g/L h, respectively. In the case where glucose was fed the respective values were higher but still low compared to other PHA producers. However, only a few microorganisms have been reported to directly utilize raw starch for PHA production [21,22]. During the experiments, the authors observed the simultaneous production of organic acids and exopolymers and this was the main reason for the low PHB accumulation. Higher PHA efficiency could be achieved by optimizing cultivation parameters to drive carbon flux towards PHA biosynthesis and also by applying genetic engineering to knock out genes responsible for the production of side products such as exopolymers.

PHA production in aquarium salt medium supplemented with 1% of different types of cellulosic substrates such as α-cellulose, avicel PH101, sigmacell 101, carboxymethyl cellulose (CMC), and cellobiose have also been studied [23]. In flask experiments, PHB production was 11.8, 14.6, 13.7, and 12.8% of the DCW respectively. Fed-batch cultivation strategy resulted in increased PHB contents reaching up to 52.8% and 19.2% of the DCW using glucose and avicel respectively, as carbon sources.

Recently, another approach towards PHA biosynthesis from *S. degradans* was proposed. During their experiments, Sawant et al. [24] observed that *Bacillus cereus* (KF801505) was growing together with *S. degradans* 2-40 as a contaminant and had the ability of producing high amounts of PHAs [25]. In addition, the viability and agar degradation potential of *S. degradans* increased with the presence of *B. cereus*. Taking those into account, they further investigated the ability of co-cultures of *S. degradans* and *B. cereus* to produce PHAs using 2% w/v agarose and xylan without any prior treatment. PHA contents obtained from agarose and xylan were 19.7% and 34.5% of the DCW respectively when co-cultures were used compared to 18.1% and 22.7% achieved by pure cultures of *S. degradans*. This study reported for the first time the production of PHAs using agarose. Moreover, according to the results, the highest PHA content from xylan was obtained using a natural isolate. So far, only recombinant *Escherichia coli* has been reported to produce 1.1% PHA from xylan, which increased to 30.3% and 40.4% upon supplementation of arabinose and xylose, respectively [26].

These unique features of *S. degradans* open up the possibility to use it as a source of carbohydrases in order to saccharify lignocellulosic materials. Thus, coupled hydrolysis and fermentation is a promising alternative for the production of PHAs using carbon sources that may derive from biomass residues of different origin (Table 2). However, saccharification and coupled PHA production need to be studied in detail in order to understand their potential and find ways to increase the rates of their processes.

2.1.2. Caldimonas Taiwanensis

Caldimonas taiwanensis is a bacterial strain isolated from hot spring water in southern Taiwan in 2004 [27]. Researchers had been searching for thermophilic amylase-producing bacteria since those enzymes are of high industrial importance for the food and pharmaceutical sector. In addition, since starch hydrolysis is known to be faster at relatively high temperatures, thermophilic amylases are usually preferred [28]. Upon morphological and physiological characterization, it was shown that this Gram-negative, aerobic, rod shaped bacteria can form PHB granules.

A few years later, Sheu et al. [22] investigated PHA production from a wide variety of carbon sources. At first cultivation of *C. taiwanensis* on a three-fold diluted Luria-Bertani medium supplemented with sodium gluconate, fructose, maltose, and glycerol as the sole carbon sources under optimal nitrogen limiting conditions, C/N = 30 was performed. PHB contents reached up to 70, 62, 60, and 52% of the CDW at 55 °C in shake flask experiments. In the next step, fatty acids were tested as sole carbon sources for growth and PHA production. It was observed that the strain did not grow at a temperature between 45 °C or 55 °C while no PHA was formed. Nevertheless, when mixtures of gluconate and valerate were provided bacterial growth was feasible and PHA cellular content reached up to 51% of its CDW. The presence of valerate induced the presence of HV units in the polymer resulting in the production of a PHBV copolymer constituting of 10–95 mol % HV depending on the relative valerate concentration in the mixture. Moreover, mixtures of commercially available starches and valerate were evaluated for PHA production at 50 °C. The carbon source mixture consisted of 1.5% starch and 0.05% valerate. Starch types examined were cassava, corn, potato, sweet potato, and wheat starch. After 32 h of cultivation PHBV copolymer was produced in all cases, composed of approximately 10 mol % HV. PHA contents of 67, 65, 55, 52, and 42% of its CDW were achieved respectively.

Despite the fact that biotechnological process using thermophilic bacteria need to be performed at high temperatures, they reduce the risk of contamination. Another advantage is the fact that thermophiles grow faster compared to mesophiles, therefore less time is needed to achieve maximum PHA accumulation [22]. Moreover, employing *C. taiwanensis* for PHA production using starch-based raw materials is extremely beneficial, in economic terms, since no prior saccharification is required. On the other hand, as mentioned before, *C. taiwanensis* cannot grow on fatty acids but when a mixture of valerate and gluconate/or starch is supplied bacterial growth and PHA accumulation occurs in the form of PHBV copolymer. However, the concentration of fatty acids may result in toxicity for bacterial cells, that up to a point can be overcome by the fast growth of cells. In addition, since amylose, amylopectin, and nitrogen contents vary between types of starch, prior characterization needs to be performed. The feasibility of enzymatic degradation of amylose and amylopectin is considered a key factor as it regulates the amount of sugars present in the medium. Last but not least, nitrogen content should be also controlled as high amounts favor biomass growth instead of PHA accumulation.

2.2. Bioremediation Technologies Allowing PHA Production

One of the major causes of environmental pollution is the presence of volatile aromatic hydrocarbons such as benzene, toluene, ethylbenzene, and xylene (BTEX) that are found in crude oil and petroleum products. In addition, huge amounts of starting materials for the production of petrochemical based plastics, such as styrene, are released annually [29]. Moreover, chemical additives in plastics, which are accumulated in the environment due to their recalcitrance, can leach out and are detectable in aquatic environments, dust and air because of their high volatility [30]. Also, textile dyes and effluents are one of the worst polluters of our precious water bodies and soils. All the above are posing mutagenic, carcinogenic, allergic, and cytotoxic threats to all life forms [31].

Since PHAs are known to have a functional role in bacterial survival under stress conditions, toxic environments characterized by poor nutrient availability are proven to be important sources of PHA producers [32]. Several attempts have been made within the last decade to explore contaminated sites as a resource of microorganisms that are expected to advance biotechnological production of PHAs.

Employment of such bacteria combines bioremediation with the production of a high value-added material. So far, bacterial strains that belong to the genus of *Sphingobacterium*, *Bacillus*, *Pseudomonas*, and *Rhodococcus* have been isolated and studied regarding their PHA production potential degrading environmental pollutants, as summarized in Table 2 [3,29,33–39].

Pseudomonas species are characterized by their ability to utilize and degrade a variety of carbon sources due to their wide catabolic versatility and genetic diversity. For these reasons, they are a natural choice regarding techniques of in situ and ex situ bioremediation [40]. Several *Pseudomonas* strains have been isolated from hydrocarbon-contaminated soils and together with other *Pseudomonas* sp. have been examined regarding their ability to produce PHA from hydrocarbons. In their study, Nikodinovic et al. [41], investigated PHA accumulation in several *Pseudomonas* strains from single BTEX aromatic substrates and mixed aromatic substrates as well as mixtures of BTEX with styrene. It was reported that when *P. putida* F1 was supplied with 350 μL of toluene, benzene, or ethylbenzene it accumulated PHA up to 22, 14, and 15% of its CDW respectively, while no growth was observed when *p*-xylene and styrene were supplied as the sole carbon source. In the case of *P. putida* mt-2 no growth was obtained with benzene, ethylbenzene, or styrene but when toluene and *p*-xylene were used its PHA content was 22% and 26% respectively. *P. putida* CA-3 efficiently degraded styrene but could not metabolize any of the other hydrocarbons investigated. However, a defined mixed culture of *P. putida* F1, mt-2, and CA-3 was successfully used for PHA production from BTEX and styrene mixtures, where the highest biomass concentration was achieved and PHA content reached up to 24% of the CDW. In another study, strains from petroleum-contaminated soil samples were screened on their ability to degrade toluene and synthesize mcl-PHA [42]. Among them *P. fluva* TY16 was selected to be further investigated. It was shown that the highest PHA content of 68.5% was achieved when decanoic acid was used as the carbon source. In the case of benzene, toluene, and ethylbenzene PHA contents reached up to 19.1, 58.9, and 28.6% respectively, using a continuous feeding strategy. *Pseudomonas* sp. TN301 was isolated from a river sediment sample from a site in a close proximity to a petrochemical industry [43]. Both monoaromatic and polyaromatic hydrocarbons were examined as PHA precursors and cellular mcl-PHA contents varied between 1.2% and 23% of its CDW, while this study was the first one on the ability of a bacterial strain to convert polyaromatic hydrocarbon compounds to mcl-PHA. Moreover, *Pseudomonas* strains isolated from contaminated soil and oily sludge samples from Iranian southwestern refineries accumulated 20–23% of their CDW to mcl-PHA using 2% v/v crude oil as the sole carbon source [32].

As mentioned before, styrene—used for the synthesis of polystyrene—is a major and toxic environmental pollutant. Ward et al. [29], has reported that *P. putida* CA-3 was capable of converting styrene, its metabolic intermediate phenylacetic acid and glucose into mcl-PHA under nitrogen limited conditions, characterized by conversion yields of 0.11, 0.17, and 0.22 g/g, respectively. However, higher cell density and PHA production, characterized by a conversion yield of 0.28 g/g, were observed when cells were supplied with nitrogen at a feeding rate of 1.5 mg/L/h [37]. Moreover, in a recent study the key challenges of improving transfer and increasing supply of styrene, without inhibiting bacterial growth, were addressed [35]. It was shown that by changing the feed from gaseous to liquid styrene, through the air sparger, release of styrene was reduced 50-fold, biomass concentration was five times higher, while PHA production was four-fold compared to previous experiments, with a PHA content reaching up to 32% in terms of CDW and a conversion yield of 0.17 g/g.

A two-step chemo-biotechnological approach has been proposed for the management of post-consumer polystyrene, involving its pyrolysis to styrene oil and subsequent conversion of the styrene oil to PHA by *P. putida CA-3* [44]. According to their results, after 48 h 1.6 g of mcl-PHA were obtained from 16 g of oil with a cellular content of 57%. Following the same approach, the solid fraction of pyrolyzed polyethylene terephthalate (PET) was used as feedstock for PHA production by bacteria isolated from soil exposed to PET granules at a PET processing plant [45]. The isolated strains were identified and designated as *P. putida* GO16, *P. putida* GO19, and *P. frederiksbergensis* GO23 and they were able to accumulate mcl-PHA up to 27, 23, and 24% of their CDW respectively, using 1.1 g/L sodium terephthalate as the sole carbon source under conditions of nitrogen limitation. Recently, conversion of polyethylene (PE) pyrolysis wax to mcl-PHAs was investigated employing *P. aeruginosa* PAO-1 [46]. Addition of rhamnolipid biosurfactants in the growth medium had a positive impact on bacterial growth and PHA accumulation. Substitution of ammonium chloride with ammonium nitrate led to faster growth and earlier PHA accumulation that reached up to 25% of its CDW.

A series of studies has been focused on the degradation of textile dyes for PHA production using *Sphingobacterium*, *Bacillus*, and *Pseudomonas* species. When the dye Direct Blue GLL (DBGLL) was used, *Sphingobacterium* sp. ATM completely decolorized 0.3 g/L in 24 h, while simultaneous polyhydroxyhexadecanoic acid (PHD) occurred reaching up to 64% of its CDW [38]. The potential of *B. odyssey* SUK3 and *P. desmolyticum* NCIM 2112 was also investigated. It was shown that both strains were able to decolorize 0.05 g/L DBGLL by 82% and 86% and produce PHD up to 61 and 52% of their CDW, respectively. In another study, 82% decolorization of 0.8 g/L of the textile dye Orange 3R was feasible, employing *Sphingobacterium* sp. ATM which resulted in the production of 3.48 g/L of PHD and a cellular PHA content of 65% after 48 h [39]. In addition, full decolorization of 0.5 g/L of the textile dye Direct Red 5B (DR5B) was accomplished when the medium was supplemented with glycerol, glucose, starch, molasses, frying oil, and cheese whey. In those cases PHD accumulation was 52, 56, 55, 64, 46, and 10% of its CDW respectively [47].

2.3. Halophiles

Halophiles are microorganisms that require salt for their growth and are categorized, according to their halotorerance, in two groups: moderate (up to 20% salt) and extreme (20–30% salt) halophiles [48]. Their name comes from the Greek word for 'salt-loving' [49] and can be found in the three domains of life: Archaea, Bacteria, and Eukarya. They thrive in marine and hypersaline environments around the globe such as the saline lakes, salt marshes, and salterns [50,51]. With the use of halophiles the risk for contamination is reduced and/or eliminated since non-halophilic microorganisms cannot grow in media containing high salt concentrations. This is of great importance since their use combines the advantages of low energy requirements under unsterile conditions, minimal fresh water consumption, due to its substitution with seawater for medium preparation, and the possibility of operating contamination free continuous fermentation processes that are much more efficient. In addition, downstream processing cost can be reduced by treating the bacterial cells with salt-deficient water in order to cause hypo-osmotic shock [52]. The above, together with the valorization of low cost substrates, bring halophilic bacteria a step closer to being used as biocatalysts for industrial PHA production.

PHA accumulation by halophilic archaea was first observed in 1972 by Kirk and Ginzburg [53]. So far, the most well-known and best PHA halophilic archaeon producer is *Haloferax mediterranei*, which was first isolated from seawater evaporation ponds near Alicante in Spain [54]. Several studies have shown their ability to accumulate high PHA contents utilizing low cost feedstocks. Among them, vinasse, a byproduct of ethanol production from sugarcane molasses, has been utilized [55]. After pre-treatment, via adsorption on activated carbon, 25% and 50% v/v of pre-treated vinasse led to cellular PHA contents of 70% and 66%, respectively. Maximum PHBV (86% HB–14% HV) concentration reached up to 19.7 g/L, characterized by a volumetric productivity of 0.21 g/L h and a conversion yield of 0.87 g/g, for the case where 25% pre-treated vinasse was used. In another study, stillage derived from a rice-based ethanol industry, was investigated [56]. PHA accumulation was 71% of its CDW that led to 16.4 g/L of PHBV (85% HB-15% HV) with a yield coefficient of 0.35 g/g and a volumetric productivity of 0.17 g/L h. Moreover, cheese whey hydrolysate—obtained upon acid pre-treatment—has been used for the production of PHBV with low HV fraction, 1.5 mol % [57]. Batch cultivation of *H. mediterranei* led to the production of 7.54 g/L of biomass, with a PHA content of 53%, and a volumetric productivity of 0.17 g/L h. Olive mill wastewater (OMW), which is a highly polluting waste, was also utilized recently as the sole carbon source for PHA production [58]. Using a medium containing 15% of OMW up to 43% of PHBV/CDW was produced consisting of 6 mol % HV in a one-stage cultivation step.

Halophilic bacteria belong to γ-Proteobacteria and they can grow on a wide range of pH, temperature, and salinity concentrations up to 30% (w/v NaCl) and possess the ability to accumulate PHA [52]. *Halomonas* TD01 has been isolated from Aydingkol Lake in China. This strain was investigated regarding its PHA production potential under unsterile and continuous conditions [59]. Glucose salt medium was used and initial fed-batch cultivation resulted in the production of 80 g/L of biomass with a PHA content of 80%, in the form of PHB, after 56 h. A continuous and unsterile cultivation process was developed that lasted for 14 days, and that allowed cells to grow to an average of 40 g/L containing 60% PHB in the first reactor. Cells were forwarded by continuous pumping from the first to the second reactor that contained nitrogen-deficient glucose salt medium. In the second reactor PHB levels ranged from 65 to 70% of its CDW and a conversion yield of 0.5 g/g was achieved. This was the first attempt for continuous PHA production under non-sterile conditions from a halophilic bacterium. In addition, Yue et al. [60] explored the potential of *Halomonas campaniensis* LS21, isolated from the Dabancheng salt lake in China, to produce PHA in a seawater-based open and continuous process. The strain utilized a mixture of substrates mainly consisting of cellulose, starch, animal fats and proteins. Instead of fresh water fermentation was performed using artificial seawater composed of 26.7 g/L NaCl, among others under a pH around 10 and 37 °C. PHB accumulation reached up to 26% during 65 days of continuous fermentation without any contamination. Through this study the benefits of long-lasting, seawater-based, and continuous processes for PHA production under unsterile conditions were demonstrated.

Bacillus megaterium has recently drawn attention since several studies have isolated such stains from salterns. *Bacillus megaterium* H16, isolated from the solar salters of Ribandar Goa in India, was shown to accumulate up to 39% PHA in the presence (5% w/v) or absence of NaCl using glucose [61]. In another study, a mangrove isolate that was found to belong to *Bacillus* sp. could tolerate salinity up to 9% w/v [62]. The certain strain was able to utilize a wide variety of carbon sources such as monosaccharides, organic acids, acid pre-treated liquor, and lignocellulosic biomass reaching cellular PHA contents of up to 73% of its CDW. Furthermore, *Bacillus megaterium uyuni* S29, isolated from Uyuni salt lake in Bolivia, was examined in terms of its salinity tolerance and impact on biomass and PHB production [63]. It was observed that the strain could grow at 10% w/v NaCl while PHB production was observed even at high salinity levels of 25% w/v. Optimum results for biomass and PHB production were achieved in medium containing 4.5% w/v NaCl and were 5.4 and 2.2 g/L characterized by a yield coefficient of 0.13 g/g and a volumetric PHB productivity of 0.10 g/L h.

The results obtained from the studies described above are very promising and demonstrate the remarkable potential of halophiles for biotechnological production of PHAs. Although, processes performed under high salinity concentrations present disadvantages such as the corrosion of stainless steel fermenters and piping systems [51,52]. However, since no sterilization is required when halophiles are used, other types of low cost materials, such as plastics and ceramics, may be used to design and construct fermentation and piping systems in order to overcome corrosion issues. In addition, the number of halophilic bacteria that their genome is being sequenced is constantly increasing throughout the years. Subsequently, in the near future, molecular biology techniques will result in metabolically engineered strains with better performances regarding their industrial application [48,51].

Table 2. Characteristic parameters describing PHA production from different types of bacteria.

Strain	Carbon Source	PHA	Cultivation Mode	DCW (g L^{-1})	PHA (g L^{-1})	PHA (%)	$Y_{P/S}$	Ref.
Lignocellulose Degraders								
S. degradans	Glucose	PHB				17.2		[20]
	Glucose	PHB	Fed-batch	12.7	2.7	21.4	0.17	[21]
	Starch		Fed-batch	11.7	2.0	17.5	0.14	
	Glucose	PHB	Flask	2.1	0.46	22.4		[23]
	Cellobiose		Flask	2.0	0.42	20.8		
	α-Cellulose		Flask	1.2	0.14	11.8		
	Avicel		Flask	1.0	0.15	14.6		
	Sigmacell		Flask	1.0	0.14	13.7		
	CMC		Flask	1.1	0.14	12.7		
	Glucose		Batch	1.6	0.40	25.3		
	Glucose		Fed-batch	4.2	2.20	52.8		
	Avicel		Fed-batch	2.1	0.40	19.2		
	Agarose	PHB	One-step batch		0.24	18.1		[24]
	Xylan		One-step batch		0.20	22.7		
	Agarose		Two-step batch		0.31	18.4		
	Xylan		Two-step batch		0.24	15.3		
Co-culture of *S. degradans* and *B. cerues*	Agarose	PHB	One-step batch		0.29	19.7		[24]
	Xylan		One-step batch		0.27	34.5		
	Agarose		Two-step batch		0.23	15.3		
	Xylan		Two-step batch		0.33	30.2		
C. taiwanensis	Propionate + Glc [a]	PHBV (88–12) [f]	Flask	2.0	1.04	52		[22]
	Valerate + Glc [a]	PHBV (49–51)		1.0	0.51	51		
	Hexanoate + Glc [a]	PHBHHx [c]		2.7	1.67	62		
	Hexanoate + Glc [a] + AA [b]	PHBHHx [d]		1.2	0.56	47		
	Heptanoate + Glc [a]	PHBV (65–35)		1.7	0.56	33		
	Heptanoate + Glc [a] + AA	PHBV (15–85)		0.3	0.05	17		
	Octanoate + Glc [a]	PHB		0.4	0.05	13		
	Cassava starch + Val [e]	PHBV (87–13)		2.8	1.88	67		
	Corn starch + Val [e]	PHBV (80–10)		3.3	2.14	65		
	Potato + Val [e]	PHBV (80–10)		2.6	1.43	55		
	Sweet potato + Val [e]	PHBV (80–10)		1.6	0.83	52		
	Wheat starch + Val [e]	PHBV (80–10)		4.1	1.72	42		
Polyhydroxyalkanoates and Bioremediation								
P. putida F1	Benzene	mcl-PHA	Flask	0.34	0.05	14		[41]
	Toluene			0.72	0.16	22		
	Ethylbenzene			0.67	0.10	15		
P. putida mt-2	Toluene	mcl-PHA	Flask	0.37	0.08	22		[41]
	p-Xylene			0.53	0.14	26		
P. putida CA-3	Styrene	mcl-PHA	Flask	0.79	0.26	33		[41]
P. fluva TY16	Benzene	mcl-PHA	Continuous feeding	2.54		19	0.03	[42]
	Toluene			3.87		59	0.11	
	Ethylbenzene			2.80		29	0.04	
P. putida CA-3	Styrene pyrolysis oil	mcl-PHA	Flask	2.80	1.60	57	0.10	[44]
Sphingobacterium sp. ATM	Orange 3R dye	PHA	Flask		3.48	65		[38]
B. odysseyi SUK3					2.10	61		
P. desmolyticim NCIM 2112					1.12	52		
Halophiles								
H. mediterranei DSM 1411	25% pre-treated vinasse	PHBV (86–14)	Flask		19.7	70	0.87	[55]
	Stillage	PHBV (85–15)			16.4	71	0.35	[56]
	Hydrolyzed cheese whey	PHBV (98.5–1.5)	Batch	7.54		54	0.78	[57]
	15% v/v olive mill wastewater	PHBV (94-6)	Flask		0.2	43		[58]
Halomonas TD01	Glucose salt medium	PHA	Continuous two-fermentor			65	0.51	[59]

Table 2. *Cont.*

Strain	Carbon Source	PHA	Cultivation Mode	DCW (g L^{-1})	PHA (g L^{-1})	PHA (%)	$Y_{P/S}$	Ref.
Halomonas campaniensis LS21	Mixed substrates (mostly comprised of kitchen waste)	PHB	Continuous pH-stat			26		[60]
B. megaterium H16	Glucose salt medium	PHB	Flask			39		[61]
B. megaterium uyuni S29	Glucose salt medium	PHB	Flask	5.42	2.22	41	0.13	[63]

[a] Mixtures consisting of 0.1% fatty acid and 1.5% gluconate; [b] 2mM acrylic acid; [c] 99.5% HB, 0.5% HHx; [d] 98.5% HB, 1.5% HHx; [e] Mixtures consisting of 1.5% starch type + 0.05% Valerate; [f] PHBV (%HB–%HV).

2.4. Synthetic Biology of PHA Producers

Synthetic biology tools may aid in developing competitive bioprocesses by engineering biocatalysts with the potential of being employed at industrial scale, producing large amounts of PHA at low prices [64]. Industrial biotechnology requires non-pathogenic, fast growing bacteria that do not produce toxins and their genome is easily manipulated. Utilization of cellulose and fast growth under a wider range of temperature and pH are considered a plus [65]. The effort of minimizing PHA production cost focuses mainly on engineering strains that show higher PHA production efficiency from raw waste material, require less energy consumption during PHA production, simplify downstream processing, and produce tailored functional polymers for high value-added applications.

In order to achieve high PHA volumetric productivities high cell densities need to be obtained of up to 200 g/L, characterized by high cellular PHA contents, above 90% g PHA/g CDW. Manipulation of genes related to the oxygen uptake, quorum sensing, and PHA biosynthetic mechanisms may enhance PHA production [65]. For example, oxygen limitation may occur, after obtaining high cell densities, in order to initiate/promote PHB production. In a relevant study, anaerobic metabolic pathways were designed in *E. coli* (over-expressing hydrogenase 3 and acetyl-CoA synthatase) to facilitate production of both hydrogen and PHB. In that way, the formation of toxic compounds such as formate and acetate was avoided by driving carbon fluxes towards the production of PHB. The engineered strain showed improved hydrogen and PHB production. In addition, PHB pathway optimization has been also investigated in *E. coli* by adjusting expression levels of the three genes *phbC*, *phbA*, and *phbB* [66]. *phbCAB* operon was cloned from the native PHA producing strain *Ralstonia eutropha*. Rational designed Ribosomal Binding Sites (RBS) libraries were constructed based on high or low copy number plasmids in a one-pot reaction by an Oligo-Linker Mediated Assembly method (OLMA). Bacterial strains accumulating cellular contents of 0 to 92% g PHB/g CDW were designed and a variety of molecular weights ranging between 2.7–6.8 $\times$ 10^6 was achieved. The certain study demonstrated that this semirational approach combining library design, construction, and proper screening is an efficient tool in order to optimize PHB production.

Another example where synthetic biology has been implemented are halophilic bacteria, which allow for PHA production under continuous mode and unsterile conditions. These features increase the competitiveness of industrial PHA production. In addition, halophilic bacteria have been proven easy for genetic manipulation, thus allowing for the construction of a hyper-producing strain [65,67]. For example, both recombinant and wild type *Halomonas campaniensis* LS21were able to grow on mixed substrates (kitchen wastes) in the presence of 26.7 g/L NaCl, at pH 10 and temperature of 37 °C continuously, for 65 days, without any contamination. Recombinant *H. campaniensis* produced almost 70% PHB compared to wild type strain that in which PHA accounted for 26% of its CDW [60].

Engineering the morphology of bacteria, in terms of cell size increase, has been recently investigated. Apart from PHA granules, several bacteria may accumulate polyphosphates, glycogen, sulfur, or proteins within their cells that limit cell space availability. In order to increase cell size, approaches such as deletion or weak expression of on actin-like protein gene mreB in recombinant

E. coli resulted in increasing PHB accumulation by 100% [68–70]. In addition, manipulating PHA granule-associated proteins leads to an increase in PHA granule size allowing for easier separation [71].

Intracellular accumulation of PHA necessitates several extraction and purification steps. Synthetic biology approaches have been developed to control and facilitate the release of PHA granules to the medium. For example, the programmed self-disruptive strain *P. putida* BXHL has been constructed in a recent study, deriving from the prototype *P. putida* KT2440 which is a well-known mcl-PHA producer [72]. This was based on a controlled autolysis system utilizing endolysin Ejl and holing Ejh isolated from EJ-1 phage and in order to improve the efficiency of the lytic system this was tested in *P. putida* tol-pal mutant strains with alterations in outer membrane integrity. According to results, it was shown that the engineered lytic system of *P. putida* BXHL provided a novel approach to inducing controlled cell lysis under PHA producing conditions, either produce PHA accumulating cells that were more susceptible to lytic treatments. The certain study demonstrated a new perspective on engineered cells facilitating PHA extraction in a more environmentally friendly and economic way.

PHA structures include PHA homopolymers, random and block copolymers, and also different monomer molar fractions in copolymers. Block copolymers have been reported regarding their resistance against polymer aging. This is of crucial importance since slower degradation of polymer occurs leading to better performance and consistent polymer properties [73]. It has been observed that downregulating of β-oxidation cycle in *P. putida* and *P. entomophila* may be used for controlling PHA structure when fatty acids are used as precursors for PHA production [73–76]. In the case of fatty acids, containing functional groups are consumed by bacteria, introduction of those functional groups into PHA polymer chains occurs [77]. In addition, recombinant strains of *E. coli* have been constructed for the synthesis of block polymers with superior properties [78–81]. PHA diversity is possible by engineering basic biosynthesis pathways (acetoacetyl-CoA pathway, in situ fatty acid synthesis, and/or β-oxidation cycles) as well as through the specificity of PHA synthase [82].

3. PHA Production by Mixed Microbial Consortia (MMC)

Currently, industrial PHA production is conducted using natural isolates or engineered strains and pure substrates [83,84]. An alternative scenario that would contribute to the reduction of the PHA production cost is to employ mixed culture biotechnology [85,86]. This approach uses open (under non-sterile conditions) mixed microbial consortia (MMC) and ecological selection principles, where microorganisms able to accumulate PHA are selected by the operational conditions imposed on the biological system. Thus, the principle is to engineer the ecosystem, rather than the strains, combining the methodology of environmental biotechnology with the goals of industrial biotechnology [87]. The cost reduction derives mainly from operations being performed under non-sterile conditions, and their consequent energy savings, and the higher adaptability of MMC to utilize waste streams as substrates.

Processes for PHA production in mixed cultures are usually performed in two steps (Figure 1). In the first step, SBR reactors (sequential batch reactors) are used to select and enrich a microbial population with high PHA production capacity by applying transient conditions. In the second step, the culture from the SBR is subjected to conditions maximizing the PHA accumulation, from where cells are harvested for PHA extraction and purification when they reach maximum PHA content [88,89].

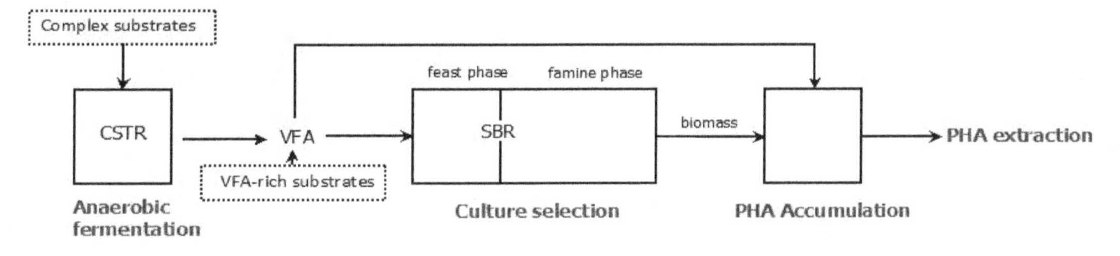

Figure 1. PHA production process by mixed microbial cultures. Modified from [88]. CSTR: continuous stirred tank reactor, SBR: Sequential Batch Reactor.

Unlike pure cultures, where glucose is mostly used as a substrate for PHA production, mixed culture biotechnology makes use of volatile fatty acids (VFA) as the precursors for PHA production [90]. The main reason is that carbohydrates in MMCs, as well as other substrates such as glycerol, tend to form glycogen besides PHA [91,92]. For those substrates, a previous step is generally included (Figure 1), during which they are fermented into volatile fatty acids (VFA) in continuous mode (CSTR). Moreover, this is also applied to complex substrates, such as olive mill wastewater [10,93,94], cheese whey [95], and other food wastes [96] in order to obtain a more homogeneous readily available feed for the PHA production. VFA conversion into PHA require few steps, and usually presents very high yields and rapid uptake rates [88,97]. This is especially the case for butyric acid, which has now been reported in many studies as the VFA presenting the highest yields (up to 0.94 Cmol PHA/Cmol S) and being the one preferably up-taken by MMCs [2,98–100]. As a matter of fact, butyric acid preference has been observed even in cultures that were not exposed to it during the enrichment [93].

It is worth mentioning that the distribution of VFA is known to affect the PHA monomer composition, where VFA with an even number of carbon atoms tend to produce PHB while VFA with odd carbon atoms tend to produce PHBV copolymers with different % HV molar fractions [101]. Based on this fact, many studies have suggested the possibility of regulating the PHA composition by manipulating the fermentation conditions in the preceding acidogenesis step [2,95,100].

This section provides an overview of the different types of existing enrichment techniques, performed within the last 10 years. Each one of the enrichment strategies presents different advantages; either related to the cost of the process or to increased cellular PHA content. Thus, they all present opportunities to further improve economic and sustainable PHA production. Such opportunities are commented for each of the enrichment techniques. This section is followed by a compilation of recent advances regarding the PHA accumulation stage, aiming at increasing the productivity. Finally, recent attempts to bring MMC to pilot scale are described, followed by a section highlighting the main challenges and bottlenecks of the MMC.

3.1. Types of Enrichments

Until the late 2000s, two types of enrichments dominated the research panorama related to PHA mixed culture biotechnological production, namely the anaerobic/aerobic selection and aerobic dynamic feeding. These types of enrichments have already been previously reviewed in other articles [88,102–104], thus apart from a general description of the mechanisms, only the recent trends are described in the respective Sections 3.1.1 and 3.1.2. The following sections are dedicated to recent (and still less widespread) types of enrichments developed within the last decade. The main characteristics of all types of enrichments described are summarized in Table 3.

Table 3. Summary on the main characteristics of the enrichment techniques applied for MMCs.

Anaerobic-Aerobic Enrichment (AN/AE) (Section 3.1.1)		
	Feast phase	**Famine phase**
Aeration	No	Yes
e^- acceptor	– (PHA)	Oxygen
Energy source	Glycogen/polyphosphate	Oxidation of PHA
Carbon source	External substrate	PHA
Driving force for PHA accumulation	• Lack of electron acceptor * • Transient presence of substrate ***	
Aerobic Dynamic Feeding (ADF) (Sections 3.1.2 and 3.1.3)		
Classical Aerobic Dynamic Feeding (Section 3.1.2)		
	Feast phase	**Famine phase**
Aeration	Yes	Yes
e^- acceptor	Oxygen	Oxygen
Energy source	Oxidation of substrate	Oxidation of PHA
Carbon source	External substrate	PHA
Nitrogen availability	Yes **	Yes **
Driving force for PHA accumulation	• Transient presence of substrate ***	
Aerobic Dynamic Feeding (ADF) with Intermediate Settling Phase (Section 3.1.3)		
	Feast phase	**Famine phase**
Aeration	Yes	Yes
e^- acceptor	Oxygen	Oxygen
Energy source	Oxidation of substrate	Oxidation of PHA
Carbon source	External substrate	PHA
Nitrogen availability	Yes	Yes
Driving force for PHA accumulation	• Transient presence of substrate *** • Higher settling capacity of PHA rich cells • Elimination of residual COD after feast phase prevents growth of non-PHA accumulating bacteria	
Aerobic Dynamic Feeding (ADF) with Nitrogen Limitation in the Feast-Phase (Section 3.1.3)		
	Feast phase	**Famine phase**
Aeration	Yes	Yes
e^- acceptor	Oxygen	Oxygen
Energy source	Oxidation of substrate	Oxidation of PHA
Carbon source	External substrate	PHA
Nitrogen availability	No	Yes
Driving force for PHA accumulation	• Transient presence of substrate *** • Nitrogen limitation during the feast phase	

Table 3. *Cont.*

	Photosynthetic Enrichment (Section 3.1.5)	
	Photosynthetic Enrichments—Illuminated SBR	
	Feast phase	**Famine phase**
Aeration	No	No
e^- acceptor	– (PHA)	Oxygen produced by algae
Energy source	Light	Oxidation of PHA + Light
Carbon source	External substrate	PHA
Driving force for PHA accumulation	• Lack of external electron acceptor with presence of light • Transient presence of substrate ***	
	Photosynthetic Enrichment—Dark Feast Phase	
	Feast phase	**Famine phase**
Aeration	No	No
e^- acceptor	– (PHA)	Oxygen produced by algae
Energy source	Glycogen	Oxidation of PHA + Light
Carbon source	External substrate	PHA
Driving force for PHA accumulation	• Lack of external electron acceptor with presence of light • Transient presence of substrate ***	
	Photosynthetic Enrichment—Permanent Feast Phase	
	Feast phase	**Famine phase**
Aeration	No	No famine phase
e^- acceptor	– (PHA)	
Energy source	Light	
Carbon source	External substrate	
Driving force for PHA accumulation	• Lack of external electron acceptor with presence of light	
	Aerobic-Anoxic Enrichment (Section 3.1.6)	
	Feast phase	**Famine phase**
Aeration	Yes	No
e^- acceptor	Oxygen	NO_3 /NO_2
Energy source	Oxidation of substrate	Oxidation of PHA
Carbon source	External substrate	PHA
Driving force for PHA accumulation	• Transient presence of substrate ***	

Table 3. *Cont.*

	Anoxic-Aerobic Enrichment (Section 3.1.7)	
	Feast phase	**Famine phase**
Aeration	No	Yes
e^- acceptor	NO_3/NO_2	Oxygen
Energy source	Oxidation of substrate	Oxidation of PHA
Carbon source	External substrate	PHA
Driving force for PHA accumulation	• Transient presence of substrate ***	
	Microaerophilic Enrichment (Section 3.1.8)	
	Feast phase	**Famine phase**
Aeration	Yes	Yes
e^- acceptor	Oxygen	Oxygen
Energy source	Oxidation of substrate	Oxidation of PHA
Carbon source	External substrate	PHA
Driving force for PHA accumulation	• Transient presence of substrate *** • Limitation of electron acceptor	

* Even though the lack of electron acceptor is the driving force of the enrichment, this limitation is not mandatory for these cultures to produce PHA, which can also be produced aerobically. ** various C/N ratios have been applied resulting in a limitation of nitrogen in the famine or late feast phase. Nevertheless, most wide-spread configuration provides nitrogen in both phases *** Transient presence of substrate leads to the following effects in all cases mentioned in the table: Growth during famine phase consuming the PHA accumulated; Limitation of internal growth factors; Higher responsiveness of PHA producers to substrate addition.

3.1.1. Anaerobic/Aerobic Enrichments (AN/AE)

PHA production from MMC was first observed in biological phosphate removal by activated sludge in wastewater treatment plants, where aerobic and anaerobic steps alternate [90]. Thus, the first attempts of enriching a PHA storing community were performed by replicating those conditions. In these cases, Polyphosphate Accumulating Organisms (PAOs) and Glycogen Accumulating Organisms (GAOs) were described to accumulate PHA during the anaerobic phase, where the electron acceptor becomes limiting. In the aerobic phase, these microbes would consume the internally stored PHA, using the available oxygen, obtaining thus a higher adenosine triphosphate (ATP) yield compared to the substrate being metabolized anaerobically [102,104]. Nevertheless, as substrate catabolism and PHA formation, during the anaerobic phase, requires ATP and reducing equivalents, these microbes would also depend on the accumulation of glycogen or polyphosphate from the stored PHA during respiration, that limit the PHA production capacity of the cultures. Maximum PHA contents of around 20% had been reported from these consortia enriched under AN/AE conditions. However, in the late 1990s the aerobic dynamic feeding (ADF) enrichment was introduced, obtaining higher PHA contents [102–104], and thus, research efforts in AN/AE remained limited. Nevertheless, recent developments in PHA production following AN/AE enrichment, demonstrated that PHA might be also accumulated aerobically, without depending on glycogen and phosphate reserves, and high PHA storage capacities (up to 60%) could be obtained [105]. Thus, further research could prove the feasibility of AN/AE that would demonstrate benefits of saving energy costs, in terms of aeration requirements, compared to ADF.

3.1.2. Aerobic Dynamic Feeding (ADF)

In enrichment under ADF conditions the limiting factor, promoting PHA accumulation, is carbon substrate availability rather than the electron acceptor [104]. This process relies on subsequent feast/famine cycles where the culture is subjected initially to an excess of carbon source, and then submitted to carbon deficiency under aerobic conditions. Bacteria that are able to convert carbon to PHA during the feast phase have a competitive advantage towards the rest of the microbial population, as they utilize PHA as a carbon and energy reserve during the famine phase, allowing them to grow over non-PHA storing microorganisms [104,105]. Moreover, a limitation of internal factors, such as RNA and enzymes required for growth, seems to be crucial [102]. In order for cells to grow, a considerable amount of RNA and enzymes are needed, which might not be available after long starvation periods. Nevertheless PHA synthase, the key enzyme for PHA polymerization, is active during PHA production and degradation, generating a futile cycle that wastes ATP but enables the PHA mechanism to be ready when a sudden addition of carbon occurs, providing them with higher responsiveness [106–108]. In this way, a new competitive advantage of PHA producers arises, given that they can use PHA to regulate substrate consumption and growth [106]. PHA contents up to 90% of the dry cell weight have been reported using this strategy [89,109], higher than the ones reached following AN/AE enrichment [102–104].

Most of the studies performed within the last 10 years have been based on ADF enrichment. Apart from investigating the feasibility of this strategy using a variety of substrates, recently reviewed by Valentino and colleagues [97], the main focus has been on evaluating the impact of different parameters during the enrichment process. An overview on how those parameters influence PHA production from MMCs has been recently reported on, including the effect of hydraulic retention time (HRT), solids retention time (SRT), pH, temperature, nitrogen concentration, dissolved oxygen concentration (DO), cycle length, influent concentration, feast/famine ratio, and food/microbe ratio [110]. Regarding process configuration, a continuous system has been proposed where instead of an SBR, the feast and the famine phases were operated in separate CSTR [111]. Although no significant improvements were observed with respect to the conventional SBR configuration, the study demonstrated that successful enrichment was also possible in continuous mode, which is considered advantageous in the case of putative coupling to the following PHA accumulation step under continuous mode.

3.1.3. Variations of the ADF Enrichments

Even though microbial communities with high PHA-storing capacity have been obtained with several parameter combinations, the presence of non-PHA accumulating microorganisms is still not completely avoided. This is partially due to the presence of organic content other than VFA present in waste streams, allowing the growth of non-PHA accumulating bacteria [112]. A possible way to overcome the presence of such bacteria has been recently proposed where the culture was settled and the supernatant was discharged just after VFA depletion. In this way, consumption of the remaining organic matter—measured as chemical oxygen demand (COD)—and growth of side-population was prevented while the fraction of PHA-producers was considerably increased as verified by molecular techniques [113]. An increase in the PHA content from 48% to 70% was observed by applying this strategy. The authors suggested that, apart from the role of the remaining COD, also the increased cell density of PHA packed cells enhanced the enrichment, as those cells would have a higher tendency to settle after the feast phase. This observation coincided with the results obtained using only acetate as a carbon source, where the settling after the feast phase also increased the PHA accumulation capacity of the culture from 41% to 64–74% [114]. As no residual COD was present in those experiments, the effect could be entirely attributed to differences in the cell density that led to an additional physical selection.

Similarly, the growth of such non-PHA accumulating bacteria was observed to be restricted by applying nitrogen limitation [112]. Following this reasoning, a variation of the ADF was recently proposed, where also a nitrogen deficiency was imposed during the feast phase, while providing nitrogen during the famine phase to enable growth from the accumulated PHA. Thus, an uncoupling of the carbon and nitrogen took place in the SBR [2,86,115]. This strategy resulted in higher PHA contents at the end of the feast phase using synthetic VFA [2,115], cheese whey [86], and 1,3-propanediol from fermented crude glycerol [116] as a substrates. When a separate PHA accumulation was performed with the cultures of such enrichments, higher PHA yields and productivities were obtained with this strategy compared to carbon-nitrogen coupled ADF [86]. This strategy also allowed for a more stable system for long term operation. Nevertheless, given that the enrichment is already performed at conditions maximizing PHA accumulation, it was also suggested that a separate PHA accumulation might no longer be required if part of the biomass is already harvested after the feast phase, something that would significantly contribute towards the reduction of the costs of the process [115]. Moreover, given that there is already a selective pressure in the feast phase, the duration of the famine phase would be less important, so as to achieve an effective selection, and this could enable a reduction of its duration leading to enhanced process productivity [86].

3.1.4. ADF Enrichments in Halophilic Conditions

As previously discussed, the use of halophilic bacteria comes with various advantages. PHA production using halophilic bacterial populations can be performed using seawater instead of fresh or distilled water or using high salinity wastewater produced by several industries, namely food processing industries. In addition, halophilic bacteria can be lysed in distilled water thus reducing downstream processing costs due to lower quantities of solvents required. Enrichment of a halophilic PHA accumulating consortium under ADF conditions has been recently investigated using different carbon sources as substrate, which resulted in cellular PHA contents reaching up to 65% and 61% PHA using acetate and glucose respectively, demonstrating the potential of this strategy in MMC [117]. Recently, a previously enriched MMC fed with a mixture of VFAs containing 0.8 g/L Na^+ was examined regarding its PHA accumulation capacity under transient concentrations of 7, 13, and 20 g/L NaCl [118]. Since the particular MMC was not adapted to saline conditions, PHA accumulation capacities and rates decreased with higher NaCl concentrations while biopolymer composition was affected in terms of HB:HV ratio.

3.1.5. Mixed Photosynthetic Consortia

A new approach relying on the photosynthetic activity of mixed consortia has been explored recently [119–122]. Based on the previous observation of PHA production in photosynthetic strains, an illuminated SBR operating without aeration was proposed, eliminating the costly need for aeration during ADF enrichments. In such a system, photosynthetic bacteria uptake an external carbon source, in the form of acetate, during the feast phase using light as an energy source. PHB was produced at the same time as a sink of NADH, given that no electron acceptor was present. During the famine phase PHB was consumed using oxygen as an electron acceptor, which was not provided though aeration but produced by algae also present in the SBR. Under these conditions, up to 20% PHB was attained during the PHA accumulation step [119], which was also possible by utilizing other VFAs such as propionate and butyrate [121]. However, it was observed that a dark feast phase could also be envisioned, given that similarly to AN/AE enrichments, glycogen accumulation occurred during the famine phase, which was subsequently used as a complementary energy source to uptake acetate. With this SBR configuration, which would considerably reduce the need of illumination, a 30% PHA was accomplished [122].

The best PHA productivity though, was obtained in a system operating in a permanent illuminated feast phase (instead of successive feast-famine cycles) without oxygen supply [120]. This was based on the fact that photosynthetic accumulating bacteria out-compete other bacteria and algae without the need of transient presence of carbon source. Therefore, productivity was significantly increased due to the elimination of the famine phase and since there was no need for a separate PHA accumulation reactor. However, considerable input of light was required in order to obtain cultures with high PHA content (up to 60% of the dry weight), so the economic advantages of such systems should be further explored. Nevertheless, this process will allow significant savings in energy, since no sterilization and aeration are required, which will have an impact on the final price of the polymer.

3.1.6. Aerobic–Anoxic Enrichment Coupled with Nitrification/Denitrification

Basset and colleagues [123] developed a novel scheme for the treatment of municipal wastewater integrating nitrification/denitrification with the selection of PHA storing biomass, under an aerobic/anoxic and feast/famine regime. The process took place in a SBR (where NH_4^+ is converted into NO_3^- with a simultaneous selection of PHA storing biomass—and with COD being converted to PHA) and the subsequent PHA accumulation in a batch reactor (where PHA is consumed to allow denitrification, under famine–anoxic conditions, without the need of external addition of organic matter). The carbon source added during the selection and accumulation steps consisted of fermentation liquid from the Organic Fraction of Municipal Solids Waste (OFMSW) and primary sludge fermentation liquor.

The advantage of this approach is that the potential for recovering biopolymers from wastewater presents particular interest, when the latter is integrated within the normal operation of the plant. An important benefit of this strategy is that anoxic denitrification usually requires a carbon source, which at that point is usually low and already consumed in a wastewater treatment plant (WWTP). In this case however, it can occur without external addition of carbon source by using those stored internally in the form of PHA.

Results showed that during SBR operation ammonium oxidation to nitrite reached on average $93.4 \pm 5.25\%$. The overall nitrogen removal was 98% (resulting in an effluent with only 0.8 mg NH_4^- N L^{-1}). Similar results were obtained by Morgan-Sagastume et al. [124]. Denitrification efficiency and rate did not seem to be affected by the carbon source. When sufficient PHA amount was available, denitrification of all available nitrate was observed. COD removal reached up to 70% when DO level was higher (2–3 mg L^{-1}). PHA content decreased during nitrification due to the lack of external COD. However, after complete nitrification, there was enough PHA to carry out the denitrification process. Even though biomass was rich in PHA storing bacteria, PHA accumulation reached only 6.2% during the feast phase (first 10 min) and it was progressively consumed before the initiation of the anoxic phase to 2.3%, which was enough to complete the subsequent denitrification. Nevertheless, the selection of PHA storing biomass under feast (aerobic)–famine (anoxic) conditions required less DO compared to the typical feast–famine regime carried out under continuous aerobic conditions, leading to a reduction of 40% of the energy demand.

The PHA accumulation capacity of the biomass, previously selected in the SBR, was further evaluated in accumulation batch reactors with the use of OFMSW, and primary sludge fermentation liquid. After 8 h of accumulation with OFMSW, the stored PHA was 10.6% (wt.). In the case where fermented sludge liquid and OFMSW was used as carbon source, the contribution to growth was higher, due to the elevated nutrient content, and the lower VFA/COD ratio but only 8.6% PHA was accumulated after 8 h. The carbon source was proven to play an important role in the PHA accumulation step as the presence of non-VFA COD contributed to the growth of non-PHA-storing biomass [124,125]. PHA storage yields could be potentially improved with a more efficient solid–liquid separation after the fermentation process.

3.1.7. Anoxic–Aerobic Strategy Coupled with Nitrification/Denitrification

Anoxic–aerobic enrichments coupled with denitrification, where nitrate is used as electron acceptor during PHA accumulation, has been already explored using synthetic VFA since early 2000s [103]. A recent study applying this strategy has been reported which investigated the use of the condensate and wash water from a sugar factory [126]. Furthermore, they considered that, in combination with harvesting enriched biomass from the process water treatment, side-streams could be exploited as a substrate for PHA accumulation. This approach (together with the aerobic–anoxic strategy shown in Section 3.1.6) would have the big advantage of significantly reducing the PHA production costs, through the integration of already existing full-scale WWTP and reduction of aeration needs.

In that study [126], they used parallel SBRs fed alternatively with condensate and wash water, developing a microbial consortium that removes inorganic nitrogen by aerobic and anoxic bioprocess steps of nitrification and denitrification. Alternating bioprocess conditions of anoxic feast (supporting denitrification) and aerobic famine (supporting nitrification) in mixed open cultures was expected to furnish a biomass with stable PHA accumulating potential characterized by its ability to remove carbon, nitrogen, and phosphorus in biological processes. In more detail, one laboratory SBR was operated with suspended activated sludge (AS) and long SRT, similar to the full-scale (SRT > 6 days), while the other SBR was a hybrid suspended activated sludge and moving bed biofilm reactor (MBBR) with short SRT of 4–6 days. MBBR technology employs thousands of polyethylene biofilm carriers operating in mixed motion within an aerated wastewater treatment basin. Therefore, the MBBR-SBR was used as a means for maintaining nitrifying activity while enabling enrichment of biomass at relatively low SRT.

The results showed a COD removal performance of 94% and 96% for AS- and MBBR-SBRs, respectively. Full nitrification was achieved in both systems, with exception of periods showing phosphorous or mineral trace element limitation. Soluble nitrogen removal reached 80 $\pm$ 21% and 83 $\pm$ 11% for AS- and MBBR-SBRs, respectively. MBBR-SBR showed more stable performance under lab-scale operation. The process achieved a PHA content of 60% g PHA/g VSS in both cases. A significant advantage was the possibility of lowering the SRT while maintaining a robust nitrification activity and improving the removal of soluble phosphorus from the process water.

3.1.8. Microaerophilic Conditions

In 1998, Satoh and colleagues [127] investigated the feasibility of activated sludge (from laboratory scale anaerobic–aerobic reactors) for the production of PHA, by optimizing the DO concentration provided to the system. They were able to obtain a PHA content of around 20% in anaerobic conditions and 33% under aerobic conditions. When applying a microaerophilic–aerobic process, by supplying a limited amount of oxygen into the anaerobic zone, they were able to increase the PHA accumulation to 62% of sludge dry weight.

PHA production using palm oil mill effluent (POME) was investigated by Din and colleagues [128], using a laboratory SBR system under aerobic feeding conditions. The microorganisms were grown in serial configuration under non-limiting conditions for biomass growth, whereas in the parallel configuration the nutrient presence was controlled so as to minimize biomass growth in favor of intracellular PHA production. PHA production under aerobic, anoxic, and microaerophilic conditions was investigated and it was shown that PHA concentration and content increased rapidly at the early stages of oxygen limitation while the production rate was reduced at a later stage implying that oxygen limitation would be more advantageous in the PHA accumulation step.

Another interesting study was published by Pratt and colleagues [129], where the effect of microaerophilic conditions was evaluated during the accumulation phase, using an enriched PHA culture, harvested from a SBR fed with fermented dairy waste. Batch experiments were conducted to examine the effect of DO on PHA storage and biomass growth. The results showed that in microaerophilic conditions a higher fraction of substrate was accumulated as PHA, compared to high DO conditions. Also, the intracellular PHA content was 50% higher during early accumulation phase. Interestingly, the accumulation capacity was not affected by the DO, despite its influence on biomass growth. The PHA content in both low and high DO concentrations reached approximately 35%. However, the time needed to achieve maximum PHA content at low DO level was three times longer than in the case of high DO concentration. The reason why PHA accumulation was proven to be less sensitive to DO, compared to its effect on biomass growth, was explained by the fact that low DO levels limit the availability of ATP, while high DO supply provides surplus ATP and high growth rates (and consequently reduced PHA yield). In addition, when MMCs were fed with multiple VFAs (acetate, propionate, butyrate, and valerate) it was also shown that, during PHA accumulation, high DO concentration is required to reach maximum PHA accumulation rates due to low specific VFA uptake rates under low DO levels [130].

The effect of dual nitrogen and DO limitation has been also investigated in MMCs fed with a VFA mixture of acetate, propionate, and butyrate and acidified OMW [10,99]. As discussed above, it was shown that during the PHA accumulation step, under batch mode, lower substrate uptake and PHA production rates were obtained compared to assays performed under nitrogen limitation. Moreover, PHA accumulation percentages and the yield coefficient $Y_{PHA/S}$ was lower in the case of dual limitation, while the accumulation of non-PHA polymers within the cells was indicated.

Those reports demonstrate that manipulating oxygen concentration could influence growth and PHA storage. Manipulating DO instead of limiting nitrogen or phosphate availability could represent a significant opportunity for PHA production processes that utilize nutrient rich feedstocks. A major advantage of operating at low DO is the reduced aeration requirements leading to reduction of operating costs. However, this advantage can be countered by the fact that PHA accumulation in low DO environments can be significantly slower.

3.1.9. PHA Accumulation without Previous Enrichment

The three-step process described in Figure 1, consisting in an enrichment step followed by an accumulation step, has been proven efficient to obtain cultures with high PHA contents. Nevertheless, several authors have put in doubt the alternative of having a separate enrichment step. The main reason is that during the enrichment, PHA is produced, but it is also allowed to be consumed to drive the selection. Thus, this step consumes substrate without leading to any net production of PHA, lowering the overall PHA/Substrate yields of the system [97,131]. Thus, skipping this step could imply considerable improvements on those parameters.

Already in 2002, PHA accumulation without a previous enrichment step using activated sludge was reported to obtain PHA contents up to 30% [132]. Later on, fed-batch cultivation under nitrogen limiting conditions was reported to obtain up to 57% PHA [133]. Cavaillé and colleagues performed fed-batch PHA accumulation experiment as well using activated sludge without previous enrichment, but applying phosphorous limitation and achieved up to 70% PHA [134]. Substrate to PHA yield reached up to 0.2 Cmol PHA/Cmol S using acetic acid. This yield was considerably lower than that obtained in PHA accumulation steps submitted to a previous enrichment step (up to 0.9 Cmol PHA/Cmol S), but comparable to the overall yields of enrichment and accumulation strategies summed up [134]. Moreover, they further developed the system into a continuous process, and attained a stable operation where the cells in the effluent contained 74% PHA [135]. Their findings also evidenced that the continuous system was not stable at severely phosphorous limited experiments, because the growth rate could not be maintained and the cells were washed out of the reactor. The key was that differently from when a separate enrichment is performed, the continuous PHA accumulation without previous enrichment relied on the occurrence of both growth and storage responses. Moreover, the authors suggested that phosphorous limitation might offer more flexibility than nitrogen limitation when both PHA formation and growth are a goal, given that phosphorous is less needed than nitrogen for growth-related metabolism [135].

3.2. PHA Accumulation

As in the enrichment step, several operational parameters such the temperature [136] and the pH have an impact on PHA accumulation [2,137–139]. Nonetheless, the most critical aspect during PHA accumulation experiments is the cultivation strategy employed. High substrate concentrations supplied under batch mode should be avoided since they can cause inhibition and thus limit PHA productivity [105,140]. In order to circumvent that, several fed-batch strategies have been suggested. Pulsed fed-batch cultivation has been suggested when synthetic VFA mixtures are used [140,141]. However, due to an increase occurring in the working volume after the addition of substrate the feed should be very concentrated. Nevertheless, this is rarely the case with fermentation effluents, since they usually do not exceed 20 g COD/L [97]. Discharge of the exhaust supernatant has been suggested as an alternative [137,142,143] yet, this approach requires a settling step between batches, which severely limits the productivity [137].

Continuous feeding processes have shown the best results until now, given that they can attain a sustained productivity [89,137,143,144]. Pulsed fed-batch production may result in high PHA productivity but as the substrate is being consumed, PHA productivity eventually decreases. This phenomenon might be avoided by supplying substrate in a continuous manner. Continuous substrate addition has been successfully performed using the pH as an indicator, given the pH increases with VFA consumption [89,137,143,144]. On the other hand, less successful results have been obtained when the substrate was supplied, taking into account previously observed substrate uptake rates resulting in either accumulation or limitation of substrate in the reactor [140,141]. Alternatively, an on-demand continuous addition of substrate, based on change of DO, has been proven efficient to maintain optimal amounts of carbon substrate in the reactor [145].

High productivities up to 1.2 g PHA/L h combined with high PHA yields, 0.8 Cmol PHA/Cmol S, have been achieved with continuous-feeding systems [137]. However, similarly to the pulsed fed-batch, such values have been reported only when synthetic substrates of high concentration were present in the feed. Much lower values are reported in real substrates due to the diluted nature of these substrates and the consequent increment in reactor volumes [97]. A way to overcome this could be the development of a continuous feeding scheme for PHA accumulation under low biomass loading rates (3.5–5.5 Cmol VFA/Cmol X/d). So far, this venture has only been investigated once using MMC [139]. The authors suggested a PHA accumulation reactor operating under continuous mode, were the effluent was allowed to settle and the resulting sludge recycled back to the reaction vessel. The system worked with a rather diluted effluent in the feed (around 100 Cmol VFA) and was shown to obtain higher specific productivities than the pulse-fed-batch. Nevertheless, overall productivity of the system was not reported in that study. It is worth noting that the system was not coupled to an SBR operating in continuous mode, so the operation of the reactor was also for a limited period of time.

Regarding the nutrient availability, nitrogen limitation or deficiency is usually reported to improve the PHA yield and content [2,105,146,147]. On the other hand, several studies have concluded that the role of nitrogen during the accumulation step is secondary since PHA storage was preferred over growth regardless of the nitrogen concentration [144,148]. Moreover, in another study it was observed that nitrogen limitation did not enhance the PHA accumulation [149]. As a matter of fact, the main reason for limiting the amount nitrogen is to prevent bacterial growth of non PHA-accumulating bacteria [112]. Hence, different observations from different cultures do not imply contradictions but highlight the fact that the requirement for nitrogen limitation, in order to obtain high PHA contents, is highly dependent on the composition of the enriched culture and the type of substrate fed. In terms of productivity though, nutrient limitation rather than deficiency was reported to show higher productivities [145]. According to a certain study, the absence of an essential nutrient for growth leads to cellular PHA saturation, while nutrient limitation allows cells to duplicate prolonging PHA accumulation without enabling excessive growth response. In addition, it was shown that the best productivities were obtained from dual limitation of nitrogen and phosphorous.

3.3. Pilot Scale Experiences

Several industrial/agro-industrial effluents and residues have been investigated so far as potential feedstocks for PHA production. Effluents, rich in sugars, glycerol, and/or fatty acids were either directly used for the selection of PHA accumulating MMCs or were previously fermented for the conversion of carbohydrates to VFAs, the preferable precursors for PHA production using MMCs. Numerous studies on PHA production from industrial effluents have been performed, that have been recently reviewed by Valentino et al. [97]. On the other hand, studies on PHA production in pilot-scale by MMCs are rather scarce and relevant efforts have recently started, in 2010s, while no full-scale production of PHA by MMC exists yet.

A common feature in all pilot-scale studies is that effluents/feedstocks were always fermented prior to PHA production. Also, most efforts on PHA production in pilot-scale focus on integrating and combining PHA production with existing processes in wastewater treatment plants so as to reduce the production cost by exploiting the available infrastructure as much as possible. In this context, anoxic/aerobic MMC selection regimes can be coupled to nitrification and denitrification activities despite the fact that the highest PHA yields and cell content in PHA are usually reported for aerobic dynamic feeding selection regimes. Another tendency in pilot-scale studies is the use of a different effluent for the MMC enrichment than the one fed during the PHA accumulation step. The first pilot-scale study, concerning MMC PHA production, was performed from pre-fermented milk and ice-cream processing wastewater, as reported by Chakravarty et al. [150], with a PHA content of 43% and a PHA yield of 0.25 kg PHA/kg COD being obtained. In 2014, Jia et al. [125] studied the production of PHA in pilot-scale with pre-hydrolyzed and fermented raw excess sludge. In both studies, activated sludge was used as the raw material for its enrichment to PHA accumulating microorganisms. A series of pilot-scale studies was conducted and published with the participation of Anoxkaldnes and Veolia Water Technologies [126,151,152]. In the study of Morgan-Sagastume et al. [151], the potential of waste sludge, generated in wastewater treatment plants as a feedstock for PHA production was evaluated. This was done in the context of integrating PHA production in existent WWTP valorizing at the same time the excess sludge that in general represents a burden for further treatment and disposal. A very interesting point in the study of Bengtsson et al. [152] is that denitrifying microbial biomass was also selected towards high PHA producing potential by applying an anoxic-feast and aerobic-famine selection pattern and therefore the process comprised nitrification and denitrification steps followed by accumulation of PHA. Tamis et al. [153] investigated PHA production from fermented wastewater deriving from a candy bar factory. Activated sludge was enriched in PHA accumulating microorganisms under aerobic feast and famine regime and the obtained PHA content was the highest reported so far in pilot-scale at 70–76%. Table 4 summarizes the main features of the pilot-scale studies published so far. Overall, as it regards pilot-scale studies, their performance cannot really be compared to respective lab-scale studies but they provide valuable information on PHA formation under variable feedstock characteristics and allow production of significant amounts of polymer that can be processed for a full characterization. The variation of feedstock composition combined with the oxygen mass transfer limitations occurring at a larger scale, could be the main reason why PHA yields and cell content in pilot-scale studies are in general lower than the ones reported in lab-scale experiments.

Table 4. Main characteristics of PHA production in pilot-scale.

Pilot Plant, Location	Feedstock	Origin of MMC and Enrichment Strategy	$Y_{P/S}$ (g/g) *	PHA % (%mol HB: %mol HV)	mg PHA/g X/h	Ref.
Nagpur, India	Pre-fermented milk and ice cream processing wastewater	Activated sludge	0.425 *	39–43		[150]
Lucun WWTP in Wuxi, China	Hydrolyzed and acidified raw excess sludge	Activated sludge/synthetic mixture of VFA, ADF feast famine with carbon limitation and inhibitor of nitrification	0.044–0.29 *		2.06–39.31	[125]
Eslöv, Sweden	Beet process water, 38% in VFA	PHA producing MMC from pre-fermented effluent of Procordia Foods		60 (85:15 HB:HV)		[126]
Brussels North WWTP (Aquiris, Belgium)	Pre-hydrolyzed and fermented WWTP sludge	Sludge fed with municipal WW under aerobic feast famine	0.25–0.38	27–38 (66–74:26–34 HB:HV)	100–140	[151]
Leeuwarden WWTP, Friesland, Netherlands	Fermented residuals from green-house tomato production	Sludge fed with municipal WW under anoxic feast/aerobic famine	0.30–0.39	34–42 (51–58:42:49 HB:HV)	28–35	[152]
Mars company, Veghel, Netherlands	Fermented wastewater from a candy bar factory	Activated sludge from a WWTP fed with the fermented wastewater under aerobic feast/famine with inhibitor of nitrification	0.30	70–76 (84:16 HB:HV)		[153]

* Yield calculated on a COD basis by using the coefficients: for HB: 1.67 g COD PHA/g PHA and for HV: 1.92 g COD PHA/g PHA.

3.4. Challenges and Perspectives Regarding PHA Production by Mixed Microbial Consortia

PHA process brought to an industrial scale using pure substrates and strains has broadly surpassed cell densities of 150 g/L with PHA contents up to 90% and productivities in the range from 1 to 3 g PHA/L h [4]. Comparable productivities, of up to 1.2 g PHA/L h, combined with high PHA yields, up to 0.8 Cmol PHA/Cmol S, have already been attained from MMCs using synthetic substrates [137]. Likewise, PHA contents up to 90% of the CDW have been reported [109]. Thus, MMC have proven to be able to achieve comparable results to pure cultures in synthetic media. Nevertheless, PHA productivities could be further increased if higher cell densities were obtained, a parameter that is usually below 10 g/L in MMC. In pure cultures, the PHA accumulation phase is preceded by a biomass growth phase in order to achieve high biomass densities [154]. In the accumulation phase, the feeding strategy is modified accordingly (usually limiting the nitrogen) to obtain a high PHA content. Nevertheless, in processes for PHA production from MMC, the biomass generation step is also the enrichment step. Thus, two objectives, which might not have the same optimal conditions, are combined in the same process unit. A future direction could be to test if adopting a microbial biomass generation step—leading to higher cell densities before the PHA accumulation—could maintain the high PHA content of the cells in MMC.

In order to achieve a sustainable production of PHA, both in economic and environmental terms, high productivities should also be obtained in waste substrates. In the current state of art, neither pure strains nor MMC have achieved high productivities when waste streams are used as substrates [83,97,155]. Thus, the challenge is common.

One of the main issues that compromises productivity when using waste streams is their diluted nature [155,156]. This applies for the case of MMC where an anaerobic fermentation step is usually performed in order to convert sugars to fatty acids. However, it is also the case with other industrial wastes used both in pure and mixed cultures, such as whey or lignocellulosic biomass, that requires pre-processing to release its sugars [83,155]. When these effluents are provided as feed in fed-batch PHA accumulations, they provoke substantial increases in the reactor operating volume, thus reducing the productivity.

A promising way to obtain high cell densities and productivities would be the use of cell recycling systems coupled to fed-batch processes [156]. This strategy has been scarcely applied to PHA production until now, with only one report using an MMC [139] and one study in pure strains [157]. The latter obtained cell densities up to 200 g/L with a productivity of 4.6 g PHA/L/h by using external cross-flow membranes to recirculate cells into the fed-batch reactor using *C. necator* [157]. Thus, the strategy seems to offer good opportunities to increase the cell density and PHA concentration. Likewise, reactor designs preventing the cells from escaping the system, while allowing supernatant removal, could result in high cell densities and reduced reactor volumes.

With the same scope, other research groups have proposed influent concentration. Although evaporation has been suggested [157], other less energy intensive methods such as forward osmosis membranes could more likely be applied [158,159]. A very interesting approach was recently published where forward osmosis Aquaporin® membranes, mimicking biological protein channels, were suggested to concentrate fermentation effluents from glycerol and wheat straw [160]. The novelty lied in the fact that the concentrated feedstock could be used as the water draw solution with the diluted fermentation effluent being the water feed solution. This enabled the recirculation of water from the effluent to the influent in an energy efficient way since the process was based solely on the use of forward osmosis membranes without the need of the costly regeneration of the draw solution (usually performed by applying the energy intensive reverse osmosis). Integration of such systems in the PHA production process could also enhance its productivity.

4. PHA Recovery

The development of new strategies and methodologies for PHA recovery is one of the main factors associated with the feasibility of a PHA production bio-refinery using microbial mixed fermentation. As demonstrated in Figure 2, PHA recovery uses a variation of different techniques. However, PHA purification generally requires five steps: biomass-harvesting, pre-treatment, PHA recovery, PHA accumulation, polishing, and drying. Biomass harvesting is the concentration of biomass using techniques such as filtration or centrifugation. As PHAs are intracellular polymers, it is necessary to concentrate the biomass prior PHA recovery. Nevertheless, some researchers have evaluated PHA recovery without biomass harvesting to facilitate process scale-up and to reduce costs. Pre-treatment's main objective is to facilitate PHA retrieval from the microbial biomass; these techniques include drying techniques (lyophilization and thermal drying), grinding, chemical and biochemical pre-treatments, etc. The pre-treatment step can combine two or more methods. PHA retrieval phase utilizes two principal methods: PHA solubilization and the disruption of non-PHA cell mass (NPCM). In some cases, NPCM disruption precedes a PHA solubilization step. The PHA accumulation step is dependent on the retrieval technique utilized. In PHA solubilization, the PHA is concentrated by using alcohols precipitation. On the other hand, in NPCM disruption, recovery is performed by collecting the PHA granules. As final steps, recovered PHAs can be polished by removing residues, from the previous steps, or can be dried; depending on the separation steps utilized. As of 2013, two reviews specialized in PHA recovery were published [161,162]. This section primarily focuses in the most recent developments in pre-treatments and PHA retrieval steps; additionally, this section describes a more industrial opinion on the PHA recovery methods.

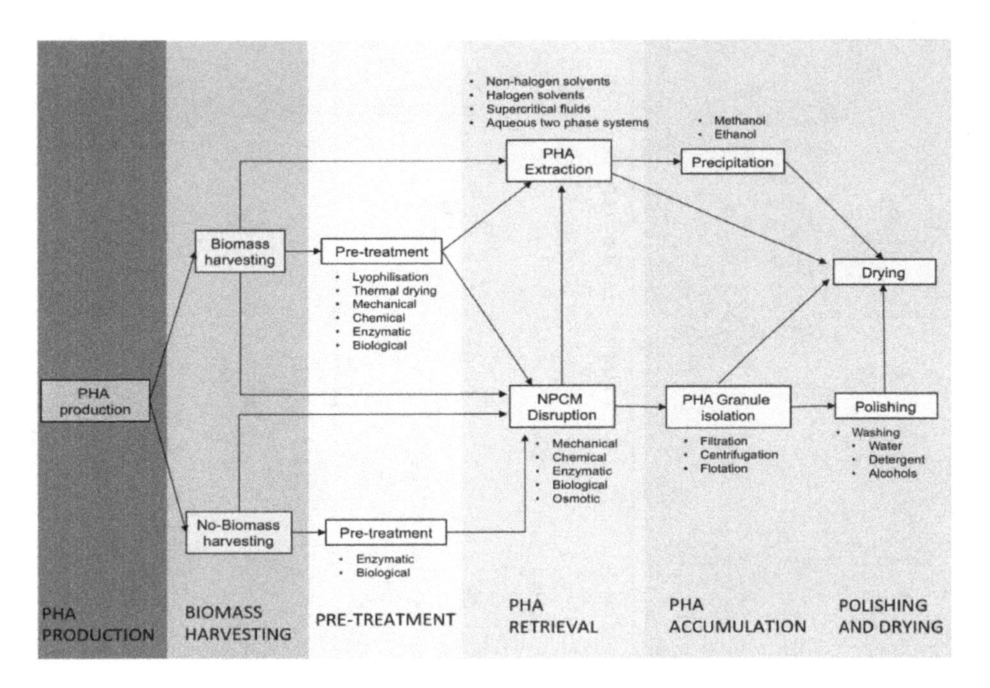

Figure 2. PHA recovery strategies.

4.1. Pre-Treatments

Pre-treatments are chemical, physical, or biological methods employed to facilitate the retrieval of PHA. These methods focus on weakening the cell structure that protects and surrounds the PHA granules. After biomass harvesting, drying is the traditional pre-treatment used in PHA recovery, which includes heat drying and lyophilization; the latter is the most employed pre-treatment in PHA recovery processes. This technique removes the majority of water molecules in the biomass facilitating the posterior PHA extraction. Although lyophilization has interesting features, it has economic and technical difficulties that reduce its future application in an industrial PHA recovery process. In recent years, these industrial difficulties increased the interest in PHA recovery from wet biomass instead of dry biomass [163–165]. Lyophilization can be a unique pre-treatment or it can precede a further pre-treatment. Lyophilization has preceded thermal, mechanical, and chemical pre-treatments (see Table 5). All these treatments included a further retrieval technique associated with solvent extraction [166,167]. Samori et al. [166], described chemical pre-treatment with NaClO as the best pre-treatment for PHA recovery compared with thermal and mechanical pre-treatments; however, this pre-treatment generated an important reduction in the molecular weight of the polymer.

Pre-treatments without lyophilization consisted of chemical and physical methods, with chemical methods including sodium chloride (NaCl) and sodium hypochlorite (NaClO). Anis et al. [163], employed NaCl as a pre-treatment for a NaClO digestion. The additional pre-treatment step generated an increment in purity and yield. NaCl pre-treatment modifies the osmotic conditions in the medium leading the cells to dehydrate and shrink, this destabilizes the cell membranes facilitating the PHA granules' liberation. Physical methods include high temperature and ultrasonication methods both methods anteceded NPCM digestion methods. Neves and Muller [168] evaluated three temperatures (121, 100, and 80 °C) during 15 min as pre-treatment conditions. The 121 °C treatment achieved the best results; whereas, 100 and 80 °C treatments recovered significantly lower amounts of PHA. The heat treatment improves the PHA removal by denaturizing proteins, DNA and RNA, destabilizing the microbial cell wall and inactivating the PHB depolymerase [168]. Ultrasonication employs sound waves to create disruption in the cell wall and open the cytosolic material to the aqueous medium. Leong et al. [164] utilized ultrasonication prior an aqueous two-phase extraction. The advantages of this method are the lack of any previous cell harvesting method and the fast pace in which is performed.

Table 5. Pre-treatment techniques applied for PHA recovery.

Pre-Treatment	Further PHA Retrieval Treatment	Pre-Treatment Conditions	Purity (%)	Recovery (%)	Ref.
Sodium chloride (NaCl)	NaOH digestion	NaCl (8 g/L), 30 °C, 3 h	97.7	97.5	[163]
Ultrasonication	Aqueous two-phase extraction	Ultrasonication at 30 kHz per cycle 15 min		-	[164]
Sodium hypochlorite (NaClO)	Non-halogenated solvent extraction	NaClO (10%), 37 °C, 1 h		-	[165]
Thermal pre-treatment	Enzymatic digestion and chloroform extraction	Autoclave, 15 min 121 °C		94.1	[168]
Thermal pre-treatment [1]	Non-halogenated solvent extraction	150 °C, 24 h		50	[166]
Ultrasonication and glass beds [1]	Non-halogenated solvent extraction	Glass beads (0.5 mm) and Ultrasonication (10 pulses of 2 min)		50	[166]
Sodium hypochlorite [1]	Non-halogenated solvent extraction	NaClO (5%), 100 °C, 15 min	93	82	[166]
Hot acetone [1]	Non-halogenated solvent extraction	Acetone, 100 °C, 30 min		-	[167]

[1] The pre-treatment included a previous lyophilization step.

Pre-treatments can increase the recovery and purity of the PHA extracted from a fermentation broth. However, their implementation in a PHA industrial recovery process needs to be evaluated using economic and technical analysis. The yield and purity increment should counterbalance with the additional cost associated with the introduction of this step in the purification line. Additionally, it is important to remark that even though several pre-treatments can be used, not all of them are suitable for industrial applications. Industrially suitable pre-treatments require the use of wet biomass or unharvested biomass in order to reduce the number of purification steps and the costs associated with the purification process. Lyophilization use should concentrate on PHA chemical analyses or be replaced with more suitable drying techniques for process scale up. The selection of a pre-treatment is dependent of the bacterial strain, fermentation broth characteristics, and further PHA application; therefore, each PHA process needs to be analyzed individually from an economic, environmental, and technical point of view.

4.2. Retrieval Techniques

4.2.1. Non-PHA Cell Mass (NPCM) Disruption

In recent years, the use of NPCM disruption as a tool for retrieving PHA from bacterial biomass has increased. This increment is associated with the necessity of environmental and safe options to replace the use of halogenated solvents used in the traditional PHA extraction methods. Additionally, some of the NPCM techniques for the PHA extraction have used wet biomass or unharvested biomass, which is an advance through the reduction of purification steps and costs. This review grouped the novel NPCM disruption techniques into chemical, enzymatic, and biological disruption.

Chemical Disruption

NPCM chemical disruption includes methods that utilize chemical compounds to disrupt bacterial cell wall and the denaturalization or degradation of cytosolic material. The three principal methods for the chemical disruption are sodium hydroxide (NaOH), sodium hypochlorite (NaClO), and sodium dodecyl sulphate (SDS). Additional chemical treatments include water and acid treatments. NaOH treatment destabilizes the cell wall by reacting with the lipid layer in a saponification reaction and increases the cell membrane permeability [163]. NaOH treatment has obtained high recovery and purification percentage using pre-treated and unpretreated biomass (see Table 6). However, pre-treated biomass aided the NaOH treatments to achieve improvements in purity and recovery [163,169]. The combination of NaCl pre-treatment and NaOH digestion increased the purity and recovery of NaOH treatment by approximately 10% [163]. Similarly, lyophilization and freezing helped to increase the PHA purity using NaOH; however, these pre-treatments did not increase the recovery percentage [169]. López-Abelairas et al. [170] described a recovery reduction in treatments with high biomass and NaOH concentrations. Biomass concentration above 2.5% created a constant reduction in recovery and purity; however, biomass concentrations of 7.5% and 10% yielded the greatest recovery reductions. NaOH concentrations over 0.5 N affected recovery, although, the purity percentage at high percentage was constant. Likewise, Villano et al. [171] achieved recovery between 80–87% utilizing high NaOH (1 M), high biomass concentration (6:1 biomass:chemical solution) and extended extraction times (6–24 h). However, the purity achieved by this treatment was below 60%.

Sodium hypochlorite (NaClO) has confirmed its positive aspects as an NCMP disruption treatment [161,162]. Therefore, in recent years, the assessments of this treatment advanced to larger scales or continuous processes [171,172]. A continuous sequential process for PHA production and recovery obtained high polymer recovery (100%) and purity (98%). This continuous process contains three steps: a production step using microbial mixed culture, a PHA accumulation step, and a PHA extraction step using NaClO (5%) disruption for 24 h. This approach produced 1.43 g PHA/L·d and was stable for four months [171].

Another chemical disruption treatment is sodium dodecyl sulphate (SDS). This surfactant is a well-known detergent used in the recovery of genetic material. SDS treatment obtained recovery and purity values comparable with other chemical disruption techniques; moreover, this disruption technique obtained similar retrieval with and without biomass pre-treatment [169,173]. The amount of SDS varied between 0.025% and 0.2%, higher concentrations of SDS generated higher purity as result of SDS micelles formation. High micelles production is also associated with the solubilization of PHA granules generating a reduction in the recovery yield. SDS has complemented NaOH disruption, this combination exhibited superior levels of purity, especially in the removal of hydrophobic impurities [169].

Besides the previous treatments, other authors have described water and acid disruption methods as effective treatments for NPCM disruption. Distilled water achieved high purity (94%) and recovery (98%) percentages; however, the process needed 18 h and lyophilized biomass to reach these high percentages. The process duration improved by adding SDS (0.1%) into the mixture [173]. In contrast, distilled water disruption treatment with wet biomass obtained recovery (80%) and purity (58%) percentages lower than lyophilized biomass. Mohammadi et al. [174], described a higher purity and recovery yield using distilled water disruption with recombinant bacteria instead of wild bacteria. Recombinant bacterial cell wall is thinner than in wild type bacteria, which facilitates the cell wall breaking by osmotic pressure. Acid treatments have demonstrated their capability to disrupt NPCM; López-Abelairas et al. [170] described a recovery and purity percentage using a sulphuric acid solution (0.64 M) similar to alkaline treatments (NaOH, NaClO). They selected acid disruption as the best recovery method focused in operational and environmental factors. The authors chose acid disruption because this process had lower cost, environmental impact (greenhouse gas emissions), and polymer degradation than alkaline treatments [170].

Chemical disruption treatments are a significant option in PHA recovery since they present environmental and economic advantages over the traditional PHA extraction using halogenated solvents. Environmentally, chemical disruption avoids the use of toxic solvents such as chloroform. Economic advantages include liquid current recycling, the use of wet biomass, and the reagent cost. The principal drawback for chemical disruption has been polymer degradation; however, the use of mixtures of chemicals and process optimization has reduced polymer degradation. The selection of a chemical disruption method for PHA recovery from mixed microbial cultures needs an all-around evaluation of technical, economic, and environmental factors that consider the positive and negative effects and how they can affect the feasibility of a PHA bio-refinery plant.

Enzymatic NPCM Disruption

Enzymatic NPCM disruption utilizes purified enzymes or crude extracts to disrupt the bacterial cell wall. Proteases are the principal enzymatic activities employed in enzymatic disruption; however, other types of enzymes or enzymatic cocktail have effectively degraded NCMP. Enzymatic disruption advantages include their low energy requirements, aqueous recovery, and low capital investment; in contrast, the enzymes production cost is the principal disadvantage for industrial implementation [175]. Gutt et al. [176], evaluated the recovery of P3HBHV from *Cupriavidus necator* by several methods including enzymatic disruption. Simple enzymatic treatment (lysozyme) obtained low recovery and purity (Table 6). The authors attributed these low percentages to the absence of additional chemical or mechanical treatments, which have proved necessary for achieving high recovery and purity. Martino et al. [177] evaluated enzymatic disruption using simultaneously enzymatic (Alcalase) and chemical treatments (SDS and EDTA). SDS and EDTA contributed to cell wall and membrane lysis whereas Alcalase solubilized the cytosolic material. This enzymatic/chemical digestion treatment eliminated the requirement of heat pre-treatment used in previous enzymatic disruption researchers [161,162]. Kachrimanidou et al. [175], developed a novel enzymatic disruption method by using crude enzymes from solid-state fermentation of *Aspergillus awamori*. This method achieved good recovery (98%) and purity (97%) without using additional chemicals; however, it required heat and lyophilization as pre-treatments. Approaches focused in the reduction of enzyme costs are necessary to facilitate the industrial application of enzymatic disruption including the use of immobilized enzymes, integration of enzymes production as part of a PHA biorefinery, and genetically engineering enzymes.

Biological NPCM Disruption

Biological disruption utilizes biological agents (virus) or organisms to liberate PHA from bacterial cells. The first biological disruption technique used viral particles to break bacterial cells. Bacteriophages were included in bacterial lines to utilize the viruses' lytic cycle to liberate PHA granules. When the lytic cycle is completed, the virus escapes from the host cell by breaking down the cell wall; this breaking down also liberates the PHA particles allowing their recovery [162]. In recent years, biological disruption methods included bacteria predators, rats, and mealworms [178–180].

Martinez et al. [181], proposed the use of obligate predatory bacteria *Bdellovibrio bacteriovorus* as an innovative cell lytic agent suitable to recover intracellular bioproducts such as PHA. *B. bacteriovorus* achieved a PHA recovery of 60%, the recovery percentage obtained was attributed to PhaZ depolymerase activity which hydrolyses PHA and expresses during all the stages of *B. bacteriovorus*'s life cycle [181]. To improve the use of *B. bacteriovorus* in PHA recovery, Martinez et al. [180], developed *B. bacteriovorus* mutant strains, one with an inactive medium-chain-length PHA depolymerase (*B. bacteriovorus* Bd3709) and another with inactive short-chain-length-PHA depolymerase (*B. bacteriovorus* Bd2637). *B. bacteriovorus* Bd3709 increased PHA recovery from 60% to 80% when predating *Pseudomonas putida*, whereas, *B. bacteriovorus* Bd2637 increased PHB recovery from 48% to 63% when predating PHB-accumulating *E. coli* ML35 [180]. Biological recovery using *B. bacteriovorus* has advantages such as avoiding the use of cell harvesting and pre-treatments. However, it has disadvantages such as the use of organic solvent steps, processing time, and low recovery. In the future, this biological method can be complemented with other recovery treatments to avoid solvents' usage and to increase PHA recovery.

In recent years, authors have used complex organisms' digestive system as a NPCM disruption technique. In these treatments, different organisms were fed with PHA-rich bacteria; afterwards, PHA granules were recovered from these organism's feces. This process selectively digested the NPCM without reducing the PHA molecular weight. Kunasundari et al. [182], fed Sprague Dawley rats with lyophilized cells of *Cupriavidus necator* in a single cell protein diet during several days. The fecal pellets were whitish and rich in PHA (82–97%). The authors also demonstrated the safety and tolerability of the Sprague Dawley rats to a *Cupriavidus necator* diet. Kunasundari et al. [179], studied the purification of the Sprague Dawley PHA-rich fecal pellets using water and surfactants. The use of SDS 2% as a further purification step increased the purity of the PHA biological recovered at levels similar to solvent extraction. Similar to Sprague Dawley rats, Murugan et al. [180] fed lyophilized bacteria to mealworms and recovered PHA granules from their fecal pellets. The PHA granules had an 89% purity when washed with water and reached almost 100% purity when treated with SDS. The authors reported higher protein content in mealworms fed with *C. necator* cells than mealworms fed with oats. Mealworms fed with *C. necator* can be an alternative protein source in aquaculture and poultry diets. Biological NPCM disruption is an alternative to other disruption methods; it does not require expensive instrumentation, solvents, or strong chemicals, and the organisms doing the PHA recovery can be a marketable product too. However, the biological recovery process takes longer than any other recovery process and needs biomass pre-treatment. Depending on the final use, PHA biological recovery can be an integrated process for renewable production of feed, food, and materials.

Table 6. NPCM chemical disruption treatments.

NPCM Digestion Type	NPCM Disruption Method	Pre-Treatment	PHA Accumulation Method	Disruption Conditions	Microbial Strain	PHA Content in Biomass (%)	Purity (%)	Recovery (%)	Ref.
Chemical	NaOH	Chemical Pre-treatment	Centrifugation	NaOH (0.1 M), 30 °C, 1 h, 350 rpm	*C. necator*	68	90.8	95.3	[163]
Chemical	NaOH		Centrifugation	NaOH (0.1 M), 30 °C, 1 h, 350 rpm	*C. necator*	68	82.7	94.4	[163]
Chemical	NaOH	Lyophilization	Centrifugation	NaOH (0.1 M), 30 °C, 1 h,	*C. necator*	68	80–90	80–90	[183]
Chemical	NaOH	Lyophilization	Centrifugation	NaOH 0.05 M, 3 h, 0 rpm, 4 °C	*C. necator*	30	98.6	96.9	[174]
Chemical	NaOH	Lyophilization and milling	Centrifugation	NaOH (0.5 N), 4 h, 37 °C, 500 rpm	*C. necator*	65	93	80	[170]
Chemical	NaOH		Centrifugation	NaOH (0.2 M), 200 rpm, 30 °C, 1 h	*Mixed Culture*	62–72	87	97	[169]
Chemical	NaClO		Centrifugation	NaClO (5%) 24 h	*Mixed Culture*	46	90	~100	[171]
Chemical	NaClO	Mechanical pre-treatment	Precipitation	NaClO 13% (v/v), room temperature, 1 h.	*Ralstonia eutropha*	65.2	95.6	91.3	[172]
Chemical	NaClO	Lyophilization and Milling	Centrifugation	NaClO (13%), 37 °C, 500 rpm,4 h.	*C. necator*	65	97	82	[170]
Chemical	NaOH and SDS		Centrifugation	NaOH (0.2 M) and SDS (0.2 %), 200 rpm, 30°C, 1 h	*Mixed Culture*	62–72	99	91	[169]
Chemical	SDS		NaClO and Centrifugation	SDS (0.1%), 24 h	*H. mediterranei*	70	~100	97	[55]
Chemical	SDS		Centrifugation	SDS (0.1%), 24 h	*H. mediterranei*	71.2	~100	97	[56]
Chemical	SDS		Centrifugation	SDS (0.1%), 60 °C, 2 h	*Halomonas* sp. SK5	48	94	98	[173]
Chemical	SDS		Centrifugation	SDS (0.2 %), 200 rpm, 30 °C, 1 h	*Mixed Culture*	62–72	79	63.5	[169]
Chemical	H_2SO_4	Lyophilization and Milling	Chemical treatment and Centrifugation	H_2SO_4 (0.64 M), 6 h, 80 °C	*C. necator*	65	98	79	[170]
Chemical	Water	Lyophilization	Centrifugation	dH_2O, 30 °C, 1 h,	*Comamonassp.*	30	80.6	96	[174]
Chemical	Water	Lyophilization	Centrifugation	dH_2O, 30 °C, 18 h	*Halomonas* sp.	48	94	98	[173]
Enzymatic	Alcalase, SDS and EDTA		Centrifugation	Alcalase (0.3 U g−1), SDS (0.3 g g−1), EDTA (0.01 g g−1). Na_2HPO_4 buffer, 150 rpm, 55 °C, 1 h	*C. necator*	37	94		[177]
Enzymatic	Crude extract	Heat treatment and lyophilization	Centrifugation	*Aspergillus oryzae* crude extract, Na_2HPO_4-citric acid buffer and 47 °C	*C. necator*	78.9	98	97	[175]
Enzymatic	Lysozyme		Centrifugation	Lysozyme solution (2 mg/mL). 1 h, 3 °C	*C. necator*	41	41	75	[176]
Biological	Mealworm (*Tenebrio molitor*)	Lyophilization	Chemical treatment, centrifugation	50 g of mealworms fed 5% of their body weight per day for 16 days.	*C. necator*	37	89%		[178]
Biological	Sprague Dawley rats	Lyophilization and grinding	Chemical treatment, centrifugation	150–200 g rats were feed 15 g/day/animal, 28 days 25 °C	*C. necator*	37	89.3	100	[179]
Biological	Sprague Dawley rats	Lyophilization	Water	150–200 g rats were feed 15 g/day/animal, 28 days 25 °C	*C. necator*	54	82–97	40–47	[182]
Biological	*Bdellovibrio bacteriovorus* HD100		Centrifugation	*P. putida* was inoculated with *B. bacteriovorus* strains 48 h. 30 °C	*P. putida*	55		60	[181]
Biological	*Bdellovibrio bacteriovorus* HD100 and Bd3709		Centrifugation	*P. putida* and *E. coli* cultures were inoculated with *B. bacteriovorus* strains 48 h. 30 °C	*P. putida*	55		80	[180]

4.2.2. PHA Extraction

As PHA are produced inside the cellular biomass, in order to be retrieved they have to be separated from the non-PHA cell mass (NPCM). The simplest, least destructive to biopolymer and most direct way for PHA is to be extracted from the biomass; significant quantities of hazardous solvents and energy input are required for this, creating a potential counterbalance to sustainability and economics towards commercialization [184].

Several studies have proposed various solvent extraction methods for PHA recovery that improve parameters such as yield, purity, and cost of extraction, while at the same time maintaining the physicochemical properties of the biopolymer [185].

Non-Halogenated Solvents

Although several types of extraction systems exist for the production of biopolymers, the majority of extraction methods for PHA still involve the use of organic solvents in which the polymer is soluble. Fei et al. [186], aimed to develop an effective and environmentally-friendly solvent system so as to extract PHB from bacterial biomass. In order to accomplish that, they used a solvent mixture of acetone/ethanol/propylene carbonate (A/E/P, 1:1:1 v/v/v) for extracting PHB from *Cupriavidus necator*. When the A/E/P mixture was used at high temperature, it could recover 92% pure PHB with 85% yield from dry biomass, and 90% purity with 83% PHB yield from wet biomass. Additionally, if hexane was added, it could further enhance the purity and recovery quantities of PHB.

Bacterial PHA could be used for medical applications due to its biocompatibility. However, using inappropriate solvents or techniques during extraction of PHA from bacterial biomass could result in contamination by pyrogenic compounds (e.g., lipopolysaccharides), which eventually leads to rejection of the material for medical use. This problem could be overcome by using a temperature-controlled method for the recovery of poly(3-hydroxyoctanoate-co-3-hydroxyhexanoate) from *Pseudomonas putida* GPo1. Non-chlorinated solvents were found to be the optimal solvents for such tests and, specifically, n-hexane and 2-propanol. The purity reached more than 97% (w/w) and the endotoxicity between 10–15 EU/g PHA. Further re-dissolution in 2-propanol at 45 °C and precipitation at 10 °C resulted in a purity of nearly 100% (w/w) and endotoxicity equal to 2 EU/g PHA [187].

Another approach, however, is that the use of aqueous solvents could benefit the integration into a biorefinery scheme. A study aimed at connecting the exploitation of a raw biowaste, such as used cooking oil (UCO), and producing a desired final product (i.e., amorphous granules of PHA) based on aqueous solvents in a way that will prove the effective reliability of an overall biotechnological approach. Used cooking oil was utilized as the only carbon source for the production of PHB by cultivating *Cupriavidus necator* DSM 428 in batch reactors. The PHB granules were extracted from the biomass using sodium dodecyl sulfate (SDS), ethylenediaminetetraacetic acid (EDTA), and the enzyme Alcalase in an aqueous medium. The PHB granule recovery reached more than 90% and highly pure amorphous polymer was finally obtained [177].

A different alternative to the use of halogen-free and environmental-friendly methods implemented the use of water and ethanol for the recovery of PHA from recombinant *Cupriavidus necator*, in comparison to the well-established chloroform extraction method. Comparing the results obtained from experiments under different incubation times (1, 3, and 5 h) and temperatures (4 and 30 °C) showed that the optimized halogen-free method produced a PHA with 81% purity and 96% recovery yield, whereas the chloroform extraction system resulted in a highly pure PHA with 95% recovery yield. This method could potentially be developed as an alternative and more environmentally-friendly method for industrial application [188].

Aqueous Two-Phase Extraction Systems (ATPS)

In addition to the conventional isolation and purification methods, such as solvent extraction, aqueous two-phase systems (ATPS) have many advantages and important characteristics that attract the attention of researchers and industries. The main advantages are that because ATPS comprise of high water content (70–90% w/w), thus they provide a beneficial environment for separation of sensitive biomaterials. Also, the materials that form the different phases/layers of ATPS are, in principle, safe and environmentally-friendly compared to conventional solvent extraction methods; additionally, an intricate ability for high capacity processing which leads to reduced purification steps; finally, large-scale purification using ATPS can be easily and reliably predicted from laboratory experimental data. ATPS is a feasible solution for industrial demand of cost-effective and highly efficient large-scale bioseparation technologies with short processing times [189,190].

Regarding PHA retrieval methods, one interesting approach is the use of aqueous two-phase systems with the aim of enhancing the accumulation of PHA in one phase using environmentally-friendly layer-forming constituents. This method offers advantages of supporting a beneficial environment for bioseparation, capability of handling high operating capacity, and reducing downstream processing volume, thus proving extractive bioconversion via ATPS can be an optimum solution. Leong et al. [189], examined the effect of pH and salts' addition in *Ralstonia eutropha* H16 cultures, using an ATPS system as a mechanism for PHA extraction. The optimum result obtained in this study was a PHA concentration of 0.139 g/L (purification factor: 1.2–1.63) and recovery yield 55–65% using ATPS of polyethylene glycol (PEG) 8000/sodium sulphate adjusted to pH 6 and the addition of 0.5 M NaCl.

In another study using ATPS, the thermal separation of the phases and how it affected the PHA extraction was studied. The most important ATPS parameters (type and concentration of thermoseparating polymer, salt addition, feedstock load, and thermoseparating temperature) were optimized in order to achieve high PHA retrieval from the bacterial lysate. By taking advantage of the properties of the thermo-responsive polymer (whose solubility decreases in its aqueous solution as the temperature increases), cloud point extraction (CPE) is an ATPS technique that offers the capability to its phase-forming component to be reused. Extraction of PHA from *Cupriavidus necator* H16 via CPE was investigated. The best conditions for PHA extraction (recovery yield of 94.8% and purification factor 1.42) were reached under the following conditions: 20% w/w ethylene oxide-propylene oxide (EOPO), 10 mM NaCl addition, and a thermoseparating temperature of 60 °C with crude feedstock limit of 37.5% w/w. Another benefit of this process is the ability to recycle and reuse EOPO 3900 at least twice, achieving a satisfying yield and purification factor [164].

5. Conclusions

Wide production, commercialization, and thus application of PHAs as a biodegradable alternative to conventional plastics is still limited due to high production cost. Bioprocess technologies are still being developed while bacterial resources are still being explored.

Pure cultures are constantly investigated for their potential to valorize waste byproducts as a low-cost feedstock. The ability of certain bacteria to directly utilize lignocellulosic biomass as the carbon source for PHA production is a huge bioprocess advantage. In this case, the need for chemicals, energy, and labor is minimized. PHA production can be also used as a tool for the bioremediation of oil contaminated sites, as bacterial strains can degrade environmental pollutants, minimizing their toxicity and environmental impact. Moreover, halophilic PHA producers combine a series of advantages, such as growth on seawater and possibility of continuous processes under non-sterile conditions. Those features will significantly contribute to the PHA cost reduction and minimize the requirements for fresh water. Synthetic biology tools are expected to aid in the enhancement of PHA production efficiency, simplify downstream process, and regulate PHA composition providing customized materials for specific applications. However, robust strains are yet to be developed.

Efforts have been also made towards the PHA production using mixed microbial cultures (MMC). MMC-based processes, apart from offering a reduction in the operational costs and the possibility to adapt to a wider range of waste substrates, they could be integrated in current wastewater treatment plants. Recent developments regarding different enrichment strategies and the PHA production step offer new opportunities to make the PHA production more feasible. Cell densities and derived productivities attained with MMC are the main current bottleneck.

The development of economic and simple downstream processes is crucial for the recovery of PHAs. Methods based on the utilization of environmentally friendlier techniques are constantly being investigated, with enzymatic methods advancing the bio-based profile of the process. Reduction of large amounts of chemicals, used per cell dry mass, is going to benefit the economics of the process as well as society and the environment.

Acknowledgments: Constantina Kourmentza would like to thank the European Commission for providing financial support through the 'SimPHAsRLs' project (Grant Agreement no. 625774) funded by the scheme FP7-PEOPLE-2013-IEF-Marie Curie Action: Intra-European Fellowships for Career Development. Jersson Plácido would like to thank the financial support provided by the European Regional Development Fund/Welsh Government funded BEACON research program (Swansea University). Anna Burniol-Figols, Christiano Varrone, and Hari Gavala thank the European Commission for the financial support of this work under FP7 Grant Agreement no. 613667 (acronym: GRAIL). Maria Reis would like to acknowledge and thank all partners and researchers working on 'INCOVER' (Grant Agreement no. 689242) and 'NoAW' (Grant Agreement no. 688338) projects and the European Union for providing funding under H2020 research and innovation program.

Author Contributions: Constantina Kourmentza conceived the concept of the study and reviewed the literature regarding PHA production using pure culture biotechnology. Anna Burniol-Figols, Christiano Varrone, and Hari Gavala prepared the section referring to PHA production by mixed microbial consortia. Nikolaos Venetsaneas and Jersson Plácido reviewed the part of PHA recovery techniques. Maria Reis revised and edited the manuscript.

References

1. Plastics Europe. *Plastics Europe Plastics—The Facts 2016*; Plastics Europe: Brussels, Belgium, 2016; pp. 1–38.
2. Kourmentza, C.; Kornaros, M. Biotransformation of volatile fatty acids to polyhydroxyalkanoates by employing mixed microbial consortia: The effect of pH and carbon source. *Bioresour. Technol.* **2016**, *222*, 388–398. [CrossRef] [PubMed]
3. Singh Saharan, B.; Grewal, A.; Kumar, P. Biotechnological Production of Polyhydroxyalkanoates: A Review on Trends and Latest Developments. *Chin. J. Biol.* **2014**, *2014*, 1–18. [CrossRef]
4. Chen, G.-Q. A microbial polyhydroxyalkanoates (PHA) based bio- and materials industry. *Chem. Soc. Rev.* **2009**, *38*, 2434–2446. [CrossRef] [PubMed]
5. López, N.I.; Pettinari, M.J.; Nikel, P.I.; Méndez, B.S. Polyhydroxyalkanoates: Much More than Biodegradable Plastics. *Adv. Appl. Microbiol.* **2015**, *93*, 73–106. [PubMed]
6. Kourmentza, C.; Koutra, E.; Venetsaneas, N.; Kornaros, M. Integrated Biorefinery Approach for the Valorization of Olive Mill Waste Streams Towards Sustainable Biofuels and Bio-Based Products. In *Microbial Appications Vol. 1—Bioremediation and Bioenergy*; Kalia, V.C., Kumar, P., Eds.; Springer: Berlin, Germany, 2017; Volume 1, pp. 211–238.
7. Plastics Technology. Available online: http://www.ptonline.com/articles/prices-bottom-out-for-polyolefins-pet-ps-pvc-move-up (accessed on 26 May 2017).
8. Eno, R.; Hill, J. Metabolix Bio-industrial Evolution. In Proceedings of the Jefferies 11th Global Clean Technology Conference, New York, NY, USA, 23–24 February 2011.
9. Markets and Markets Polyhydroxyalkanoate (PHA) Market by Type (Monomers, Co-Polymers, Terpolymers), Manufacturing Technology (Bacterial Fermentation, Biosynthesis, Enzymatic Catalysis), Application (Packaging, Bio Medical, Food Services, Agriculture)-Global Forecast to 202. Available online: http://www.marketsandmarkets.com/Market-Reports/pha-market-395.html (accessed on 26 May 2017).
10. Kourmentza, C.; Ntaikou, I.; Lyberatos, G.; Kornaros, M. Polyhydroxyalkanoates from *Pseudomonas* sp. using synthetic and olive mill wastewater under limiting conditions. *Int. J. Biol. Macromol.* **2015**, *74*, 202–210. [CrossRef] [PubMed]
11. Fraiberg, M.; Borovok, I.; Weiner, R.M.; Lamed, R. Discovery and characterization of cadherin domains in *Saccharophagus degradans* 2-40. *J. Bacteriol.* **2010**, *192*, 1066–1074. [CrossRef] [PubMed]
12. Weiner, R.M.; Taylor, L.E.; Henrissat, B.; Hauser, L.; Land, M.; Coutinho, P.M.; Rancurel, C.; Saunders, E.H.; Longmire, A.G.; Zhang, H.; et al. Complete genome sequence of the complex carbohydrate-degrading marine bacterium, *Saccharophagus degradans* strain 2-40T. *PLoS Genet.* **2008**, *4*, e1000087. [CrossRef] [PubMed]

13. Ekborg, N.A.; Taylor, L.E.; Longmire, A.G.; Henrissat, B.; Weiner, R.M.; Steven, W.; Hutcheson, S.W. Genomic and Proteomic Analyses of the Agarolytic System Expressed by *Saccharophagus degradans* 2-40. *Appl. Environ. Microbiol.* **2006**, *72*, 3396–3405. [CrossRef] [PubMed]
14. Andrykovitch, G.; Marx, I. Isolation of a new polysaccharide-digesting bacterium from a salt marsh. *Appl. Environ. Microbiol.* **1988**, *54*, 1061–1062. [PubMed]
15. Taylor, L.E.; Henrissat, B.; Coutinho, P.M.; Ekborg, N.A.; Hutcheson, S.W.; Weiner, R.M. Complete cellulase system in the marine bacterium *Saccharophagus degradans* strain 2-40T. *J. Bacteriol.* **2006**, *188*, 3849–3861. [CrossRef] [PubMed]
16. Howard, M.B.; Ekborg, N.A.; Taylor, L.E.; Weiner, R.M.; Hutcheson, S.W. Genomic Analysis and Initial Characterization of the Chitinolytic System of *Microbulbifer degradans* Strain 2-40. *J. Bacteriol* **2003**, *185*, 3352–3360. [CrossRef] [PubMed]
17. Howard, M.B.; Ekborg, N.A.; Taylor, L.E.; Weiner, R.M.; Hutcheson, S.W. Chitinase B of *"Microbulbifer degradans"* 2-40 Contains Two Catalytic Domains with Different Chitinolytic Activities. *J. Bacteriol.* **2004**, *186*, 1297–1303. [CrossRef] [PubMed]
18. Suvorov, M.; Kumar, R.; Zhang, H.; Hutcheson, S. Novelties of the cellulolytic system of a marine bacterium applicable to cellulosic sugar production. *Biofuels* **2011**, *2*, 59–70. [CrossRef]
19. Munoz, L.E.A.; Riley, M.R. Utilization of cellulosic waste from tequila bagasse and production of polyhydroxyalkanoate (pha) bioplastics by *Saccharophagus degradans*. *Biotechnol. Bioeng.* **2008**, *100*, 882–888. [CrossRef] [PubMed]
20. González-García, Y.; Nungaray, J.; Córdova, J.; González-Reynoso, O.; Koller, M.; Atlic, A.; Braunegg, G. Biosynthesis and characterization of polyhydroxyalkanoates in the polysaccharide-degrading marine bacterium *Saccharophagus degradans* ATCC 43961. *J. Ind. Microbiol. Biotechnol.* **2008**, *35*, 629–633. [CrossRef] [PubMed]
21. González-García, Y.; Rosales, M.A.; González-Reynoso, O.; Sanjuán-Dueñas, R.; Córdova, J. Polyhydroxybutyrate production by *Saccharophagus degradans* using raw starch as carbon source. *Eng. Life Sci.* **2011**, *11*, 59–64. [CrossRef]
22. Sheu, D.S.; Chen, W.M.; Yang, J.Y.; Chang, R.C. Thermophilic bacterium *Caldimonas taiwanensis* produces poly(3-hydroxybutyrate-co-3-hydroxyvalerate) from starch and valerate as carbon sources. *Enzyme Microb. Technol.* **2009**, *44*, 289–294. [CrossRef]
23. Sawant, S.S.; Tran, T.K.; Salunke, B.K.; Kim, B.S. Potential of *Saccharophagus degradans* for production of polyhydroxyalkanoates using cellulose. *Process Biochem.* **2017**. [CrossRef]
24. Sawant, S.S.; Salunke, B.K.; Taylor, L.E.; Kim, B.S. Enhanced agarose and xylan degradation for production of polyhydroxyalkanoates by co-culture of marine bacterium, *Saccharophagus degradans* and its contaminant, *Bacillus cereus*. *Appl. Sci.* **2017**, *7*, 225. [CrossRef]
25. Sawant, S.S.; Salunke, B.K.; Kim, B.S. A Laboratory Case Study of Efficient Polyhydoxyalkonates Production by *Bacillus cereus*, a Contaminant in *Saccharophagus degradans* ATCC 43961 in Minimal Sea Salt Media. *Curr. Microbiol.* **2014**, *69*, 832–838. [CrossRef] [PubMed]
26. Salamanca-Cardona, L.; Ashe, C.S.; Stipanovic, A.J.; Nomura, C.T. Enhanced production of polyhydroxyalkanoates (PHAs) from beechwood xylan by recombinant *Escherichia coli*. *Appl. Microbiol. Biotechnol.* **2014**, *98*, 831–842. [CrossRef] [PubMed]
27. Chen, W.M.; Chang, J.S.; Chiu, C.H.; Chang, S.C.; Chen, W.C.; Jiang, C.M. *Caldimonas taiwanensis* sp. nov., a amylase producing bacterium isolated from a hot spring. *Syst. Appl. Microbiol.* **2005**, *28*, 415–420. [CrossRef] [PubMed]
28. Elleuche, S.; Antranikian, G. Starch-Hydrolyzing Enzymes from Thermophiles. In *Thermophilic Microbes in Environmental and Industrial Biotechnology: Biotechnology of Thermophiles*; Satyanarayana, T., Littlechild, J., Kawarabayasi, Y., Eds.; Springer: Dordrecht, The Netherlands, 2013; pp. 509–533.
29. Ward, P.G.; De Roo, G.; O'Connor, K.E. Accumulation of polyhydroxyalkanoate from styrene and phenylacetic acid by *Pseudomonas putida* CA-3. *Appl. Environ. Microbiol.* **2005**, *71*, 2046–2052. [CrossRef] [PubMed]
30. Thompson, R.C.; Moore, C.J.; Vom Saal, F.S.; Swan, S.H. Plastics, the environment and human health: Current consensus and future trends. *Philos. Trans. R. Soc. B* **2009**, *364*, 2153–2166. [CrossRef] [PubMed]
31. Khandare, R.V.; Govindwar, S.P. Phytoremediation of textile dyes and effluents: Current scenario and future prospects. *Biotechnol. Adv.* **2015**, *33*, 1697–1714. [CrossRef] [PubMed]

32. Goudarztalejerdi, A.; Tabatabaei, M.; Eskandari, M.H.; Mowla, D.; Iraji, A. Evaluation of bioremediation potential and biopolymer production of pseudomonads isolated from petroleum hydrocarbon-contaminated areas. *Int. J. Environ. Sci. Technol.* **2015**, *12*, 2801–2808. [CrossRef]
33. Hori, K.; Abe, M.; Unno, H. Production of triacylglycerol and poly(3-hydroxybutyrate-co-3-hydroxyvalerate) by the toluene-degrading bacterium *Rhodococcus aetherivorans* IAR1. *J. Biosci. Bioeng.* **2009**, *108*, 319–324. [CrossRef] [PubMed]
34. Hori, K.; Kobayashi, A.; Ikeda, H.; Unno, H. *Rhodococcus aetherivorans* IAR1, a new bacterial strain synthesizing poly(3-hydroxybutyrate-co-3-hydroxyvalerate) from toluene. *J. Biosci. Bioeng.* **2009**, *107*, 145–150. [CrossRef] [PubMed]
35. Nikodinovic-Runic, J.; Casey, E.; Duane, G.F.; Mitic, D.; Hume, A.R.; Kenny, S.T.; O'Connor, K.E. Process analysis of the conversion of styrene to biomass and medium chain length polyhydroxyalkanoate in a two-phase bioreactor. *Biotechnol. Bioeng.* **2011**, *108*, 2447–2455. [CrossRef] [PubMed]
36. Tan, G.-Y.A.; Chen, C.-L.; Ge, L.; Li, L.; Tan, S.N.; Wang, J.-Y. Bioconversion of styrene to poly(hydroxyalkanoate) (PHA) by the new bacterial strain *Pseudomonas putida* NBUS12. *Microbes Environ.* **2015**, *30*, 76–85. [CrossRef] [PubMed]
37. Goff, M.; Ward, P.G.; O'Connor, K.E. Improvement of the conversion of polystyrene to polyhydroxyalkanoate through the manipulation of the microbial aspect of the process: A nitrogen feeding strategy for bacterial cells in a stirred tank reactor. *J. Biotechnol.* **2007**, *132*, 283–286. [CrossRef] [PubMed]
38. Tamboli, D.P.; Kurade, M.B.; Waghmode, T.R.; Joshi, S.M.; Govindwar, S.P. Exploring the ability of *Sphingobacterium* sp. ATM to degrade textile dye Direct Blue GLL, mixture of dyes and textile effluent and production of polyhydroxyhexadecanoic acid using waste biomass generated after dye degradation. *J. Hazard. Mater.* **2010**, *182*, 169–176. [CrossRef] [PubMed]
39. Tamboli, D.P.; Gomare, S.S.; Kalme, S.S.; Jadhav, U.U.; Govindwar, S.P. Degradation of Orange 3R, mixture of dyes and textile effluent and production of polyhydroxyalkanoates from biomass obtained after degradation. *Int. Biodeterior. Biodegrad.* **2010**, *64*, 755–763. [CrossRef]
40. Kahlon, R.S. *Pseudomonas: Molecular and Applied Biology*; Springer: Berlin, Germany, 2016.
41. Nikodinovic, J.; Kenny, S.T.; Babu, R.P.; Woods, T.; Blau, W.J.; O'Connor, K.E. The conversion of BTEX compounds by single and defined mixed cultures to medium-chain-length polyhydroxyalkanoate. *Appl. Microbiol. Biotechnol.* **2008**, *80*, 665–673. [CrossRef] [PubMed]
42. Ni, Y.Y.; Kim, D.Y.; Chung, M.G.; Lee, S.H.; Park, H.Y.; Rhee, Y.H. Biosynthesis of medium-chain-length poly(3-hydroxyalkanoates) by volatile aromatic hydrocarbons-degrading *Pseudomonas fulva* TY16. *Bioresour. Technol.* **2010**, *101*, 8485–8488. [CrossRef] [PubMed]
43. Narancic, T.; Kenny, S.T.; Djokic, L.; Vasiljevic, B.; O'Connor, K.E.; Nikodinovic-Runic, J. Medium-chain-length polyhydroxyalkanoate production by newly isolated *Pseudomonas* sp. TN301 from a wide range of polyaromatic and monoaromatic hydrocarbons. *J. Appl. Microbiol.* **2012**, *113*, 508–520. [CrossRef] [PubMed]
44. Ward, P.G.; Goff, M.; Donner, M. A Two Step Chemo—Biotechnological Conversion of Polystyrene to a Biodegradable Thermoplastic. *Environ. Sci. Technol.* **2006**, *40*, 2433–2437.
45. Kenny, S.T.; Runic, J.N.; Kaminsky, W.; Woods, T.; Babu, R.P.; Keely, C.M.; Blau, W.; O'Connor, K.E. Up-cycling of PET (Polyethylene Terephthalate) to the biodegradable plastic PHA (Polyhydroxyalkanoate). *Environ. Sci. Technol.* **2008**, *42*, 7696–7701. [CrossRef] [PubMed]
46. Guzik, M.W.; Kenny, S.T.; Duane, G.F.; Casey, E.; Woods, T.; Babu, R.P.; Nikodinovic-Runic, J.; Murray, M.; O'Connor, K.E. Conversion of post consumer polyethylene to the biodegradable polymer polyhydroxyalkanoate. *Appl. Microbiol. Biotechnol.* **2014**, *98*, 4223–4232. [CrossRef] [PubMed]
47. Tamboli, D.P.; Kagalkar, A.N.; Jadhav, M.U.; Jadhav, J.P.; Govindwar, S.P. Production of polyhydroxyhexadecanoic acid by using waste biomass of *Sphingobacterium* sp. ATM generated after degradation of textile dye Direct Red 5B. *Bioresour. Technol.* **2010**, *101*, 2421–2427. [CrossRef] [PubMed]
48. Maheshwari, D.K.; Saraf, M. *Halophiles: Biodiversity and Sustainable Exploitation*; Springer: Berlin, Germany, 2015; Volume 6.
49. Konstantinidis, G. *Elsevier'S Dictionary of Medicine and Biology Greek German Italian Latin*; Elsevier: Amsterdam, The Netherlands, 2005.
50. Setati, M. Diversity and industrial potential of hydrolaseproducing halophilic/halotolerant eubacteria. *Afr. J. Biotechnol.* **2010**, *9*, 1555–1560.

51. Yin, J.; Chen, J.C.; Wu, Q.; Chen, G.Q. Halophiles, coming stars for industrial biotechnology. *Biotechnol. Adv.* **2015**, *33*, 1433–1442. [CrossRef] [PubMed]
52. Quillaguamán, J.; Guzmán, H.; Van-Thuoc, D.; Hatti-Kaul, R. Synthesis and production of polyhydroxyalkanoates by halophiles: Current potential and future prospects. *Appl. Microbiol. Biotechnol.* **2010**, *85*, 1687–1696. [CrossRef] [PubMed]
53. Kirk, R.G.; Ginzburg, M. Ultrastructure of two species of halobacterium. *J. Ultrastruct. Res.* **1972**, *41*, 80–94. [CrossRef]
54. Rodriguez-Valera, F.; Ruiz-Berraquero, F.; Ramos-Cormenzana, A. Isolation of Extremely Halophilic Bacteria Able to Grow in Defined Inorganic Media with Single Carbon Sources. *Microbiology* **1980**, *119*, 535–538. [CrossRef]
55. Bhattacharyya, A.; Pramanik, A.; Maji, S.K.; Haldar, S.; Mukhopadhyay, U.K.; Mukherjee, J. Utilization of vinasse for production of poly-3-(hydroxybutyrate-co-hydroxyvalerate) by *Haloferax mediterranei*. *AMB Express* **2012**, *2*, 34. [CrossRef] [PubMed]
56. Bhattacharyya, A.; Saha, J.; Haldar, S.; Bhowmic, A.; Mukhopadhyay, U.K.; Mukherjee, J. Production of poly-3-(hydroxybutyrate-co-hydroxyvalerate) by *Haloferax mediterranei* using rice-based ethanol stillage with simultaneous recovery and re-use of medium salts. *Extremophiles* **2014**, *18*, 463–470. [CrossRef] [PubMed]
57. Pais, J.; Serafim, L.S.; Freitas, F.; Reis, M.A.M. Conversion of cheese whey into poly(3-hydroxybutyrate-co-3-hydroxyvalerate) by *Haloferax mediterranei*. *New Biotechnol.* **2016**, *33*, 224–230. [CrossRef] [PubMed]
58. Alsafadi, D.; Al-Mashaqbeh, O. A one-stage cultivation process for the production of poly-3-(hydroxybutyrate-co-hydroxyvalerate) from olive mill wastewater by *Haloferax mediterranei*. *New Biotechnol.* **2017**, *34*, 47–53. [CrossRef] [PubMed]
59. Tan, D.; Xue, Y.-S.; Aibaidula, G.; Chen, G.-Q. Unsterile and continuous production of polyhydroxybutyrate by *Halomonas* TD01. *Bioresour. Technol.* **2011**, *102*, 8130–8136. [CrossRef] [PubMed]
60. Yue, H.; Ling, C.; Yang, T.; Chen, X.; Chen, Y.; Deng, H.; Wu, Q.; Chen, J.; Chen, G.-Q. A seawater-based open and continuous process for polyhydroxyalkanoates production by recombinant *Halomonas campaniensis* LS21 grown in mixed substrates. *Biotechnol. Biofuels* **2014**, *7*, 108. [CrossRef]
61. Salgaonkar, B.B.; Mani, K.; Braganca, J.M. Characterization of polyhydroxyalkanoates accumulated by a moderately halophilic salt pan isolate *Bacillus megaterium* strain H16. *J. Appl. Microbiol.* **2013**, *114*, 1347–1356. [CrossRef] [PubMed]
62. Moorkoth, D.; Nampoothiri, K.M. Production and characterization of poly(3-hydroxy butyrate-co-3 hydroxyvalerate) (PHBV) by a novel halotolerant mangrove isolate. *Bioresour. Technol.* **2016**, *201*, 253–260. [CrossRef] [PubMed]
63. Rodríguez-Contreras, A.; Koller, M.; Braunegg, G.; Marqués-Calvo, M.S. Poly[(*R*)-3-hydroxybutyrate] production under different salinity conditions by a novel *Bacillus megaterium* strain. *New Biotechnol.* **2016**, *33*, 73–77. [CrossRef] [PubMed]
64. Chen, G.-Q. New challenges and opportunities for industrial biotechnology. *Microb. Cell Fact.* **2012**, *11*, 111. [CrossRef] [PubMed]
65. Wang, Y.; Yin, J.; Chen, G.Q. Polyhydroxyalkanoates, challenges and opportunities. *Curr. Opin. Biotechnol.* **2014**, *30*, 59–65. [CrossRef] [PubMed]
66. Li, T.; Ye, J.; Shen, R.; Zong, Y.; Zhao, X.; Lou, C.; Chen, G.-Q. Semirational Approach for Ultrahigh Poly(3-hydroxybutyrate) Accumulation in *Escherichia coli* by Combining One-Step Library Construction and High-Throughput Screening. *ACS Synth. Biol.* **2016**, *5*, 1308–1317. [PubMed]
67. Fu, X.-Z.; Tan, D.; Aibaidula, G.; Wu, Q.; Chen, J.-C.; Chen, G.-Q. Development of *Halomonas* TD01 as a host for open production of chemicals. *Metab. Eng.* **2014**, *23*, 78–91. [CrossRef] [PubMed]
68. Wang, Y.; Wu, H.; Jiang, X.; Chen, G.-Q. Engineering *Escherichia coli* for enhanced production of poly(3-hydroxybutyrate-co-4-hydroxybutyrate) in larger cellular space. *Metab. Eng.* **2014**, *25*, 183–193. [CrossRef] [PubMed]
69. Jiang, X.-R.; Wang, H.; Shen, R.; Chen, G.-Q. Engineering the bacterial shapes for enhanced inclusion bodies accumulation. *Metab. Eng.* **2015**, *29*, 227–237. [CrossRef] [PubMed]
70. Jiang, X.-R.; Chen, G.-Q. Morphology engineering of bacteria for bio-production. *Biotechnol. Adv.* **2016**, *34*, 435–440. [CrossRef] [PubMed]

71. Pfeiffer, D.; Jendrossek, D. Localization of poly(3-Hydroxybutyrate) (PHB) granule-associated proteins during PHB granule formation and identification of two new phasins, phap6 and phap7, in *Ralstonia eutropha* H16. *J. Bacteriol.* **2012**, *194*, 5909–5921. [CrossRef] [PubMed]
72. Martínez, V.; García, P.; García, J.L.; Prieto, M.A. Controlled autolysis facilitates the polyhydroxyalkanoate recovery in *Pseudomonas putida* KT2440. *Microb. Biotechnol.* **2011**, *4*, 533–547. [CrossRef] [PubMed]
73. Tripathi, L.; Wu, L.-P.; Chen, J.; Chen, G.-Q. Synthesis of Diblock copolymer poly-3-hydroxybutyrate -block-poly-3-hydroxyhexanoate [PHB-b-PHHx] by a β-oxidation weakened *Pseudomonas putida* KT2442. *Microb. Cell Fact.* **2012**, *11*, 44. [CrossRef] [PubMed]
74. Tripathi, L.; Wu, L.-P.; Dechuan, M.; Chen, J.; Wu, Q.; Chen, G.-Q. *Pseudomonas putida* KT2442 as a platform for the biosynthesis of polyhydroxyalkanoates with adjustable monomer contents and compositions. *Bioresour. Technol.* **2013**, *142*, 225–231. [CrossRef] [PubMed]
75. Li, S.; Cai, L.; Wu, L.; Zeng, G.; Chen, J.; Wu, Q.; Chen, G.-Q. Microbial Synthesis of Functional Homo-, Random, and Block Polyhydroxyalkanoates by β-Oxidation Deleted *Pseudomonas entomophila*. *Biomacromolecules* **2014**, *15*, 2310–2319. [CrossRef] [PubMed]
76. Shen, R.; Cai, L.W.; Meng, D.C.; Wu, L.P.; Guo, K.; Dong, G.X.; Liu, L.; Chen, J.C.; Wu, Q.; Chen, G.Q. Benzene containing polyhydroxyalkanoates homo- and copolymers synthesized by genome edited *Pseudomonas entomophila*. *Sci. China Life Sci.* **2014**, *57*, 4–10. [CrossRef] [PubMed]
77. Meng, D.C.; Shen, R.; Yao, H.; Chen, J.C.; Wu, Q.; Chen, G.Q. Engineering the diversity of polyesters. *Curr. Opin. Biotechnol.* **2014**, *29*, 24–33. [CrossRef] [PubMed]
78. Tripathi, L.; Wu, L.-P.; Meng, D.; Chen, J.; Chen, G.-Q. Biosynthesis and Characterization of Diblock Copolymer of P(3-Hydroxypropionate)-block-P(4-hydroxybutyrate) from Recombinant *Escherichia coli*. *Biomacromolecules* **2013**, *14*, 862–870. [CrossRef] [PubMed]
79. Zhuang, Q.; Wang, Q.; Liang, Q.; Qi, Q. Synthesis of polyhydroxyalkanoates from glucose that contain medium-chain-length monomers via the reversed fatty acid β-oxidation cycle in *Escherichia coli*. *Metab. Eng.* **2014**, *24*, 78–86. [CrossRef] [PubMed]
80. Wang, Q.; Luan, Y.; Cheng, X.; Zhuang, Q.; Qi, Q. Engineering of *Escherichia coli* for the biosynthesis of poly(3-hydroxybutyrate-co-3-hydroxyhexanoate) from glucose. *Appl. Microbiol. Biotechnol.* **2015**, *99*, 2593–2602. [CrossRef] [PubMed]
81. Meng, D.-C.; Wang, Y.; Wu, L.-P.; Shen, R.; Chen, J.-C.; Wu, Q.; Chen, G.-Q. Production of poly(3-hydroxypropionate) and poly(3-hydroxybutyrate-co-3-hydroxypropionate) from glucose by engineering *Escherichia coli*. *Metab. Eng.* **2015**, *29*, 189–195. [CrossRef] [PubMed]
82. Chen, G.Q.; Hajnal, I.; Wu, H.; Lv, L.; Ye, J. Engineering Biosynthesis Mechanisms for Diversifying Polyhydroxyalkanoates. *Trends Biotechnol.* **2015**, *33*, 565–574. [CrossRef] [PubMed]
83. Koller, M.; Maršálek, L.; de Sousa Dias, M.M.; Braunegg, G. Producing microbial polyhydroxyalkanoate (PHA) biopolyesters in a sustainable manner. *New Biotechnol.* **2017**, *37*, 24–38. [CrossRef] [PubMed]
84. Możejko-Ciesielska, J.; Kiewisz, R. Bacterial polyhydroxyalkanoates: Still fabulous? *Microbiol. Res.* **2016**, *192*, 271–282. [CrossRef] [PubMed]
85. Bugnicourt, E.; Cinelli, P.; Lazzeri, A.; Alvarez, V. Polyhydroxyalkanoate (PHA): Review of synthesis, characteristics, processing and potential applications in packaging. *Express Polym. Lett.* **2014**, *8*, 791–808. [CrossRef]
86. Oliveira, C.S.S.; Silva, C.E.; Carvalho, G.; Reis, M.A. Strategies for efficiently selecting PHA producing mixed microbial cultures using complex feedstocks: Feast and famine regime and uncoupled carbon and nitrogen availabilities. *New Biotechnol.* **2016**, *37*, 69–79. [CrossRef] [PubMed]
87. Kleerebezem, R.; van Loosdrecht, M.C. Mixed culture biotechnology for bioenergy production. *Curr. Opin. Biotechnol.* **2007**, *18*, 207–212. [CrossRef] [PubMed]
88. Serafim, L.S.; Lemos, P.C.; Albuquerque, M.G.E.; Reis, M.A.M. Strategies for PHA production by mixed cultures and renewable waste materials. *Appl. Microbiol. Biotechnol.* **2008**, *81*, 615–628. [CrossRef] [PubMed]
89. Johnson, K.; Jiang, Y.; Kleerebezem, R.; Muyzer, G.; Van Loosdrecht, M.C.M. Enrichment of a mixed bacterial culture with a high polyhydroxyalkanoate storage capacity. *Biomacromolecules* **2009**, *10*, 670–676. [CrossRef] [PubMed]
90. Van Loosdrecht, M.C.M.; Pot, M.A.; Heijnen, J.J. Importance of bacterial storage polymers in bioprocesses. *Water Sci. Technol.* **1997**, *35*, 41–47. [CrossRef]

91. Dircks, K.; Beun, J.J.; Van Loosdrecht, M.; Heijnen, J.J.; Henze, M. Glycogen metabolism in aerobic mixed cultures. *Biotechnol. Bioeng.* **2001**, *73*, 85–94. [CrossRef] [PubMed]
92. Moralejo-Gárate, H.; Mar'Atusalihat, E.; Kleerebezem, R.; Van Loosdrecht, M.C.M. Microbial community engineering for biopolymer production from glycerol. *Appl. Microbiol. Biotechnol.* **2011**, *92*, 631–639. [CrossRef] [PubMed]
93. Dionisi, D.; Carucci, G.; Petrangeli Papini, M.; Riccardi, C.; Majone, M.; Carrasco, F. Olive oil mill effluents as a feedstock for production of biodegradable polymers. *Water Res.* **2005**, *39*, 2076–2084. [CrossRef] [PubMed]
94. Ntaikou, I.; Valencia Peroni, C.; Kourmentza, C.; Ilieva, V.I.; Morelli, A.; Chiellini, E.; Lyberatos, G. Microbial bio-based plastics from olive-mill wastewater: Generation and properties of polyhydroxyalkanoates from mixed cultures in a two-stage pilot scale system. *J. Biotechnol.* **2014**, *188C*, 138–147. [CrossRef] [PubMed]
95. Duque, A.F.; Oliveira, C.S.S.; Carmo, I.T.D.; Gouveia, A.R.; Pardelha, F.; Ramos, A.M.; Reis, M.A.M. Response of a three-stage process for PHA production by mixed microbial cultures to feedstock shift: Impact on polymer composition. *New Biotechnol.* **2014**, *31*, 276–288. [CrossRef] [PubMed]
96. Amulya, K.; Jukuri, S.; Venkata Mohan, S. Sustainable multistage process for enhanced productivity of bioplastics from waste remediation through aerobic dynamic feeding strategy: Process integration for up-scaling. *Bioresour. Technol.* **2015**, *188*, 231–239. [CrossRef] [PubMed]
97. Valentino, F.; Morgan-Sagastume, F.; Campanari, S.; Villano, M.; Werker, A.; Majone, M. Carbon recovery from wastewater through bioconversion into biodegradable polymers. *New Biotechnol.* **2016**, *37*, 9–23. [CrossRef] [PubMed]
98. Kourmentza, C.; Mitova, E.; Stoyanova, N.; Ntaikou, I.; Kornaros, M. Investigation of PHAs production from acidified olive oil mill wastewater (OOMW) by pure cultures of *Pseudomonas* spp. strains. *New Biotechnol.* **2009**, *25*, S269. [CrossRef]
99. Kourmentza, C.; Ntaikou, I.; Kornaros, M.; Lyberatos, G. Production of PHAs from mixed and pure cultures of *Pseudomonas* sp. using short-chain fatty acids as carbon source under nitrogen limitation. *Desalination* **2009**, *248*, 723–732. [CrossRef]
100. Marang, L.; Jiang, Y.; van Loosdrecht, M.C.M.; Kleerebezem, R. Butyrate as preferred substrate for polyhydroxybutyrate production. *Bioresour. Technol.* **2013**, *142*, 232–239. [CrossRef] [PubMed]
101. Shen, L.; Hu, H.; Ji, H.; Cai, J.; He, N.; Li, Q.; Wang, Y. Production of poly(hydroxybutyrate-hydroxyvalerate) from waste organics by the two-stage process: Focus on the intermediate volatile fatty acids. *Bioresour. Technol.* **2014**, *166*, 194–200. [CrossRef] [PubMed]
102. Dias, J.M.L.; Lemos, P.C.; Serafim, L.S.; Oliveira, C.; Eiroa, M.; Albuquerque, M.G.E.; Ramos, A.M.; Oliveira, R.; Reis, M.A.M. Recent advances in polyhydroxyalkanoate production by mixed aerobic cultures: From the substrate to the final product. *Macromol. Biosci.* **2006**, *6*, 885–906. [CrossRef] [PubMed]
103. Salehizadeh, H.; Van Loosdrecht, M.C.M. Production of polyhydroxyalkanoates by mixed culture: Recent trends and biotechnological importance. *Biotechnol. Adv.* **2004**, *22*, 261–279. [CrossRef] [PubMed]
104. Reis, M.A.M.; Serafim, L.S.; Lemos, P.C.; Ramos, A.M.; Aguiar, F.R.; Van Loosdrecht, M.C.M. Production of polyhydroxyalkanoates by mixed microbial cultures. *Bioprocess Biosyst. Eng.* **2003**, *25*, 377–385. [CrossRef] [PubMed]
105. Albuquerque, M.G.E.; Eiroa, M.; Torres, C.; Nunes, B.R.; Reis, M.A.M. Strategies for the development of a side stream process for polyhydroxyalkanoate (PHA) production from sugar cane molasses. *J. Biotechnol.* **2007**, *130*, 411–421. [CrossRef] [PubMed]
106. Ren, Q.; De Roo, G.; Ruth, K.; Witholt, B.; Zinn, M. Simultaneous Accumulation and Degradation of Polyhydroxyalkanoates: Futile Cycle or Clever Regulation? *Biomacromolecules* **2009**, *10*, 916–922. [CrossRef] [PubMed]
107. Frigon, D.; Muyzer, G.; Van Loosdrecht, M.; Raskin, L. rRNA and poly-β-hydroxybutyrate dynamics in bioreactors subjected to feast and famine cycles. *Appl. Environ. Microbiol.* **2006**, *72*, 2322–2330. [CrossRef] [PubMed]
108. Prieto, A.; Escapa, I.F.; Martínez, V.; Dinjaski, N.; Herencias, C.; de la Peña, F.; Tarazona, N.; Revelles, O. A holistic view of polyhydroxyalkanoate metabolism in *Pseudomonas putida*. *Environ. Microbiol.* **2016**, *18*, 341–357. [CrossRef] [PubMed]
109. Jiang, Y.; Marang, L.; Kleerebezem, R.; Muyzer, G.; van Loosdrecht, M.C.M. Polyhydroxybutyrate production from lactate using a mixed microbial culture. *Biotechnol. Bioeng.* **2011**, *108*, 2022–2035. [CrossRef] [PubMed]

110. Pardelha, F.A. *Constraint-Based Modelling of Mixed Microbial Populations: Application to Polyhydroxyalkanoates Production*; Faculdade de Ciências e Tecnologia: Caparica, Portugal, 2013.
111. Albuquerque, M.G.E.; Concas, S.; Bengtsson, S.; Reis, M.A.M. Mixed culture polyhydroxyalkanoates production from sugar molasses: The use of a 2-stage CSTR system for culture selection. *Bioresour. Technol.* **2010**, *101*, 7112–7122. [CrossRef] [PubMed]
112. Marang, L.; Jiang, Y.; Van Loosdrecht, M.C.M.; Kleerebezem, R. Impact of non-storing biomass on PHA production: An enrichment culture on acetate and methanol. *Int. J. Biol. Macromol.* **2014**, *71*, 74–80. [CrossRef] [PubMed]
113. Korkakaki, E.; van Loosdrecht, M.C.M.; Kleerebezem, R. Survival of the fastest: Selective removal of the side population for enhanced PHA production in a mixed substrate enrichment. *Bioresour. Technol.* **2016**, *216*, 1022–1029. [CrossRef] [PubMed]
114. Chen, Z.; Guo, Z.; Wen, Q.; Huang, L.; Bakke, R.; Du, M. A new method for polyhydroxyalkanoate (PHA) accumulating bacteria selection under physical selective pressure. *Int. J. Biol. Macromol.* **2015**, *72*, 1329–1334. [CrossRef] [PubMed]
115. Silva, F.; Campanari, S.; Matteo, S.; Valentino, F.; Majone, M.; Villano, M. Impact of nitrogen feeding regulation on polyhydroxyalkanoates production by mixed microbial cultures. *New Biotechnol.* **2017**, *37*, 90–98. [CrossRef] [PubMed]
116. Burniol-Figols, A.; Varrone, C.; Daugaard, A.E.; Skiadas, I.V.; Gavala, H.N. Polyhydroxyalkanoates (PHA) production from fermented crude glycerol by mixed microbial cultures. In Proceedings of the Sustain ATV Conference, Book of Abstracts, Kobenhavn, Denmark, 30 November 2016.
117. Cui, Y.-W.; Zhang, H.-Y.; Lu, P.-F.; Peng, Y.-Z. Effects of carbon sources on the enrichment of halophilic polyhydroxyalkanoate-storing mixed microbial culture in an aerobic dynamic feeding process. *Sci. Rep.* **2016**, *6*, 30766. [CrossRef] [PubMed]
118. Palmeiro-Sánchez, T.; Fra-Vázquez, A.; Rey-Martínez, N.; Campos, J.L.; Mosquera-Corral, A. Transient concentrations of NaCl affect the PHA accumulation in mixed microbial culture. *J. Hazard. Mater.* **2016**, *306*, 332–339. [CrossRef] [PubMed]
119. Fradinho, J.C.; Domingos, J.M.B.; Carvalho, G.; Oehmen, A.; Reis, M.A.M. Polyhydroxyalkanoates production by a mixed photosynthetic consortium of bacteria and algae. *Bioresour. Technol.* **2013**, *132*, 146–153. [CrossRef] [PubMed]
120. Fradinho, J.C.; Reis, M.A.M.; Oehmen, A. Beyond feast and famine: Selecting a PHA accumulating photosynthetic mixed culture in a permanent feast regime. *Water Res.* **2016**, *105*, 421–428. [CrossRef] [PubMed]
121. Fradinho, J.C.; Oehmen, A.; Reis, M.A.M. Photosynthetic mixed culture polyhydroxyalkanoate (PHA) production from individual and mixed volatile fatty acids (VFAs): Substrate preferences and co-substrate uptake. *J. Biotechnol.* **2014**, *185*, 19–27. [CrossRef] [PubMed]
122. Fradinho, J.C.; Oehmen, A.; Reis, M.A.M. Effect of dark/light periods on the polyhydroxyalkanoate production of a photosynthetic mixed culture. *Bioresour. Technol.* **2013**, *148*, 474–479. [CrossRef] [PubMed]
123. Basset, N.; Katsou, E.; Frison, N.; Malamis, S.; Dosta, J.; Fatone, F. Integrating the selection of PHA storing biomass and nitrogen removal via nitrite in the main wastewater treatment line. *Bioresour. Technol.* **2016**, *200*, 820–829. [CrossRef] [PubMed]
124. Morgan-Sagastume, F.; Karlsson, A.; Johansson, P.; Pratt, S.; Boon, N.; Lant, P.; Werker, A. Production of polyhydroxyalkanoates in open, mixed cultures from a waste sludge stream containing high levels of soluble organics, nitrogen and phosphorus. *Water Res.* **2010**, *44*, 5196–5211. [CrossRef] [PubMed]
125. Jia, Q.; Xiong, H.; Wang, H.; Shi, H.; Sheng, X.; Sun, R.; Chen, G. Production of polyhydroxyalkanoates (PHA) by bacterial consortium from excess sludge fermentation liquid at laboratory and pilot scales. *Bioresour. Technol.* **2014**, *171*, 159–167. [CrossRef] [PubMed]
126. Anterrieu, S.; Quadri, L.; Geurkink, B.; Dinkla, I.; Bengtsson, S.; Arcos-Hernandez, M.; Alexandersson, T.; Morgan-Sagastume, F.; Karlsson, A.; Hjort, M.; et al. Integration of biopolymer production with process water treatment at a sugar factory. *New Biotechnol.* **2014**, *31*, 308–323. [CrossRef] [PubMed]
127. Satoh, H.; Iwamoto, Y.; Mino, T.; Matsuo, T. Activated sludge as a possible source of biodegradable plastic. *Water Sci. Technol.* **1998**, *38*, 103–109. [CrossRef]

128. Din, M.F.; Mohanadoss, P.; Ujang, Z.; van Loosdrecht, M.; Yunus, S.M.; Chelliapan, S.; Zambare, V.; Olsson, G. Development of Bio-PORec®system for polyhydroxyalkanoates (PHA) production and its storage in mixed cultures of palm oil mill effluent (POME). *Bioresour. Technol.* **2012**, *124*, 208–216. [CrossRef] [PubMed]
129. Pratt, S.; Werker, A.; Morgan-Sagastume, F.; Lant, P. Microaerophilic conditions support elevated mixed culture polyhydroxyalkanoate (PHA) yields, but result in decreased PHA production rates. *Water Sci. Technol.* **2012**, *65*, 243–246. [CrossRef]
130. Wang, X.; Oehmen, A.; Freitas, E.B.; Carvalho, G.; Reis, M.A.M. The link of feast-phase dissolved oxygen (DO) with substrate competition and microbial selection in PHA production. *Water Res.* **2017**, *112*, 269–278. [CrossRef] [PubMed]
131. Gurieff, N.; Lant, P. Comparative life cycle assessment and financial analysis of mixed culture polyhydroxyalkanoate production. *Bioresour. Technol.* **2007**, *98*, 3393–3403. [CrossRef] [PubMed]
132. Takabatake, H.; Satoh, H.; Mino, T.; Matsuo, T. PHA (polyhydroxyalkanoate) production potential of activated sludge treating wastewater. *Water Sci. Technol.* **2002**, *45*, 119–126. [PubMed]
133. Mengmeng, C.; Hong, C.; Qingliang, Z.; Shirley, S.N.; Jie, R. Optimal production of polyhydroxyalkanoates (PHA) in activated sludge fed by volatile fatty acids (VFAs) generated from alkaline excess sludge fermentation. *Bioresour. Technol.* **2009**, *100*, 1399–1405. [CrossRef] [PubMed]
134. Cavaillé, L.; Grousseau, E.; Pocquet, M.; Lepeuple, A.S.; Uribelarrea, J.L.; Hernandez-Raquet, G.; Paul, E. Polyhydroxybutyrate production by direct use of waste activated sludge in phosphorus-limited fed-batch culture. *Bioresour. Technol.* **2013**, *149*, 301–309. [CrossRef] [PubMed]
135. Cavaillé, L.; Albuquerque, M.; Grousseau, E.; Lepeuple, A.S.; Uribelarrea, J.L.; Hernandez-Raquet, G.; Paul, E. Understanding of polyhydroxybutyrate production under carbon and phosphorus-limited growth conditions in non-axenic continuous culture. *Bioresour. Technol.* **2016**, *201*, 65–73. [CrossRef] [PubMed]
136. Johnson, K.; van Geest, J.; Kleerebezem, R.; van Loosdrecht, M.C.M. Short- and long-term temperature effects on aerobic polyhydroxybutyrate producing mixed cultures. *Water Res.* **2010**, *44*, 1689–1700. [CrossRef] [PubMed]
137. Albuquerque, M.G.E.; Martino, V.; Pollet, E.; Avérous, L.; Reis, M.A.M. Mixed culture polyhydroxyalkanoate (PHA) production from volatile fatty acid (VFA)-rich streams: Effect of substrate composition and feeding regime on PHA productivity, composition and properties. *J. Biotechnol.* **2011**, *151*, 66–76. [CrossRef] [PubMed]
138. Villano, M.; Beccari, M.; Dionisi, D.; Lampis, S.; Miccheli, A.; Vallini, G.; Majone, M. Effect of pH on the production of bacterial polyhydroxyalkanoates by mixed cultures enriched under periodic feeding. *Process Biochem.* **2010**, *45*, 714–723. [CrossRef]
139. Chen, Z.; Huang, L.; Wen, Q.; Guo, Z. Efficient polyhydroxyalkanoate (PHA) accumulation by a new continuous feeding mode in three-stage mixed microbial culture (MMC) PHA production process. *J. Biotechnol.* **2015**, *209*, 68–75. [CrossRef]
140. Serafim, L.S.; Lemos, P.C.; Oliveira, R.; Reis, M.A.M. Optimization of polyhydroxybutyrate production by mixed cultures submitted to aerobic dynamic feeding conditions. *Biotechnol. Bioeng.* **2004**, *87*, 145–160. [CrossRef] [PubMed]
141. Moita, R.; Freches, A.; Lemos, P.C. Crude glycerol as feedstock for polyhydroxyalkanoates production by mixed microbial cultures. *Water Res.* **2014**, *58*, 9–20. [CrossRef] [PubMed]
142. Pardelha, F.; Albuquerque, M.G.E.; Reis, M.A.M.; Dias, J.M.L.; Oliveira, R. Flux balance analysis of mixed microbial cultures: Application to the production of polyhydroxyalkanoates from complex mixtures of volatile fatty acids. *J. Biotechnol.* **2012**, *162*, 336–345. [CrossRef] [PubMed]
143. Chen, H.; Meng, H.; Nie, Z.; Zhang, M. Polyhydroxyalkanoate production from fermented volatile fatty acids: Effect of pH and feeding regimes. *Bioresour. Technol.* **2013**, *128*, 533–538. [CrossRef]
144. Jiang, Y.; Marang, L.; Tamis, J.; van Loosdrecht, M.C.M.; Dijkman, H.; Kleerebezem, R. Waste to resource: Converting paper mill wastewater to bioplastic. *Water Res.* **2012**, *46*, 5517–5530. [CrossRef] [PubMed]
145. Valentino, F.; Karabegovic, L.; Majone, M.; Morgan-Sagastume, F.; Werker, A. Polyhydroxyalkanoate (PHA) storage within a mixed-culture biomass with simultaneous growth as a function of accumulation substrate nitrogen and phosphorus levels. *Water Res.* **2015**, *77*, 49–63. [CrossRef]
146. Johnson, K.; Kleerebezem, R.; van Loosdrecht, M.C.M. Influence of ammonium on the accumulation of polyhydroxybutyrate (PHB) in aerobic open mixed cultures. *J. Biotechnol.* **2010**, *147*, 73–79. [CrossRef] [PubMed]

147. Venkateswar Reddy, M.; Venkata Mohan, S. Effect of substrate load and nutrients concentration on the polyhydroxyalkanoates (PHA) production using mixed consortia through wastewater treatment. *Bioresour. Technol.* **2012**, *114*, 573–582. [CrossRef] [PubMed]
148. Moralejo-Garate, H.; Palmeiro-Sanchez, T.; Kleerebezem, R.; Mosquera-Corral, A.; Campos, J.L.; van Loosdrecht, M.C.M. Influence of the cycle lenght on the production of PHA from Glycerol by Bacterial Enrichments in Sequencing Batch Reactors. *Biotechnol. Bioeng.* **2013**, *110*, 3148–3155. [CrossRef] [PubMed]
149. Dionisi, D.; Majone, M.; Vallini, G.; Di Gregorio, S.; Beccari, M. Effect of the applied organic load rate on biodegradable polymer production by mixed microbial cultures in a sequencing batch reactor. *Biotechnol. Bioeng.* **2006**, *93*, 76–88. [CrossRef]
150. Chakravarty, P.; Mhaisalkar, V.; Chakrabarti, T. Study on poly-hydroxyalkanoate (PHA) production in pilot scale continuous mode wastewater treatment system. *Bioresour. Technol.* **2010**, *101*, 2896–2899. [CrossRef] [PubMed]
151. Morgan-Sagastume, F.; Hjort, M.; Cirne, D.; Gérardin, F.; Lacroix, S.; Gaval, G.; Karabegovic, L.; Alexandersson, T.; Johansson, P.; Karlsson, A.; et al. Integrated production of polyhydroxyalkanoates (PHAs) with municipal wastewater and sludge treatment at pilot scale. *Bioresour. Technol.* **2015**, *181*, 78–89. [CrossRef] [PubMed]
152. Bengtsson, S.; Karlsson, A.; Alexandersson, T.; Quadri, L.; Hjort, M.; Johansson, P.; Morgan-Sagastume, F.; Anterrieu, S.; Arcos-Hernandez, M.; Karabegovic, L.; et al. A process for polyhydroxyalkanoate (PHA) production from municipal wastewater treatment with biological carbon and nitrogen removal demonstrated at pilot-scale. *New Biotechnol.* **2017**, *35*, 42–53. [CrossRef] [PubMed]
153. Tamis, J.; Lužkov, K.; Jiang, Y.; van Loosdrecht, M.C.M.; Kleerebezem, R. Enrichment of *Plasticicumulans acidivorans* at pilot-scale for PHA production on industrial wastewater. *J. Biotechnol.* **2014**, *192*, 161–169. [CrossRef] [PubMed]
154. Lee, S.Y. Bacterial polyhydroxyalkanoates. *Biotechnol. Bioeng.* **1996**, *49*, 1–14. [CrossRef]
155. Obruca, S.; Benesova, P.; Marsalek, L.; Marova, I. Use of Lignocellulosic Materials for PHA Production. *Chem. Biochem. Eng. Q.* **2015**, *29*, 135–144. [CrossRef]
156. Ienczak, J.L.; Schmidell, W.; De Aragão, G.M.F. High-cell-density culture strategies for polyhydroxyalkanoate production: A review. *J. Ind. Microbiol. Biotechnol.* **2013**, *40*, 275–286. [CrossRef] [PubMed]
157. Ahn, W.S.; Park, S.J.; Lee, S.Y. Production of poly (3-hydroxybutyrate) from whey by cell recycle fed-batch culture of recombinant *Escherichia coli*. *Biotechnol. Lett.* **2001**, *23*, 235–240. [CrossRef]
158. Shaffer, D.L.; Werber, J.R.; Jaramillo, H.; Lin, S.; Elimelech, M. Forward osmosis: Where are we now? *Desalination* **2015**, *356*, 271–284. [CrossRef]
159. Jung, K.; Choi, J.D.R.; Lee, D.; Seo, C.; Lee, J.; Lee, S.Y.; Chang, H.N.; Kim, Y.C. Permeation characteristics of volatile fatty acids solution by forward osmosis. *Process Biochem.* **2015**, *50*, 669–677. [CrossRef]
160. Kalafatakis, S.; Braekevelt, S.; Vilhelmsen Carlsen, N.S.; Lange, L.; Skiadas, I.V.; Gavala, H.N. On a novel strategy for water recovery and recirculation in biorefineries through application of forward osmosis membranes. *Chem. Eng. J.* **2017**, *311*, 209–216. [CrossRef]
161. Koller, M.; Niebelschütz, H.; Braunegg, G. Strategies for recovery and purification of poly[(*R*)-3-hydroxyalkanoates] (PHA) biopolyesters from surrounding biomass. *Eng. Life Sci.* **2013**, *13*, 549–562. [CrossRef]
162. Madkour, M.H.; Heinrich, D.; Alghamdi, M.A.; Shabbaj, I.I.; Steinbüchel, A. PHA recovery from biomass. *Biomacromolecules* **2013**, *14*, 2963–2972. [CrossRef] [PubMed]
163. Anis, S.N.S.; Md Iqbal, N.; Kumar, S.; Amirul, A.A. Effect of different recovery strategies of P(3HB-co-3HHx) copolymer from *Cupriavidus necator* recombinant harboring the PHA synthase of Chromobacterium sp. USM2. *Sep. Purif. Technol.* **2013**, *102*, 111–117. [CrossRef]
164. Leong, Y.K.; Lan, J.C.-W.; Loh, H.-S.; Ling, T.C.; Ooi, C.W.; Show, P.L. Cloud-point extraction of green-polymers from *Cupriavidus necator* lysate using thermoseparating-based aqueous two-phase extraction. *J. Biosci. Bioeng.* **2017**, *123*, 370–375. [CrossRef] [PubMed]
165. Aramvash, A.; Gholami-Banadkuki, N.; Seyedkarimi, M.-S. An efficient method for the application of PHA-poor solvents to extract polyhydroxybutyrate from *Cupriavidus necator*. *Biotechnol. Prog.* **2016**, *32*, 1480–1486. [CrossRef] [PubMed]

166. Samorì, C.; Abbondanzi, F.; Galletti, P.; Giorgini, L.; Mazzocchetti, L.; Torri, C.; Tagliavini, E. Extraction of polyhydroxyalkanoates from mixed microbial cultures: Impact on polymer quality and recovery. *Bioresour. Technol.* **2015**, *189*, 195–202. [CrossRef] [PubMed]
167. Yang, Y.H.; Jeon, J.M.; Yi, D.H.; Kim, J.H.; Seo, H.M.; Rha, C.K.; Sinskey, A.J.; Brigham, C.J. Application of a non-halogenated solvent, methyl ethyl ketone (MEK) for recovery of poly(3-hydroxybutyrate-co-3-hydroxyvalerate) [P(HB-co-HV)] from bacterial cells. *Biotechnol. Bioprocess Eng.* **2015**, *20*, 291–297. [CrossRef]
168. Neves, A.; Müller, J. Use of enzymes in extraction of polyhydroxyalkanoates produced by *Cupriavidus necator*. *Biotechnol. Prog.* **2012**, *28*, 1575–1580. [CrossRef] [PubMed]
169. Jiang, Y.; Mikova, G.; Kleerebezem, R.; van der Wielen, L.A.; Cuellar, M.C. Feasibility study of an alkaline-based chemical treatment for the purification of polyhydroxybutyrate produced by a mixed enriched culture. *AMB Express* **2015**, *5*, 5. [CrossRef] [PubMed]
170. López-Abelairas, M.; García-Torreiro, M.; Lú-Chau, T.; Lema, J.M.; Steinbüchel, A. Comparison of several methods for the separation of poly(3-hydroxybutyrate) from *Cupriavidus necator* H16 cultures. *Biochem. Eng. J.* **2015**, *93*, 250–259. [CrossRef]
171. Villano, M.; Valentino, F.; Barbetta, A.; Martino, L.; Scandola, M.; Majone, M. Polyhydroxyalkanoates production with mixed microbial cultures: From culture selection to polymer recovery in a high-rate continuous process. *New Biotechnol.* **2014**, *31*, 289–296. [CrossRef] [PubMed]
172. Heinrich, D.; Madkour, M.H.; Al-Ghamdi, M.A.; Shabbaj, I.I.; Steinbüchel, A. Large scale extraction of poly(3-hydroxybutyrate) from *Ralstonia eutropha* H16 using sodium hypochlorite. *AMB Express* **2012**, *2*, 59. [CrossRef] [PubMed]
173. Rathi, D.-N.; Amir, H.G.; Abed, R.M.M.; Kosugi, A.; Arai, T.; Sulaiman, O.; Hashim, R.; Sudesh, K. Polyhydroxyalkanoate biosynthesis and simplified polymer recovery by a novel moderately halophilic bacterium isolated from hypersaline microbial mats. *J. Appl. Microbiol.* **2013**, *114*, 384–395. [CrossRef] [PubMed]
174. Mohammadi, M.; Hassan, M.A.; Phang, L.Y.; Shirai, Y.; Man, H.C.; Ariffin, H. Intracellular polyhydroxyalkanoates recovery by cleaner halogen-free methods towards zero emission in the palm oil mill. *J. Clean. Prod.* **2012**, *37*, 353–360. [CrossRef]
175. Kachrimanidou, V.; Kopsahelis, N.; Vlysidis, A.; Papanikolaou, S.; Kookos, I.K.; Monje Martínez, B.; Escrig Rondán, M.C.; Koutinas, A.A. Downstream separation of poly(hydroxyalkanoates) using crude enzyme consortia produced via solid state fermentation integrated in a biorefinery concept. *Food Bioprod. Process.* **2016**, *100*, 323–334. [CrossRef]
176. Gutt, B.; Kehl, K.; Ren, Q.; Boesel, L.F. Using ANOVA Models to Compare and Optimize Extraction Protocols of P3HBHV from *Cupriavidus necator*. *Ind. Eng. Chem. Res.* **2016**, *55*, 10355–10365. [CrossRef]
177. Martino, L.; Cruz, M.V.; Scoma, A.; Freitas, F.; Bertin, L.; Scandola, M.; Reis, M.A.M. Recovery of amorphous polyhydroxybutyrate granules from *Cupriavidus necator* cells grown on used cooking oil. *Int. J. Biol. Macromol.* **2014**, *71*, 117–123. [CrossRef] [PubMed]
178. Murugan, P.; Han, L.; Gan, C.Y.; Maurer, F.H.J.; Sudesh, K. A new biological recovery approach for PHA using mealworm, *Tenebrio molitor*. *J. Biotechnol.* **2016**, *239*, 98–105. [CrossRef] [PubMed]
179. Kunasundari, B.; Arza, C.R.; Maurer, F.H.J.; Murugaiyah, V.; Kaur, G.; Sudesh, K. Biological recovery and properties of poly(3-hydroxybutyrate) from *Cupriavidus necator* H16. *Sep. Purif. Technol.* **2017**, *172*, 1–6. [CrossRef]
180. Martínez, V.; Herencias, C.; Jurkevitch, E.; Prieto, M.A. Engineering a predatory bacterium as a proficient killer agent for intracellular bio-products recovery: The case of the polyhydroxyalkanoates. *Sci. Rep.* **2016**, *6*, 24381. [CrossRef] [PubMed]
181. Martínez, V.; Jurkevitch, E.; García, J.L.; Prieto, M.A. Reward for *Bdellovibrio bacteriovorus* for preying on a polyhydroxyalkanoate producer. *Environ. Microbiol.* **2013**, *15*, 1204–1215. [CrossRef] [PubMed]
182. Kunasundari, B.; Murugaiyah, V.; Kaur, G.; Maurer, F.H.J.; Sudesh, K. Revisiting the Single Cell Protein Application of *Cupriavidus necator* H16 and Recovering Bioplastic Granules Simultaneously. *PLoS ONE* **2013**, *8*, 1–15. [CrossRef] [PubMed]
183. Anis, S.N.S.; Iqbal, N.M.; Kumar, S.; Al-Ashraf, A. Increased recovery and improved purity of PHA from recombinant *Cupriavidus necator*. *Bioengineered* **2013**, *4*, 115–118. [CrossRef] [PubMed]

184. Koller, M.; Bona, R.; Chiellini, E.; Braunegg, G. Extraction of short-chain-length poly-acetone under elevated temperature and pressure. *Biotechnol. Lett.* **2013**, *35*, 1023–1028. [CrossRef] [PubMed]
185. Aramvash, A.; Gholami-Banadkuki, N.; Moazzeni-Zavareh, F.; Hajizadeh-Turchi, S. An Environmentally Friendly and Efficient Method for Extraction of. *J. Microbiol. Biotechnol.* **2015**, *25*, 1936–1943. [CrossRef] [PubMed]
186. Fei, T.; Cazeneuve, S.; Wen, Z.; Wu, L.; Wang, T. Effective Recovery of Poly-β-Hydroxybutyrate (PHB) Biopolymer from *Cupriavidus necator* Using a Novel and Environmentally Friendly Solvent System. *Biotechnol. Prog.* **2016**, *32*, 678–685. [CrossRef] [PubMed]
187. Furrer, P.; Panke, S.; Zinn, M. Efficient recovery of low endotoxin medium-chain-length poly ([*R*]-3-hydroxyalkanoate) from bacterial biomass. *J. Microbiol. Methods* **2007**, *69*, 206–213. [CrossRef] [PubMed]
188. Mohammadi, M.; Ali Hassan, M.; Phang, L.-Y.; Ariffin, H.; Shirai, Y.; Ando, Y. Recovery and purification of intracellular polyhydroxyalkanoates from recombinant *Cupriavidus necator* using water and ethanol. *Biotechnol. Lett.* **2012**, *34*, 253–259. [CrossRef] [PubMed]
189. Leong, Y.K.; Koroh, F.E.; Show, P.L.; Lan, C.-W.J.; Loh, H.-S. Optimisation of Extractive Bioconversion for Green Polymer via Aqueous Two-Phase System Optimisation of Extractive Bioconversion for Green Polymer. *Chem. Eng. Trans.* **2015**, *45*, 1495–1500.
190. Iqbal, M.; Tao, Y.; Xie, S.; Zhu, Y.; Chen, D.; Wang, X.; Huang, L.; Peng, D.; Sattar, A.; Shabbir, M.A.B.; et al. Aqueous two-phase system (ATPS): An overview and advances in its applications. *Biol. Proced. Online* **2016**, *18*, 18. [CrossRef] [PubMed]

8

Recent Advances in the use of Polyhydroxyalkanoates in Biomedicine

Alejandra Rodriguez-Contreras

Department of Materials Science and Metallurgical Engineering, Universitat Politècnica de Catalunya (UPC), Escola d'Enginyeria de Barcelona Est (EEBE), Eduard Maristany 10-14, 08930 Barcelona, Spain; sandra8855@hotmail.com

Abstract: Polyhydroxyalkanoates (PHAs), a family of natural biopolyesters, are widely used in many applications, especially in biomedicine. Since they are produced by a variety of microorganisms, they possess special properties that synthetic polyesters do not have. Their biocompatibility, biodegradability, and non-toxicity are the crucial properties that make these biologically produced thermoplastics and elastomers suitable for their applications as biomaterials. Bacterial or archaeal fermentation by the combination of different carbohydrates or by the addition of specific inductors allows the bioproduction of a great variety of members from the PHAs family with diverse material properties. Poly(3-hydroxybutyrate) (PHB) and its copolymers, such as poly(3-hydroxybutyrate-co-3-hydroxyvalerate) (PHVB) or poly(3-hydroxybutyrate-co-4-hydroxybutyrate) (PHB4HB), are the most frequently used PHAs in the field of biomedicine. PHAs have been used in implantology as sutures and valves, in tissue engineering as bone graft substitutes, cartilage, stents for nerve repair, and cardiovascular patches. Due to their good biodegradability in the body and their breakdown products being unhazardous, they have also been remarkably applied as drug carriers for delivery systems. As lately there has been considerable and growing interest in the use of PHAs as biomaterials and their application in the field of medicine, this review provides an insight into the most recent scientific studies and advances in PHAs exploitation in biomedicine.

Keywords: polyhydroxyalkanoates; biomedicine; biomaterials; Poly(3-hydroxybutyrate); tissue engineering; wound healing; delivery system; poly(3-hydroxybutyrate-co-3-hydroxyvalerate) (PHVB); poly(3-hydroxybutyrate-co-4-hydroxybutyrate)

---

1. Introduction

Synthetic plastics are used in many different applications, as they are a family of versatile materials. However, there is a global awareness of the environmental impact of these fossil-based polymers. At the same time, there is growing recognition that organic matter of biological origin can be a worthy alternative [1]. In this regard, natural polymers or biopolymers show many advantages relative to petrochemical materials, as they are biodegradable and produced from renewable sources. Furthermore, due to their similarity to the native natural environment, their biopolymer functions show good biological performance and adaptability, and adequate body reaction [2]. This makes them very attractive for their application not only in biomedicine but also in other fields such as pharmacology and biotechnology [1]. In biomedicine as the theoretical branch of medicine that applies the principles of biology, biochemistry, and biophysics to medical research and practice, the combination of synthetic and natural polymers is frequently used [3–6].

Polyhydroxyalkanoates (PHAs) are a big family of naturally produced polyesters. Chemically, they are linear polymers composed of hydroxyalkanoate units as their basic structure (Figure 1a).

These biopolymers are accumulated within the cytoplasm of diverse microorganisms under conditions of nutrient depletion and in the presence of an excess of carbon source [7–9]. They appear as granules and function as carbohydrate and energy storage (Figure 1c). PHAs can be produced by biotechnological processes via bacterial and archaeal fermentation. The members of the PHA family differ widely in their structure and properties (Figure 1a,d,e) depending on the producing microorganism, biosynthesis conditions, and type of carbon source used in the production process [8–10]. In general, PHAs are thermoplastic or elastomeric, and their sufficiently high molecular mass provides them with properties similar to those of conventional petrochemical polymers (Figure 1d) [11–14]. More specifically, they can be classified depending on their monomeric composition: short-chain-length PHA (*scl*-PHA), consisting of 3 to 5 carbon atoms per monomer; medium-chain-length PHA (*mcl*-PHA), with 6 to 14 carbon atoms; and the rather rare group of long-chain-length PHA (*lcl*-PHA), which presents more than 14 carbon atoms [15,16]. The vast majority of microorganisms synthesize either *scl*-PHAs containing primarily 3-hydroxybutyrate (3HB) units or *mcl*-PHAs containing 3-hydroxyhexanoate (3HHx), 3-hydroxyoctanoate (3HO), 3-hydroxydecanoate (3HD), and 3-hydroxydodecanoate (3HHD) as the major monomers [7,15]. While *scl*-PHAs are crystalline and feature typical thermoplastic properties, *mcl*-PHA resins resemble elastomers and latex-like materials with typically low glass transition temperature and lower molecular mass if compared to *scl*-PHA [15,17,18].

Poly(3-hydroxybutyrate) (PHB) is the most frequently occurring PHA member and is a linear, unbranched homopolymer consisting of (R)-3-hydroxybutyric acid units. When extracted from bacterial biomass, PHB tends to crystallize [19]. Although its applications are limited mainly by its high crystallinity and brittleness, which reduce its flexibility and ductility, PHB can be modified by simply physical blending or chemical alteration to fine-tune its mechanical properties [10,20]. Another strategy to modify its mechanical properties is by copolymerization via bacterial fermentation using different precursors (Figure 1d). For instance, a common PHB copolymer, poly(3-hydroxybutyrate-co-3-hydroxyvalerate) (PHBV), is characterized as less crystalline and more flexible than PHB itself [21], and its properties can be varied according to the 3-hydroxyvalerate (3HV) content in the structure (Figure 1d) [22]. PHBV is usually produced by adding valeric acid to the fermentation medium [23]. Poly(3-hydroxybutyrate-co-4-hydroxybutyrate) (PHB4HB) copolymer is another of the most well-known members of the PHA family. With higher 4-hydroxybutyric acid (4HB) content, PHA is more elastomeric and with outstanding elongation at break (Figure 1d) [24]. These PHAs members together with poly(3-hydroxybutyrate-co-3-hydroxyvalerate-co-3-hydroxyhexanoate) (PHBVHHx) represent the most commonly applied PHAs in biomedicine [25].

Current research on PHAs focuses on subjects such as gaining a better understanding of the mechanisms related to their biosynthesis, or how to modulate PHAs properties for different applications. The development of natural and recombinant microorganisms to efficiently produce PHAs and the finding of alternative raw materials that lead their production to more competitive costs are also important research topics [26,27].

PHAs show major advantages compared with traditional synthetic polymers. However, it is because of their biodegradability, biocompatibility, and non-toxicity that they are especially appealing materials for biomedical applications. Furthermore, an additional benefit is their unchanged local pH value during degradation. This makes them well tolerated by cells and the immune system compared to other polymers clinically used such as poly(lactide-co-glycolide) (PLGA), poly(ε-caprolactone) (PCL), poly(glycolic acid) (PGA), and poly(lactic acid) (PLA) [20]. In the last decades, there has been an increase in PHAs exploitation in biomedicine. Therefore, this review is an attempt to summarize the most important advances published in the last few years on the use of microbially originated PHAs used in this field.

a)

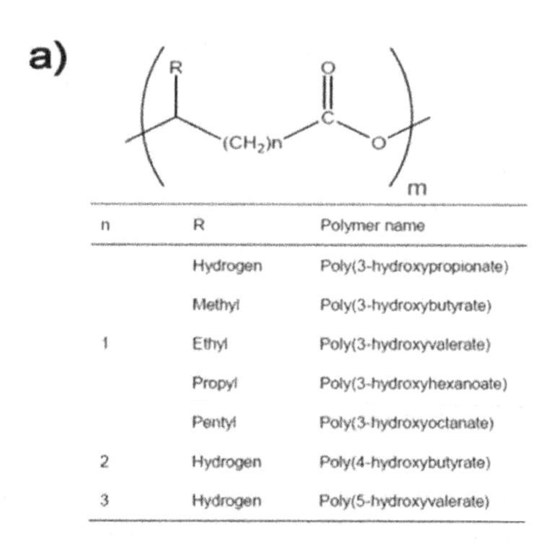

n	R	Polymer name
	Hydrogen	Poly(3-hydroxypropionate)
	Methyl	Poly(3-hydroxybutyrate)
1	Ethyl	Poly(3-hydroxyvalerate)
	Propyl	Poly(3-hydroxyhexanoate)
	Pentyl	Poly(3-hydroxyoctanate)
2	Hydrogen	Poly(4-hydroxybutyrate)
3	Hydrogen	Poly(5-hydroxyvalerate)

b)

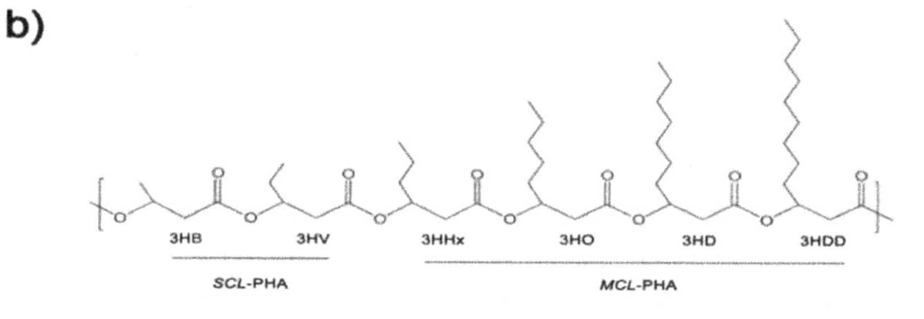

c)

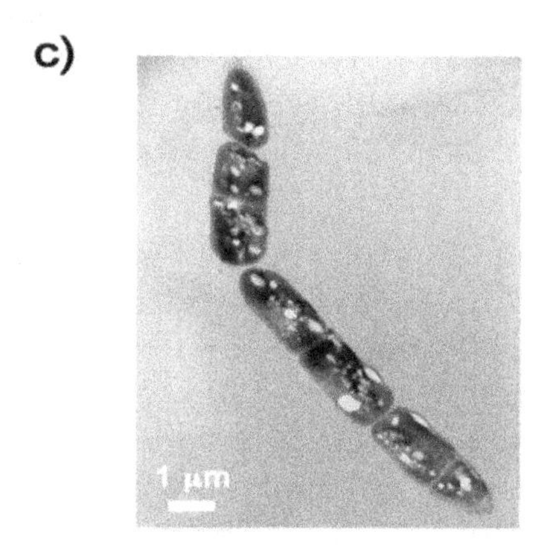

d)

Properties	PHB	P(3HB-3HV) for 4 – 95% (mol/mol) 3HV	PP
Molecular mass (10^5 g/mol)	1 – 8	3	2.2 – 7
Density (kg/m^3)	1.25	1.20	0.905
T_m (°C)	171 – 182	75 – 172	176
Crystallinity degree (X_c) (%)	80	55 – 70	70
T_g (°C)	4 – 10	-13 – +8	-10
O_2 Permeability [(cm^3/(m^{2}atm)]	45	n. d.	1700
UV Resistance	Good	Good	Bad
Resistance to solvents	Bad	Bad	Good
Tensile strength (Mpa)	40	25 – 30	38
Elongation to break (%)	6	8 – 1200	400
Young's modulus (GPa)	3.5	2.9 (3% 3HV); 0.7 (25% 3HV)	1.7
Biodegradability	Yes	Yes	No

e)

Parameter	Values for					
	PHB	P(3HB-3% 3HV)	P(3HB-20%3HV)	P4HB	P(3HB-16%4HB)	P(3HO-3HHx)
Melting temperature (T_m) (°C)	177	170	145	60	152	61
Glass transition temperature (T_g) (°C)	4	-	-1	-50	-8	-36
Tensile strength (MPa)	40	38	32	104	26	9
Young´s modulus (GPa)	3.5	2.9	1.2	149	-	0.008
Elongation at break (%)	6	-	50	1000	444	380

Figure 1. (**a**) Chemical structure of the polyhydroxyalkanoates (PHA) biopolymer family, the monomer number m range from 100 to 30,000 [12]. (**b**) Some commonly synthesized scl-PHA monomers (scl-HA) and mcl-PHA monomers (mcl-HA). 3HB: 3-hydroxybutyrate, 3HV: 3-hydroxyvalerate, 3HHx: 3-hydroxyhexanoate, 3HO: 3-hydroxyoctanoate, 3HD: 3-hydroxydecanoate, 3HDD: 3-hydroxydodecanoate. (**c**) Transmission electron microscopy micrograph of *Bacillus megaterium* uyuni S29 after 4 h of fermentation showing PHB granules as refractile inclusion bodies [28]. (**d**) Some physical, thermal, chemical, and mechanical properties of PHB and poly(3-hydroxybutyrate-co-3-hydroxyvalerate) (PHBV) compared to those of the petrol-based polypropylene (PP) [12,29]. (**e**) Table of properties of some PHAs members and copolymers [12,29].

2. Tissue Engineering

Tissue engineering is an interdisciplinary field of research focused on the creation of vital tissues by a combination of biomaterials, cells, and bioactive molecules, aiming to repair damaged or diseased tissues and organs [30]. Tissues can be classified as hard tissue substitutes, such as bone and cartilage, or soft tissues, such as vascular and skin grafts [31]. The biomaterial used must have two crucial features to function as tissue repairer: to possess mechanical properties for supporting the organ during new tissue regeneration, and enhanced surface topography to allow efficient cell adhesion and proliferation. In this regard, engineered scaffolds are designed to closely mimic the topography, spatial distribution, and chemical environment corresponding to the native extracellular matrix of the intended tissue in order to support cell growth and differentiation [32]. PHAs constitute a great alternative for tissue engineering due to their versatility regarding their mechanical properties, combined with great biocompatibility with minimal tissue toxicity and degradability. Thus, PHAs have been exploited for the replacement and healing of both hard and soft tissues in tissue engineering to repair cartilage, cardiovascular tissues, skin, bone marrow, and nerve conduits [22,33–35].

2.1. *Hard Tissue*

2.1.1. Bone Tissue Engineering

Bone tissue engineering refers to the regeneration of new bone by providing mechanical support while inducing cell growth. For this application, hydroxyapatite (HA), inorganic substances, hydrogels, and even other biocompatible polymers are used to blend with PHAs to optimize their compressive elastic modulus and maximum stress. For instance, Degli Esposti et al. [36] very recently published the exploitation of a mixture of PHB with HA particles for the development of bio-resorbable porous scaffolds for bone tissue regeneration. The osteoinductivity and osteoconductivity of the bioactive scaffolds were attained mainly due to the incorporation of HA. By combining $CaCO_3$-mineralized piezoelectric with PHB- and PHBV-based scaffolds, Chernozem et al. [37] elaborated PHA biocomposites that provided biodegradability and stimulated bone tissue repair. The presence of mineral led to a pronounced apatite-forming behavior of the biodegradable PHAs scaffolds, and this turned out to stimulate the growth of the bone tissue. A more complex system is the one produced by Meischel et al. [38], who evaluated the response of bone to PHA composite implants in the femora of growing rats. Composites were constituted by PHB with zirconium dioxide, Herafill® (calcium sulfate, calcium carbonate, triglycerides, and gentamicin; produced by Hereus), and Mg-alloy WZ21. Longitudinal observation of the bone reaction at the implant site and resorption of the implanted pins were monitored, and the results showed that PHB composited with zirconium dioxide and 30% Herafill possessed the highest values of bone accumulation. The authors concluded that the mechanical properties (elastic modulus, tensile strength, and strain properties) of PHB composites in these conditions were close to that of bone.

Hydrogels can be used to create scaffolds with a well-interconnected porous structure. However, they provide poor mechanical stability and very low bioactivity, failing to create suitable constructs for bone tissue engineering. In order to improve the mechanical stability of hydrogels, Sadat et al. [39] developed a scaffold system based on combining a mix of biodegradable PHB and HA with a protein-based hydrogel in a single tri-layered scaffold. These scaffolds provided high strength, had the ability to encapsulate cells, and enhanced bone cell adaptability (Figure 2a).

In their study, Ding et al. [40] mixed the natural polyester PHB with the synthetic polyester PCL. They fabricated PHB/PCL/58S sol-gel bioactive glass hybrid scaffolds by electrospinning of the polymers and inorganic substances. The combination of the high stiffness of PHB, the flexibility of PCL, and the bioactivity of 58S bioactive glass in one single fibrous structure showed potential for using in bone tissue engineering integration. The composite enhanced the primary biological response of osteoblast-like cells and their viability, and significantly increased alkaline phosphatase enzyme activity.

2.1.2. Cartilage

Tissue engineering of cartilage provides promising strategies for the regeneration of damaged articular cartilage. There are significant challenges, since current surgical procedures are unable to restore normal cartilage function. It is important to create an alternative that matches the long-term mechanical stability and durability of this native hard tissue [41]. Some recent studies demonstrated that the use of PHAs can be a solution. Ching et al. [41] produced diverse blends of PHB with poly(3-hydroxyoctanoate) (P3HO) as biodegradable polymer scaffolds. By studding different ratios of both polymers, they optimized their structure, stiffness, degradation rates, and biocompatibility. At a polymer rate (PHB/P3HO) of 1:0.25, the blend closely mimicked the collagen fibrillar meshwork of native cartilage and attained the stiffness of native articular cartilage (Figure 2b). They concluded that by fine tuning the ultrastructure and mechanical properties using different blends, these two polymers allowed the production of a cartilage repair kit for clinical use and the reduction of the risk of developing secondary osteoarthritis. More recently, Toloue et al. [42] evaluated the mechanical properties and cell viability of a mix of PHB with 3% chitosan reinforced with alumina as a scaffold for cartilage reparation. The presence of alumina nanowires significantly increased the tensile strength of PHB and PHB/chitosan scaffolds. In vitro studies showed that chondrocyte cells spread more on the composite than on pure PHB scaffolds. The authors concluded that the electrospun scaffold of PHB with chitosan and 3% alumina had the potential to be applied in cartilage tissue engineering.

2.2. *Soft Tissue*

2.2.1. Cardiac Tissue Engineering

Cardiac tissue engineering is currently a prime focus of research because of an enormous clinical need. *Mcl*-PHAs have demonstrated exceptional properties for cardiac tissue engineering applications. They are more elastic than other members of their family, showing an elastomeric nature, higher glass transition temperatures, and the potential to integrate with the myocardial network and be conjugated with bioactive molecules, such as vascular endothelial growth factor, to further increase cellular attachment, viability, and proliferation [18].

Guo et al. [35] summarized the recent use of P4HB as a promising biomaterial for applications in cardiac tissue engineering such as congenital heart defects, heart valves, and vascular grafts. The versatile material is also used in other applications as an absorbable monofilament for sutures, and hernia, tendon, and ligament repair, among others. Bagdadi et al. [43] used P3HO as a potential material for cardiac tissue engineering. They fabricated P3HO-based multifunctional cardiac patches with mechanical properties that were close to those of cardiac muscle. Furthermore, they were shown to be as good as collagen in terms of cell viability, proliferation, and adhesion. Likewise, Constantinides et al. [18] used *mcl*-PHAs for this application. They first produced the *mcl*-PHAs by bacterial fermentation with *Pseudomonas mendocina* CH50 using glucose as the sole carbon source under nitrogen limiting conditions. Then, the obtained *mcl*-PHAs were reinforced with PCL (5%) to produce thin films. The blended structures were implanted in post mortem murine heart in situ. The composites demonstrated possessing a great potential for maximizing tissue regeneration in myocardial infarction. Besides this study, there was research using PHBVHHx [25] in the form of membranes and PHB4HB [44] for the production of cardiac patches. These studies were carried out with stem cells of different origin.

Valvular heart diseases are the third leading cause of cardiovascular disease. Thus, heart valve tissue engineering (HVTE) has appeared as an important strategy to treat these disorders. Ideally, a designed construct should withstand the native dynamic mechanical environment, guide the regeneration of the diseased tissue, and more importantly, have the ability to grow with the patient's heart [45]. Xue et al. [45] summarized different types of synthetic biodegradable elastomers that have been explored for HVTE. Referring to a published work of Chen et al. [46], they specify that this class of elastomers, the PHAs, are generally stronger than polyurethane-based elastomers and more suitable to work under dynamic conditions such as those of cardiovascular tissue.

2.2.2. Wound Healing

The need for novel materials in the effective regeneration of injured skin is a serious concern in reconstructive medicine [47]. Many natural (collagen, alginic acid, hyaluronic acid, chitosan, fucoidan) and synthetic (teflon, polyurethanes, methyl methacrylate) polymers are being used in the preparation of artificial dressing materials for wound healing applications [48]. This complex application requires that the biomaterial fulfills the functions of healthy skin, which has an antimicrobial effect, promotes moist wound environment, permits gaseous exchange, provides mechanical protection, and is sufficiently elastic to fit the wound shape [47]. The PHA family of biopolymers has also extended in this novel medical area. One major factor inhibiting natural wound-healing processes is bacterial infection, especially in chronic wounds [49]. There are studies on wound healing with antibiotic delivery systems and applying PHAs as a remedy. For instance, Marcano et al. [49] optimized the micro/nano-structure of a wound dressing in order to obtain a more efficient antibiofilm protein-release profile for biofilm inhibition and/or detachment. Thus, they developed a three-dimensional (3D) substrate based on asymmetric PHA membranes to entrap an antibiofilm protein (Figure 2c). Similarly, the team of Volova [47] constructed wound dressings from PHB4HB membranes for skin wound repair and evaluated their effectiveness in experiments with laboratory animals. The nonwoven membranes of PHB4HB carried the culture of allogenic fibroblasts. The use of the biopolymer reduced inflammation, enhanced the angiogenic properties of the skin, and facilitated the wound healing process.

2.2.3. PHAs for Organ Tissues

PHBVHHx is considered a promising PHA member for the growth of stem cells, and certain studies utilized it as biomaterial for the preparation of three-dimensional supportive scaffolds for organ tissue. In some works [50,51], PHBVHHx films and scaffolds were developed and loaded with mesenchymal stem cells from human umbilical cord (UC-MSCs) to recover injured liver. Biopolymer scaffolds were transplanted into liver-injured mice, and the results demonstrated that the PHA scaffold significantly promoted the recovery of injured liver and could be used for liver tissue engineering. In the case of the work by Li et al. [50], differences between PHBVHHx and some other commonly used biopolymers such as PLA, PHB4HB, and PHBHHx were examined by loading them with stem cells into their scaffolds (Figure 2d). They concluded that the PHBVHHx structures exhibited the highest cell attachment and, when loaded with mesenchymal stem cells, significantly improved the recovery of injured liver.

PHAs have also been used in tendon healing. In order to improve the initial biomechanical repair strength of tendon tears at risk of failure, Tashjian et al. [52] produced a bioresorbable scaffold to reinforce the suture-tendon interface in rotator cuff repairs. A study of cyclic and ultimate failure properties of PHA mesh was conducted, obtaining better mechanical results than in the control condition (without the reinforcement).

As a hard tissue, Findrik et al. [31] exploited a blend of PLA and PHB to use it as a tubular substitute for urethra replacement. They dealt with the combination of both polymers to provide stabile conditions during the engineering of the replacement by adjusting material degradation and viscosity. By using a 3D printing process, a cubic sample representing basic scaffold structures and a tubular one serving as urethra substitution were designed.

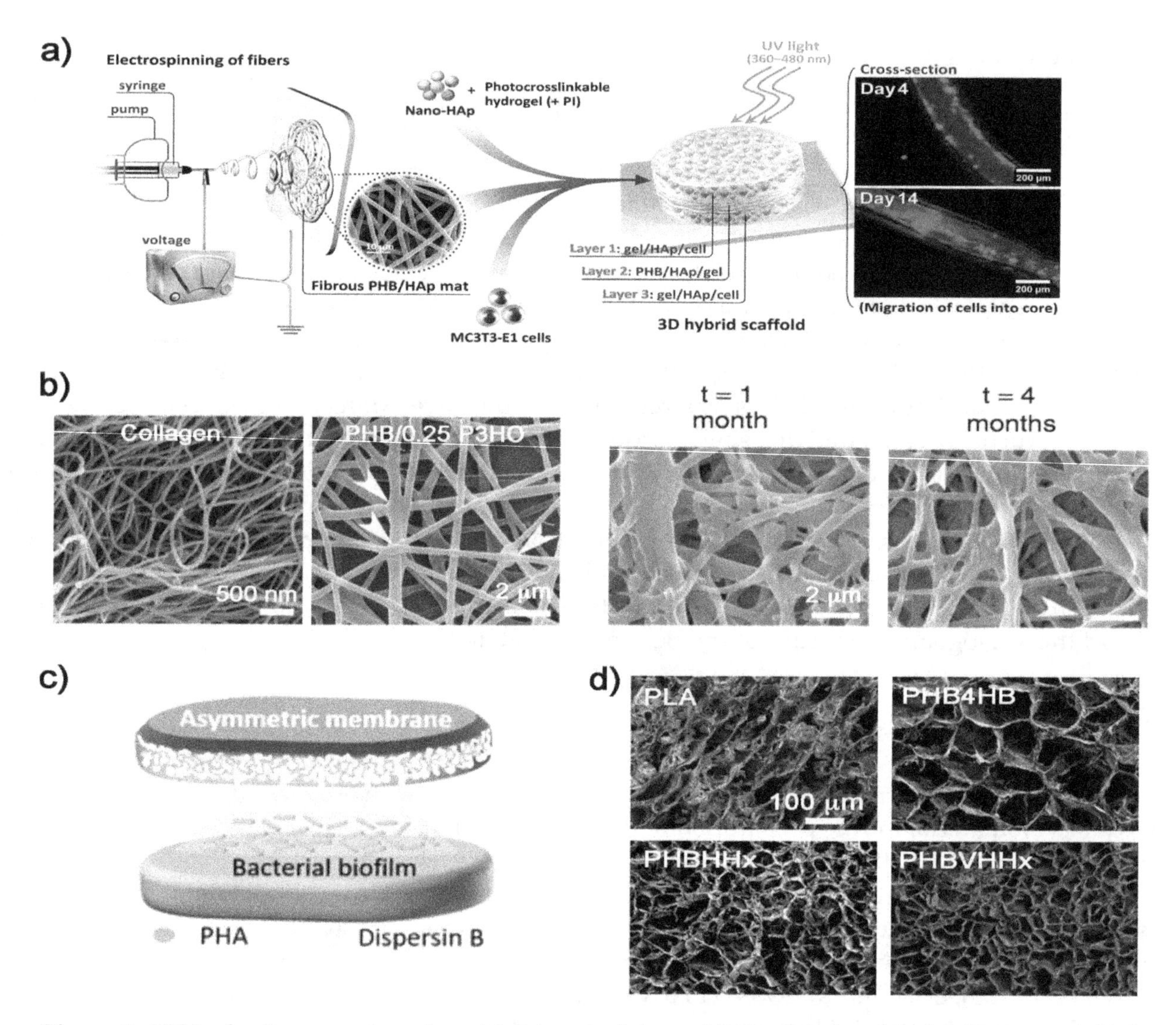

Figure 2. PHAs for tissue engineering: (**a**) Scheme of the published study of Sadat-Shojai et al. [39] where a cell-laden tri-layered scaffold of PHB with hydroxyapatite (HA) was performed to enhance bone regeneration in vivo. (**b**) Scanning electron microscopy (SEM) micrographs of PHB/P3HO scaffolds where the electrospun fibers with a ratio blend of 1:0.25 provided structures more similar to collagen natural fibers. Biopolymeric fibers after hydrolytic degradation [41]. (**c**) Scheme of the asymmetric PHA membranes entrapping an anti-biofilm protein (dispersin B) for wound healing [49]. (**d**) SEM micrographs of biopolymers scaffolds from Li et al. [50]. The biopolymer structures displayed different pore sizes where stem cells were loaded into, and the PHBVHHx ones exhibited the highest cell attachment.

3. Drug Delivery Systems

One of the key reasons for the common use of the PHA biopolymer family as drug carriers is their biodegradability under different environments. A vast number of microorganisms secrete extracellular PHA-hydrolyzing enzymes (PHA depolymerases and other enzymes) to degrade PHA polymers into oligomers and monomers, which subsequently act as nutrients inside the cells [53–55]. PHAs typically degrade by hydrolytic and bacterial depolymerase mechanisms over 52-plus weeks in vivo [56]. Furthermore, there are studies that compare PHAs biodegradability with that of other synthetic or semisynthetic polymers. Gil-Castell et al. [57] compared the durability of PLGA, polydioxanone (PDO), polycaprolactone (PCL), and PHB scaffolds. Results showed that for long-term applications, PCL and PHB were more appropriate materials than PLGA and PDO, which could be used in short-term applications. Regarding their biodegradability in ultra-pure water and phosphate buffer solution at 37 °C, the PHB molar mass progressively decreased, reaching almost 50% after 650 days of immersion. However, PHAs' biodegradability depends on different factors such as the composition of the biopolymer, its stereo regularity, crystallinity (degradability decreases as the overall crystallinity increases), molecular mass (biopolymers are generally biodegraded more rapidly when their molecular mas is lower), and environmental conditions (temperature, moisture level, pH, and nutrient supply) [58]. This makes this biopolymer family especially appealing for delivery systems, since the controllable retarding properties of systems based on PHAs can be modulated mainly by their molecular mass and copolymer composition. Moreover, PHAs have already demonstrated a significant impact on the drug bioavailability, better encapsulation, and less toxicity of biodegradable polymers [59].

In the literature, several reviews of the use of PHAs as carriers in biomedine can be found. They embrace the shape of particles, spheres, micelles, liposomes, vesicles, or capsules as therapeutic delivery carriers. For instance, Masood et al. [60] reviewed the current implications of encapsulation of anticancer agents within PHAs, PLGA, and cyclodextrin-based nanoparticles to precisely target the tumor site. The recent scientific developments in the preparation of functionalized PHAs, PHA-drug and PHA-protein conjugates, multifunctional PHA nanoparticles, and micelles as well as biosynthetic PHA particles for drug delivery were reviewed by Michalak et al. [61]. The recent advances of using PHA-based nano-vehicles as therapeutic delivery carriers were summarized by Li and Loh [59]. Pramual et al. [62] developed and investigated nanoparticles of PHAs as carriers of a hydrophobic photosensitizer for photodynamic therapy. Besides these reviews, a patent has been published on the fabrication of a delivery system comprising *scl*-PHA nanoparticles having an anticancer drug encapsulated for oral administration [63]. Also, a similar study on the production of PHB and PHB/poly(ethylene glycol) (PEG)-based microparticles loaded with antitumor drugs by the spray-drying technique was recently published [64]. Apart from these, there are not many more new studies in which PHAs spherical shape structures are considered for drug delivery.

Manero's group has been working in the exploitation of PHAs in biomedicine, and they have recently published some studies focused on the application of PHAs as therapeutic delivery carriers. They produced antibiotic (doxycycline)-loaded micro- and nano-particles of PHB with different methodologies [65]. The produced carriers were capable of diffusing the active principle from the material to the media, creating a bacteria-free protective region. Later, new strategies for combining the antibacterial properties of doxycycline-loaded PHB micro- and nano-spheres on titanium (Ti) were developed to obtain implant surfaces with antibacterial activity [66]. Furthermore, they studied a novel approach to benefit the synergistic effects of antifouling PEG together with doxycycline-loaded PHB spheres (Figure 3a,b).

The use of PHAs for delivery systems has been studied with different structures. For instance, Lee et al. [67] developed a system of drug-containing PHA fibers that can be electrospun directly onto a metal stent in order to form a biocompatible coating (Figure 3c). PHAs have been used as matrixes to construct release formulations of antibiotic delivery, providing them with antimicrobial, antifungal, anti-biofilm, anti-inflammatory and virucidal properties dependent on the conjugated/enclosed therapeutic agent. Manero's group has exploited PHAs matrixes as coatings with an antibacterial delivery effect [68–70]. Aiming to obtain antimicrobial surfaces to prevent implant infections, they studied different strategies for developing antibacterial coatings on Ti and Tantalum (Ta). The surface of the biometals was coated with different PHAs (PHB, PHBV, and PHB4HB) using a dip-coating technique. Water-in-oil PHAs emulsions with the bioactive agents were produced to use them as coating fluids. The systems designed for drug delivery not only proved to assure the elimination of the first stage of bacterial biofilm formation (bacterial adhesion), but also their proliferation, since the biopolymer coating with antibiotic was able to degrade with time under physiological conditions, thus guaranteeing a controlled drug release over time (Figure 3d).

Complex systems for drug delivery, such as the one published by Timin et al. [71], have also been developed. In this example, the authors deposited polymer and hybrid microcapsules, which were used as drug carriers, onto polymer microfiber scaffolds of PCL, PHB, and PHB doped with the conductive polyaniline (PANi). The immobilization of the microcapsules (loaded with bioactive molecules) onto the scaffold surfaces enabled multimodal triggering by physical and biological stimuli, providing the controllable release of the drug from the scaffolds. PHB and PHB-PANi scaffolds promoted the adhesion of mesenchymal stem cells compared to that of the PCL scaffolds. With this methodology, they provided a way to incorporate bioactive compounds onto polymer scaffolds, which makes these multimodal materials suitable for personalized drug therapy and bone tissue engineering.

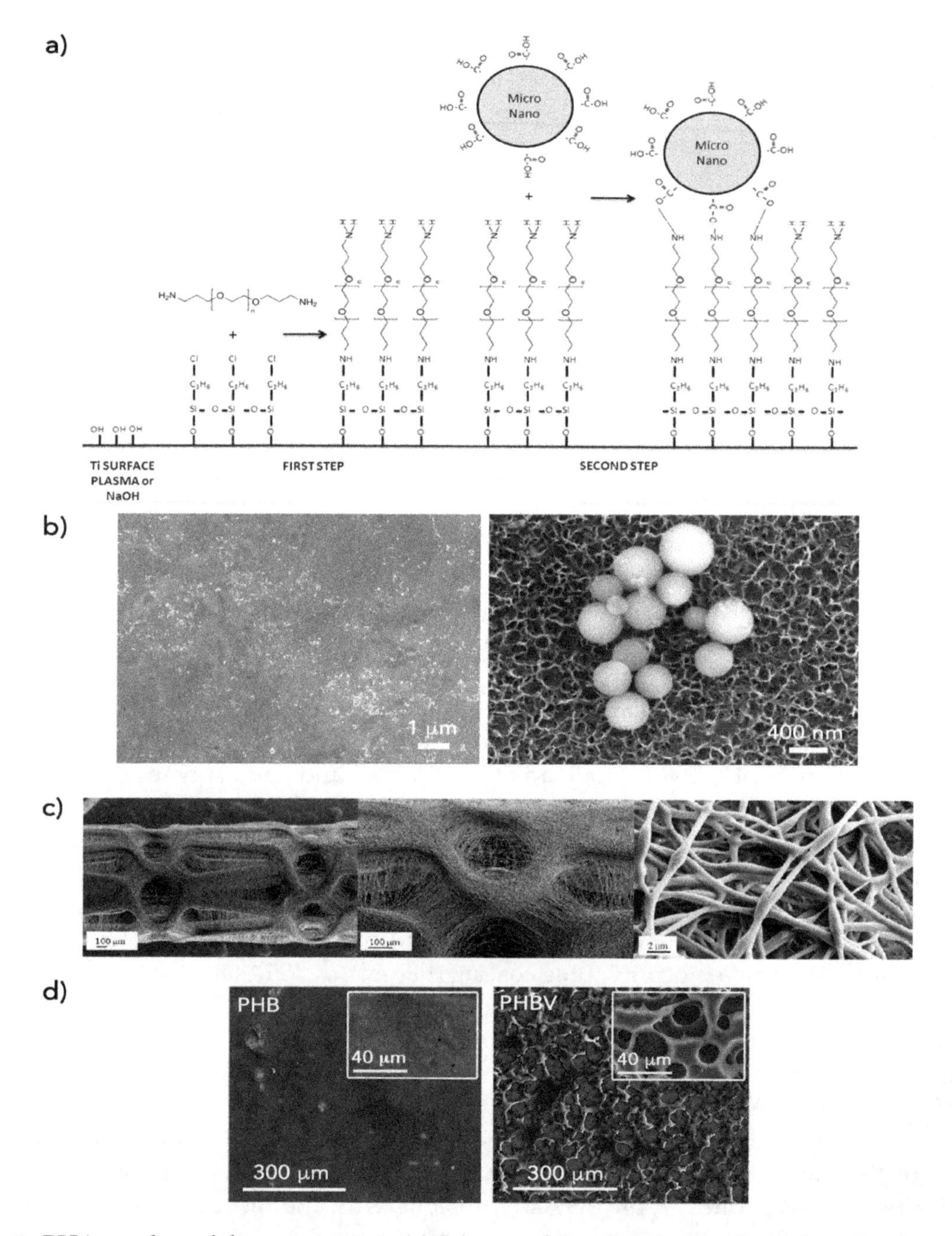

Figure 3. PHAs as drug delivery systems. (**a**) Scheme of the chemical reaction for adhesion of PHB micro- and nano-spheres on Ti surfaces: activation of the Ti surfaces (by plasma or NaOH treatment), silanization with the alkoxysilane 3-chloropropyltriethoxysilane (CPTES), covalent bounding with difunctionalized poly(ethylene glycol) (PEG), and covalent bonding with doxycycline-loaded PHB-spheres. (**b**) Field emission scanning electron microscopy micrographs of Ti surfaces with doxycycline-loaded spheres of PHB [66]. (**c**) SEM images of direct coatings of paclitaxel loaded P(3HB-co-95 mol% 4HB) nanofibers onto a metal stent (40×, 100×, and 5000×) [67]. (**d**) FESEM images at different magnifications of PHB and PHBV matrixes that totally coated Ti surfaces [69].

4. Conclusions

In the field of biomedicine, biopolymers show many advantages that make them superior to synthetic polymers, predominately because of their natural origin. PHAs represent a big family of biologically produced polymers that show common properties such as biocompatibility, biodegradability, and non-toxicity. These properties together with the ease of PHAs for tuning and adapting their mechanical properties, either by combination with other substances or by copolymerization in the biotechnological production process, make them very attractive for their application in different sectors, especially biomedicine. In the last few years, PHAs have been studied to be used in tissue engineering for hard and soft tissue replacement, and as therapeutic delivery carriers. According to the studies presented in this review and the successful results discussed, given the versatile properties that can be provided by PHAs and the need to continue improving biomedical solutions, PHAs will most likely continue to be investigated as an appealing alternative, penetrating the biomedical market in a not-too-distant future.

References

1. Rodríguez-Contreras, A. Concepts and Recent Advances on Biopolymers for Biomedical Applications: Special Mention to the PHAs Family. *Adv. Biotechnol.* **2019**, *IV*, 1–27.
2. Mavelil-Sam, R.; Pothan, L.A.; Thomas, S. Polyssacharide and Protein BsedAerogels: An Introductory Outlook. In *Biobased Aerogels: Polysaccharide and Protein-Based Materials*; Thomas, S., Pothan, L.A., Mavelil-Sam, R., Eds.; Royal Society of Chemistry: London, UK, 2018.
3. Won, J.-E.; El-Fiqi, A.; Jegal, S.-H.; Han, C.-M.; Lee, E.-J.; Knowles, J.C.; Kim, H.-W. Gelatin-apatite bone mimetic co-precipitates incorporated within biopolymer matrix to improve mechanical and biological properties useful for hard tissue repair. *J. Biomater. Appl.* **2013**, *28*, 1213–1225. [CrossRef] [PubMed]
4. Augustine, R. Skin bioprinting: A novel approach for creating artificial skin from synthetic and natural building blocks. *Prog. Biomater.* **2018**, *7*, 77–92. [CrossRef] [PubMed]
5. Park, H.-H.; Ko, S.-C.; Oh, G.-W.; Heo, S.-J.; Kang, D.-H.; Bae, S.-Y.; Jung, W.-K. Fabrication and characterization of phlorotannins/poly (vinyl alcohol) hydrogel for wound healing application. *J. Biomater. Sci. Polym. Ed.* **2018**, *29*, 972–983. [CrossRef] [PubMed]
6. Caddeo, S.; Mattioli-Belmonte, M.; Cassino, C.; Barbani, N.; Dicarlo, M.; Gentile, P.; Baino, F.; Sartori, S.; Vitale-Brovarone, C.; Ciardelli, G. Newly-designed collagen/polyurethane bioartificial blend as coating on bioactive glass-ceramics for bone tissue engineering applications. *Mater. Sci. Eng. C* **2019**, *96*, 218–233. [CrossRef] [PubMed]
7. Lee, S.Y. Bacterial polyhydroxyalkanoates. *Biotechnol. Bioeng.* **1996**, *49*, 1–14. [CrossRef]
8. Valappil, S.P.; Peiris, D.; Langley, G.J.; Herniman, J.M.; Boccaccini, A.R.; Bucke, C.; Roy, I. Polyhydroxyalkanoate (PHA) biosynthesis from structurally unrelated carbon sources by a newly characterized Bacillus spp. *J. Biotechnol.* **2007**, *127*, 475–487. [CrossRef] [PubMed]
9. Chen, G. *Plastics from Bacteria: Natural Functions and Applications, Microbiology Monographs*; Chen, G.G.-Q., Ed.; Springer: Berlin/Heidelberg, Germany, 2010; Volume 14.
10. Chen, G.-Q. A microbial polyhydroxyalkanoates (PHA) based bio- and materials industry. *Chem. Soc. Rev.* **2009**, *38*, 2434–2446. [CrossRef]
11. Koller, M. Production of Poly Hydroxyalkanoate (PHA) biopolyesters by extremophiles? *MOJ Polym. Sci.* **2017**, *1*, 69–85.
12. Sudesh, K.; Abe, H.; Doi, Y. Synthesis, structure and properties of polyhydroxyalkanoates: Biological polyesters. *Prog. Polym. Sci.* **2000**, *25*, 1503–1555. [CrossRef]
13. Yadav, P.; Yadav, H.; Shah, V.G.; Shah, G.; Dhaka, G. Biomedical Biopolymers, their Origin and Evolution in Biomedical Sciences: A Systematic Review. *J. Clin. Diagn. Res.* **2015**, *9*, ZE21–ZE25. [CrossRef] [PubMed]
14. Koller, M. Polyhydroxyalkanoate Biosynthesis at the Edge of Water Activity-Haloarchaea as Biopolyester Factories. *Bioengineering* **2019**, *6*, 34. [CrossRef] [PubMed]
15. Reddy, C.S.K.; Rashmi Ghai, R.; Kalia, V.C. Polyhydroxyalkanoates: An overview. *Bioresour. Technol.* **2003**, *87*, 137–146. [CrossRef]
16. Koller, M. Chemical and Biochemical Engineering Approaches in Manufacturing Polyhydroxyalkanoate

(PHA) Biopolyesters of Tailored Structure with Focus on the Diversity of Building Blocks. *Chem. Biochem. Eng. Q.* **2019**, *32*, 413–438. [CrossRef]

17. Koller, M.; Salerno, A.; Dias, M.; Reiterer, A.; Braunegg, G. Modern Biotechnological Polymer Synthesis: A Review. *Food Technol. Biotechnol.* **2010**, *48*, 255–269.
18. Constantinides, C.; Basnett, P.; Lukasiewicz, B.; Carnicer, R.; Swider, E.; Majid, Q.A.; Srinivas, M.; Carr, C.A.; Roy, I. In Vivo Tracking and 1H/19F Magnetic Resonance Imaging of Biodegradable Polyhydroxyalkanoate/Polycaprolactone Blend Scaffolds Seeded with Labeled Cardiac Stem Cells. *ACS Appl. Mater. Interfaces* **2018**, *10*, 25056–25068. [CrossRef]
19. Fei, T.; Cazeneuve, S.; Wen, Z.; Wu, L.; Wang, T. Effective recovery of poly-β-hydroxybutyrate (PHB) biopolymer from Cupriavidus necator using a novel and environmentally friendly solvent system. *Biotechnol. Prog.* **2016**, *32*, 678–685. [CrossRef]
20. Koller, M. Biodegradable and Biocompatible Polyhydroxy-alkanoates (PHA): Auspicious Microbial Macromolecules for Pharmaceutical and Therapeutic Applications. *Molecules* **2018**, *23*, 362. [CrossRef]
21. Köse, G.T.; Kenar, H.; Hasırcı, N.; Hasırcı, V. Macroporous poly(3-hydroxybutyrate-co-3-hydroxyvalerate) matrices for bone tissue engineering. *Biomaterials* **2003**, *24*, 1949–1958. [CrossRef]
22. Grigore, M.E.; Grigorescu, R.M.; Iancu, L.; Ion, R.-M.; Zaharia, C.; Andrei, E.R. Methods of synthesis, properties and biomedical applications of polyhydroxyalkanoates: A review. *J. Biomater. Sci. Polym. Ed.* **2019**, *30*, 695–712. [CrossRef]
23. Berezina, N. Enhancing the 3-hydroxyvalerate component in bioplastic PHBV production by Cupriavidus necator. *Biotechnol. J.* **2012**, *7*, 304–309. [CrossRef] [PubMed]
24. Rahayu, A.; Zaleha, Z.; Yahya, A.R.M.; Majid, M.I.A.; Amirul, A.A. Production of copolymer poly(3-hydroxybutyrate-co-4-hydroxybutyrate) through a one-step cultivation process. *World J. Microbiol. Biotechnol.* **2008**, *24*, 2403–2409. [CrossRef]
25. Shijun, X.; Junsheng, M.; Jianqun, Z.; Ping, B. In vitro three-dimensional coculturing poly3-hydroxybutyrate-co-3-hydroxyhexanoate with mouse-induced pluripotent stem cells for myocardial patch application. *J. Biomater. Appl.* **2016**, *30*, 1273–1282. [CrossRef] [PubMed]
26. Zheng, Y.; Chen, J.-C.; Ma, Y.-M.; Chen, G.-Q. Engineering biosynthesis of polyhydroxyalkanoates (PHA) for diversity and cost reduction. *Metab. Eng.* **2019**. [CrossRef] [PubMed]
27. Brigham, C.J.; Riedel, S.L. The Potential of Polyhydroxyalkanoate Production from Food Wastes. *Appl. Food Biotechnol.* **2019**, *6*. [CrossRef]
28. Rodriguez-Contreras, A.; Koller, M.; Miguel, M.-d.S.D.; Calafell, M.; Braunegg, G.; Marqués-Calvo, M.S. Novel Poly[(R)-3-hydroxybutyrate]-producing bacterium isolated from a Bolivian hypersaline lake. *Food Technol. Biotechnol.* **2013**, *51*, 123–130.
29. Akaraonye, E.; Keshavarz, T.; Roy, I. Production of polyhydroxyalkanoates: The future green materials of choice. *J. Chem. Technol. Biotechnol.* **2010**, *85*, 732–743. [CrossRef]
30. Waghmare, V.S.; Wadke, P.R.; Dyawanapelly, S.; Deshpande, A.; Jain, R.; Dandekar, P. Starch based nanofibrous scaffolds for wound healing applications. *Bioact. Mater.* **2018**, *3*, 255–266. [CrossRef] [PubMed]
31. Findrik Balogová, A.; Hudák, R.; Tóth, T.; Schnitzer, M.; Feranc, J.; Bakoš, D.; Živčák, J. Determination of geometrical and viscoelastic properties of PLA/PHB samples made by additive manufacturing for urethral substitution. *J. Biotechnol.* **2018**, *284*, 123–130. [CrossRef] [PubMed]
32. Lizarraga-Valderrama, L.R.; Taylor, C.S.; Claeyssens, F.; Haycock, J.W.; Knowles, J.C.; Roy, I. Unidirectional neuronal cell growth and differentiation on aligned polyhydroxyalkanoate blend microfibres with varying diameters. *J. Tissue Eng. Regen. Med.* **2019**. [CrossRef] [PubMed]
33. Butt, F.I.; Muhammad, N.; Hamid, A.; Moniruzzaman, M.; Sharif, F. Recent progress in the utilization of biosynthesized polyhydroxyalkanoates for biomedical applications—Review. *Int. J. Biol. Macromol.* **2018**, *120*, 1294–1305. [CrossRef] [PubMed]
34. Singh, A.K.; Srivastava, J.K.; Chandel, A.K.; Sharma, L.; Mallick, N.; Singh, S.P. Biomedical applications of microbially engineered polyhydroxyalkanoates: An insight into recent advances, bottlenecks, and solutions. *Appl. Microbiol. Biotechnol.* **2019**, *103*, 2007–2032. [CrossRef] [PubMed]
35. Kai, G.; Martin, D.P. Chapter 7: Poly-4-hydroxybutyrate (P4HB) in Biomedical Applications and Tissue Engineering. In *Biodegradable Polymers, Volume 2: New Biomaterials Advancement and Challenges*; Chu, C.-C., Ed.; Nova Sience: Hauppauge, NY, USA, 2015; pp. 199–231.

36. Degli Esposti, M.; Chiellini, F.; Bondioli, F.; Morselli, D.; Fabbri, P. Highly porous PHB-based bioactive scaffolds for bone tissue engineering by in situ synthesis of hydroxyapatite. *Mater. Sci. Eng. C* **2019**, *100*, 286–296. [CrossRef] [PubMed]
37. Chernozem, R.V.; Surmeneva, M.A.; Shkarina, S.N.; Loza, K.; Epple, M.; Ulbricht, M.; Cecilia, A.; Krause, B.; Baumbach, T.; Abalymov, A.A.; et al. Piezoelectric 3-D Fibrous Poly(3-hydroxybutyrate)-Based Scaffolds Ultrasound-Mineralized with Calcium Carbonate for Bone Tissue Engineering: Inorganic Phase Formation, Osteoblast Cell Adhesion, and Proliferation. *ACS Appl. Mater. Interfaces* **2019**, *11*, 19522–19533. [CrossRef] [PubMed]
38. Meischel, M.; Eichler, J.; Martinelli, E.; Karr, U.; Weigel, J.; Schmöller, G.; Tschegg, E.K.; Fischerauer, S.; Weinberg, A.M.; Stanzl-Tschegg, S.E. Adhesive strength of bone-implant interfaces and in-vivo degradation of PHB composites for load-bearing applications. *J. Mech. Behav. Biomed. Mater.* **2016**, *53*, 104–118. [CrossRef] [PubMed]
39. Sadat-Shojai, M.; Khorasani, M.-T.; Jamshidi, A. A new strategy for fabrication of bone scaffolds using electrospun nano-HAp/PHB fibers and protein hydrogels. *Chem. Eng. J.* **2016**, *289*, 38–47. [CrossRef]
40. Ding, Y.; Li, W.; Müller, T.; Schubert, D.W.; Boccaccini, A.R.; Yao, Q.; Roether, J.A. Electrospun Polyhydroxybutyrate/Poly(ε-caprolactone)/58S Sol–Gel Bioactive Glass Hybrid Scaffolds with Highly Improved Osteogenic Potential for Bone Tissue Engineering. *ACS Appl. Mater. Interfaces* **2016**, *8*, 17098–17108. [CrossRef]
41. Ching, K.Y.; Andriotis, O.G.; Li, S.; Basnett, P.; Su, B.; Roy, I.; Tare, R.S.; Sengers, B.G.; Stolz, M. Nanofibrous poly(3-hydroxybutyrate)/poly(3-hydroxyoctanoate) scaffolds provide a functional microenvironment for cartilage repair. *J. Biomater. Appl.* **2016**, *31*, 77–91. [CrossRef]
42. Toloue, E.B.; Karbasi, S.; Salehi, H.; Rafienia, M. Evaluation of Mechanical Properties and Cell Viability of Poly (3-Hydroxybutyrate)-Chitosan/Al(2)O(3) Nanocomposite Scaffold for Cartilage Tissue Engineering. *J. Med. Signals Sens.* **2019**, *9*, 111–116. [CrossRef]
43. Bagdadi, A.V.; Safari, M.; Dubey, P.; Basnett, P.; Sofokleous, P.; Humphrey, E.; Locke, I.; Edirisinghe, M.; Terracciano, C.; Boccaccini, A.R.; et al. Poly(3-hydroxyoctanoate), a promising new material for cardiac tissue engineering. *J. Tissue Eng. Regen. Med.* **2018**, *12*, e495–e512. [CrossRef]
44. Ma, Y.-X.; Mu, J.-S.; Zhang, J.-Q.; Bo, P. Myocardial Patch Formation by Three-Dimensional 3-Hydroxybutyrate-co-4-Hydroxybutyrate Cultured with Mouse Embryonic Stem Cells. *J. Biomater. Tissue Eng.* **2016**, *6*, 629–634. [CrossRef]
45. Xue, Y.; Sant, V.; Phillippi, J.; Sant, S. Biodegradable and biomimetic elastomeric scaffolds for tissue-engineered heart valves. *Acta Biomater.* **2017**, *48*, 2–19. [CrossRef]
46. Chen, Q.; Liang, S.; Thouas, G.A. Elastomeric biomaterials for tissue engineering. *Prog. Polym. Sci.* **2013**, *38*, 584–671. [CrossRef]
47. Shishatskaya, E.I.; Nikolaeva, E.D.; Vinogradova, O.N.; Volova, T.G. Experimental wound dressings of degradable PHA for skin defect repair. *J. Mater. Sci. Mater. Med.* **2016**, *27*, 165. [CrossRef] [PubMed]
48. Sezer, A.D.; Cevher, E. Biopolymers as Wound Healing Materials: Challenges and New Strategies. In *Biomaterials Applications for Nanomedicine*; Rosario, P., Ed.; IntechOpen Limited: London, UK, 2011; pp. 383–414.
49. Marcano, A.; Bou Haidar, N.; Marais, S.; Valleton, J.-M.; Duncan, A.C. Designing Biodegradable PHA-Based 3D Scaffolds with Antibiofilm Properties for Wound Dressings: Optimization of the Microstructure/Nanostructure. *ACS Biomater. Sci. Eng.* **2017**, *3*, 3654–3661. [CrossRef]
50. Li, P.; Zhang, J.; Liu, J.; Ma, H.; Liu, J.; Lie, P.; Wang, Y.; Liu, G.; Zeng, H.; Li, Z.; et al. Promoting the recovery of injured liver with poly (3-hydroxybutyrate-co-3-hydroxyvalerate-co-3-hydroxyhexanoate) scaffolds loaded with umbilical cord-derived mesenchymal stem cells. *Tissue Eng Part A* **2015**, *21*, 603–615. [CrossRef] [PubMed]
51. Su, Z.; Li, P.; Wu, B.; Ma, H.; Wang, Y.; Liu, G.; Zeng, H.; Li, Z.; Wei, X. PHBVHHx scaffolds loaded with umbilical cord-derived mesenchymal stem cells or hepatocyte-like cells differentiated from these cells for liver tissue engineering. *Mater. Sci. Eng. C* **2014**, *45*, 374–382. [CrossRef]
52. Tashjian, R.; Kolz, C.; Suter, T.; Henninger, H. Biomechanics of Polyhydroxyalkanoate Mesh-Augmented Single-Row Rotator Cuff Repairs. *Am. J. Orthop.* **2016**, *45*, E527–E533.
53. Yutaka, T.; Buenaventurada, C. Degradation of Microbial Polyesters. *Biotechnol. Lett.* **2004**, *26*, 1181–1189.

54. Rodríguez-Contreras, A.; Calafell-Monfort, M.; Marqués-Calvo, M.S. Enzymatic degradation of poly(3-hydroxybutyrate) by a commercial lipase. *Polym. Degrad. Stab.* **2012**, *97*, 2473–2476. [CrossRef]
55. Rodríguez-Contreras, A.; Calafell-Monfort, M.; Marqués-Calvo, M.S. Enzymatic degradation of poly(3-hydroxybutyrate-co-4-hydroxybutyrate) by commercial lipases. *Polym. Degrad. Stab.* **2012**, *97*, 597–604. [CrossRef]
56. Misra, S.K.; Valappil, S.P.; Roy, I.; Boccaccini, A.R. Polyhydroxyalkanoate (PHA)/Inorganic Phase Composites for Tissue Engineering Applications. *Biomacromolecules* **2006**, *7*, 2249–2258. [CrossRef]
57. Gil-Castell, O.; Badia, J.D.; Bou, J.; Ribes-Greus, A. Performance of Polyester-Based Electrospun Scaffolds under In Vitro Hydrolytic Conditions: From Short-Term to Long-Term Applications. *Nanomaterials* **2019**, *9*, 786. [CrossRef]
58. Errico, C.; Bartoli, C.; Chiellini, F.; Chiellini, E. Poly(hydroxyalkanoates)-Based Polymeric Nanoparticles for Drug Delivery. *J. Biomed. Biotechnol.* **2009**, *2009*, 10. [CrossRef]
59. Li, Z.; Loh, X.J. Recent advances of using polyhydroxyalkanoate-based nanovehicles as therapeutic delivery carriers. *Nanomed. Nanobiotechnology* **2017**, *9*, e1429. [CrossRef]
60. Masood, F. Polymeric nanoparticles for targeted drug delivery system for cancer therapy. *Mater. Sci. Eng. C* **2016**, *60*, 569–578. [CrossRef]
61. Michalak, M.; Kurcok, P.; Hakkarainen, M. Polyhydroxyalkanoate-based drug delivery systems. *Polym. Int.* **2017**, *66*, 617–622. [CrossRef]
62. Pramual, S.; Assavanig, A.; Bergkvist, M.; Batt, C.A.; Sunintaboon, P.; Lirdprapamongkol, K.; Svasti, J.; Niamsiri, N. Development and characterization of bio-derived polyhydroxyalkanoate nanoparticles as a delivery system for hydrophobic photodynamic therapy agents. *J. Mater. Sci. Mater. Med.* **2016**, *27*, 40. [CrossRef]
63. Masood, F.; Chen, P.; Yasin, T.; Hameed, A. Novel Delivery System for Anticancer Drug Based on Short-Chain-Length Polyhydroxyalkanoate Nanoparticles. U.S. Patent US2015/0118293A1, 30 April 2015.
64. Murueva, A.V.; Shershneva, A.M.; Abanina, K.V.; Prudnikova, S.V.; Shishatskaya, E.I. Development and characterization of ceftriaxone-loaded P3HB-based microparticles for drug delivery. *Dry. Technol.* **2019**, *37*, 1131–1142. [CrossRef]
65. Rodríguez-Contreras, A.; Canal, C.; Calafell-Monfort, M.; Ginebra, M.-P.; Julio-Moran, G.; Marqués-Calvo, M.-S. Methods for the preparation of doxycycline-loaded phb micro- and nano-spheres. *Eur. Polym. J.* **2013**, *49*, 3501–3511. [CrossRef]
66. Rodríguez-Contreras, A.; Marqués-Calvo, M.S.; Gil, F.J.; Manero, J.M. Modification of titanium surfaces by adding antibiotic-loaded PHB spheres and PEG for biomedical applications. *J. Mater. Sci. Mater. Med.* **2016**, *27*, 124. [CrossRef]
67. Lee, Y.-F.; Sridewi, N.; Ramanathan, S.; Sudesh, K. The Influence of Electrospinning Parameters and Drug Loading on Polyhydroxyalkanoate (PHA) Nanofibers for Drug Delivery. *Int. J. Biotechnol. Wellness Ind.* **2015**, *4*, 103–113.
68. Rodríguez-Contreras, A.; Rupérez, E.; Marqués-Calvo, M.S.; Manero, J.M. Chapter 7—PHAs as matrices for drug delivery. In *Materials for Biomedical Engineering*; Holban, A.-M., Grumezescu, A.M., Eds.; Elsevier: Amsterdam, The Netherlands, 2019; pp. 183–213.
69. Rodríguez-Contreras, A.; García, Y.; Manero, J.M.; Rupérez, E. Antibacterial PHAs coating for titanium implants. *Eur. Polym. J.* **2017**, *90*, 66–78. [CrossRef]
70. Rodríguez-Contreras, A.; Guillem-Marti, J.; Lopez, O.; Manero, J.M.; Ruperez, E. Antimicrobial PHAs coatings for solid and porous Tantalum implants. *Colloids Surf. B Biointerfaces* **2019**, *182*, 110317. [CrossRef]
71. Timin, A.S.; Muslimov, A.R.; Zyuzin, M.V.; Peltek, O.O.; Karpov, T.E.; Sergeev, I.S.; Dotsenko, A.I.; Goncharenko, A.A.; Yolshin, N.D.; Sinelnik, A.; et al. Multifunctional Scaffolds with Improved Antimicrobial Properties and Osteogenicity Based on Piezoelectric Electrospun Fibers Decorated with Bioactive Composite Microcapsules. *ACS Appl. Mater. Interfaces* **2018**, *10*, 34849–34868. [CrossRef]

9

Fed-Batch Synthesis of Poly(3-Hydroxybutyrate) and Poly(3-Hydroxybutyrate-*co*-4-Hydroxybutyrate) from Sucrose and 4-Hydroxybutyrate Precursors by *Burkholderia sacchari* Strain DSM 17165

Miguel Miranda de Sousa Dias [1], Martin Koller [2,3,*], Dario Puppi [4], Andrea Morelli [4], Federica Chiellini [4] and Gerhart Braunegg [3]

[1] Université Pierre et Marie Curie UPMC, Institut national de la santé et de la recherche médicale INSERM, Centre national de la recherche scientifique CNRS, Institut de la Vision, Sorbonne Universités, 17 rue Moreau, 75012 Paris 06, France; migueldias670@hotmail.com
[2] Institute of Chemistry, University of Graz, NAWI Graz, Heinrichstrasse 28/III, 8010 Graz, Austria
[3] ARENA—Association for Resource Efficient and Sustainable Technologies, Inffeldgasse 21b, 8010 Graz, Austria; g.braunegg@tugraz.at
[4] BIOLab Research Group, Department of Chemistry & Industrial Chemistry, University of Pisa, UdR INSTM Pisa, Via Moruzzi, 13, 56124 Pisa, Italy; d.puppi@dcci.unipi.it (D.P.); a.morelli@dcci.unipi.it (A.M.); federica.chiellini@unipi.it (F.C.)
* Correspondence: martin.koller@uni-graz.at

Academic Editor: Anthony Guiseppi-Elie

Abstract: Based on direct sucrose conversion, the bacterium *Burkholderia sacchari* is an excellent producer of the microbial homopolyester poly(3-hydroxybutyrate) (PHB). Restrictions of the strain's wild type in metabolizing structurally related 3-hydroxyvalerate (3HV) precursors towards 3HV-containing polyhydroxyalkanoate (PHA) copolyester calls for alternatives. We demonstrate the highly productive biosynthesis of PHA copolyesters consisting of 3-hydroxybuytrate (3HB) and 4-hydroxybutyrate (4HB) monomers. Controlled bioreactor cultivations were carried out using saccharose from the Brazilian sugarcane industry as the main carbon source, with and without co-feeding with the 4HB-related precursor γ-butyrolactone (GBL). Without GBL co-feeding, the homopolyester PHB was produced at a volumetric productivity of 1.29 g/(L·h), a mass fraction of 0.52 g PHB per g biomass, and a final PHB concentration of 36.5 g/L; the maximum specific growth rate μ_{max} amounted to 0.15 1/h. Adding GBL, we obtained 3HB and 4HB monomers in the polyester at a volumetric productivity of 1.87 g/(L·h), a mass fraction of 0.72 g PHA per g biomass, a final PHA concentration of 53.7 g/L, and a μ_{max} of 0.18 1/h. Thermoanalysis revealed improved material properties of the second polyester in terms of reduced melting temperature T_m (161 °C vs. 178 °C) and decreased degree of crystallinity X_c (24% vs. 71%), indicating its enhanced suitability for polymer processing.

Keywords: 4-hydroxybutyrate; biopolymers; *Burkholderia sacchari*; copolyester; poly(3-hydroxybutyrate-*co*-4-hydroxybutyrate); polyhydroxyalkanoate (PHA); saccharose; sucrose; sugarcane

---

1. Introduction

Polyhydroxyalkanoates (PHA) are a versatile group of microbial biopolyesters with properties mimicking those of petrol-based plastics. A growing number of described bacterial and archaeal prokaryotic species accumulate PHA as refractive granular inclusion bodies in the cell's cytoplasm. PHA granules are surrounded by a complex membrane of proteins and lipids; these functional "carbonosomes" are typically accumulated under conditions of an excess exogenous carbon source in parallel with the limitation of a growth-essential component like the nitrogen source or phosphate [1–4]. Playing a major biological role, the presence of intracellular PHA supports bacterial survival under the conditions of carbon starvation. Moreover, PHA has pivotal functions in protecting cells against environmental stress conditions such as extreme temperature [5,6], exposure to oxidants [5,7], organic solvents [7], and UV-irradiation [6]. Depending on their composition, we distinguish homopolyesters, consisting of only one type of monomer, from heteropolyesters, composed of two or more types of monomers differing in their side chains (copolyesters) or both in their side chains and backbones (terpolyesters). In this context, the best known member of the PHA family, namely the homopolyester poly(3-hydroxybutyrate) (PHB), has restricted processability due to its high brittleness and crystallinity if compared to heteropolyesters consisting of different monomers such as 3-hydroxybutyrate (3HB), 3-hydroxyvalerate (3HV), 4-hydroxybutyrate (4HB), or 3-hydroxyhexanoate (3HHx) [8]. Changing PHA's composition on the monomeric level offers the possibility to fine-tune the polymer properties (melting temperature T_m, glass transition temperature T_g, degree of crystallinity X_c, degradability, elongation at break, or tensile strength) according to the customer's demands [9]. Apart from utilization in its crude form, PHA can be processed together with compatible organic or inorganic materials to make various composites and blends with tailored properties in terms of density, permeability, tensile strength, (bio)degradability, crystallinity, etc. [10–12]. To an increasing extent, the processing of PHA with nanoparticles is reported to generate novel designer bio-plastics especially useful for, *inter alia*, "smart packaging" [13,14].

Nowadays, there is an emerging trend of substituting petrol-based plastics with sustainable "bio-alternatives" with low environmental impact, that are biodegradable and bio-based in their nature [15,16]. Nevertheless, PHA production is still challenged by cost-decisive factors which make them considerably more expensive than their petrochemical counterparts; in order to optimize PHA production economically, all single process steps have to be taken into account [4,17]. Enhanced downstream processing to recover intracellular PHA from the biomass [18–21], bioreactor design and process regime [22–25], and in-depth understanding of the kinetics of the bioprocess [26] are crucial factors when designing a new PHA production process. Nevertheless, the selection of the most suitable carbonaceous raw materials to be used as feedstocks for PHA biosynthesis is the issue that is most difficult to solve. In this context, there is an increasing trend towards the application of carbon-rich (agro) industrial waste materials to produce the so called "2nd generation PHA" [4]. Among these materials, the current literature familiarizes us with PHA production based on surplus whey [27], abundant lignocelluloses [28–30], waste lipids from animal processing [31–33], used plant-and cooking oils [34–36], crude glycerol from biodiesel production [37–40], plant root hydrolysates [30], extracts and hydrolysates of spent coffee ground [41,42], and molasses [43]. Such waste materials already performed well as substrates on the laboratory scale, but are still awaiting their implementation in industrial-scale PHA production processes. This is mainly due to problems associated with upstream processing, insecure supply chains, presence of inhibitory compounds, or fluctuating composition of the industrial waste streams [4]. An emerging trend in using industrial waste streams is recognized in the direct conversion of CO_2 from industrial effluent gases [44]; here, cyanobacteria [45,46] or "Knallgasbacteria" [47] are potential cellular factories used to convert CO_2 to "3rd generation PHA" and additional valued products. Although also promising on the laboratory scale, development of these processes to industrial maturity has hitherto not been reached [44–47].

Apart from 2nd and 3rd generation PHA, the production of PHA based on materials relevant for food and feed purposes ("1st generation PHA") can also become economically viable given the integration of PHA-production facilities into existing production lines, where the raw material is generated [48]. This is successfully demonstrated at PHB Industrial SA (PHBISA), a company located in the Brazilian state of São Paolo. PHBISA is involved in the cane sugar business, predominantly fermenting hydrolyzed sucrose to bioethanol, and selling sucrose in its native form; a small part of sucrose is currently converted to PHA in a pilot plant with 100 ton annual capacity, and marketed under the trade mark Biocycle^{TM} [48]. Remarkably, this bio-refinery process works energetically autarkic by the thermal conversion of surplus sugarcane bagasse to generate steam and electrical energy, which are used in the bioprocesses and the distillation for ethanol recovery. Moreover, distillative ethanol recovery generates a mixture of medium-chain-length alcohols (butanol, pentanol, etc.), which are used by the company for extractive PHA recovery from microbial biomass. This strategy saves expenses for the typically applied and often halogenated extraction solvents, which considerably contribute to the entire PHA production costs [48]. Currently, PHA production at PHBISA is carried out using the well-known production strain *Cupriavidus necator*, a eubacterial organism lacking the enzymatic activity for sucrose cleavage; hence, sucrose hydrolysis of the monomeric sugars (glucose and fructose) is a needed laborious operation step during upstream processing. For further optimization of this sucrose-based PHA production process, the assessment of alternative production strains appears reasonable. Such new whole-cell biocatalysts should fulfill some requirements: Growth rate and volumetric PHA productivity that are competitive with the data known for *C. necator*; direct sucrose conversion without the need for hydrolysis; temperature optima in the slightly thermophile range (in order to save cooling costs, a decisive cost factor under the climatic conditions prevailing in São Paolo); and last but not least, the strain should be able to produce copolyesters with advanced material properties.

A strain that appears promising in all these criteria is *Burkholderia sacchari* IPT 101 (DSM 17165), originally isolated from the soil of Brazilian sugarcane fields and investigated by Brämer and colleagues [49]. The strain is reported to accumulate high amounts of PHA inter alia from glucose [39,50], sucrose [49,50], glycerol [39,50], organic acids [51], pentose-rich substrate cocktails mimicking hydrolysates of bagasse [52], and hydrolyzed straw [53]. Aimed at the optimized utilization of lignocellulose hydrolysate, efforts are currently devoted to further improve the strain's substrate conversion ability in terms of xylose uptake [54]. PHA production by this organism and its mutant strains was demonstrated both in mechanically stirred tank bioreactors [52,53,55,56] and in airlift bioreactors [57]. As a drawback, the wild type strain displays insufficient ability for 3HV formation from structurally related precursors such as propionic acid, which is in contrast to pronounced 3HV formation by its mutant strain *B. sacchari* IPT 189 [54,56,58,59]. Formation of copolyesters consisting of 3HB and 4HB, hence P(3HB-*co*-4HB), was successfully demonstrated by co-feeding glucose or wheat straw hydrolysate (WSH) and the 4HB-related precursor compound γ-butyrolactone (GBL) [49]. Only recently has the production of copolyesters of 3HB and 3-hydroxyhexanoate (3HHx) by genetically engineered *B. sacchari* been reported [60]. In the present study, we demonstrate for the first time the feasibility of high-cell density production of PHB and P(3HB-*co*-4HB) by *B. sacchari* based on saccharose from PHBISA and the 4HB-precursor GBL, and for the first time, GBL's saponified form, 4-hydroxybutyrate sodium salt (Na-4HB). Furthermore, by addressing the contradictory literature information on the optimum temperature at which this organism thrives [50–54], we adapted the strain to an elevated cultivation temperature of 37 °C according to the requirements at the Brazilian production site [48,61]. Detailed kinetic data under controlled conditions in laboratory bioreactors, and an in-depth comparison of the polymer data of PHB and P(3HB-*co*-4HB), respectively, are provided.

2. Materials and Methods

2.1. Strain Maintenance and Adaptation to Elevated Temperature

Burkholderia sacchari DSM 17165 was purchased from DSMZ, Germany, and were grown on solid media plates (medium according to Küng [62] with 10 g/L of sucrose as the carbon source and 2 g/L ammonium sulfate as the nitrogen source). In two-week intervals, single colonies were transferred to new plates and incubated at 37 °C. All mineral components of the medium were purchased in p.a. quality (Company Roth, Graz, Austria), whereas sugarcane sucrose was obtained as unrefined saccharose directly from PHBISA.

2.2. Shaking Flask Cultivation to Assess Production of 4HB-Containing PHA

For preparation of pre-cultures, fresh single colonies from solid media were transferred to 100 mL of a liquid mineral medium containing the following components (g/L): KH_2PO_4, 9.0; $Na_2HPO_4 \cdot 2H_2O$,

3.0; $(NH_4)_2SO_4$, 2.0; $MgSO_4 \cdot 7H_2O$, 0.2 g; $CaCl_2 \cdot 2H_2O$, 0.02; $NH_4Fe(III)$citrate, 0.03; SL6, 1.0 (mL/L); sucrose, 15.0. These pre-cultures were incubated at 37 °C under continuous shaking; after 24 h, 5 mL of these pre-cultures were used for inoculation of four flasks each containing 100 mL of the minimal medium. The pH-value was adjusted to 7.0. After 8 h of incubation at 37 °C, 4HB-precursors were added to the cultures as follows: Two of the flasks were supplied with a solution of GBL, and two cultures with a solution of Na-4HB. Both solutions were added in a quantity to achieve a final precursor (GBL or the anion of 4HB, respectively) concentration of 1.5 g/L each. 15 h later, the re-feed of 4HB precursors was accomplished using the same quantity (1.5 g/L). After 47 h of cultivation, the experiment was stopped and the fermentation broth was analyzed for cell dry mass (CDM), PHA mass fraction in CDM, and PHA composition (fractions of 3HB and 4HB) (analytical methods *vide infra*).

2.3. Bioreactor Cultivations

2.3.1. PHB Production

Single colonies of *B. sacchari* were used to inoculate 100 mL (pre-cultures) of the medium according to Küng as described above. These pre-cultures were incubated (37 °C) for 36 h; then, 5 mL each of these pre-cultures were used for the inoculation of seven shaking flasks each containing 250 mL of the minimal medium. These cultures were incubated under continuous shaking at 37 °C for 36 h, until high cell densities (8–9 g/L) were reached, and two of them were used to inoculate a Labfors 3 bioreactor (Infors, CH) with an initial working volume of 1.5 L (1.0 L fresh medium with compounds calculated for 1.5 L plus 0.5 L inoculum). At the start of the cultivation, sucrose and $(NH_4)_2SO_4$ amounted to 15 g/L and 2.5 g/L, respectively. The set point for dissolved oxygen concentration (DOC) was 40% of the air saturation during the growth phase, and 20% during nitrogen-limited conditions; DOC was controlled by automatic adjustment of the stirrer speed and aeration rate. The pH-value was set to 7.0 and controlled automatically by the addition of H_2SO_4 (10%) to decrease the pH-value, and ammonia solution (25%) during the growth phase or NaOH (10%) during the accumulation phase to increase the pH-value. Hence, during the growth phase, the addition of the nitrogen source was coupled with pH-value correction. The cultivation was carried out at 37 °C. The time points of sugar addition (50% w/w aqueous solution of Brazilian sugarcane saccharose) are indicated in Figure 2 by arrows; the total amount of sucrose solution refeed amounted to 360 g.

2.3.2. P(3HB-*co*-4HB) Production:

This process was based on inoculum preparation according to the previous experiment. Cultivation in the bioreactor was performed using a minimal medium identical to the process at the company PHBISA (g/L): KH_2PO_4, 5.0; $(NH_4)_2SO_4$, 2.5; $MgSO_4 \cdot 7H_2O$, 0.8; NaCl; 1.0; $CaCl_2 \cdot 2H_2O$, 0.02; $NH_4Fe(III)$citrate, 0.05; trace element solution SL6 2.5 mL/L; sucrose 30; and the 4HB-precursor 4HB was provided by dropwise addition during the accumulation phase (total addition of GBL 15.5 g/L). Also in this case, a Labfors 3 bioreactor with an initial working volume of 1.5 L (1.0 L fresh medium with compounds calculated for 1.5 L plus 0.5 L inoculum) was used with the same basic parameters (DOC, T, pH-value) as described for the previous fermentation. The time points of sugar addition are indicated in Figure 7 by the arrows; the total amount of sucrose refeed amounted to 207 g of solution.

2.4. Cell Dry Mass (CDM) Determination

A gravimetric method was used to determine CDM in the fermentation samples. Five mL of the culture broth was centrifuged in pre-weighed glass screw-cap tubes for 10 min at 10 °C and 4000 rpm in a Heraeus Megafuge 1.0 R refrigerated centrifuge (Heraeus, Hanau, Germany). The supernatant was decanted, and subsequently used for substrate analysis. The cell pellets were washed with distilled water, re-centrifuged, frozen, and lyophilized (freeze-dryer Christ Alpha 1-4 B, Martin Christ Gefriertrocknungsanlagen GmbH, Osterode am Harz, Germany) to constant mass. CDM was expressed as the mass difference between the tubes containing cell pellets minus the mass of the empty tubes. The determination was done in duplicate. The lyophilized pellets were subsequently used for determination of intracellular PHA as described in the next paragraph.

2.5. Analysis of PHA Content in Biomass and Monomeric PHA Composition

For the analysis of PHA, standards of P(3HB-*co*-5.0%-3HV) (Biopol^{TM}, ICI, London, UK) were used for determination of the 3HB content; for determination of 4HB, "self-made" Na-4HB (next paragraph) was used as the reference material. Intracellular PHA in lyophilized biomass samples was transesterificated to volatile methyl esters of hydroxylkanoic acids via Braunegg's acidic methanolysis method [63]. Analyses were carried out with an Agilent Technologies 6850 gas chromatograph (30-m HP5 column, Hewlett-Packard, Palo Alto, CA, USA; Agilent 6850 Series Autosampler). The compounds were detected by a flame ionization detector; the split ratio was 1:10.

2.6. Preparation of Na-4HB

Na-4HB was synthesized by manually dropping a defined quantity of GBL into an equimolar aqueous solution of NaOH under continuous stirring and cooling. The obtained solution of Na-4HB was further frozen and lyophilized (freeze-dryer Christ Alpha 1–4 B) to obtain Na-4HB as a white powder. This powder was applied as a reference material for the analysis and as a co-substrate.

2.7. Substrate Analysis

The determination of carbon sources (sucrose and its hydrolysis products glucose, fructose, Na-4HB, and GBL) was accomplished by HPLC-RI using an Aminex HPX 87H column (thermostated at 75 °C, Biorad, Hercules, CA, USA), a LC-20AD pump, a SIC-20AC autosampler, a RID-10A refractive index detector, and a CTO-20AC column oven. Pure sucrose, glucose, fructose, Na-4HB, and GBL were used as standards for external calibration. Isocratic elution was carried out with 0.005 M H_2SO_4 at a flow rate of 0.6 mL/min.

2.8. Analysis of Nitrogen Source (NH_4^+)

The determination of the nitrogen source was done using an ammonium electrode (Orion) with ammonium sulfate solution standards (300–3000 ppm) as described previously [39].

2.9. PHA Recovery

After the end of the experiments, the fermentation broth was *in situ* pasteurized (80 °C, 30 min). Afterwards, the biomass was separated from the liquid supernatant via centrifugation (12,000 g; Sorvall® RC-5B Refrigerated Superspeed centrifuge, DuPont Instruments, Wilmington, NC, USA), frozen, and lyophilized (freeze-dryer Christ Alpha 1-4 B). Dry biomass was decreased by overnight stirring with a 10-fold mass of ethanol; after drying, PHA was extracted from the degreased, dried biomass by continuous overnight stirring in a 25-fold mass of chloroform in light-protected glass vessels. The solution containing the PHA was separated by vacuum-assisted filtration, and concentrated by evaporation of the major part of the solvent (Büchi Rotavapor® R-300). This concentrated PHA solution was dropped into permanently stirred ice-cooled ethanol. Precipitated PHA filaments of high purity were obtained by vacuum-assisted filtration, dried, and subjected to polymer characterization (*vide infra*).

2.10. Polymer Characterization

2.10.1. Molecular Mass Distribution

Gel Permeation Chromatography (GPC) analysis was carried out on a Waters 600 model (Waters Corporation, Milford, MA, USA) equipped with a Waters 410 Differential Refractometer and two PLgel 5 μm mixed-C columns (7.8 × 300 mm^2). The mobile phase constituted by chloroform (CHROMASOLV®

for HPLC amylene stabilized, Sigma-Aldrich, Milan, Italy) was eluted at a flow rate of 1 mL/min. Monodisperse polystyrene standards were used for calibration (range 500–1.800,000 g/mol). Samples were prepared at a concentration of ca. 0.5% (*w*/*v*).

2.10.2. Thermoanalysis

Differential Scanning Calorimetry (DSC) analysis was performed using a Mettler DSC-822E instrument (Mettler Toledo, Novate Milanese, Italy) under a nitrogen flow rate of 80 mL/min. The analysis was carried out in the range from −20 to 200 °C at a heating and cooling rate of 10 °C/min. By considering the second heating cycles in the thermograms, the glass transition temperature (T_g) was evaluated by analyzing the inflection point, while the melting temperature (T_m) and crystallinity (X_c) was evaluated by analyzing the endothermic peak. X_c was determined by considering the value of the melting enthalpy of 146 J/g for the 100% crystalline PHB. Both characterization tests were carried out on five replicates for each kind of sample and the data were presented as mean ± standard deviation. Statistical differences were analyzed using one-way analysis of variance (ANOVA), and a Tukey test was used for post hoc analysis. A p-value < 0.05 was considered statistically significant.

3. Results

3.1. Impact of 4HB-Precursors GBL and Na-4HB on Poly-(3-hydroxybutyrate-co-4-hydroxybutyrate) (P(3HB-co-4HB)) Biosynthesis by Burkholderia sacchari DSM 17165 on Sucrose

Figure 1 illustrates the outcomes of the shaking flask experiment comparing the effect of adding 4HB-precursors GBL and Na-4HB to *B. sacchari* cultivated on sucrose as main carbon source. After 47 h of incubation, the CDM concentration was in the range of 5 g/L in all experimental setups. Final PHA concentrations amounted to 1–2 g/L without significant differences between the individual cultivation setups. Using GBL as the 4HB-related precursor, PHA fractions in the CDM were slightly lower than in the case of using Na-4HB, but almost identical to the setups without precursor addition (ca. 30% vs. ca. 35%, respectively). The 4HB fractions in PHA (4HB/PHA) differ in dependence on the applied precursor; using GBL, this value amounted to 20.8%, while it was only 14.1% when using Na-4HB. As expected, the setups cultivated on sucrose as the sole carbon source (no addition of 4HB-related precursors) resulted in the generation of the PHB homopolyester. Here, it has to be emphasized that it is not clear from the available data if the generated polyester is definitely a P(3HB-*co*-4HB) copolyester with random distribution of the individual building blocks, a blend of homopolymers consisting of 3HB or 4HB, respectively, or a blend of different P(3HB-*co*-4HB) copolyesters with different 4HB fractions.

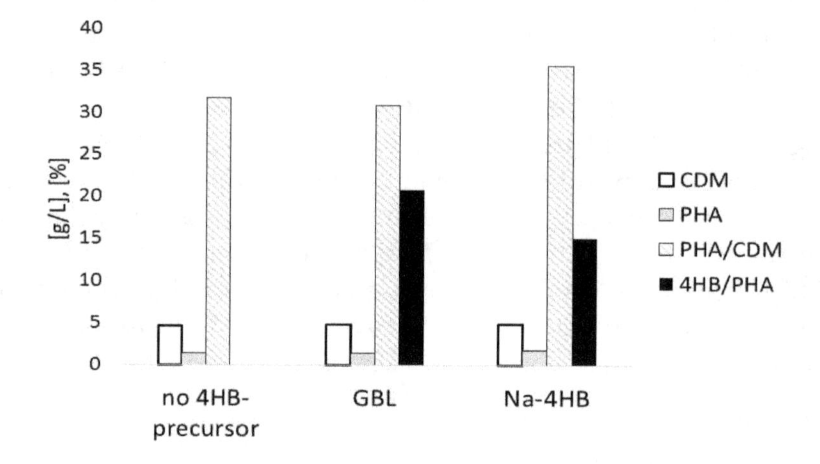

Figure 1. Cell dry mass (CDM) (g/L), polyhydroxyalkanoate (PHA) (g/L), mass fraction of PHA in CDM (%), and mass fraction of 4-hydroxybutyrate (4HB) in PHA (%): *B. sacchari* after 47 h of cultivation on 15 g/L sucrose and 4HB-precursors γ-butyrolactone (GBL) or Na-4HB (precursor addition: 1.5 g/L after 8 h, refeed of 1.5 g/L after 15 h).

3.2. Poly(3-hydroxybutyrate) (PHB) Production with Burkholderia sacchari on the Bioreactor Scale; Sucrose as the Sole Carbon Source

3.2.1. Bioprocess

This experiment aimed to test a medium similar to the one used at the industrial company PHBISA for sucrose-based PHA production by *C. necator*, and to study its influence on the kinetic data and on the polymer production (*cf.* Materials and Methods section). Of major importance, it was intended to considerably increase the concentration of the residual biomass and to achieve higher productivities for PHA. This was accomplished using an advanced strategy for adding the nitrogen source (NH_4^+) during the microbial growth phase by coupling the addition of the nitrogen source with the correction of the pH-value. Instead of a periodic re-feed of $(NH_4)_2SO_4$ solution to maintain the nitrogen concentration at the desired level, NH_4OH was used as a base for correction of the pH-value and, at the same time, to provide the nitrogen needed by the strain to grow. Hence, the addition of the nitrogen source was directly coupled to the excretion of acidic metabolites during the growth phase. After 12.5 h of fermentation, the NH_4OH solution as the pH-correction agent was replaced by NaOH solution (20%) in order to provoke a nutritional stress by limitation of the nitrogen source to stop the biomass formation and to enhance PHA production; this time point is marked by a full line in Figure 3. The depletion of the nitrogen source occurred after 19 h of cultivation.

Figure 2 illustrates the time curves of the sugar concentrations (sucrose, glucose, and fructose). It is easily seen that the strain possesses the metabolic ability to rapidly hydrolyze the disaccharide sucrose to its monomeric sugars by the excretion of an extracellular invertase enzyme. Immediately after inoculation, hydrolysis started, resulting in about 9 g/L sucrose and 6 g/L of monomers (glucose plus fructose) already present in the first sample taken at $t = 0$ h. The time points of sucrose additions are marked by arrows in Figure 2. Remarkably, the concentrations of the two monosaccharides do not follow the same trend with time, which might be due to the changing conversion rates of the individual monomers (glucose or fructose, respectively) with the changing environmental (nutritional) conditions during the cultivation. Mathematical modelling of the data to elucidate the metabolic processes should therefore be performed in follow-up experiments by specialists in the field of metabolic flux analysis. A total quantity of 360 g sucrose solution was added during the process. A total sugar consumption of 29.14 g/(L·h) was observed, and a conversion yield of sugar to CDM of 0.18 g/g (calculated for the entire sugar addition and also encompassing the not utilized sugar in the spent fermentation broth) (Table 1). Limitation of the carbon source was avoided during the entire cultivation period by permanent monitoring (HPLC) and re-feeding (Figure 2).

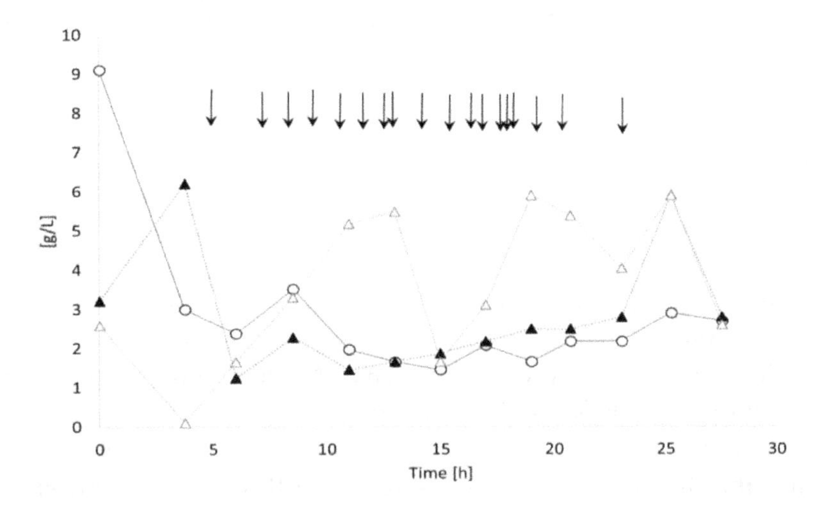

Figure 2. Substrate time curves: *B. sacchari* on sucrose without supplementation of 4HB-precursors. Open spheres: sucrose; black triangles: glucose; open triangles: fructose. Arrows indicate the time points of pulse feedings of the sucrose solution.

Table 1. Results of the bioreactor fermentations.

Kinetic Parameter	PHB Production Process (1st Bioreactor Cultivation)	P(3HB-*co*-4HB) Production Process (2nd Bioreactor Cultivation)
$\mu_{max.}$ (1/h)	0.41 (t = 3.75–6 h)	0.23 (t = 6–8 h)
max. CDM (g/L)	70.0 (t = 25.25 h)	78.6 (t = 32 h)
max. PHA concentration (g/L)	36.8 (t = 27.5 h)	55.8 (t = 29 h)
max. fraction of PHA in CDM (% w/w)	53.0 (t = 27.5 h)	72.6 (t = 29 h)
max. fraction of 4HB in PHA (% mol/mol)	-	1.6 (t = 39 h)
Volumetric productivity for PHA (g/L·h)	1.29 (t = 0–27.5 h)	1.87 (t = 0–39 h)
$\text{Yield}_{\text{CDM/sucorse}}$ (g/g)	0.18	0.38
Yield 4HB/GBL (g/g)	-	0.05
max. specific productivity q_P (g/(g·h))	0.19 (t = 7.25 h)	0.17 (t = 17.75 h)
Material Characterization		
Weight average molecular mass Mw (kDa)	627 ± 13	315 ± 24
Polydispersity P_i (Mw/Mn)	2.66 ± 0.13	2.51 ± 0.15
Glass transition temperature T_g (°C)	1.0 ± 0.6	1.8 ± 0.2
Melting point T_m (°C)	177.6 ± 0.6	160.9 ± 0.8
Degree of crystallinity X_c (%)	70.9 ± 0.9	24.0 ± 3.6

Figure 3 illustrates the time curves of the CDM, residual biomass, and PHA during the process. After the onset of nitrogen limitation after 19 h (indicated by a dashed line in Figure 3), the concentration of the residual biomass remained constant (35 g/L), whereas the PHA concentration increased, reaching a maximum concentration of 36.5 g/L at the end of the fermentation. This corresponds to a final CDM concentration of 70 g/L. Due to the fact that no 4HB-related precursors were supplied, homopolyester PHB was accumulated. The volumetric productivity for PHB, calculated for the entire process, amounted to 1.29 g/(L·h). For the entire process (t = 0 to 27.5 h), the yield for the conversion of sugars to CDM amounted to 0.18 g/g, whereas during the nitrogen-limited phase of cultivation, a conversion yield for sugars to PHB of 0.08 g/g was evidenced (Table 1).

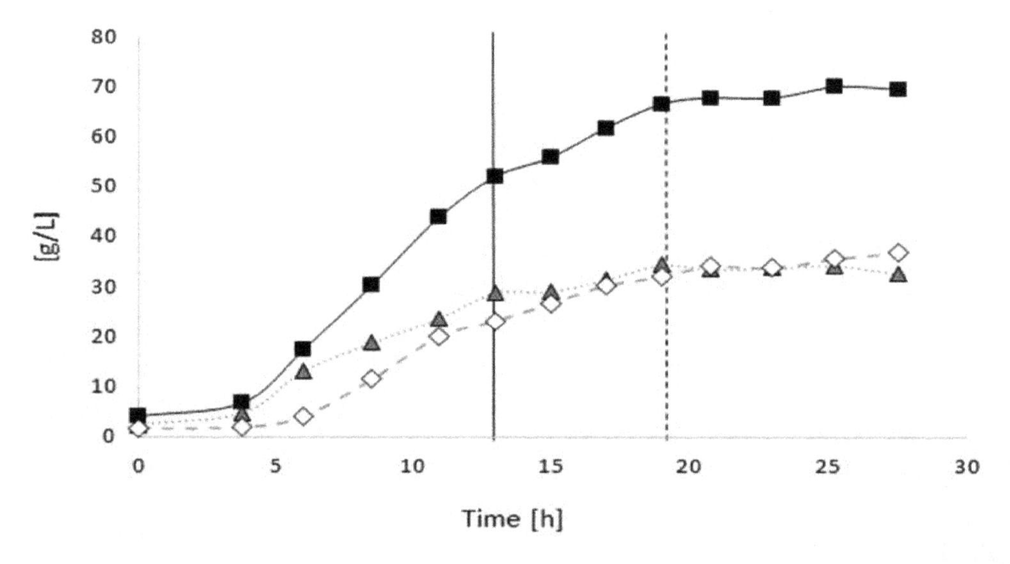

Figure 3. Time curves of CDM, residual biomass, and PHB concentration: *B. sacchari* on sucrose without supplementation of 4HB-precursors. Black squares: CDM; open rhombi: PHA; grey triangles: residual biomass. Full black line: Exchange of NH_4OH solution by NaOH solution as the pH-corrective agent, dashed line: start of nitrogen depletion in the medium.

Figure 4 illustrates the time curves of the specific growth rate μ and the specific product (PHB) formation rate q_P for the entire process. Here, it is visible that the maximum specific growth (μ_{max} = 0.41 1/h) was monitored at around 5 h of cultivation. For the entire growth phase (t = 3.75–13 h), μ_{max} was determined with 0.15 1/h by plotting the natural logarithm LN of the residual biomass concentration vs. time. After the exchange of NH_4OH by the NaOH solution and the resulting depletion of the nitrogen source, the specific growth tremendously decreased, and a slight decrease of the residual biomass concentration, indicated by the negative values for μ in Figure 4, was observed. The highest specific PHB production was observed starting from the onset of the exponential growth phase ($t = 5$ h) until the start of nitrogen depletion at $t = 12$ h; a q_P of about 0.19 g/(g·h) was measured for the period between the two subsequent samplings at $t = 6$ and 8.5 h. In later periods of the process, only a slight increase of PHB production, manifested by low values for q_P, was observed.

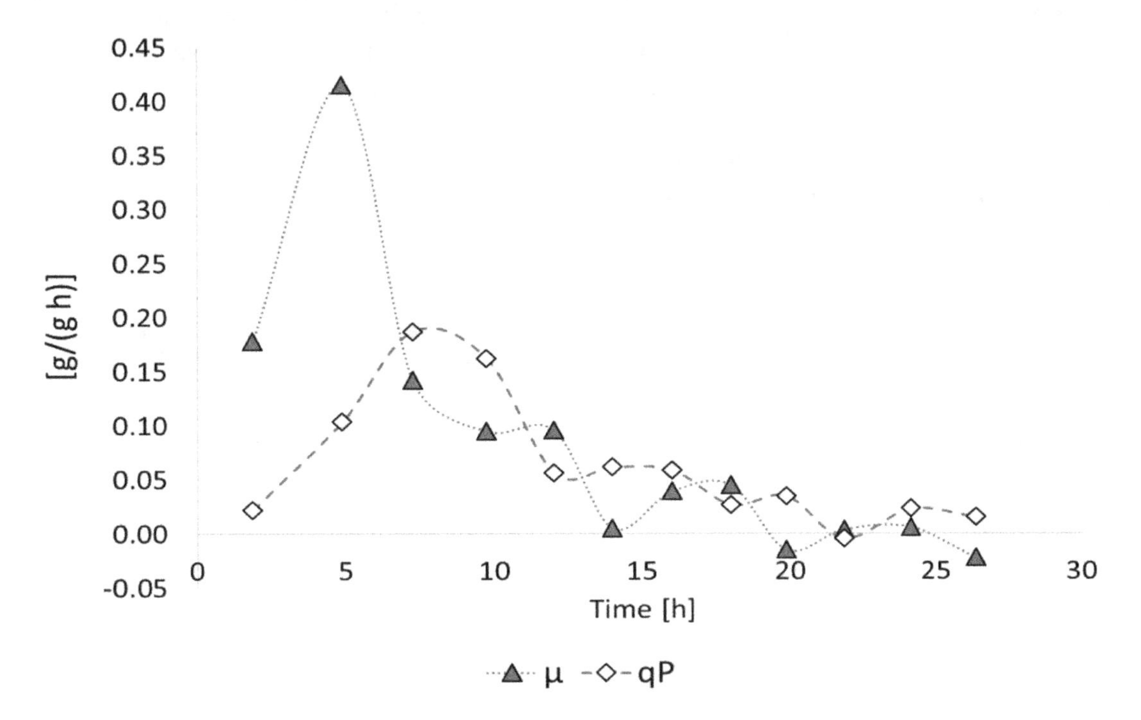

Figure 4. Time course of the specific growth rate μ and specific PHA production rate q_P: *B. sacchari* on sucrose without supplementation of 4HB-precursors. Full black line: Exchange of NH_4OH solution by NaOH solution as the pH-corrective agent; dashed line: start of nitrogen depletion in the medium.

3.2.2. Polymer Characterization:

After the end of the experiment, the biomass was separated from the liquid supernatant via centrifugation, and was frozen and lyophilized. The dry biomass was decreased with ethanol and the polymer was extracted using chloroform. The weight average molecular mass (Mw) and the polydispersity (P_i; Mw/Mn) values of the extracted homopolymer were determined by gel permeation chromatography (GPC). The Mw was 627 ± 13 kDa and P_i was 2.66 ± 0.13 kDa (Table 1). Differential scanning calorimetry (DSC) analysis was carried out to determine the glass transition temperature (T_g), melting temperature (T_m), and crystallinity (X_c) of the PHB samples. Analysis of the obtained data showed that the T_g of the produced PHB was 1.0 ± 0.6 °C and the T_m was 177.6 ± 0.6 °C, while X_c was 70.9% ± 0.9%.

3.3. Controlled Poly(3-Hydroxybutyrate-co-4-Hydroxybutyrate) (P(3HB-co-4HB)) Production with Burkholderia sacchari on the Bioreactor Scale: Sucrose plus GBL as Carbon Subsubstrates.

3.3.1. Bioprocess

Based on the results from the shaking flask scale reported in this study and previous findings which confirmed *B. sacchari*'s potential to produce PHA containing 4HB by co-feeding sucrose and 4HB-related precursor compounds, this material was produced under controlled conditions at the bioreactor scale. It was aimed at generating a residual biomass concentration of about 20 g/L and a

PHA mass fraction in CDM exceeding 60 g/L in order to be competitive with the *C. necator*—mediated sucrose-based PHA production process at PHBISA.

Figure 5 shows the time curves of the CDM, PHA, and residual biomass, whereas Figure 6 illustrates the corresponding time curves of the sugar concentrations; again, arrows mark the time points of sucrose addition. Also in this cultivation, the nitrogen source (NH_4^+) served as the growth-limiting factor. NH_4^+ was added continuously during the growth phase as aqueous NH_4OH solution (25%) according to the response of the pH-electrode. The maximum specific growth rate μ_{max} measured between two subsequent samplings (t = 6–8 h) amounted to 0.23 1/h for the entire growth phase (t = 0–10 h), and the μ_{max} for the entire exponential growth phase was determined to be 0.18 1/h. About 21 g/L of catalytically active residual biomass was produced until the onset of nitrogen depletion. Figure 7 shows the time curve of the main carbon source sucrose and its hydrolysis products glucose and fructose, which are produced by the extracellular invertase excreted by the organism; again, the rapid hydrolysis of sucrose is evident.

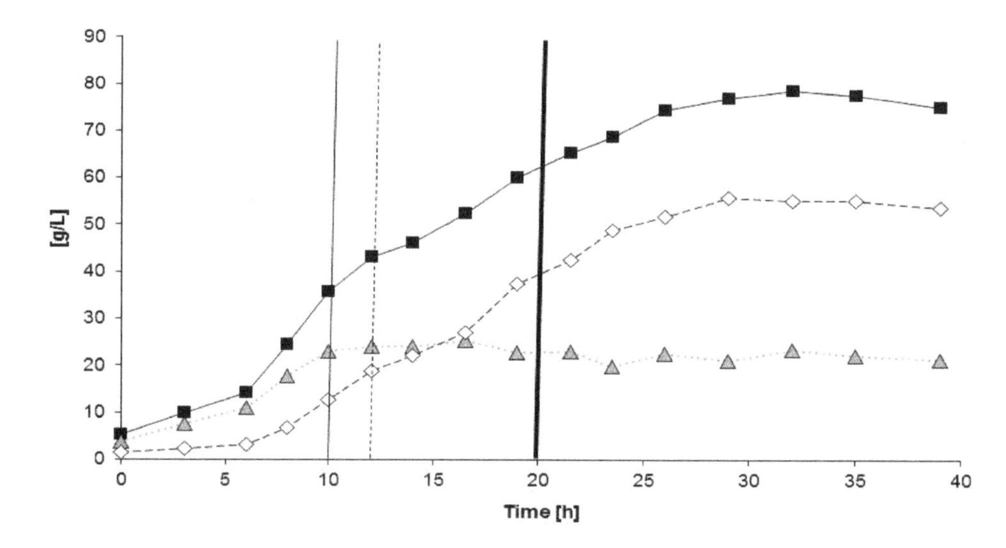

Figure 5. Time curves of product concentrations: *B. sacchari* on sucrose and the addition of γ-butyrolactone (GBL) as 4HB precursor. Black squares: CDM; open rhombi: PHA; grey triangles: residual biomass. Thin black line: Exchange of NH_4OH solution by NaOH solution as the pH-corrective agent; dash line: start of nitrogen depletion in the medium; bold black line: start of GBL feed.

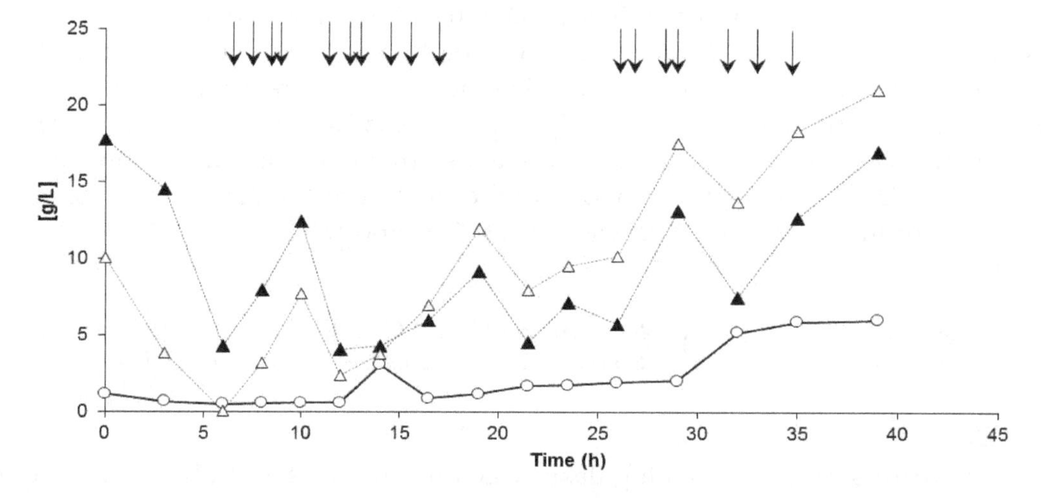

Figure 6. Actual concentrations of sugars: *B. sacchari* on sucrose and the addition of GBL as 4HB precursor. Arrows indicate the refeed with sucrose solution. Open spheres: sucrose; black triangles: glucose; open triangles: fructose.

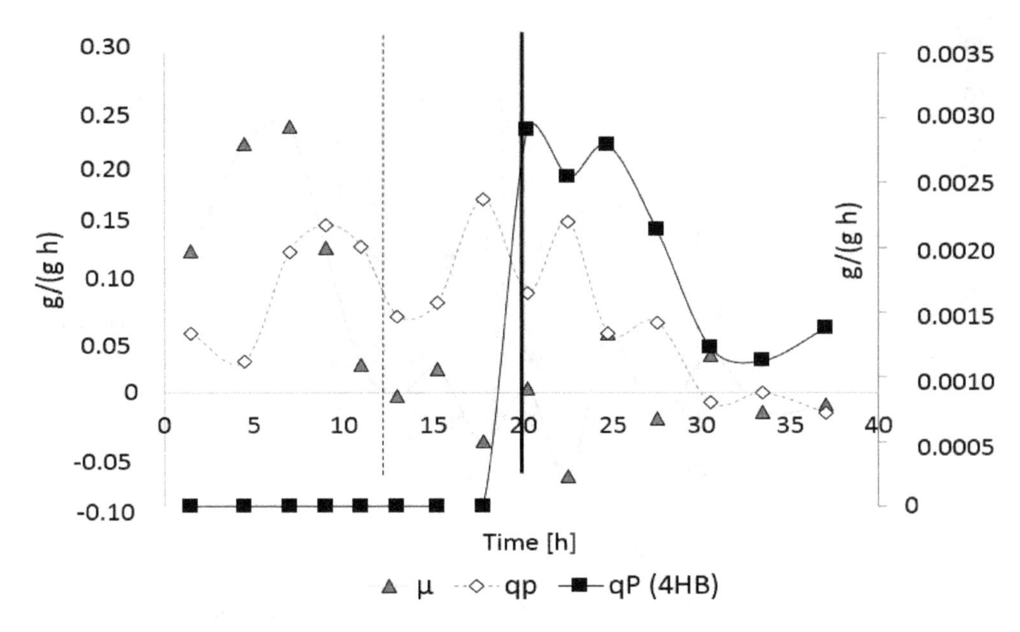

Figure 7. Time course of the specific growth rate μ and specific PHA production rate q_P (left axis): *B. sacchari* on sucrose and the addition of GBL as 4HB precursor. Thin black line: Exchange of NH_4OH solution by NaOH solution as the pH-corrective agent; dashed line: start of nitrogen depletion in the medium; bold black line: start of GBL feed.

After 10 h of fermentation, the nitrogen source supply was stopped by exchanging NH_4OH with NaOH as the pH-value correction agent; now, the second phase of the process was initiated (accumulation phase). During this phase, the time curve of the residual biomass was constant and the increase of CDM until the end of the experiment was only due to the increasing intracellular concentration of PHA (see Figure 5). It is visible that already during the exponential phase of the microbial growth (t = 7–10 h) that considerable amounts of PHA were produced ("growth associated product formation"). During the phase of product formation, GBL was added dropwise in order to not move into inhibiting concentration ranges. The actual GBL concentration was always below the detection limit when analyzing the samples; hence, GBL was completely converted by the cells. During the process, a total of 15.5 g/L GBL was added to the culture, distributed to a total of ten pulses of the substrate feed.

At the end of the process, the final concentrations of CDM and PHA of 75.1 g/L and 53.7 g/L, respectively, were achieved, corresponding to a PHA mass fraction in CDM of 71.5%. The total PHA concentration remained constant from t = 27.5 h. The volumetric productivity of PHA for the entire process and the conversion yield of sugar to CDM were calculated as 1.87 g/(L·h) and 0.38 g/g, respectively, which signifies an enormous enhancement in comparison to the previous experiment (Table 1).

Figure 7 illustrates the time curves of the specific growth rate μ, the specific PHA production rate q_P, and the specific 4HB production rate for the entire process. Again, starting with nitrogen limitation at about t = 12 h, the values for μ drastically decreased, whereas the specific PHA productivity q_P reached its highest values under nitrogen limited conditions; the maximum value for q_P was reached between t = 16.5 and 19 h, and amounted to 0.17 g/(g·h). Maximum specific 4HB production occurred between t = 20 and 35 h, and was calculated with about 0.003 g/(g h).

Co-feeding of GBL started after 20 h; until this time, the PHB homopolyester was produced (Figures 7 and 8). Starting with the sample taken at t = 23.5 h, 4HB-building blocks were detected in the polymer. The achieved 4HB fraction in PHA at the end of the fermentation was determined with 1.6% (mol/mol). The time curve of the polyester composition is illustrated in Figure 8. The essential process results are collected in Table 1 and directly compared with the outcomes of the previous process for the PHB production.

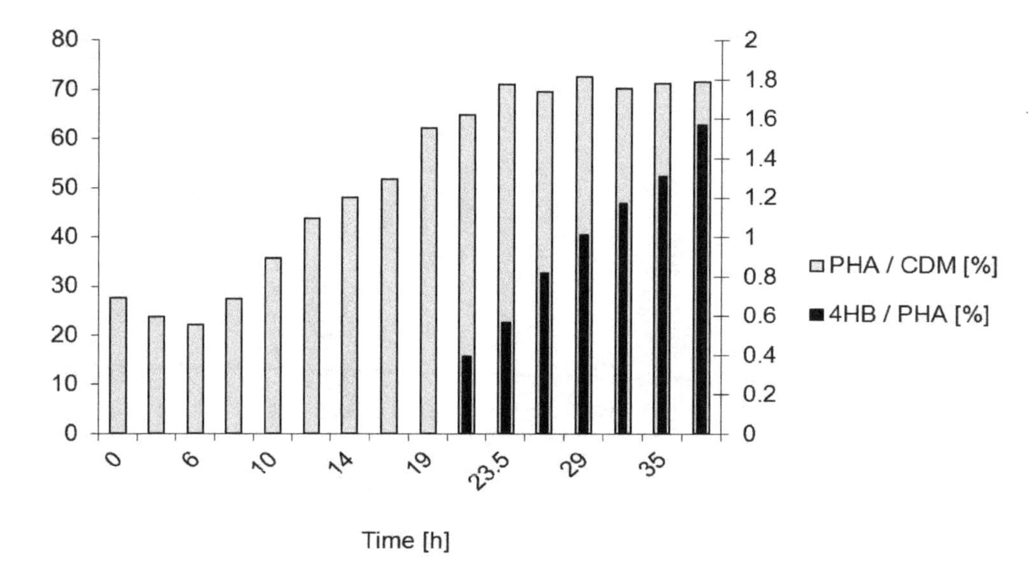

Figure 8. *B. sacchari* on sucrose and the addition of GBL as 4HB precursor. Composition of PHA during the process. Grey bars: Mass fraction of PHA in CDM (left axis). Black bars: Molar 4HB fraction on PHA (right axis). GBL addition started at $t = 20$ h.

3.3.2. Polymer Characterization:

After the end of the experiment, the biomass was separated from the liquid supernatant via centrifugation, and was frozen and lyophilized. The dry biomass was decreased with ethanol and the polymer was extracted using chloroform. The Mw and P_i values of the extracted copolymer, determined by GPC, were 315 ± 24 kDa and 2.51 ± 0.15 kDa, respectively (Table 1). Statistical differences analyses showed that the Mw of P(3HB-*co*-4HB) was significantly lower than that of PHB. In addition, analysis of the DSC data showed that P(3HB-*co*-4HB) had significantly lower X_c (24.0% ± 3.6%) and T_m (160.9 ± 0.8 °C) than PHB, while T_g was in the same range (1.8 ± 0.2 °C).

Table 1 compares both kinetic data and data from polymer characterization of both bioprocesses on the bioreactor scale.

4. Discussion

4.1. Bioprocess

The organism *B. sacchari* DSM 17165 possesses the desired ability to produce 4HB-containing PHA from sucrose plus both investigated 4HB precursors GBL and Na-4HB. The successful conversion of GBL towards 4HB building blocks is in agreement to previous findings reported by Cesário, who used glucose or WSH plus GBL for P(3HB-*co*-4HB) biosynthesis by this strain. These authors also tested P(3HB-*co*-4HB) production by this organism by using 1,4-butanediole as the 4HB-related precursor, revealing the incorporation of 4HB by GBL supplementation and the strain's inability to utilize 1,4-butanediole. No reports were previously available on the utilization of Na-4HB by this strain. The results reported by Cesário et al. show varying PHA fractions in CDM for the fed-batch cultivation of *B. sacchari* on glucose/GBL mixtures, dependent on the ratio of glucose/GBL. Cultivation on pure glucose resulted in 49.2% PHB in CDM; this value decreased with increasing GBL portions in the feed stream to only 7.1% using GBL as the sole carbon source [28]. In our shaking flask setups, the rather modest precursor supplementation of 1.5 g/L neither significantly impacted the CDM production or the PHA fraction in CDM compared to the precursor-free setups (sucrose as the sole carbon source). Remarkably, the application of the GBLs saponified from Na-4HB resulted in considerably lower 4HB fractions in PHA than observed when using the annular lactone (GBL) (21% vs. 14%). As assumed for *C. necator* [64] and *Hydrogenophaga pseudoflava* [65], GBL is imported into the cells as an intact lactone ring, which is opened only intracellularly. According to Valentin et al., only a part of 4HB is converted to 4-hydroxybutyryl-CoA (4HB-CoA) in the cells, whereas 4HB's major share is converted to succinic acid semialdehyde and succinic acid, which finally undergo conversion to the 3-hydroxybutyryl-CoA (3HB-CoA) precursor acetyl-CoA. PHA synthase in turn polymerizes 3HB-CoA and 4HB-CoA to P(3HB-*co*-4HB) [66].

As shown previously [39,52,53,55,57] and confirmed by the present work, nitrogen limitation is a suitable approach to boost PHA biosynthesis by *B. sacchari*. Generally, the strategy to constantly supply a nitrogen source by coupling the NH_4OH supply to microbial growth by automatically responding to the signal of the pH-electrode was performed successfully to rapidly generate a high concentration of catalytically active biomass at a high specific growth rate. Only about 9 h (PHB production) or 12 h (production of 4HB-containing PHA) were needed to boost the concentration of the residual biomass above 20 g/L. This shows significant progress to comparable experiments carried out by Rocha and colleagues, who used the same strategy and achieved a maximum residual biomass of about 16 g/L after 24 h of cultivation using the mutant *B. sacchari* IPT 189 [55]. The maximum growth rates μ_{max} obtained in our experiments (0.15 1/h for the first, 0.18 1/h for the second bioreactor cultivation; calculated for the entire growth phase; 0.41 and 0.23 1/h maximum valued between two subsequent samplings) can be compared to related reports found in the literature; for shaking flask cultivations of *B. sacchari* LFM 101 on sucrose, Nascimento et al. report a μ_{max} of 0.544 and 0.546 1/h at 30 and 35 °C, respectively [50]. At the bioreactor scale, Rocha et al. obtained a μ_{max} of 0.4 1/h for the first 10 hours of continuous cultivation of *B. sacchari* IPT 189 [55]; this value was also obtained by da Cruz Pradella with *B. sacchari* IPT 189 by using a fedbatch feeding regime in an airlift reactor [57]. Reliable μ_{max} values from the bioreactor scale cultivations of our production strain *B. sacchari* IPT 101 (DSM 17165) are available for xylose-based experiments, where μ_{max} amounted to 0.07–0.21 1/h with dependence on the initial xylose concentration [52]. Using glucose during the growth phase, Rodriguez-Contreras obtained a μ_{max} of 0.42 1/h [39]. Testing the effect of GBL on the growth of *B. sacchari* in shaking flask setups, Cesário et al. noticed a continued decrease of μ_{max} from 0.32 to 0.19 1/h with GBL concentrations increasing from 5 to 40 g/L, with 40 g/L glucose as the main carbon source. In this study, μ_{max} was unfortunately not reported for the fedbatch cultivations in the bioreactors for the production of PHB and P(3HB-*co*-4HB) [53].

Furthermore, we demonstrated that the organism can successfully be cultivated at an elevated temperature of 37 °C, which is beneficial for large scale operation in reactors integrated into the production facilities of the Brazilian sugarcane industry [48,61]. The cultivation temperature of 37 °C is in contrast to previous literature reports for this organism and its close relatives. Generally, 30 °C is reported as the optimum temperature to efficiently thrive most *B. sacchari* sp. [50]. In a mechanically stirred tank bioreactor, Raposo and colleagues cultivated the same strain for the production of PHB, xylitol, and xylonic acid at a temperature of 32 °C [61], whereas 30–32 °C was used by da Cruz Pradella et al. to culture its mutant strain *B. sacchari* IPT 189 for PHB biosynthesis in an airlift reactor [57], or by Rocha and colleagues in continuously operated bioreactor cultivations [55]. *B. sacchari* LFM 101, a strain that is most likely closely related to our production strain, was only recently tested by Nascimento et al. for PHA production on sucrose, glucose, and glycerol at both 30 and 35 °C. These authors report higher volumetric productivity and PHA fractions in CDM, and unaltered specific growth rates for cultivations carried out on glucose or sucrose at 35 °C or 30 °C, respectively. When using glycerol as the carbon source, no biomass formation or significant substrate consumption was observed, probably due to the lack of energy needed to convert the glycerol molecules [50]. As demonstrated by Rodriguez-Contreras et al. who operated a *B. sacchari*-mediated PHB production process at 37 °C, this problem can be overcome by feeding the cells with energy-rich carbohydrates like glucose or sucrose in the first stage (growth phase), and subsequently switching to glycerol feeding in the second phase (PHA accumulation) [39].

Values of 1.29 g/(L·h) (PHB) and 1.87 g/(L·h) (4HB-containing PHA) were achieved for the volumetric PHA productivity in the two conducted bioreactor experiments. These values are considerably higher than that reported for comparable experiments by Rodriguez-Contreras et al., who reported a volumetric productivity of 0.08 g/(L·h) for a two-stage process based on the co-feeding of *B. sacchari* with glucose and glycerol [39], and by Cesário and colleagues, who obtained 0.7 g/(L·h) for fed-batch cultures supplied with glucose and GBL, and 0.5 g/(L·h) when using WSH plus GBL for fed-batch P(3HB-*co*-4HB) production [53]. Here, it has to be emphasized that Cesário et al. [53] used considerably higher GBL dosage than we did in the study at hand; this, on the one hand, resulted in tripling the molar fractions of 4HB in PHA in comparison to our results, but, on the other hand, negatively influenced the overall volumetric PHA productivity as the fundamental economic parameter in PHA production. Regarding the obtained PHA contents in the biomass, our results show final PHA fractions in CDM of 52.4% for PHB, and 71.5% for P(3HB-*co*-4HB), respectively. The results by Cesário and colleagues report 73% PHB in CDM in fed-batch cultures with glucose as the sole carbon source, and 45% P(3HB-*co*-4HB) in CDM with pulse feeding 8 g/L GBL in the accumulation phase followed by continuously feeding GBL at a rate of 2.3 g/h. Fed-batch cultures of *B. sacchari* on WSH plus GBL reported in the same study resulted in a P(3HB-*co*-4HB) fraction in CDM of 27%. Interestingly, the authors found that in *B. sacchari*, the conversion yield of GBL towards 4HB can considerably be improved by supplementing acetate or propionate as additional "stimulants" for the 4HB biosynthesis [53]. Based on the works carried out by Lee et al. with *C. necator*, it was known previously that an increased acetyl-CoA pool from acetate conversion or from propionate ketolysis, respectively, inhibits the conversion of 4HB-CoA to acetyl-CoA, thus preserving a high 4HB-CoA pool available for the P(3HB-*co*-4HB) biosynthesis [67]. Using the mutant strain *B. sacchari* IPT 189, PHA copolyesters consisting of 3HB and 3HV were produced by Rocha et al. by co-feeding sucrose and propionic acid in two-stage bioreactor setups at a volumetric productivity of 1 g/(L·h); in these experiments, the biomass contained a PHA mass fraction of up to 60%, which is higher than in our PHB production process (52.4%), but lower than the value obtained in the present study for P(3HB-*co*-4HB) production (71.5%) [55]. The two-stage co-feeding experiments with *B. sacchari* carried out by Rodriguez-Contreras et al. on glucose and glycerol generated a PHA fraction in CDM that hardly exceeded 10% [39]. Using mixtures of xylose and glucose to mimic differently composed lignocellulosic hydrolysates, Raposo and associates produced PHB by fed-batch cultivations of *B. sacchari* in laboratory bioreactors. Changing the pulse size, feeding rate, and glucose/xylose ratio, the volumetric productivities decreased from 2.7 g/(L·h) (73% PHB in CDM) for pure glucose feeding to 0.07 (11% PHB in CDM) for xylose as the sole carbon source, indicating the inhibitory effect of this pentose sugar [52].

4.2. Polymer Characterization:

The obtained data for polymer characterization were in the same range as the results provided by Cesário and colleagues, who extracted PHB and P(3HB-*co*-4HB) from *B. sacchari* biomass, cultivated on WSH, via the same method used in the present study. These authors describe a Mw for PHB of 790 kDa, and between 450 and 590 kDa for P(3HB-*co*-4HB); higher 4HB fractions gradually decreased the Mw values [29]. Our results report a M_W of 627 kDa for PHB, and 315 kDa for P(3HB-*co*-4HB). The P_i of our sucrose-based polyester samples was higher than the values reported for WSH-based PH. For PHB, we obtained a P_i of 2.66, which is similar to the value obtained for the P(3HB-*co*-4HB) sample (2.51). For comparison, the PHB and P(3HB-*co*-4HB) samples produced by Cesário and colleagues had significantly lower P_i, ranging from 1.4 to 1.7 [29]. Other comparable results were provided by Rosengart et al., who reported a P_i of 2.33 for a *B. sacchari*-based PHB [68]. A considerably lower Mw (200 kDa) was described by Rodriguez-Contreras et al. for PHB obtained by co-feeding *B. sacchari* with glucose and glycerol; in this study, a P_i of 2.5 was reported [39]. Here, it should be noted that glycerol feeding generally results in low molecular mass PHA if compared to sugar-based PHA production, as reported elsewhere [37,68]. This is due to the "endcapping effect", a phenomenon describing the termination of the *in vivo* PHA chain propagation in the presence of glycerol and other polyols [69]. The melting temperature T_m reported by Cesário and colleagues amounted to 171.7 °C for PHB, and to 158.8 and 164.3 °C for P(3HB-*co*-4HB) with 7.6 or 4.6 mol% of 4HB, respectively [29]. In our case, the T_m for PHB amounted to 177.6 °C, whereas for P(3HB-*co*-4HB) was only 160.9 °C, which matches well with the cited literature data. Our PHB displayed an X_c of 70.9 °C, which is slightly higher than that reported for the WSH-based material (64.8%) [29]. A remarkably low X_c of 24.0% was measured for our P(3HB-*co*-4HB), which is considerably lower than the value reported for P(3HB-*co*-4HB) based on WSH (between 47.2% and 52.3%) [29]. The PHB produced by Rodriguez-Contreras et al. on glucose plus glycerol displayed an X_c of 72.8% and a T_m of 163.3°C [39]. Using PHB-rich biomass from a cultivation of *B. sacchari* on glucose, Rosengart et al. [21] compared the extraction performance of unusual extraction solvents (anisol, phenetole, and cyclohexanone) with the performance of classical chloroform extraction as used in our study, and by Cesário and colleagues [53]. As an outcome, the thermal properties (T_m, T_g, X_c) and molecular mass were fully comparable to the values obtained via chloroform extraction, thus demonstrating the feasibility of switching to sustainable, non-chlorinated alternatives to chloroform [21].

5. Conclusions

The highest (up to now) reported productivity for *B. sacchari*-mediated biosynthesis of PHA with building blocks differing from 3HB is described in the present work. Adaptation of the production strain to an elevated temperature optimum of 37 °C makes it a feasible candidate for cost-efficient on-site PHB and P(3HB-*co*-4HB) production starting from cane sugar on the industrial scale. In any case, PHA production facilities should also in future be integrated into the existing production lines for sucrose-based bioethanol production in order to profit from reduced transportation costs, energetic autarky, and in-house availability of extraction solvents for PHA recovery from the biomass. Further efforts should be devoted to high-throughput continuous PHA production by this organism in a chemostat ("chemical environment is static") process regime. Similar to the results recently obtained by other production strains [70], the application of a multistep-continuous production in a bioreactor cascade displays a viable process-engineering tool to further increase volumetric productivity, and to trigger the distribution of 3HB and 4HB monomers in tailor-made copolyesters. Moreover, the highly effective invertase enzyme excreted by this strain deserves in-depth characterization and might be of interest for applications in food technology. Together with PHA production and other metabolites generated by this strain, such as xylitol or xylonic acid [52], this might open the door to implementing *B. sacchari* as a versatile platform to catalyze a bio-refinery plant starting from inexpensive feedstocks.

Acknowledgments: ARENA is grateful for the funding received by PHBISA for the industrial project "Production of the copolyester Poly(3-hydroxybutyrate-*co*-3-hydroxyvalerate) from cane sugar by fermentation with special regard to new strains adapted to high temperature and direct use of sucrose".

Author Contributions: Miguel Miranda de Sousa Dias, Gerhart Braunegg, and Martin Koller conceived and designed the biotechnological experiments; Miguel Miranda de Sousa Dias performed the biotechnological experiments; Dario Puppi, Andrea Morelli, and Federica Chiellini analyzed and interpreted the polymer data; Martin Koller wrote the predominant portion of the paper. All authors read, edited, and approved the final manuscript.

References

1. Jendrossek, D.; Pfeiffer, D. New insights in the formation of polyhydroxyalkanoate granules (carbonosomes) and novel functions of poly(3-hydroxybutyrate). *Environ. Microbiol.* **2014**, *16*, 2357–2373. [CrossRef] [PubMed]
2. Chen, G.Q.; Hajnal, I. The 'PHAome'. *Trends Biotechnol.* **2015**, *33*, 559–564. [CrossRef] [PubMed]
3. Tan, G.Y.A.; Chen, C.L.; Li, L.; Ge, L.; Wang, L.; Razaad, I.M.N.; Li, Y.; Zhao, L.; Mo, Y.; Wang, J.Y. Start a research on biopolymer polyhydroxyalkanoate (PHA): A review. *Polymers* **2014**, *6*, 706–754. [CrossRef]
4. Koller, M.; Maršálek, L.; de Sousa Dias, M.; Braunegg, G. Producing microbial polyhydroxyalkanoate (PHA) biopolyesters in a sustainable manner. *New Biotechnol.* **2017**, *37*, 24–38. [CrossRef] [PubMed]
5. Obruca, S.; Sedlacek, P.; Mravec, F.; Samek, O.; Marova, I. Evaluation of 3-hydroxybutyrate as an enzyme-protective agent against heating and oxidative damage and its potential role in stress response of poly(3-hydroxybutyrate) accumulating cells. *Appl. Microbiol. Biotechnol.* **2016**, *100*, 1365–1376. [CrossRef] [PubMed]

6. Ayub, N.D.; Pettinari, M.J.; Ruiz, J.A.; López, N.I. A polyhydroxybutyrate-producing *Pseudomonas* sp. isolated from Antarctic environments with high stress resistance. *Curr. Microbiol.* **2004**, *49*, 170–174. [CrossRef] [PubMed]
7. Obruca, S.; Marova, I.; Stankova, M.; Mravcova, L.; Svoboda, Z. Effect of ethanol and hydrogen peroxide on poly(3-hydroxybutyrate) biosynthetic pathway in *Cupriavidus necator* H16. *World J. Microbiol. Biotechnol.* **2010**, *26*, 1261–1267. [CrossRef] [PubMed]
8. Steinbüchel, A.; Valentin, H.E. Diversity of bacterial polyhydroxyalkanoic acids. *FEMS Microbiol. Lett.* **1995**, *128*, 219–228. [CrossRef]
9. Zinn, M.; Witholt, B.; Egli, T. Occurrence, synthesis and medical application of bacterial polyhydroxyalkanoate. *Adv. Drug Deliv. Rev.* **2001**, *53*, 5–21. [CrossRef]
10. Pérez Amaro, L.; Chen, H.; Barghini, A.; Corti, A.; Chiellini, E. High performance compostable biocomposites based on bacterial polyesters suitable for injection molding and blow extrusion. *Chem. Biochem. Eng. Q.* **2015**, *29*, 261–274. [CrossRef]
11. Kovalcik, A.; Machovsky, M.; Kozakova, Z.; Koller, M. Designing packaging materials with viscoelastic and gas barrier properties by optimized processing of poly(3-hydroxybutyrate-*co*-3-hydroxyvalerate) with lignin. *React. Funct. Polym.* **2015**, *94*, 25–34. [CrossRef]
12. Koller, M. Poly(hydroxyalkanoates) for food packaging: Application and attempts towards implementation. *Appl. Food Biotechnol.* **2014**, *1*, 3–15.
13. Khosravi-Darani, K.; Bucci, D.Z. Application of poly(hydroxyalkanoate) in food packaging: Improvements by nanotechnology. *Chem. Biochem. Eng. Q.* **2015**, *29*, 275–285. [CrossRef]
14. Martínez-Sanz, M.; Villano, M.; Oliveira, C.; Albuquerque, M.G.; Majone, M.; Reis, M.A.M.; Lopez-Rubio, A.; Lagaron, J.M. Characterization of polyhydroxyalkanoates synthesized from microbial mixed cultures and of their nanobiocomposites with bacterial cellulose nanowhiskers. *New Biotechnol.* **2014**, *31*, 364–376. [CrossRef] [PubMed]
15. Narodoslawsky, M.; Shazad, K.; Kollmann, R.; Schnitzer, H. LCA of PHA production–Identifying the ecological potential of bio-plastic. *Chem. Biochem. Eng. Q.* **2015**, *29*, 299–305. [CrossRef]
16. Dietrich, K.; Dumont, M.J.; Del Rio, L.F.; Orsat, V. Producing PHAs in the bioeconomy—Towards a sustainable bioplastic. *Sustain. Prod. Consum.* **2017**, *9*, 58–70. [CrossRef]
17. Koller, M. Poly(hydroxyalkanoate) (PHA) biopolyesters: Production, Performance and processing aspects. *Chem. Biochem. Eng. Q.* **2015**, *29*, 261.
18. Koller, M.; Niebelschütz, H.; Braunegg, G. Strategies for recovery and purification of poly[(*R*)-3-hydroxyalkanoates] (PHA) biopolyesters from surrounding biomass. *Eng. Life Sci.* **2013**, *13*, 549–562. [CrossRef]
19. Madkour, M.H.; Heinrich, D.; Alghamdi, M.A.; Shabbaj, I.I.; Steinbüchel, A. PHA recovery from biomass. *Biomacromolecules* **2013**, *14*, 2963–2972. [CrossRef] [PubMed]
20. Murugan, P.; Han, L.; Gan, C.Y.; Maurer, F.H.; Sudesh, K. A new biological recovery approach for PHA using mealworm, *Tenebrio molitor*. *J. Biotechnol.* **2016**, *239*, 98–105. [CrossRef] [PubMed]
21. Rosengart, A.; Cesário, M.T.; de Almeida, M.C.M.; Raposo, R.S.; Espert, A.; de Apodaca, E.D.; da Fonseca, M.M.R. Efficient P(3HB) extraction from *Burkholderia sacchari* cells using non-chlorinated solvents. *Biochem. Eng. J.* **2015**, *103*, 39–46. [CrossRef]
22. Kaur, G.; Roy, I. Strategies for large-scale production of polyhydroxyalkanoates. *Chem. Biochem. Eng. Q.* **2015**, *29*, 157–172. [CrossRef]
23. Koller, M.; Muhr, A. Continuous production mode as a viable process-engineering tool for efficient poly(hydroxyalkanoate) (PHA) bio-production. *Chem. Biochem. Eng. Q.* **2014**, *28*, 65–77.
24. Sindhu, R.; Pandey, A.; Binod, P. Solid-state fermentation for the production of poly(hydroxyalkanoates). *Chem. Biochem. Eng. Q.* **2015**, *29*, 173–181. [CrossRef]
25. Haas, C.; El-Najjar, T.; Virgolini, N.; Smerilli, M.; Neureiter, M. High cell-density production of poly(3-hydroxybutyrate) in a membrane bioreactor. *New Biotechnol.* **2017**, *37*, 117–122. [CrossRef] [PubMed]
26. Novak, M.; Koller, M.; Braunegg, M.; Horvat, P. Mathematical modelling as a tool for optimized PHA production. *Chem. Biochem. Eng. Q.* **2015**, *29*, 183–220. [CrossRef]
27. Koller, M.; Bona, R.; Chiellini, E.; Fernandes, E.G.; Horvat, P.; Kutschera, C.; Hesse, P.; Braunegg, G. Polyhydroxyalkanoate production from whey by *Pseudomonas hydrogenovora*. *Bioresour. Technol.* **2008**, *99*, 4854–4863. [CrossRef] [PubMed]
28. Obruca, S.; Benesova, P.; Marsalek, L.; Marova, I. Use of lignocellulosic materials for PHA production. *Chem. Biochem. Eng. Q.* **2015**, *29*, 135–144. [CrossRef]

29. Cesário, M.T.; Raposo, R.S.; de Almeida, M.C.M.; Van Keulen, F.; Ferreira, B.S.; Telo, J.P.; da Fonseca, M.M.R. Production of poly(3-hydroxybutyrate-*co*-4-hydroxybutyrate) by *Burkholderia sacchari* using wheat straw hydrolysates and gamma-butyrolactone. *Int. J. Biol. Macromol.* **2014**, *71*, 59–67. [CrossRef] [PubMed]
30. Haas, C.; Steinwandter, V.; Diaz De Apodaca, E.; Maestro Madurga, B.; Smerilli, M.; Dietrich, T.; Neureiter, M. Production of PHB from chicory roots—Comparison of three *Cupriavidus necator* strains. *Chem. Biochem. Eng. Q.* **2015**, *29*, 99–112. [CrossRef]
31. Muhr, A.; Rechberger, E.M.; Salerno, A.; Reiterer, A.; Schiller, M.; Kwiecień, M.; Adamus, G.; Kowalczuk, M.; Strohmeier, K.; Schober, S.; et al. Biodegradable latexes from animal-derived waste: Biosynthesis and characterization of mcl-PHA accumulated by *Ps. citronellolis*. *React. Funct. Polym.* **2013**, *73*, 1391–1398. [CrossRef]
32. Muhr, A.; Rechberger, E.M.; Salerno, A.; Reiterer, A.; Malli, K.; Strohmeier, K.; Schober, S; Mittelbach, M.; Koller, M. Novel description of mcl-PHA biosynthesis by *Pseudomonas chlororaphis* from animal-derived waste. *J. Biotechnol.* **2013**, *165*, 45–51. [CrossRef] [PubMed]
33. Titz, M.; Kettl, K.H.; Shahzad, K.; Koller, M.; Schnitzer, H.; Narodoslawsky, M. Process optimization for efficient biomediated PHA production from animal-based waste streams. *Clean Technol. Environ. Policy* **2012**, *14*, 495–503. [CrossRef]
34. Obruca, S.; Marova, I.; Snajdar, O.; Mravcova, L.; Svoboda, Z. Production of poly(3-hydroxybutyrate-*co*-3-hydroxyvalerate) by *Cupriavidus necator* from waste rapeseed oil using propanol as a precursor of 3-hydroxyvalerate. *Biotechnol. Lett.* **2010**, *32*, 1925–1932. [CrossRef] [PubMed]
35. Obruca, S.; Snajdar, O.; Svoboda, Z.; Marova, I. Application of random mutagenesis to enhance the production of polyhydroxyalkanoates by *Cupriavidus necator* H16 on waste frying oil. *World J. Microbiol. Biotechnol.* **2013**, *29*, 2417–2428. [CrossRef] [PubMed]
36. Walsh, M.; O'Connor, K.; Babu, R.; Woods, T.; Kenny, S. Plant oils and products of their hydrolysis as substrates for polyhydroxyalkanoate synthesis. *Chem. Biochem. Eng. Q.* **2015**, *29*, 123–133. [CrossRef]
37. Hermann-Krauss, C.; Koller, M.; Muhr, A.; Fasl, H.; Stelzer, F.; Braunegg, G. Archaeal production of polyhydroxyalkanoate (PHA) co-and terpolyesters from biodiesel industry-derived by-products. *Archaea* **2013**, *2013*. [CrossRef] [PubMed]
38. Cavalheiro, J.M.; Raposo, R.S.; de Almeida, M.C.M.; Cesário, M.T.; Sevrin, C.; Grandfils, C.; Da Fonseca, M.M.R. Effect of cultivation parameters on the production of poly(3-hydroxybutyrate-*co*-4-hydroxybutyrate) and poly(3-hydroxybutyrate-4-hydroxybutyrate-3-hydroxyvalerate) by *Cupriavidus necator* using waste glycerol. *Bioresour. Technol.* **2012**, *111*, 391–397. [CrossRef] [PubMed]
39. Rodríguez-Contreras, A.; Koller, M.; Miranda-de Sousa Dias, M.; Calafell-Monfort, M.; Braunegg, G.; Marqués-Calvo, M.S. Influence of glycerol on poly(3-hydroxybutyrate) production by *Cupriavidus necator* and *Burkholderia sacchari*. *Biochem. Eng. J.* **2015**, *94*, 50–57. [CrossRef]
40. Koller, M.; Marsalek, L. Potential of diverse prokaryotic organisms for glycerol-based polyhydroxyalkanoate production. *Appl. Food Biotechnol.* **2015**, *2*, 3–15.
41. Obruca, S.; Petrik, S.; Benesova, P.; Svoboda, Z.; Eremka, L.; Marova, I. Utilization of oil extracted from spent coffee grounds for sustainable production of polyhydroxyalkanoates. *Appl. Microbiol. Biotechnol.* **2014**, *98*, 5883–5890. [CrossRef] [PubMed]
42. Obruca, S.; Benesova, P.; Petrik, S.; Oborna, J.; Prikryl, R.; Marova, I. Production of polyhydroxyalkanoates using hydrolysate of spent coffee grounds. *Process Biochem.* **2014**, *49*, 1409–1414. [CrossRef]
43. Carvalho, G.; Oehmen, A.; Albuquerque, M.G.; Reis, M.A. The relationship between mixed microbial culture composition and PHA production performance from fermented molasses. *New Biotechnol.* **2014**, *31*, 257–263. [CrossRef] [PubMed]
44. Khosravi-Darani, K.; Mokhtari, Z.B.; Amai, T.; Tanaka, K. Microbial production of poly(hydroxybutyrate) from C1 carbon sources. *Appl. Microbiol. Biotechnol.* **2013**, *97*, 1407–1424. [CrossRef] [PubMed]
45. Drosg, B.; Fritz, I.; Gattermayr, F.; Silvestrini, L. Photo-autotrophic production of poly(hydroxyalkanoates) in cyanobacteria. *Chem. Biochem. Eng. Q.* **2015**, *29*, 145–156. [CrossRef]
46. Koller, M.; Marsalek, L. Cyanobacterial Polyhydroxyalkanoate Production: *Status Quo* and *Quo Vadis*? *Curr. Biotechnol.* **2015**, *4*, 464–480. [CrossRef]
47. Tanaka, K.; Miyawaki, K.; Yamaguchi, A.; Khosravi-Darani, K.; Matsusaki, H. Cell growth and P(3HB) accumulation from CO_2 of a carbon monoxide-tolerant hydrogen-oxidizing bacterium, *Ideonella* sp. O-1. *Appl. Microbiol. Biotechnol.* **2011**, *92*, 1161–1169. [CrossRef] [PubMed]

48. Nonato, R.; Mantelatto, P.; Rossell, C. Integrated production of biodegradable plastic, sugar and ethanol. *Appl. Microbiol. Biotechnol.* **2001**, *57*, 1–5. [PubMed]
49. Brämer, C.O.; Vandamme, P.; da Silva, L.F.; Gomez, J.G.; Steinbüchel, A. Polyhydroxyalkanoate-accumulating bacterium isolated from soil of a sugar-cane plantation in Brazil. *Int. J. Syst. Evol. Microbiol.* **2001**, *51*, 1709–1713. [CrossRef] [PubMed]
50. Nascimento, V.M.; Silva, L.F.; Gomez, J.G.C.; Fonseca, G.G. Growth of *Burkholderia sacchari* LFM 101 cultivated in glucose, sucrose and glycerol at different temperatures. *Sci. Agric.* **2016**, *73*, 429–433. [CrossRef]
51. Alexandrino, P.M.R.; Mendonça, T.T.; Bautista, L.P.G.; Cherix, J.; Lozano-Sakalauskas, G.C.; Fujita, A.; Ramos Filho, E.; Long, P.; Padilla, G.; Taciro, M.K.; et al. Draft genome sequence of the polyhydroxyalkanoate-producing bacterium *Burkholderia sacchari* LMG 19450 isolated from Brazilian sugarcane plantation soil. *Genome Announc.* **2015**, *3*, e00313-15. [CrossRef] [PubMed]
52. Raposo, R.S.; de Almeida, M.C.M.; de Oliveira, M.D.C.M.; da Fonseca, M.M.; Cesário, M.T. A *Burkholderia sacchari* cell factory: Production of poly-3-hydroxybutyrate, xylitol and xylonic acid from xylose-rich sugar mixtures. *New Biotechnol.* **2017**, *34*, 12–22. [CrossRef] [PubMed]
53. Cesário, M.T.; Raposo, R.S.; de Almeida, M.C.M.; van Keulen, F.; Ferreira, B.S.; da Fonseca, M.M.R. Enhanced bioproduction of poly-3-hydroxybutyrate from wheat straw lignocellulosic hydrolysates. *New Biotechnol.* **2014**, *31*, 104–113. [CrossRef] [PubMed]
54. Lopes, M.S.G.; Gomez, J.G.C.; Silva, L.F. Cloning and overexpression of the xylose isomerase gene from *Burkholderia sacchari* and production of polyhydroxybutyrate from xylose. *Can. J. Microbiol.* **2009**, *55*, 1012–1015. [CrossRef] [PubMed]
55. Rocha, R.C.; da Silva, L.F.; Taciro, M.K.; Pradella, J.G. Production of poly(3-hydroxybutyrate-*co*-3-hydroxyvalerate) P(3HB-*co*-3HV) with a broad range of 3HV content at high yields by *Burkholderia sacchari* IPT 189. *World J. Microbiol. Biotechnol.* **2008**, *24*, 427–431. [CrossRef]
56. Mendonça, T.T.; Gomez, J.G.C.; Buffoni, E.; Sánchez Rodriguez, R.J.; Schripsema, J.; Lopes, M.S.G.; Silva, L.F. Exploring the potential of *Burkholderia sacchari* to produce polyhydroxyalkanoates. *J. Appl. Microbiol.* **2014**, *116*, 815–829. [CrossRef] [PubMed]
57. Da Cruz Pradella, J.G.; Taciro, M.K.; Mateus, A.Y.P. High-cell-density poly(3-hydroxybutyrate) production from sucrose using *Burkholderia sacchari* culture in airlift bioreactor. *Bioresour. Technol.* **2010**, *101*, 8355–8360. [CrossRef] [PubMed]
58. Brämer, C.O.; Silva, L.F.; Gomez, J.G.C.; Priefert, H.; Steinbüchel, A. Identification of the 2-methylcitrate pathway involved in the catabolism of propionate in the polyhydroxyalkanoate-producing strain *Burkholderia sacchari* IPT101T and analysis of a mutant accumulating a copolyester with higher 3-hydroxyvalerate content. *Appl. Environ. Microbiol.* **2002**, *68*, 271–279. [CrossRef] [PubMed]
59. Silva, L.F.; Gomez, J.G.C.; Oliveira, M.S.; Torres, B.B. Propionic acid metabolism and poly-3-hydroxybutyrate-*co*-3-hydroxyvalerate (P3HB-*co*-3HV) production by *Burkholderia* sp. *J. Biotechnol.* **2000**, *76*, 165–174. [CrossRef]
60. Mendonça, T.T.; Tavares, R.R.; Cespedes, L.G.; Sánchez-Rodriguez, R.J.; Schripsema, J.; Taciro, M.K.; Gomez, J.G.C.; Silva, L.F. Combining molecular and bioprocess techniques to produce poly(3-hydroxybutyrate-*co*-3-hydroxyhexanoate) with controlled monomer composition by *Burkholderia sacchari*. *Int. J. Biol. Macromol.* **2017**, *98*, 654–663. [CrossRef] [PubMed]
61. Silva, L.F.; Taciro, M.K.; Raicher, G.; Piccoli, R.A.M.; Mendonça, T.T.; Lopes, M.S.G.; Gomez, J.G.C. Perspectives on the production of polyhydroxyalkanoates in biorefineries associated with the production of sugar and ethanol. *Int. J. Biol. Macromol.* **2014**, *71*, 2–7. [CrossRef] [PubMed]
62. Küng, W. Wachstum und Poly-D-(-)-3-Hydroxybuttersäure-Akkumulation bei *Alcaligenes latus*. Diploma Thesis, Graz University of Technology, Graz, Austria, 1982.
63. Braunegg, G.; Sonnleitner, B.; Lafferty, R. A rapid gaschromatographic method for the determination of poly-β-hydroxybutyric acid in microbial biomass. *Eur. J. Appl. Microbiol.* **1978**, *6*, 29–37. [CrossRef]
64. Kunioka, M.; Kawaguchi, Y.; Doi, Y. Production of biodegradable copolyesters of 3-hydroxybutyrate and 4-hydroxybutyrate by *Alcaligenes eutrophus*. *Appl. Microbiol. Biotechnol.* **1989**, *30*, 569–573. [CrossRef]
65. Choi, M.H.; Yoon, S.C.; Lenz, R.W. Production of poly(3-hydroxybutyric acid-co-4-hydroxybutyric acid) and poly(4-hydroxybutyric acid) without subsequent degradation by *Hydrogenophaga pseudoflava*. *Appl. Environ. Microbiol.* **1999**, *65*, 1570–1577. [PubMed]

66. Valentin, H.E.; Zwingmann, G.; Schönebaum, A.; Steinbüchel, A. Metabolic pathway for biosynthesis of poly(3-hydroxybutyrate-*co*-4-hydroxybutyrate) from 4-hydroxybutyrate by *Alcaligenes eutrophus*. *Eur. J. Biochem.* **1995**, *227*, 43–60. [CrossRef] [PubMed]
67. Lee, Y.H.; Kang, M.S.; Jung, Y.M. Regulating the molar fraction of 4-hydroxybutyrate in poly(3-hydroxybutyrate-4-hydroxybutyrate) biosynthesis by *Ralstonia eutropha* using propionate as a stimulator. *J. Biosci. Bioeng.* **2000**, *89*, 380–383. [CrossRef]
68. Koller, M.; Bona, R.; Braunegg, G.; Hermann, C.; Horvat, P.; Kroutil, M.; Martinz, J.; Neto, J.; Pereira, L.; Varila, P. Production of polyhydroxyalkanoates from agricultural waste and surplus materials. *Biomacromolecules* **2005**, *6*, 561–565. [CrossRef] [PubMed]
69. Madden, L.A.; Anderson, A.J.; Shah, D.T.; Asrar, J. Chain termination in polyhydroxyalkanoate synthesis: Involvement of exogenous hydroxy-compounds as chain transfer agents. *Int. J. Biol. Macromol.* **1999**, *25*, 43–53. [CrossRef]
70. Atlić, A.; Koller, M.; Scherzer, D.; Kutschera, C.; Grillo-Fernandes, E.; Horvat, P.; Chiellini, E.; Braunegg, G. Continuous production of poly([*R*]-3-hydroxybutyrate) by *Cupriavidus necator* in a multistage bioreactor cascade. *Appl. Microbiol. Biotechnol.* **2011**, *91*, 295–304. [CrossRef] [PubMed]

10

Tequila Agave Bagasse Hydrolysate for the Production of Polyhydroxybutyrate by *Burkholderia sacchari*

Yolanda González-García [1,*], Janessa Grieve [1], Juan Carlos Meza-Contreras [1], Berenice Clifton-García [2] and José Antonio Silva-Guzman [1]

1 Department of Wood, Cellulose and Paper, University of Guadalajara, 45020 Zapopan, Mexico; janessagrieve@me.com (J.G.); jcmezac@gmail.com (J.C.M.-C.); jasilva@dmcyp.cucei.udg.mx (J.A.S.-G.)
2 Department of Chemical Engineering, University of Guadalajara, 44430 Guadalajara, Mexico; bere_clifton@hotmail.com
* Correspondence: yolanda.ggarcia@academicos.udg.mx

Abstract: Tequila agave bagasse (TAB) is the fibrous waste from the Tequila production process. It is generated in large amounts and its disposal is an environmental problem. Its use as a source of fermentable sugars for biotechnological processes is of interest; thus, it was investigated for the production of polyhydroxybutyrate (PHB) by the xylose-assimilating bacteria *Burkholderia sacchari*. First, it was chemically hydrolyzed, yielding 20.6 $g \cdot L^{-1}$ of reducing sugars, with xylose and glucose as the main components (7:3 ratio). Next, the effect of hydrolysis by-products on *B. sacchari* growth was evaluated. Phenolic compounds showed the highest toxicity (> 60% of growth inhibition). Then, detoxification methods (resins, activated charcoal, laccases) were tested to remove the growth inhibitory compounds from the TAB hydrolysate (TABH). The highest removal percentage (92%) was achieved using activated charcoal (50 $g \cdot L^{-1}$, pH 2, 4 h). Finally, detoxified TABH was used as the carbon source for the production of PHB in a two-step batch culture, reaching a biomass production of 11.3 $g \cdot L^{-1}$ and a PHB accumulation of 24 g PHB g^{-1} dry cell (after 122 h of culture). The polymer structure resulted in a homopolymer of 3-hydroxybutyric acid. It is concluded that the TAB could be hydrolyzed and valorized as a carbon source for producing PHB.

Keywords: polyhydroxybutyrate; tequila bagasse; hydrolysate detoxification; activated charcoal; phenolic compounds

---

1. Introduction

Polyhydroxyalkanoate (PHA) is a family of biodegradable and biocompatible polymers produced by microorganisms. They have great potential as a substitute for petroleum-based plastics as packing material, disposable items, and biomedical devices [1]:

Polyhydroxybutyrate (PHB) is the main representative of these polymers. Nowadays, some manufacturers such as Metabolix (USA), Tianjin Green Bioscience Co., Ltd. (China), and Biocycle PHB Industrial S.A (Brazil) commercialize this biopolymer, which is produced from raw materials such as food crops and sugarcane [1]. Nevertheless, its production cost is 5 to 10 times higher than that of conventional plastics; in particular, the substrate cost for PHA production represents almost half of the production cost [2].

Thus, the use of low-cost substrates for producing PHAs is a matter of interest. Among various alternative substrates, lignocellulosic wastes have gained considerable attention since they could be hydrolyzed to yield sugars (i.e., glucose, xylose, and arabinose) for fermentation processes [3]. Examples of this type of waste, studied for PHA production include sugar cane bagasse, rice, wheat straw, and corn stover [4].

Tequila agave bagasse (TAB) is a residual fibrous material from the Tequila production process. In 2018, 346,700.00 tons of TAB were generated. This waste did not have any specific use, but since it contains large amounts of cellulose and hemicellulose [5], it might be used as a low-cost substrate for fermentation processes. It has been hydrolyzed and used as a carbon source in submerged fermentation for the production of bioethanol, organic acids, and lipids [6,7]. Nevertheless, the research about the use of TAB for the production of PHAs is scarce [8] as well as the effect of the hydrolysis byproducts generated from its chemical saccharification on the growth and PHB accumulation by any PHA-producing microorganism.

Lignocellulosic wastes conversion to PHAs generally requires a hydrolysis step to obtain fermentable sugars and then a detoxification process to remove inhibitory compounds produced during hydrolysis [9]. Different detoxification methodologies (activated charcoal, ionic resins, and enzymes) have been studied in bagasse hydrolysates, such as sugar cane bagasse hydrolysate, but there are not reports about their use for detoxification of TABH.

This research aimed to evaluate the effect of growth-inhibition compounds (present in the TABH) on the growth of the xylose-utilizing bacteria *B. sacchari*, their removal using activated charcoal, resins, and laccases, and the use of the detoxified TABH for the production of PHB.

2. Materials and Methods

2.1. Characterization TAB

TAB fibers were washed (Sprout-Waldron refiner, D2A509NH) and then centrifuged to remove water, sun-dried for 48 h, ground, and sieved (60 mesh). Cellulose, hemicellulose, lignin, ashes, extractable, and humidity content were determined by the Technical Association of the Pulp and Paper Industry (TAPPI) standards (T203, T222, and T257).

2.2. Chemical Hydrolysis of TAB and Hydrolysate Characterization

Acid hydrolysis of TAB fibers was performed in 250 mL capped flasks using the following conditions: H_2SO_4, 2.3% (*w*/*v*); TAB: acid solution, 1:10 (*w*/*v*); temperature, 130 °C (autoclave); time, 20 min. Sixty mesh sieved TAB (250 μm fiber size), and not sieved TAB (mixed fiber sizes, from 125–420 μm) were used. After hydrolysis, the remaining fibers were separated by filtration (filter paper Whatman 2). Next, pH was adjusted to 5.5 by the addition of $Ca(OH)_2$ (constant agitation), and the precipitated solids were removed by filtration (filter paper Whatman 2). The TABH was analyzed for determining: total sugars, reducing sugars, and total phenolic compounds (Phenol-sulfuric [10], DNS (dinitrosalicylic acid) [11], and Folin–Ciocalteu method [12], respectively). Monomeric sugars were analyzed by HPLC (Waters) with an IR detector and using an Aminex 87P column: mobile phase, water; flow 0.6 mL min^{-1}; temperature, 80 °C; sample volume, 20 μL).

2.3. Microorganism and Culture Media

The strain *B. sacchari* 17165 was purchased from DSMZ, reactivated according to DSMZ instructions in R2A medium (in $g{\cdot}L^{-1}$): yeast extract, 0.5; meat peptone 0.5; casein peptone, 0.5; glucose, 0.5; soluble starch, 0.5; sodium pyruvate, 0.3; K_2HPO_4, 0.3; and $MgSO_4{\cdot}7H_2O$, 0.05; pH 7. The strain was incubated at 30 °C and 150 rpm for 48 h and then frozen at −20 °C (2 mL microtubes with 0.8 mL of bacterial suspension and 0.2 mL of glycerol).

Two culture media were studied for *B. sacchari* growth and PHB production: control medium and Detoxified TABH medium. Control medium composition was (in $g{\cdot}L^{-1}$): xylose, 14; glucose, 6; KH_2PO_4, 2.5; Na_2HPO_4, 5.5; $(NH_4)_2SO_4$, 6; $MgSO_4{\cdot}7H_2O$, 0.5; $CaCl_2{\cdot}2H_2O$, 0.02; ammonium ferric citrate, 0.1; peptone, 1; and yeast extract, 1. Trace elements solution (1 $mL{\cdot}L^{-1}$) with the following composition ($g{\cdot}L^{-1}$) was also added: H_3BO_3, 0.30; $CoCl_2{\cdot}6H_2O$, 0.20; $ZnSO_4{\cdot}7H_2O$, 0.1; $MnCl_2{\cdot}4H_2O$, 0.03; $NaMoO_4{\cdot}2H_2O$, 0.03; $NiCl_2{\cdot}6H_2O$, 0.02; and $CuSO_4{\cdot}5H_2O$, 0.01.

The Detoxified TABH medium was based on the hydrolysate (20 $g \cdot L^{-1}$ of reducing sugars, xylose: glucose ratio of 7:3) with the addition of all the components above mentioned except for the sugars.

pH was adjusted to 6.5, and culture media were autoclaved (peptone and yeast extract were autoclaved separately) before inoculation with 10% (*v*/*v*) of *B. sacchari* pre-culture ($OD_{600nm} = 0.5$). The incubation conditions used were 30 °C and 150 rpm.

2.4. *B. sacchari Growth Inhibition by TABH*

TABH was used as a culture medium, without diluting and diluted with water (5 times), in order to determine the microorganism's tolerance to it. The sugars content of the diluted TABH were adjusted to the same concentration and proportion of the concentrated TABH, and the mineral salts, peptone, and yeast extract (as mentioned in Materials and Methods section) were added to both hydrolysates. After sterilization, 125 mL glass vials media (22 mL of TABH) were inoculated with 3 mL of *B. sacchari* pre-culture (control medium, $OD_{600nm} = 0.5$) and incubated under the conditions previously mentioned.

Microbial growth (Dry cell mass) at 0, 24, 48, and 72 h was estimated by measuring the optical density of the sample at 600 nm and by its correlation with the dry cell weight, obtained gravimetrically.

2.5. *B. sacchari Growth Inhibition by Model Toxic Compounds*

Acid hydrolysis by-products reported in other hydrolysates (inhibitory compounds) were added separately to the control medium (concentrated solutions previously sterilized), in order to evaluate their particular effect on *B. sacchari* growth. The following compounds (0.5–12 $g \cdot L^{-1}$) were studied: furfural, hydroxymethylfurfural, vanillin, coumaric acid, and ferulic acid. Glass vials of 125 mL containing 22 mL of control medium (added with each inhibitory compound) were inoculated with 3 mL of *B. sacchari* pre-culture (control medium, $OD_{600nm} = 0.5$), and the medium without inhibitory compounds was used as control.

Microbial growth was measured as previously mentioned at 0, 24, 48, and 72 h, and maximum biomass production was used for determining growth inhibition percentage I% as follows:

$$I = (X_{max}C - X_{max}I) \times 100 / X_{max}C$$

where,

$X_{max}C$ = maximum biomass produced in the control medium,

$X_{max}I$ = maximum biomass produced in the control medium with the inhibitory compound.

2.6. *Detoxification of TABH by Different Methods*

Elimination of growth inhibitory compounds, specifically total phenolic compounds, from the TABH (without diluting) was studied using activated charcoal (Golden bell) and resins (Sigma & Aldrich): DOWEX 1 × 8 (chloride form); DOWEX G26 (hydrogen form); Amberlite XAD7 (hydrophobic). Before use, resins were washed with distilled water (agitation, 5 min) and activated according to manufacturer indications (agitation, 1 h): DOWEX G26, HCl 0.4 N; DOWEX 1 × 8 NaOH 0.1 N; Amberlite XAD7, methanol. Finally, resins were washed three times with distilled water [13].

Each adsorbent (50 $mg \cdot mL^{-1}$) was added separately into 10 mL glass vials with 2 mL of TABH at pH 2 or 7 (previously adjusted with NaOH or H_2SO_4) and kept in agitation (150 rpm) in a shaker at 25 °C (1 and 4 h). Next, the adsorbent was separated by filtration, and the total phenolic compounds were quantified by the Folin–Ciocalteu method. The phenolic compound elimination percentage E% was calculated as follows:

$$E = (P_{t0} - P_{tf}) \times 100 / P_{t0}$$

where,

P_{t0} = Total phenolic compound before the detoxification process,

Ptf = Total phenolic compound after the detoxification process.

Detoxification of TABH using enzymes (*Trametes versicolor* laccase, Sigma & Aldrich 38429) was also studied. The pH was first adjusted to 5.5, and the effect of enzyme concentration (1 and 10 U) and contact time (1 and 5 h) on phenolic compounds elimination was investigated (2 mL microtubes, 50 °C, 150 rpm).

2.7. PHB Production from Detoxified TABH by B. sacchari

The detoxified TABH medium and the control medium were used for the production of PHB by *B. sacchari* in 500 mL Erlenmeyer flasks containing 90 mL of culture medium, and inoculated with 10 mL of bacterial pre-culture (DO_{600nm} = 0.5, at 30 °C, and 150 rpm).

A two-step batch culture was performed as follows: (1) Biomass production (72 h) in medium (TABH or control) with nitrogen sources ($(NH_4)_2SO_4$, peptone and yeast extract) as described in the Materials and Methods Section; (2) PHB accumulation (48 h) in nitrogen-limited medium (TABH or control with 0.6 $g \cdot L^{-1}$ of $(NH_4)_2SO_4$, without either peptone or yeast extract). For PHB accumulation, cells from the biomass production step were aseptically recovered by centrifugation and resuspended in nitrogen-limited culture media. Samples (5 mL) were withdrawn periodically and analyzed for quantifying biomass (dry cell weight), reducing sugars (DNS), pH, and PHB (dry weight) after solvent extraction. For polymer extraction, biomass was lyophilized, weighted, suspended in ethanol, and kept in constant agitation (24 h). Next, it was air-dried, suspended in chloroform, and kept in constant agitation (24 h). Cell debris was removed by filtration, and the organic phase was added to cold methanol (1:10 *v*/*v*) and kept in the freezer for 24 h for PHB precipitation. The polymer was recovered by centrifugation, air-dried, and weighed.

2.8. PHB Characterization

The polymer extracted from *B. sacchari* cells was analyzed by ATR-FTIR (Fourier Transform Infrared-Attenuated Total Reflectance) using a Perkin Elmer Spectrum Two FTIR spectrometer. For determining its monomeric composition, the polymer was subjected to acid methanolysis [14]. The resulting methyl esters were analyzed using a Perkin Elmer XL gas chromatograph, equipped with a CP-Wax 52 CB capillary column (25 m × 0.32 mm) and a flame ionization detector. The chromatographic conditions used were: sample injection volume, 1 μL; gas carrier, nitrogen; flow rate, 20 $cm \cdot s^{-1}$; injector and detector temperatures, 210 and 220 °C, respectively. A temperature ramp was used as follows: 50 °C for 1 min, incrementing by 8 °C min^{-1}, and 160 °C for 5 min. Methyl benzoate and polyhydroxybutyrate from Fluka (after methanolysis) were used as internal and external standards, respectively.

3. Results and Discussions

3.1. Characterization of TAB

To know the cellulose, hemicellulose, lignin, extractable, ashes, and humidity content of the TAB to be used, it was characterized by TAPPI standards. Table 1 presents the data obtained, as expected, cellulose was found as the primary polymer followed by hemicellulose and lignin. Compared to other TAB characterizations previously reported, there was no significant variation in respect to the proportions of the structural components of the bagasse used in this research: cellulose, 20–50%; hemicellulose, 19–27%; and lignin, 15–20% [5,15].

Table 1. TAB characterization.

Component	Content (%)
Cellulose	50.1 ± 2.1
Hemicellulose	21.1 ± 2.4
Lignin	13.1 ± 1.3
Extractable	8.0 ± 1.1
Ashes	0.8 ± 0.1
Humidity	7.0 ± 0.9

3.2. Chemical Hydrolysis of TAB

The effect of the TAB fiber size on the concentration of total and reducing sugars, xylose, glucose, and total phenolic compounds was investigated. The results are presented in Table 2. A slightly higher concentration of reducing sugars was obtained from the mixed TAB fiber sizes (20.6 vs. 19.14 $g{\cdot}L^{-1}$), but generally, the effect of the fiber size on the hydrolysates composition was not significant.

Table 2. Tequila agave bagasse hydrolysate (TABH) composition after acid hydrolysis of different Tequila agave bagasse (TAB) fiber size.

Compound	Fiber Size	
	Mixed (125–420 μm)	60 mesh (250 μm)
Total sugars ($g{\cdot}L^{-1}$)	25.5 ± 1.5	23.9 ± 1.9
Reducing sugars ($g{\cdot}L^{-1}$)	20.61 ± 0.92	19.14 ± 1.03
Xylose (%) [1]	72	71
Glucose (%)	28	29
Total phenolic compounds ($g{\cdot}L^{-1}$)	1.7 ± 0.12	1.6 ± 0.13

[1] Percentage concerning the total amount of reducing sugars.

The factors that usually influence the effectivity of the acid hydrolysis of lignocellulosic materials are the temperature and time of the process [16]. In some investigations, the influence of the fiber size had also been studied [17]. The benefit of grinding the fibers before performing the chemical hydrolysis is to increase the exposure area of the lignocellulosic material to the acid, as well as to reduce the crystallinity of cellulose, allowing it to hydrolyze with little generation of total phenolic compounds. Nevertheless, since sieving (60 mesh) the ground TAB to obtain a particular fiber size (250 μm) did not had a significant impact in the amount of reducing sugar or phenolic compounds obtained, this step could be avoided, representing a saving in time and energy.

TAB hydrolysis under similar conditions to those used in this investigation was reported previously [5], and a concentration of 24.9 $g{\cdot}L^{-1}$ of reducing sugars was achieved (Table 3). Similar results obtained in hydrolysis studies of different materials are presented in Table 3.

Table 3. Chemical composition of hydrolysates obtained from different lignocellulosic materials.

Lignocellulosic Material	Reducing Sugars ($g{\cdot}L^{-1}$)	Phenolic Compounds ($g{\cdot}L^{-1}$)	Reference
TAB	24.9	n.r.	[5]
Sugarcane bagasse	25.38	n.r.	[18]
Sugarcane bagasse	n.r.	2.86	[19]
Sago trunk cortex	29.46	2.15	[20]
Sugarcane bagasse	30.29	2.75	[13]
Saccharum spontaneum	32.15	2.01	[21]

n.r.—not reported.

The determination of monomeric sugars was performed using HPLC, finding that the xylose to glucose ratio was around 7:3. Xylose was expected to be found in higher concentrations than glucose since xylose is the main product of the hydrolysis of hemicellulose. Hemicellulose, because of its branched chemical structure, results in being more easily hydrolyzed in comparison to cellulose, whose principal degradation product is glucose. The high crystallinity of cellulose makes it challenging to hydrolyze into monomers of glucose [22]. This tendency has been reported in other investigations using sugar cane bagasse [23,24] and wheat straw [25]. Generally, the high percentage of xylose present in the TABH would be a negative aspect, given that it is not an easily assimilated substrate for the majority of the PHA-producing microorganisms. However, *B. sacchari* is capable of metabolizing xylose [23] therefore, this should not be a limiting factor for the growth and synthesis of PHB by it.

3.3. *B. sacchari Growth Inhibition by Toxic Compounds Present in the TABH*

TABH was used as a culture medium, concentrated and diluted with water (5 times), in order to determine the microorganism's tolerance to it. It was found that only the diluted hydrolysate supported growth. On the other hand, the regular hydrolysate strongly inhibited the growth of *B. sacchari* (Figure 1), which evidences its toxic nature due to the presence of inhibitory compounds from the acid hydrolysis process.

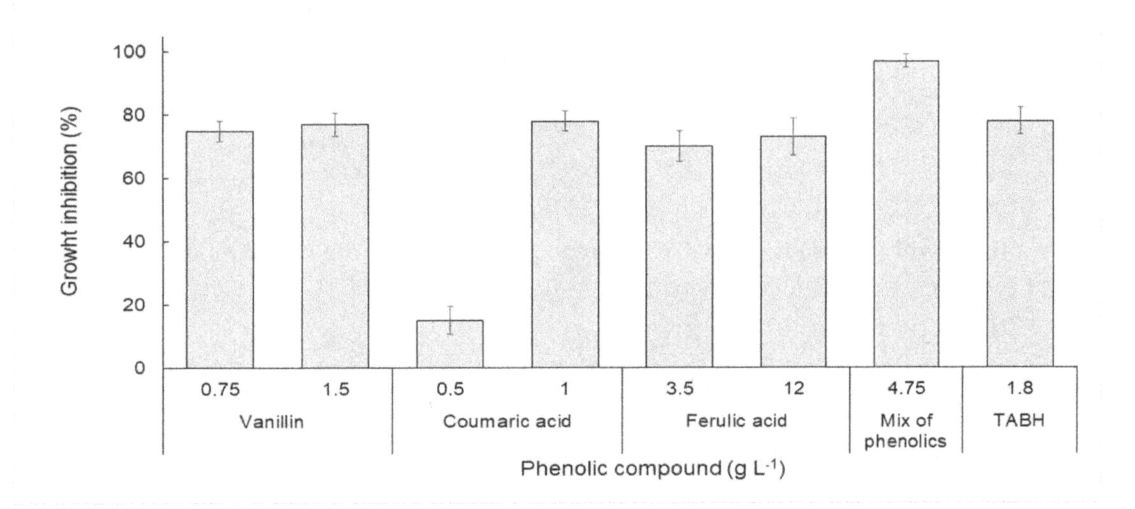

Figure 1. Inhibition of *B. sacchari* growth by model phenolic compounds associated with the acid hydrolysis of lignocellulosic materials.

The effect of pure inhibitory compounds such as furfural, hydroxymethylfurfural, acetic acid, levulinic acid, and phenolic compounds (vanillin, coumaric, and ferulic acids) on *B. sacchari* growth was investigated. It was found that the phenolic compounds caused the highest growth inhibition (above 60%), rather than the other inhibitory compounds investigated. The inhibitory effect of phenolic compounds (lowest and higher concentration used) are presented in Figure 1. Except for coumaric acid at 0.5 $g{\cdot}L^{-1}$, all the phenolic compounds tested strongly inhibited *B. sacchari* growth, while their mixture was even highly inhibitory (97% inhibition). Such toxic synergistic behavior has been previously described for other bacteria [26].

As previously mentioned, various types of bagasse hydrolysates presented concentrations of total phenolic compounds ranging from 0.13 $g{\cdot}L^{-1}$ to 3 $g{\cdot}L^{-1}$ with different levels of toxicity to the microorganism used [27,28]. The amount of total phenolic compounds produced during acid hydrolysis of the TAB was significant (1.6–1.7 $g{\cdot}L^{-1}$). Such compounds are known for decreasing the microbial growth rate associated with the loss of integrity of the cell membranes [26].

The evident growth inhibition of *B. sacchari* by the TABH confirmed the necessity of a detoxification treatment before using it as a cultivation media.

3.4. Elimination of Growth Inhibitory Compounds from TABH

Different detoxification methods were used in order to eliminate growth inhibitors (with emphasis on phenolic compounds) from the TABH, and the results are presented in Figure 2.

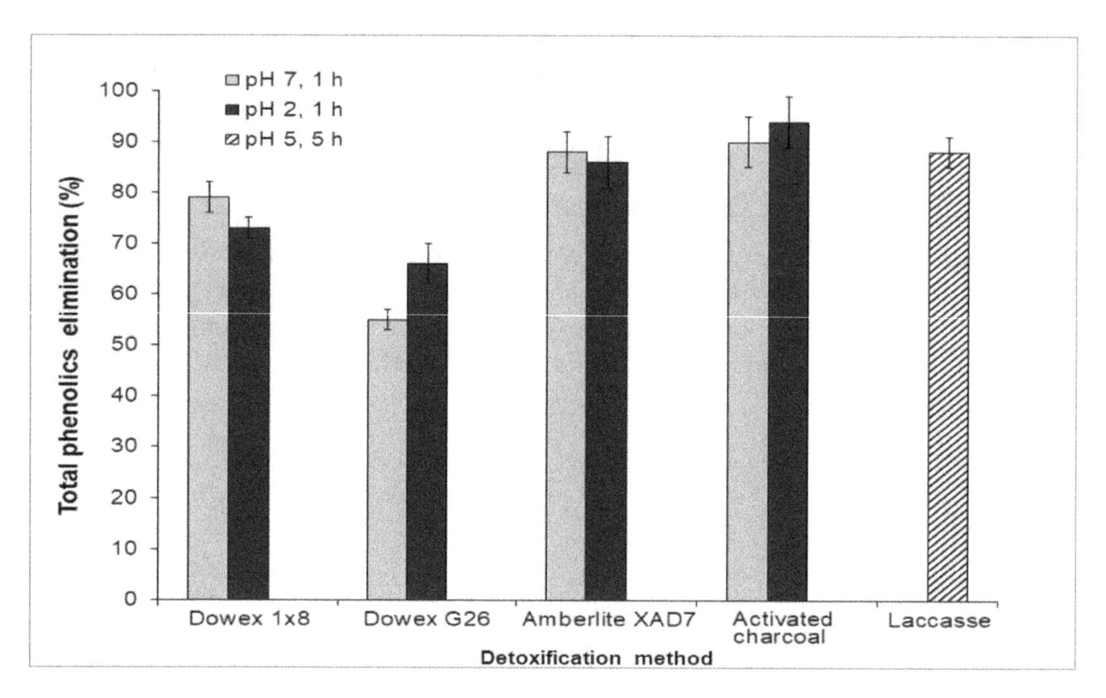

Figure 2. Total phenolic compounds removed from the TABH by different detoxification methods. For the treatment with resins and activated charcoal, 50 mg of adsorbent per mL of TABH were used. For laccase treatment, 1 U·mL^{-1} was used.

The detoxification with activated charcoal is a treatment frequently used to purify or to recover certain compounds from hydrolysates (lignin, tannin, furan derivatives, aromatic monomers, and phenolic acids) [29,30]. The pH of the hydrolysate, the concentration of activated carbon used, and the contact time are known factors that can influence the effectivity of the detoxification process with this adsorbent [31,32]. The phenolics elimination percentage using this method was around 90%, which is similar to values reported for other hydrolysates (sugarcane bagasse, 94%; olive tree pruning residue, 98%) [23,31]. The adsorptive behavior observed for the phenolic compounds on activated carbon could be explained by their polarity (electron distribution), hydrophobicity, and chemical structure [33]. The pH can affect its absorption capacity because it can influence the nature of functional groups for both adsorbent and adsorbate. Weak organic acids and phenolic compounds are better absorbed when they are in a non-ionized state ($pH < 4$) [34], this could be a possible explanation for the slight increase in the elimination of phenolic compounds at pH 2.

Regarding the elimination of phenolic compounds from TABH by the resins, it was from 55% to 88%; this result is similar to other values reported in the literature (64–94%) [35–38]. Using the Dowex 1×8 resin at acid pH, the amount of phenolic compounds adsorbed was lower than at pH 7; a comparable result was observed by Martos et al. [39]. However, the elimination of phenolic compounds using this adsorbent was higher (73–78%) than the achieved with the cation-exchange resin Dowex G26 (55–65%), which might be less efficient to adsorb phenolic compounds due to its overall negative charge. The same behavior was observed previously [36] using a similar resin (AG 50W-X8). Concerning the use of the Amberlite XAD7 resin, the elimination percentages of phenolic compounds were high (86–88%). This resin is an acrylic ester that is slightly polar and has been used for eliminating phenolic compounds from olive mill wastewater [40].

The last method assayed for detoxifying the TABH was the use of laccases. The elimination of total phenolic compounds by this method was 88%. The action mechanism of the laccase enzyme has been previously elucidated. The enzyme oxidizes phenolic compounds that lead to the formation of phenoxy-type free radicals, which are unstable and polymerize into aromatic compounds that are less toxic [41]. Other compounds present in the hydrolysate (such as salts) might inhibit to a certain extent the enzyme activity. This is a probable reason why the elimination of phenolic compounds did not reach above 90% [42]. Similar results (70–75% of phenolic compounds elimination) were obtained using laccases for detoxifying wheat straw hydrolysate (0.5 U·mL^{-1} of laccase at pH 5 for 2 h) [42].

3.5. *B. sacchari* Growth in TABH Detoxified by Different Methods

B. sacchari growth was evaluated in the detoxified TAB hydrolysates, and it was found that some hydrolysates stimulated the growth, with respect to the control medium (Figure 3A), while others were still inhibitory (Figure 3B). The TABH treated with activated charcoal allowed the highest biomass production, 16% more biomass (Figure 3A) than the presented in the control medium, followed by the TABH detoxified with enzymes (10% more biomass than the produced in the control medium). Concerning the TABH treated with resins, the growth of *B. sacchari* with respect to the control medium varied: it was slightly inhibited in TABH detoxified with XAD7 or G26 resin (3 and 7% less biomass, respectively) (Figure 3B), but enhanced (8%) in TABH treated with 1 × 8 resin (Figure 3A).

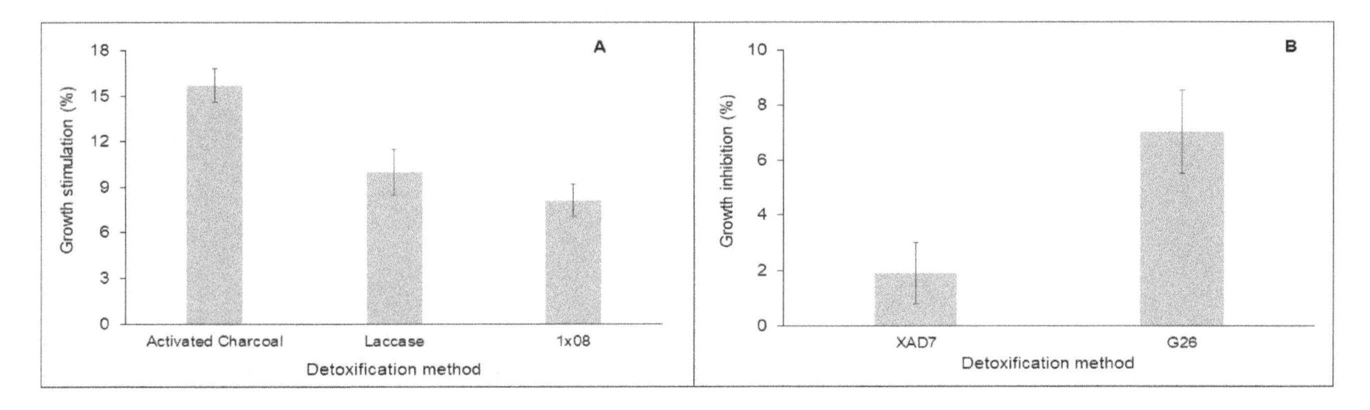

Figure 3. *B. sacchari* growth stimulation (**A**) or inhibition (**B**) by TABH detoxified using different methods.

Thus, the TABH treated with activated charcoal, resin 1 × 8, and enzymes not only supported the growth of *B. sacchari* but stimulated it. It is possible that such detoxification methods removed toxic compounds but left tolerable concentrations of some others that could act as growth factors, an effect that has been previously observed [26]. It has been mentioned that some organic acids and phenol-type compounds (including formic, acetic, levulinic, 4-hydroxybenzoic and gallic, and vanillic acid) at low concentration (below 10 mM) could stimulate the cell growth instead of suppressing it [43,44].

On the other hand, detoxification of TABH with XAD7 and G26 resins might not absorb other known toxic compounds such as furfural and HMF, which acted as growth inhibitors [35].

Activated charcoal eliminated more than 90% of total phenolic compounds from the TABH and the detoxified TABH not only supported but promoted the growth of *B. sacchari*, thus this treatment (50 g·L^{-1}, pH 2, and a contact time of 4 h) was selected for detoxifying the hydrolysate used in the PHB production experiment.

3.6. PHB Production from TABH Detoxified with Activated Charcoal

The production of PHB using *B. sacchari* from TABH detoxified with activated carbon, was investigated. After detoxification, TABH pH was adjusted to 7. Synthetic media with the same sugar concentration (20 g·L^{-1}) and xylose to glucose proportion (7:3) as the TABH was used as a control medium. Kinetics of *B. sacchari* growth, sugars consumption, and PHB production are depicted in Figure 4.

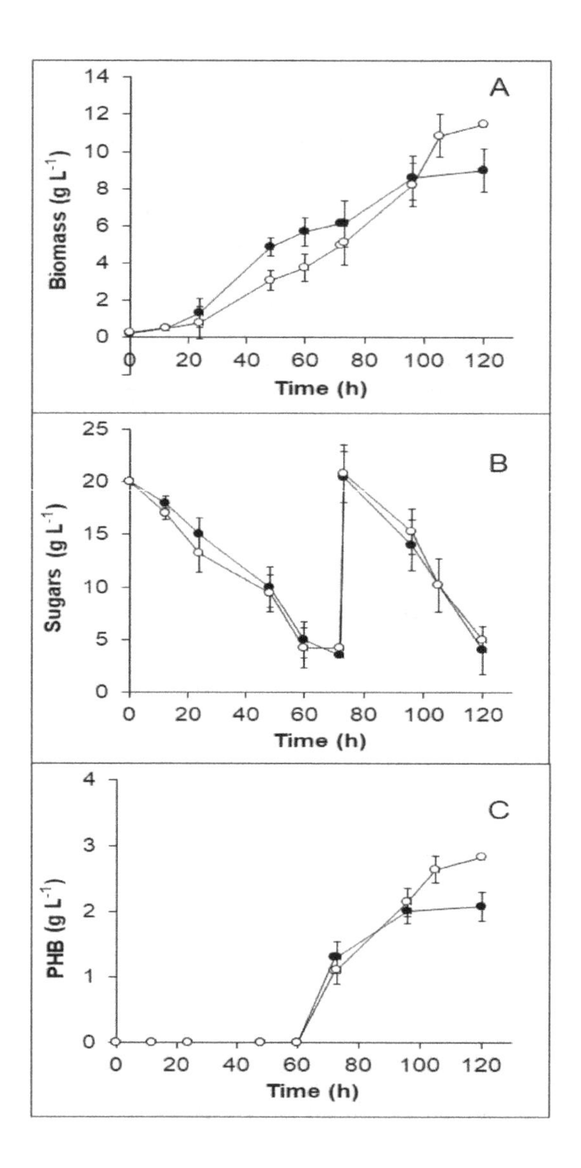

Figure 4. Kinetic profile of *B. sacchari* growing in TABH detoxified with activated charcoal (○) and control medium (●). (**A**) Biomass production. (**B**) Substrate consumption. (**C**) Polyhydroxybutyrate (PHB) production.

In both media, the lag phase was not evident, while exponential growth occurred from 12 to 72 h. Diauxic growth due to the presence of at least two carbon sources (xylose and glucose) was almost imperceptible, although a slight shift in the growth curve is observed at 40 h in both media used. At 72 h cells were harvested and resuspended in nitrogen-limited medium, cell growth continued until 96 h (control medium), and 105 h (TABH medium), although it represents both the PHB and non-PHB biomass (residual biomass). This is consistent with the PHB accumulation profile (Figure 4C) since the polymer production occurred from 72 to 96 h in the control medium, and from 72 to 105 h in TABH medium. The sugar consumption profile was similar in both culture media (Figure 4B).

The results of the kinetic parameters are presented in Table 4. Biomass and PHB production, as well as the maximum growth rate and PHB accumulation, were slightly higher in TABH medium (1.05–1.2 times) than in the control medium. Such results have been observed in other PHA producing bacteria growing in hydrolysates, and it has been hypothesized and researched that certain phenolic compounds and organic acids present in the hydrolysates in minimal concentrations, can stimulate growth and production of PHB [23].

Table 4. Biomass and PHB production by *B. sacchari* from activated charcoal detoxified TABH and mineral medium (120 h).

Parameter	Control Medium (CM)	TABH (Detoxified)
Total biomass ($g \cdot L^{-1}$)	8.78 ± 1.04	11.03 ± 1.14
Residual biomass ($g \cdot L^{-1}$) [a]	6.77 ± 1.09	8.36 ± 0.91
PHB ($g \cdot L^{-1}$)	2.01 ± 0.86	2.67 ± 0.96
PHB (%) [b]	22.91 ± 1.18	24.20 ± 1.26
μ_{max} (h^{-1})	0.08 ± 0.01	0.11 ± 0.02
$Y_{X/S}$ ($g \cdot g^{-1}$) [c]	0.23 ± 0.02	0.25 ± 0.02
$Y_{P/S}$ ($g \cdot g^{-1}$) [d]	0.10 ± 0.01	0.10 ± 0.01

[a] Total Biomass—PHB. [b] g of PHB g^{-1} total biomass × 100. [c] g of residual biomass g^{-1} reducing sugar consumed. [d] g of PHB g^{-1} reducing sugar consumed.

The PHB accumulation percentage achieved by *B. sacchari* from TABH, compares with values that have been reported from other hydrolysates (in shake flasks) by different strains: 34%, *Halomonas boliviensis* (wheat bran) [45]; 31.9% of PHB, *Ralstonia eutropha* (pulp fiber sludge); and 32% of PHB, *Sphingobium scionense* (softwood) [9].

B. sacchari has been previously used for PHB production from other lignocellulosic materials hydrolysates. The results of biomass production and PHB accumulation obtained from TABH are similar to those values reported from sugar cane bagasse hydrolysate (shaken flasks) by the same strain: 6.13 $g \cdot L^{-1}$, and 23.22%, respectively [23].

The biomass and PHB yields (on the substrate) obtained in the control medium and in the TABH medium were similar (0.23 and 0.10 $g \cdot g^{-1}$, respectively). Specifically, $Y_{P/S}$ value obtained for *B. sacchari* growing in the TABH medium is low compared to those reported for other hydrolysates (0.11 to 0.46 $g \cdot g^{-1}$) [23].

Although flask fermentations are very useful to study fermentation processes, they are restricted due to the incapacity to be controlling variables such as pH and dissolved oxygen. These are essential factors to optimize microbial growth and PHB accumulation. During the flask fermentation, the pH dropped from 7 to 5 and therefore affected the accumulation of the biopolymer [23]; thus, the production of PHB from TABH could be further optimized by using an automatized bioreactor and implementing a fed-batch system.

3.7. PHB Characterization

An FTIR analysis was performed on the polymer produced from detoxified TABH (Figure 5A), and it was compared against a Fluka ™ PHB standard. The peak around 2900 cm^{-1} is characteristic of carbon to hydrogen bonds, which are a part of the general structure of PHAs. The zone between 1700 and 1750 cm^{-1} relates to the stretching carbonyl C=O group, and the set of peaks from 1300 to 1000 cm^{-1} to the stretching of C–O bonds, both signals correspond to the ester bonds present in the PHAs structure. The peak around 1450 cm^{-1} originates from the asymmetric deformation and stretching of the bonds of methyl groups C–H, the same as the peak approximately at 1380 cm^{-1} [46].

The monomeric composition of the synthesized PHA was investigated by gas chromatography (GC). As depicted in Figure 5B, the biopolymer produced by *B. sacchari*, using the TABH as carbon source, is a homopolymer of 3-hydroxybutyric acid.

In previous experiments with *B. sacchari* using mixtures of glucose and xylose, as well as hydrolysates (wheat straw, sugar cane bagasse) as carbon sources, the polymer produced was also composed of repeating units of 3-hydroxybutyric acid [23,47].

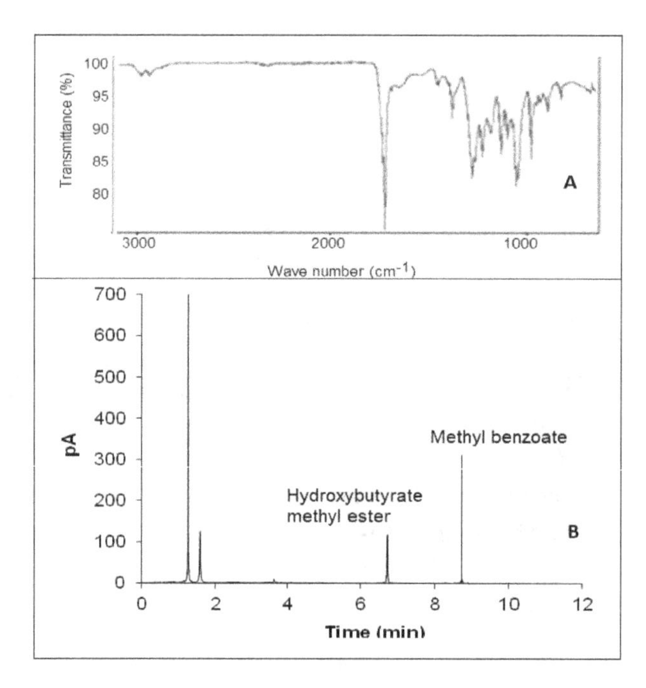

Figure 5. Characterization of the PHB produced by *B. sacchari* from TABH detoxified with activated charcoal. FTIR spectrum (**A**); GC chromatogram (**B**).

Author Contributions: Conceptualization, Y.G.-G.; formal analysis, Y.G.-G. and J.C.M.-C.; funding acquisition, J.A.S.-G.; investigation, Y.G.-G., J.G., J.C.M.-C., B.C.-G., and J.A.S.-G.; methodology, Y.G.-G., J.G., J.C.M.-C., and B.C.-G.; project administration, Y.G.-G.; supervision, Y.G.-G.; writing—original draft, Y.G.-G., J.G., J.C.M.-C., and B.C.-G..

References

1. Tsang, Y.F.; Kumar, V.; Samadar, P.; Yang, Y.; Lee, J.; Ok, Y.S.; Song, H.; Kim, K.-H.; Kwon, E.E.; Jeon, Y.J. Production of bioplastic through food waste valorization. *Environ. Int.* **2019**, *127*, 625–644. [CrossRef]
2. Salgaonkar, B.B.; Bragança, J.M. Utilization of Sugarcane Bagasse by *Halogeometricum borinquense* Strain E3 for Biosynthesis of Poly(3-hydroxybutyrate-co-3-hydroxyvalerate). *Bioengineering* **2017**, *4*, 50. [CrossRef]
3. Sandhya, M.; Aravind, J.; Kanmani, P. Production of polyhydroxyalkanoates from *Ralstonia eutropha* using paddy straw as cheap substrate. *Int. J. Environ. Sci. Technol.* **2012**, *10*, 47–54. [CrossRef]
4. Nielsen, C.; Rahman, A.; Rehman, A.U.; Walsh, M.K.; Miller, C.D. Food waste conversion to microbial polyhydroxyalkanoates. *Microb. Biotechnol.* **2017**, *10*, 1338–1352. [CrossRef] [PubMed]
5. Saucedo-Luna, J.; Castro-Montoya, A.J.; Rico, J.L.; Campos-García, J. Optimization of acid hydrolysis of bagasse from *Agave tequilana* Weber. *Rev. Mex. Ing. Quim.* **2010**, *9*, 91–97.
6. Aguilar, D.L.; Rodríguez-Jasso, R.M.; Zanuso, E.; de Rodríguez, D.J.; Amaya-Delgado, L.; Sanchez, A.; Ruiz, H.A. Scale-up and evaluation of hydrothermal pretreatment in isothermal and non-isothermal regimen for bioethanol production using agave bagasse. *Bioresour. Technol.* **2018**, *263*, 112–119. [CrossRef] [PubMed]
7. Niehus, X.; Crutz-Le Coq, A.-M.; Sandoval, G.; Nicaud, J.-M.; Ledesma-Amaro, R. Engineering *Yarrowia lipolytica* to enhance lipid production from lignocellulosic materials. *Biotechnol. Biofuels* **2018**, *11*, 11. [CrossRef]
8. Alva Munoz, L.E.; Riley, M.R. Utilization of cellulosic waste from tequila bagasse and production of polyhydroxyalkanoate (PHA) bioplastics by *Saccharophagus degradans*. *Biotechnol. Bioeng.* **2008**, *100*, 882–888. [CrossRef]
9. Obruca, S.; Benesova, P.; Marsalek, L.; Marova, I. Use of lignocellulosic materials for PHA production. *Chem. Biochem. Eng. Q.* **2015**, *29*, 135–144. [CrossRef]
10. Dubois, M.; Gilles, K.; Hamilton, J.; Rebers, P.; Smith, F. Colorimetric method based on phenol sulfuric acid. *Anal. Chem.* **1956**, *28*, 356.

11. Miller, G.L. Use of dinitrosalicylic acid reagent for determination of reducing sugar. *Anal. Chem.* **1956**, *31*, 426–428. [CrossRef]
12. Singleton, V.L.; Orthofer, R.; Lamuela-Raventos, R.M. Analysis of total phenols and other oxidation substrates and antioxidants by means of Folin–Ciocalteau reagent. *Method. Enzymol.* **1999**, *299*, 152–178.
13. Chandel, A.K.; Kapoor, R.K.; Singh, A.; Kuhad, R.C. Detoxification of sugarcane bagasse hydrolysate improves ethanol production by *Candida shehatae* NCIM 3501. *Bioresour. Technol.* **2007**, *98*, 1947–1950. [CrossRef] [PubMed]
14. Linton, E.; Rahman, A.; Viamajala, S.; Sims, R.C.; Miller, C.D. Polyhydroxyalkanoate quantification in organic wastes and pure cultures using a single-step extraction and 1H NMR analysis. *Water Sci. Technol.* **2012**, *66*, 1000–1006. [CrossRef] [PubMed]
15. Núñez, H.M.; Rodríguez, L.F.; Khanna, M. Agave for tequila and biofuels: An economic assessment and potential opportunities. *GCB Bioenergy* **2011**, *3*, 43–57. [CrossRef]
16. Mussatto, S.I.; Roberto, I.C. Alternatives for detoxification of diluted-acid lignocellulosic hydrolyzates for use in fermentative processes: A review. *Bioresour. Technol.* **2004**, *93*, 1–10. [CrossRef] [PubMed]
17. Zamudio-Jaramillo, M.A.; Castro-Montoya, A.J.; Yescas, R.M.; Parga, M.D.C.C.; Hernández, J.C.G.; Luna, J.S. Optimization of particle size for hydrolysis of pine wood polysaccharides and its impact on milling energy. *IJRER* **2014**, *4*, 338–348.
18. Laopaiboon, P.; Thani, A.; Leelavatcharamas, V.; Laopaiboon, L. Acid hydrolysis of sugarcane bagasse for lactic acid production. *Bioresour. Technol.* **2010**, *101*, 1036–1043. [CrossRef]
19. Martinez, A.; Rodriguez, M.E.; Wells, M.L.; York, S.W.; Preston, J.F.; Ingram, L.O. Detoxification of dilute acid hydrolysates of lignocellulose with lime. *Biotechnol. Prog.* **2001**, *17*, 287–293. [CrossRef]
20. Kamal, S.M.M.; Mohamad, N.L.; Abdullah, A.G.L.; Abdullah, N. Detoxification of sago trunk hydrolysate using activated charcoal for xylitol production. *Procedia Food Sci.* **2011**, *1*, 908–913. [CrossRef]
21. Chandel, A.K.; Singh, O.V.; Rao, L.V.; Chandrasekhar, G.; Narasu, M.L. Bioconversion of novel substrate *Saccharum spontaneum*, a weedy material, into ethanol by *Pichia stipitis* NCIM3498. *Bioresour. Technol.* **2011**, *102*, 1709–1714. [CrossRef] [PubMed]
22. Palmqvist, E.; Hahn-Hägerdal, B. Fermentation of lignocellulosic hydrolysates. I: Inhibition and detoxification. *Bioresour. Technol.* **2000**, *74*, 17–24. [CrossRef]
23. Silva, L.; Taciro, M.; Ramos, M.; Carter, J.; Pradella, J.; Gomez, G. Poly-3-hydroxybutyrate (P3HB) production by bacteria from xylose, glucose, and sugarcane bagasse hydrolysate. *J. Ind. Microbiol. Biot.* **2004**, *31*, 245–254. [CrossRef] [PubMed]
24. Liu, R.; Liang, L.; Cao, W.; Mingke, W.; Chen, K.; Ma, J.; Jiang, M.; Wei, P.; Ouyang, P. Succinate production by metabolically engineered *Escherichia coli* using sugarcane bagasse hydrolysate as the carbon source. *Bioresour. Technol.* **2012**, *135*.
25. Nigam, J.N. Ethanol production from wheat straw hemicellulose hydrolysate by *Pichia stipitis*. *J. Biotechnol.* **2001**, *87*, 17–27. [CrossRef]
26. Palmqvist, E.; Hahn-Hägerdal, B. Fermentation of lignocellulosic hydrolysates. II: Inhibitors and mechanisms of inhibition. *Bioresour. Technol.* **2000**, *74*, 25–33. [CrossRef]
27. Millati, R.; Niklasson, C.; Taherzadeh, M.J. Effect of pH, time and temperature of overliming on detoxification of dilute-acid hydrolyzates for fermentation by *Saccharomyces cerevisiae*. *Proc. Biochem.* **2002**, *38*, 515–522. [CrossRef]
28. Monlau, F.; Sambusiti, C.; Barakat, A.; Quéméneur, M.; Trably, E.; Steyer, J.-P.; Carrere, H. Do furanic and phenolic compounds of lignocellulosic and algae biomass hydrolyzate inhibit anaerobic mixed cultures? A comprehensive review. *Biotechnol. Adv.* **2014**, *32*, 934–951. [CrossRef]
29. Liu, X.; Fatehi, P.; Ni, Y. Removal of inhibitors from pre-hydrolysis liquor of kraft-based dissolving pulp production process using adsorption and flocculation processes. *Bioresour. Technol.* **2012**, *116*, 492–496. [CrossRef]
30. Zhang, Y.; Xia, C.; Lu, M.; Tu, M. Effect of overliming and activated carbon detoxification on inhibitors removal and butanol fermentation of poplar prehydrolysates. *Biotechnol. Biofuels.* **2018**, *11*, 178. [CrossRef]
31. Mateo, S.; Roberto, I.C.; Sánchez, S.; Moya, A.J. Detoxification of hemicellulosic hydrolyzate from olive tree pruning residue. *Ind. Crop. Prod.* **2013**, *49*, 196–203. [CrossRef]

32. Sarawan, C.; Suinyuy, T.; Sewsynker-Sukai, Y.; Kana, E.B. Optimized activated charcoal detoxification of acid-pretreated lignocellulosic substrate and assessment for bioethanol production. *Bioresour. Technol.* **2019**, *286*, 121403. [CrossRef] [PubMed]
33. Michailof, C.; Stavropoulos, G.G.; Panayiotou, C. Enhanced adsorption of phenolic compounds, commonly encountered in olive mill wastewaters, on olive husk derived activated carbons. *Bioresour. Technol.* **2008**, *99*, 6400–6408. [CrossRef] [PubMed]
34. Costa, T.D.S.; Rogez, H.; Pena, R.D.S. Adsorption capacity of phenolic compounds onto cellulose and xylan. *Food Sci. Technol.* **2015**, *35*, 314–320. [CrossRef]
35. Carvalheiro, F.; Duarte, L.C.; Lopes, S.; Parajó, J.C.; Pereira, H.; Gírio, F.M. Evaluation of the detoxification of brewery's spent grain hydrolysate for xylitol production by *Debaryomyces hansenii* CCMI 941. *Proc. Biochem.* **2005**, *40*, 1215–1223. [CrossRef]
36. Nilvebrant, N.-O.; Reimann, A.; Larsson, S.; Jönsson, L.J. Detoxification of lignocellulose hydrolysates with ion-exchange resins. *Appl. Biochem. Biotechnol.* **2001**, *91*, 35–49. [CrossRef]
37. Mota, M.I.F.; Barbosa, S.; Pinto, P.C.R.; Ribeiro, A.M.; Ferreira, A.; Loureiro, J.M.; Rodrigues, A.E. Adsorption of vanillic and syringic acids onto a macroporous polymeric resin and recovery with ethanol:water (90:10 %V/V) solution. *Sep. Purif. Technol.* **2019**, *217*, 108–117. [CrossRef]
38. Nitzsche, R.; Gröngröft, A.; Kraume, M. Separation of lignin from beech wood hydrolysate using polymeric resins and zeolites—Determination and application of adsorption isotherms. *Sep. Purif. Technol.* **2019**, *209*, 491–502. [CrossRef]
39. Martos, N.; Sánchez, A.; Molina-Díaz, A. Comparative study of the retention of nine phenolic compounds on anionic exchanger resins. *Chem. Pap.* **2005**, *59*, 161.
40. Bertin, L.; Ferri, F.; Scoma, A.; Marchetti, L.; Fava, F. Recovery of high added value natural polyphenols from actual olive mill wastewater through solid phase extraction. *Chem. Eng. J.* **2011**, *171*, 1287–1293. [CrossRef]
41. Moreno, A.D.; Ibarra, D.; Fernández, J.L.; Ballesteros, M. Different laccase detoxification strategies for ethanol production from lignocellulosic biomass by the thermotolerant yeast *Kluyveromyces marxianus* CECT 10875. *Bioresour. Technol.* **2012**, *106*, 101–109. [CrossRef] [PubMed]
42. Jurado, M.; Prieto, A.; Martínez-Alcalá, A.; Martínez, A.T.; Martínez, M.J. Laccase detoxification of steam-exploded wheat straw for second generation bioethanol. *Bioresour. Technol.* **2009**, *100*, 6378–6384. [CrossRef] [PubMed]
43. Huang, C.; Wu, H.; Liu, Q.; Li, Y.; Zong, M. Effects of aldehydes on the growth and lipid accumulation of oleaginous yeast *Trichosporon fermentans*. *J. Agric. Food Chem.* **2011**, *59*, 4606–4613. [CrossRef] [PubMed]
44. Guo, Z.; Olsson, L. Physiological response of Saccharomyces cerevisiae to weak acids present in lignocellulosic hydrolysate. *FEMS Yeast Res.* **2014**, *14*, 1234–1248. [CrossRef] [PubMed]
45. Nikodinovic-Runic, J.; Guzik, M.; Kenny, S.T.; Babu, R.; Werker, A.; O Connor, K.E. Carbon-rich wastes as feedstocks for biodegradable polymer (Polyhydroxyalkanoate) production suing bacteria. In *Advances in Applied Microbiology*; Elsevier: Amsterdam, The Netherlands, 2013; Volume 84, pp. 139–200. ISBN 978-0-12-407673-0.
46. Arcos-Hernandez, M.V.; Gurieff, N.; Pratt, S.; Magnusson, P.; Werker, A.; Vargas, A.; Lant, P. Rapid quantification of intracellular PHA using infrared spectroscopy: An application in mixed cultures. *J. Biotechnol.* **2010**, *150*, 372–379. [CrossRef]
47. Lopes, M.S.G.; Gosset, G.; Rocha, R.C.S.; Gomez, J.G.C.; Ferreira da Silva, L. PHB biosynthesis in catabolite repression mutant of *Burkholderia sacchari*. *Curr. Microbiol.* **2011**, *63*, 319–326. [CrossRef] [PubMed]

11

The Evolution of Polymer Composition during PHA Accumulation: The Significance of Reducing Equivalents

Liliana Montano-Herrera [1], Bronwyn Laycock [1], Alan Werker [2] and Steven Pratt [1,*]

[1] School of Chemical Engineering, University of Queensland, St Lucia QLD 4072, Australia; liliana.montano@usys.ethz.ch (L.M.-H.); b.laycock@uq.edu.au (B.L.)

[2] Veolia Water Technologies AB—AnoxKaldnes, Klosterängsvägen 11A SE-226 47 Lund, Sweden; alan@werker.se

* Correspondence: s.pratt@uq.edu.au

Academic Editor: Martin Koller

Abstract: This paper presents a systematic investigation into monomer development during mixed culture Polyhydroxyalkanoates (PHA) accumulation involving concurrent active biomass growth and polymer storage. A series of mixed culture PHA accumulation experiments, using several different substrate-feeding strategies, was carried out. The feedstock comprised volatile fatty acids, which were applied as single carbon sources, as mixtures, or in series, using a fed-batch feed-on-demand controlled bioprocess. A dynamic trend in active biomass growth as well as polymer composition was observed. The observations were consistent over replicate accumulations. Metabolic flux analysis (MFA) was used to investigate metabolic activity through time. It was concluded that carbon flux, and consequently copolymer composition, could be linked with how reducing equivalents are generated.

Keywords: PHA; monomer evolution; mixed culture; modeling; polymer composition; biopolymer

---

1. Introduction

Polyhydroxyalkanoates (PHAs) are biobased and biodegradable polyesters. PHA copolymers, such as poly(3-hydroxybutyrate-*co*-3-hydroxyvalerate) (PHBV), are of particular interest as they are the basis for biomaterials with desirable mechanical properties. These copolymers can be produced in mixed microbial cultures [1]. However, predicting and controlling the copolymer composition can be challenging.

PHAs are most typically synthesized in mixed microbial cultures from volatile fatty acids (VFAs), through well-described metabolic pathways [2,3]. In the specific case of PHBV, short chain acids such as acetic and propionic acids are transported though the cell membrane and converted into acetyl-CoA and propionyl-CoA respectively. PHA synthesis then takes place in three steps. Firstly, two acyl-CoA molecules are condensed in a reaction catalyzed by a thiolase to produce various intermediates. For example, two acetyl-CoA monomers form acetoacetyl-CoA (a 3-hydroxybutyrate (3HB) precursor), while one acetyl-CoA and one propionyl-CoA combine to form ketovaleryl-CoA (a 3-hydroxyvalerate (3HV) precursor) [4], Escapa et al. 2012). In addition, it has been observed that a portion of the propionyl-CoA produced is converted into acetyl-CoA though different pathways [5]. Secondly, a reduction, catalyzed by a reductase, produces 3-hydroxyalkanoate (3HA) monomers, with the reducing power to support PHA production being generated during anabolic pathways for cell growth, as well as in reactions related to the tricarboxylic acid (TCA) cycle [6]. Finally, a polymerase adds 3HA monomers to the PHA polymer. As such, the flux of carbon through the acyl-CoA intermediates influences the resulting polymer composition.

The fraction of 3HV monomer units in the final PHBV copolymer can be manipulated by adjusting the proportion of even-chain (i.e., acetic acid) to odd-chain (i.e., propionic acid) fatty acids in the feed composition [7–9], since odd-chain fatty acids are generally required for the formation of propionyl-CoA, which is the precursor of 3HV monomer. Diverse monomer compositions and sequence distributions of PHBV copolymers produced by mixed microbial cultures have been achieved using different feeding strategies with acetic and propionic acid mixtures as model substrates [10,11].

Most mixed microbial culture accumulation studies have been applied under conditions of some form of nutrient starvation to inhibit cell growth and favor PHA synthesis [12–14]. In contrast, a recent study has shown that PHA storage can occur concurrently with active biomass growth. Valentino et al. [15] achieved a consistent improvement of PHA productivity when N and P were supplied in an optimal C:N:P ratio. It is important to consider that shifts in the active biomass growth rates may influence carbon flux through the acyl-CoA intermediates and the availability of reducing equivalents for PHA synthesis, and therefore affect polymer production and composition.

Literature on mixed microbial culture PHA production coupled with high rate cell growth is scant. Simulations of existing models have successfully fitted data on PHA productivity and even monomer composition evolution in some cases [4,16,17]; however, these models apply only for scenarios of negligible growth. In addition, in these experiments the feedstock composition was kept constant during the accumulation process resulting in a polymer with a constant ratio of 3HB:3HV. These existing model frameworks are in contrast to some published experimental data that do show shifts in copolymer composition during fed-batch mixed culture PHA accumulation, even under non-growing conditions [11]. Such data indicate that the 3HB:3HV ratio during accumulation is not simply dependent on the feedstock but is also affected by the history of the accumulation and the resulting metabolic activity in the biomass. The potential for biomass growth and other processes to directly influence the composition of intracellular acyl-CoA reservoirs and hence copolymer composition has not been examined.

The aim of this paper is to examine 3HB and 3HV monomer evolution through PHA accumulation, giving consideration to the effect of biomass growth and alternating feedstocks on this process. To this end, monomer development through four sets of PHA accumulation experiments (based on the feeding regime) is investigated: Set 1: acetic acid (HAc) feed; Set 2: propionic acid (HPr) feed; Set 3: mixed HAc and HPr feed; and Set 4: alternating HAc/HPr feed. Concurrent biomass growth and carbon storage is encouraged in each set. Metabolic Flux Analysis (MFA) is used to quantify metabolic pathway activity through the accumulations.

2. Materials and Methods

2.1. Experimental Set-Up

PHA was produced at pilot scale at AnoxKaldnes AB (Lund, Sweden) using a three stage process that had been in continuous operation from 2008 to 2013 [11]. The first stage (acidogenic fermentation) was performed in a 200 L continuous stirred tank reactor under anaerobic conditions and fed with cheese whey permeate, producing a mixture comprising 35% ± 4% acetic, 4% ± 1% propionic, 49% ± 4% butyric, 4% ± 1% valeric and 8% ± 3% caproic acids. The second stage was carried out in a sequential batch reactor (SBR) operated under Aerobic Dynamic Feeding (ADF) conditions with nutrient addition (COD:N of 200:5). The excess biomass with a high PHA storage capacity (as enriched in stage two) was used to produce PHA-rich biomass in the third stage, in a reactor operated in fed batch mode. The details of this process and analytical methods can be found in Janarthanan et al. [18].

2.2. Fed-Batch PHA Production

PHA was accumulated in batches of 100 L harvested SBR mixed liquor by means of a 150 L (working volume) aerated reactor. Aeration provided mixing as well as oxygen supply. Acetic and propionic acids (HAc and HPr, respectively) were fed using different HAc:HPr ratios and feeding strategies. The microbial community was dominated by the genera *Flavisolibacter* and *Zoogloea* [18].

The carbon source concentration for pulse-wise substrate addition was ~100 gCOD/L (see Table 1), with pH adjusted to 4 and additions of nitrogen and phosphorus for nutrient limitation to give COD:N:P of 200:2:1 [15]. N and P additions were 3.82 g/L NH_4Cl and 0.22 g/L KH_2PO_4, respectively. For the fed-batch accumulations, a pulse-wise feedstock addition was applied for feed-on-demand [19] controlled by the biomass respiration response as measured by dissolved oxygen (DO) trends [7,11]. Semi-continuous (pulse-wise) additions of feedstock aliquots were made targeting peak COD concentrations of between 100 and 200 mg-COD/L. Feedstock additions were triggered by a measured relative decrease in biomass respiration rate [7]. pH was monitored but not controlled. The fed-batch accumulations were run over 20–25 h with samples taken at selected times for analyses, including VSS (volatile suspended solids), TSS (total suspended solids), PHA content and composition, soluble COD, volatile fatty acids (VFAs), and nutrients (nitrogen and phosphorus).

Table 1. Experimental conditions in the PHA (Polyhydroxyalkanoates) fed-batch accumulations.

Experiment set	Substrate Composition and Feeding Strategy (gCOD Basis)	Experiment Label	Process Time (h)	Initial VSS ($g \cdot L^{-1}$)	Total substrate added (gCOD)		Feed Concentration ($gCOD \cdot L^{-1}$)	Total Number of Pulses	
					HAc	HPr		HAc	HPr
1	100% Acetic acid	Exp 1 Exp 1′	23.2 21.9	1.2 1.4	1374 1684	- -	96 98	138 147	- -
2	100% Propionic acid	Exp 2 Exp 2′	23.3 24.6	1.2 1.7	- -	939 1439	106 102	- -	77 137
3	50% Acetic/50% propionic acid	Exp 3 Exp 3′	24.2 20.4	1.5 1.9	600 817	600 817	98 98	95 140	
4	100% Acetic acid—100% propionic acid (alternating)	Exp 4 Exp 4′	22.3 22.5	1.4 1.7	1429 570	982 552	101/103 94/97	106 57	105 56

2.3. Analytical Methods

Total concentrations were analyzed from well-mixed grab samples and soluble concentrations were analyzed after filtering the aqueous samples with 1.6 μm pore size (Ahlstrom Munktell, Falun, Sweden) filters. Volatile fatty acid concentrations were quantified by gas chromatography [20]. Solids analyses (total and volatile suspended solids, or TSS/VSS) were performed according to Standard Methods [21].

Hach Lange™ kits were used for the determination of soluble COD (sCOD) (LCK 114), NH_4–N (LCK 303), NO_3–N (LCK 339), soluble total phosphorus (LCK 349) and soluble total nitrogen (LCK 138). PHA content and monomeric composition (3HB and 3HV) of samples was determined using the gas chromatography method described in [11] using a Perkin-Elmer gas chromatograph (GC) (Perkin Elmer, Inc., Waltham, MA, USA). Quantitative ^{13}C high resolution NMR spectra were acquired on a Bruker Avance 500 spectrometer (Bruker, Billerica, MA, USA) as described by Arcos-Hernandez et al. [11] to determine polymer microstructure (details of polymer structure can be found in Supplementary Materials).

2.4. Experimental Design for PHA Accumulations

The full set of PHA accumulation (third stage) experiments are summarized in Table 1. For this work, four experiments (replicated, with replicates denoted using the symbol ′) were considered. Experiment set 1 used a single acetic acid (HAc) substrate; Experiment set 2 used a single propionic acid (HPr) substrate; Experiment set 3 used a mixed HAc and HPr substrate fed simultaneously in equal COD ratios; while in Experiment set 4 the acids were supplied in alternating pulses.

2.5. Rate and Yield Calculations

For PHA concentration at a given time (in $g\ PHA/L$), only the PHA produced during the accumulation process was considered. Therefore, the measured PHA concentration (PHA) was corrected by subtracting the initial measured PHA content (PHA_0). Typically the initial biomass PHA content ($\%PHA_0$ in wt %) was between 0% and 4%. Active biomass ($CH_{1.4}O_{0.4}N_{0.2}$) at a given time (X, recorded in g/L) was determined from the total concentration of biomass, measured as volatile suspended solids (VSS in $g\ VSS/L$), subtracting the produced PHA concentration (PHA):

$$X = VSS\ (g\ VSS/L)\ -\ PHA\ (g\ PHA/L)$$

PHA intracellular content ($\%PHA$) was calculated as the PHA concentration divided by the volatile suspended solids concentrations on a mass basis.

$$\%PHA\ (gPHA/g\ VSS)\ = \frac{PHA\ (g\ PHA/L)}{VSS\ (g\ VSS/L)}$$

The PHA fraction (f_{PHA}) was measured as PHA concentration divided by active biomass concentration on a COD basis.

$$f_{PHA} = \frac{PHA\ (in\ gCOD\ PHA/L)}{X\ (in\ mCOD\ X/L)}$$

Experimental data for the total amount of VFA consumed, PHA polymer (PHA) produced and active biomass (X) produced were fitted using global nonlinear regression in GraphPad Prism (v.6.0.5). This analysis was performed using an exponential growth model (one phase association) [22]. The batch process mass balance accounted for input feed dosing volumes as well as sampling withdrawal volumes. Kinetic rates and yields were calculated from fitted data as follows:

Acetic (q_{HAc}) and propionic acid (q_{HPr}) specific consumption rates and specific monomer 3HB and 3HV production rates: q_{HB} and q_{HV}, respectively, for the ith uptake of each acid or production of each monomer were calculated with reference to active biomass (X) concentration:

$$q_{HAc} = \frac{(HAc_i - HAc_{i-1})}{(t_i - t_{i-1})\cdot X_i}\ q_{HPr} = \frac{(HPr_i - HPr_{i-1})}{(t_i - t_{i-1})\cdot X_i}$$

$$q_S\left(Cmol\ VFA\cdot Cmol\ X^{-1}\cdot h^{-1}\right) = q_{HAc} + q_{HPr}$$

$$q_{HB} = \frac{(3HB_i - 3HB_{i-1})}{(t_i - t_{i-1})\cdot X_i}\ q_{HV} = \frac{(3HV_i - 3HV_{i-1})}{(t_i - t_{i-1})\cdot X_i}$$

$$q_{PHA}\left(Cmol\ PHA\cdot Cmol\ X^{-1}\cdot h^{-1}\right) = q_{HB} + q_{HV}$$

where t is time; HAc and HPr are the moles of acetic and propionic acids in solution; $3HB$ and $3HV$ are the moles of 3HB and 3HV, respectively; and q_S is the specific consumption rate of substrate (S). The instantaneous relative rate change in 3HV monomers ($\%3HV^{inst}$) was calculated relative to the total PHA specific production rate on a mole basis.

$$\%3HV^{inst} = \frac{q_{HV}\ (mol\ 3HV\cdot h^{-1}\cdot X^{-1})}{q_{PHA}\ (mol\ PHA\cdot h^{-1}\cdot X^{-1})}$$

As previously reported by Janarthanan et al. [18], a linear correlation was obtained between gCOD PHA produced versus total substrate consumed (also in gCOD) and the yield ($Y_{PHA/S}$) in gCOD PHA/gCOD S at 20 h was determined ($0.968 < r^2 < 0.998$). This time point was selected for consistent comparison between runs as all accumulations had reached at least 98% of plateau PHA content by this time. Likewise, plots of active biomass (in gCOD X) versus time were represented by linear regression to a linear quadratic equation, and the yield ($Y_{X/S}$) in gCOD X/gCOD S at 20 h was determined. The 95% confidence intervals associated with all the determined stoichiometric and kinetic parameters were estimated using error propagation formulae. The values were also converted to Cmol basis.

Maximum specific growth rate (μ_{max}) was calculated according to the re-parameterization of the empirical expression applied to growth curves developed by Gompertz [23]. Analysis was performed in SigmaPlot (Systat Software, v.12) plotting $\ln(X/X_0)$ versus time ($0.938 < r^2 < 0.989$).

The maximum specific VFA consumption rate (q_S, Cmol VFA/(Cmol X·h)) and maximum specific PHA storage rate (q_{PHA}, Cmol VFA/(Cmol X·h)), were determined from the trends in the experimental data during the exponential growth phase. The ratio of PHA concentration and total VFA consumed divided by the active biomass concentration at that time were plotted over time, calculating the first derivative.

2.6. Metabolic Flux Analysis (MFA)

MFA was performed in order to investigate the effect of VFA composition and the feeding strategy on active biomass growth and PHA (3HV and 3HB) monomer formation kinetics assuming a pseudo-steady state. The metabolic network used in this work is based on previously published models [4,17] and summarized in Figure 1. The reactions R_9 and R_{10} (Figure 1) describe the conversion of acetyl-CoA and propionyl-CoA into PHA precursors, where acetyl-CoA* and propionyl-CoA* are representations of molecules which have undergone the first two steps of PHA synthesis (condensation and reduction) [2,4]. Subsequently, PHA precursors are polymerized to form the biopolymer (PHB and PHV), with two units of acetyl-CoA* forming one 3HB molecule, and one unit of acetyl-CoA* and one of propionyl-CoA* forming one molecule of 3HV. The cells obtain energy from adenosine triphosphate (ATP), which is generated by the oxidation of NADH, and the efficiency of ATP production is represented by the phosphorylation efficiency (P/O) ratio (δ). The maximum theoretical P/O ratio is 3 mol-ATP/mol-$NADH_2$ in bacteria growing under aerobic conditions [24].

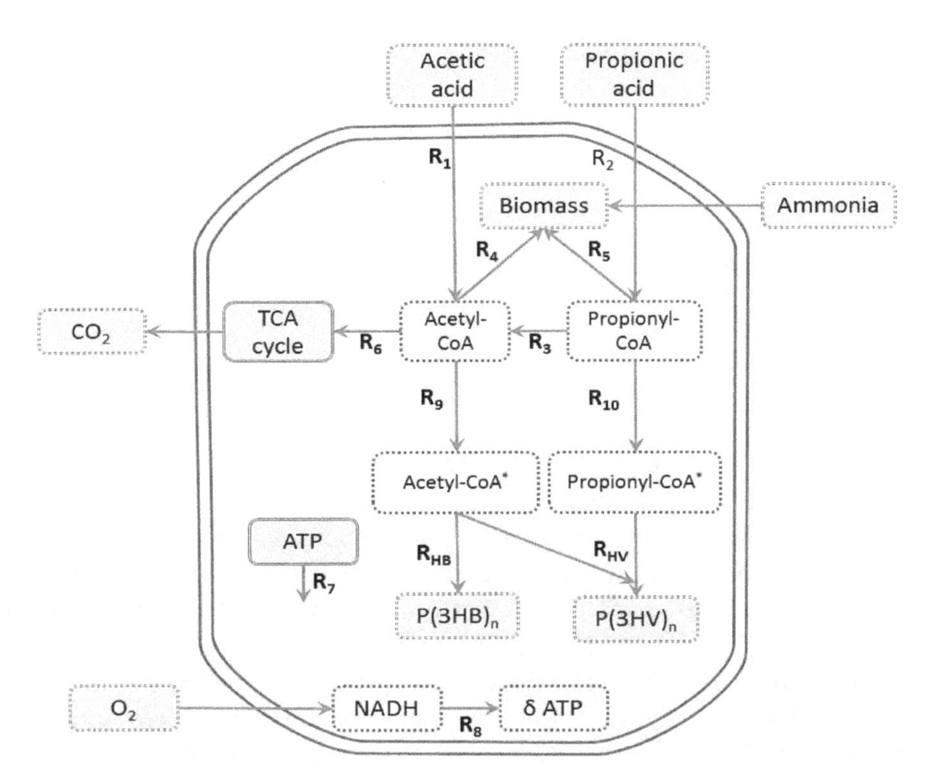

Figure 1. Metabolic network for PHBV synthesis and biomass production, adapted from [17], with permission from © 2013 Elsevier. Light blue dotted squares represent external metabolites; white dotted squares represent internal metabolites.

The metabolic model consists of 12 reactions, 6 intracellular metabolites (acetyl-CoA, propionyl-CoA, acetyl-CoA*, propionyl-CoA*, ATP, and NADH), 4 substrates (HAc, HPr, O_2, and NH_4), and 4 end products (3HB, 3HV, X, and CO_2). The system of equations has six degrees of freedom [25], and a total of seven rates were measured (VFA consumption rates, PHA monomers storage rate, oxygen uptake rate, active biomass synthesis rate, and ammonium consumption). Therefore, the system is overdetermined, with one degree of redundancy, which made it possible to estimate the experimental errors in measurements.

The following constraints and assumptions were set for MFA:

- Active biomass can be formed either from acetyl-CoA or propionyl-CoA. Previous models were performed under ammonia limiting conditions with negligible cellular growth [4,13]. In the present work, we assumed that fluxes of acetyl-CoA and propionyl-CoA used for active biomass (v_4, v_5) synthesis are proportional to the consumption rate of acetic and propionic acid (v_{HAc}, v_{HPr}) [16].

$$f_{Pr} = \frac{v_{HPr}}{v_{HAc} + v_{HPr}}$$

$$\frac{v_4}{v_5} = \frac{1 - f_{Pr}}{f_{Pr}}$$

- Reactions R_9 and R_{10} are reversible, the rest are irreversible reactions.
- The maintenance requirement (v_7) was an estimated flux while the P/O ratio was fixed ($\delta = 3$).
- PHA depolymerization was not considered.

MFA was performed using the CellNetAnalyzer (v. 2014.1, Max Planck Institute, Magdeburg, Germany) toolbox for Matlab [26]. To evaluate the consistency of experimental data with the assumed biochemistry and the pseudo-steady state assumption a *chi-squares-test* was carried out. The flux distributions calculated were found to be reliable given that the consistency index (h) values were below a reference chi-squared test function ($\chi^2 = 3.84$ for a 95% confidence level and 1 degree of redundancy) [27]. The stoichiometry of the metabolic reactions is provided in the Supplementary Materials.

3. Results and Discussion

3.1. Biomass Growth and PHA Content

The experiments were designed to follow the time evolution of PHA storage and active biomass growth during the third stage of the PHA-production system and representative Experiment sets 3 and 4 are shown in Figure 2 (sets 1 and 2 can be seen in the Supplementary Materials). The extent of production of active biomass was variable between the experiments, but higher maximum specific growth rates were achieved for accumulations where acetic acid was fed (Experiment set 1 with 100% HAc, Experiment set 3 with 50%/50% HAc/HPr, and Experiment set 4 with alternating substrates) (Table 2). However, active biomass growth rates attenuated sooner for those accumulations where acetic acid was present at all times (Experiment set 1 and Experiment set 3), while the highest biomass production (X/X_0) was achieved in Exp 4 (Figure 2, Table 2). With regard to PHA fraction evolution, a similar PHA content at plateau was achieved for all experiments. However, PHA content and yield tended to be higher in those accumulations with alternating substrates (Experiment set 4), with one experiment (Exp 4) maintaining an increasing PHA fraction even after 22 h of accumulation. This observation fits with the interpretations from other works that it is possible to stimulate PHA storage with concurrent cellular growth by supplying an optimal nutrient ratio [6,15,18].

Table 2. PHA accumulation yields and kinetic parameters.

Experiment Label	f_{Ac} Consumed (mol HAc/mol VFA)	f_{Pr} Consumed (mol HPr/mol VFA)	%PHA Plateau (gPHA/gVSS)	%3HV (mol 3HV/mol PHA) at 20 h	$Y_{PHA/S}$ (gCOD PHA/gCOD VFA)	$Y_{X/S}$ (gCOD X/gCOD VFA)	μ_{max} (h^{-1})	Final X/X_0 (gCOD/gCOD)	$-q_{Smax}$ (Cmol VFA/Cmol $X \cdot h$)	q_{PHAmax} (Cmol PHA/Cmol $X \cdot h$)
Exp 1	1.0	0	0.56 ± 0.04	0	0.48 ± 0.02	0.17 ± 0.03	0.27 ± 0.03	2.20	0.75	0.33
Exp 1′	1.0	0	0.48 ± 0.06	0	0.38 ± 0.04	0.15 ± 0.01	0.21 ± 0.02	2.28	0.80	0.36
Exp 2	0	1.0	0.40 ± 0.04	74	0.31 ± 0.03	0.16 ± 0.03	0.18 ± 0.02	2.13	0.33	0.11
Exp 2′	0	1.0	0.48 ± 0.03	80	0.40 ± 0.07	0.18 ± 0.03	0.13 ± 0.02	2.09	0.33	0.18
Exp 3	0.64	0.36	0.48 ± 0.06	40	0.39 ± 0.03	0.17 ± 0.05	0.35 ± 0.05	1.88	0.63	0.19
Exp 3′	0.64	0.36	0.52 ± 0.03	42	0.45 ± 0.03	0.12 ± 0.05	0.27 ± 0.05	1.75	0.40	0.19
Exp 4	0.72	0.28	0.59 ± 0.03	34	0.52 ± 0.03	0.18 ± 0.02	0.22 ± 0.02	3.08	0.37	0.23
Exp 4′	0.64	0.36	0.52 ± 0.06	36	0.49 ± 0.03	0.20 ± 0.02	0.19 ± 0.02	1.98	0.45	0.20

All data in table recorded as ± 95% confidence interval where possible.

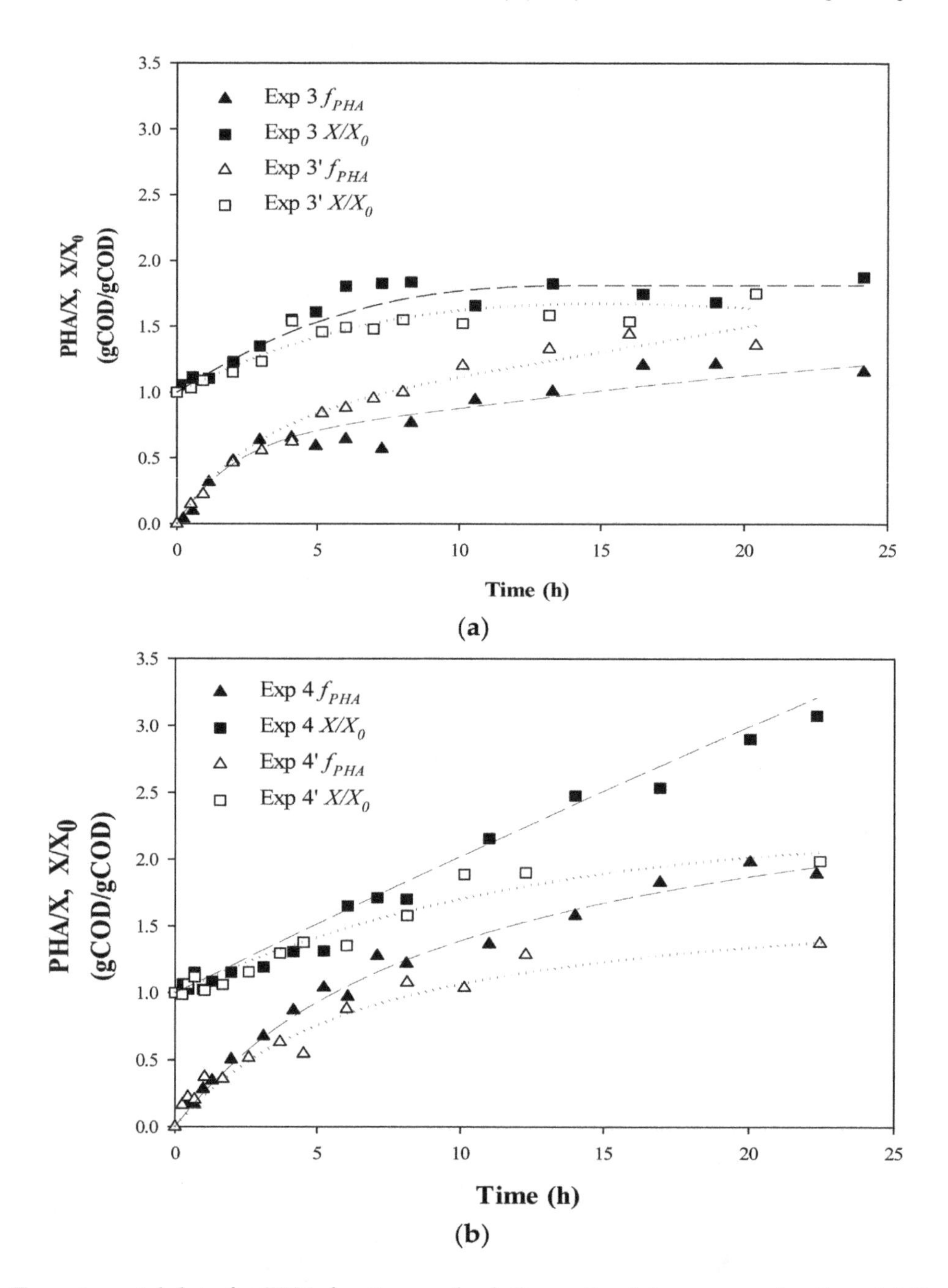

Figure 2. Experimental data for PHA fraction and relative active biomass production: (**a**) Experiment set 3: 50% acetic acid and 50% propionic acid fed simultaneously; and (**b**) Experiment set 4: 100% acetic acid alternating with 100% propionic acid. ▲, Δ, PHA fraction respect to active biomass concentration f_{PHA} (PHA/X); ■, □, Active biomass concentration respect to the initial biomass concentration (X/X_0). Dotted lines represent fitted data. Coefficients are given on gCOD basis.

3.2. Monomer Development

The polymer composition over time during the accumulations is shown in Figure 3a, while the flow of carbon to 3HV relative to PHA overall at each time point (the instantaneous 3HV fraction) is shown in Figure 3b. The trends of replicate runs all matched well with the originals in terms of monomer development, although the final 3HV content differed slightly from run to run. The highest values of 3HV content were achieved in accumulations with propionic acid present at all times; Experiment set 2 (100% HPr) and Experiment set 3 (50%/50% HAc/HPr) reached a maximum %3HV content of 0.90 and 0.72 (on a mole % basis) at 6 and 2 h, respectively (Figure 3a). Although the formation rates of 3HV units relative to the formation of PHA dominated in the early stages of the accumulation for Sets 2 and 3, a sharp decrease in the instantaneous 3HV fraction was identified (Figure 3b). In contrast, the instantaneous 3HV fraction (%3HV^{inst}) in Experiment set 4, which followed an alternating pulse feeding strategy of acetic and propionic acids, did not show any remarkable change with time. However, it should be noted that %3HV^{inst} steadily decreased during Exp 4′ but gradually increased during Exp 4 (Figure 3b). The trend in %3HV^{inst} in Exp 4 coincided with high production of active biomass in that system (Figure 3a). Overall, concurrent PHA storage and active biomass growth resulted in a dynamic trend in polymer composition, except for the alternating feeding strategy.

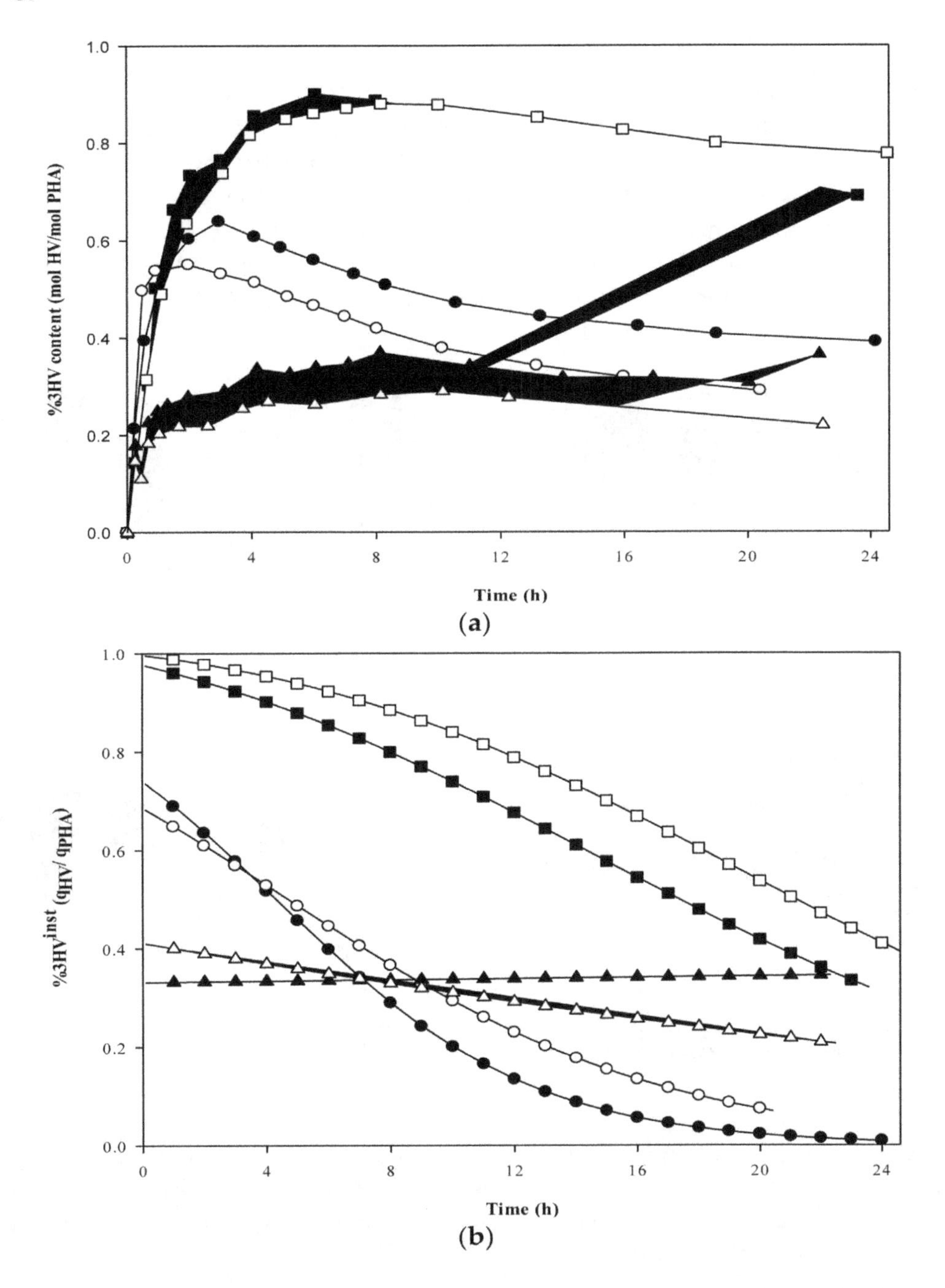

Figure 3. Evolution of 3HV fraction during PHA accumulation using different feeding strategies: (**a**) accumulated 3HV fraction; and (**b**) instantaneous 3HV fraction calculated from data regression. Experiment set 2: 100% propionic acid (■, Exp 2; and □, Exp 2′). Experiment set 3: 50% acetic acid and 50% propionic acid fed simultaneously (●, Exp 3; and ○, Exp 3′). Experiment set 4: 100% acetic acid alternating with 100% propionic acid (▲, Exp 4; and Δ, Exp 4′).

Current models for mixed culture PHA production using acetic and propionic acids as substrates consider that the proportion of 3HV monomer units in the copolymer obtained changes in proportion to the relative composition of the carbon feeds. This assumption has often adequately predicted 3HV:3HB molar composition in cultures with negligible cellular growth. However, when a shifting substrate strategy was applied, and moreover when cellular growth was maintained, even while maintaining a constant feed composition, then the published models cannot predict the observations of the present investigation. Metabolic analysis of the carbon flux distribution through time reveals why this would be so.

3.3. Carbon Flux to PHA, Biomass and CO_2

Figure 4 shows the calculated carbon flux distribution to PHA monomers, active biomass and carbon dioxide for the four experiment sets at various time points. Of note is an increase in the proportion of carbon flux directed to carbon dioxide (CO_2) production over time, mostly due to the carbon flux through the TCA cycle (R_6) in all experiments, while VFAs were being consumed despite a reduction of active biomass synthesis rate and PHA production rate. According to Escapa et al. [28], the uptake of carbon source in cultures of *Pseudomonas putida* with no PHA synthase activity remains active and the excess is directed to the TCA cycle which produces CO_2, as a way to dissipate the carbon surplus.

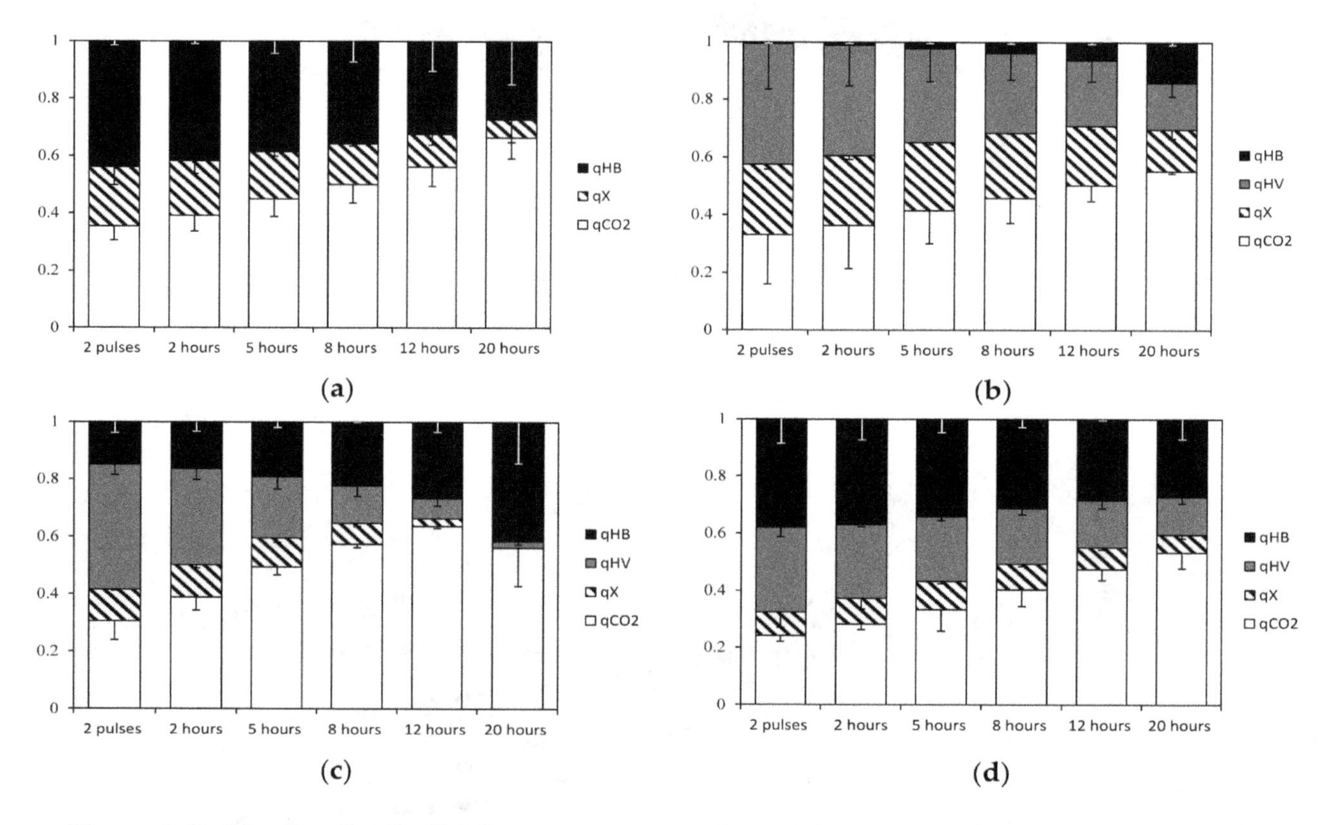

Figure 4. Carbon flux distribution (q_{PHB}, q_{PHV}, q_X, q_{CO2}) normalised respect to substrate uptake rate: (**a**) Experiment set 1: 100% acetic acid; (**b**) Experiment set 2: 100% propionic acid; (**c**) Experiment set 3: 50% acetic acid and 50% propionic acid fed simultaneously; and (**d**) Experiment set 4: 100% acetic acid alternating with 100% propionic acid.

In the model used in the present study, it was considered that the extra carbon consumed was passed to the TCA cycle for production of high energy molecules (ATP and NADH), and dissipated as ATP (Table 3). When the VFA consumption rate exceeded the respiratory capacity, a slightly decreasing trend in the respiratory quotient was observed. It should also be noted that no pathways for polymer consumption were included, although it is known that polymerization and PHA consumption can occur simultaneously [29]. It has been demonstrated that PHA operon proteins, including PHA depolymerase, are expressed from the start of the growth phase in *Pseudomonas putida* [30]. The present model was found to be feasible, if this extra carbon either goes to a depolymerization pathway or if it is spilled to produce ATP in R_7 which accounts for non-growth associated ATP maintenance. In both scenarios the same flux of CO_2 is predicted. Further model improvements that measure probable depolymerization subproducts and CO_2 production rates would be necessary to confirm the balances.

Table 3. Propionyl-CoA decarboxylation fraction, respiratory quotient and energy dissipated estimated by MFA (with standard deviation in brackets).

Experiment Set	Elapsed Duration	%Conversion PrCoA to AcCoA (mmol/mmol)		RQ (Cmmol CO_2/mol O_2)		ATP Dissipated (molATP/Cmmol VFA Consumed)	
1	2 pulses	-	-	1.19	(0.03)	0.57	(0.45)
	2 h	-	-	1.16	(0.03)	0.87	(0.46)
	5 h	-	-	1.13	(0.03)	1.32	(0.46)
	8 h	-	-	1.10	(0.04)	1.71	(0.45)
	12 h	-	-	1.07	(0.03)	2.34	(0.24)
	20 h	-	-	1.05	(0.04)	3.11	(0.44)
2	2 pulses	0.49	(0.11)	0.79	(0.03)	1.83	(1.36)
	2 h	0.51	(0.10)	0.80	(0.02)	2.09	(1.19)
	5 h	0.55	(0.07)	0.81	(0.01)	2.50	(0.90)
	8 h	0.59	(0.05)	0.81	(0.01)	2.84	(0.67)
	12 h	0.65	(0.03)	0.82	(0.01)	3.21	(0.43)
	20 h	0.75	(0.01)	0.83	(0.01)	3.68	(0.08)
3	2 pulses	0.27	(0.04)	1.08	(0.06)	0.62	(0.38)
	2 h	0.39	(0.02)	1.01	(0.03)	1.17	(0.26)
	5 h	0.55	(0.02)	0.99	(0.02)	1.99	(0.18)
	8 h	0.69	(0.02)	0.95	(0.01)	2.66	(0.08)
	12 h	0.85	(0.03)	0.95	(0.00)	3.37	(0.04)
	20 h	0.97	(0.01)	1.01	(0.08)	0.46	(0.28)
4	2 pulses	0.44	(0.05)	1.14	(0.11)	0.27	(0.04)
	2 h	0.46	(0.01)	1.10	(0.10)	0.46	(0.19)
	5 h	0.48	(0.07)	1.05	(0.08)	0.76	(0.59)
	8 h	0.54	(0.07)	1.01	(0.05)	1.30	(0.49)
	12 h	0.60	(0.07)	0.99	(0.03)	1.83	(0.35)
	20 h	0.67	(0.05)	0.97	(0.01)	2.22	(0.68)

Comparing experiments with single substrates, more CO_2 is generated when acetic acid is the only substrate compared with when propionic acid is used. The TCA cycle was calculated to be more active in Experiment set 1 (100% acetic acid), because more energy is needed to metabolize acetic acid (given 1 mol ATP is necessary to produce 1 Cmol acetyl-CoA, while activating 1 Cmol of propionic acid consumes only 0.67 mol ATP, see stoichiometry in the Supplementary Materials). On the other hand, propionic acid was found to have a higher oxygen demand when acetyl-CoA is formed from propionyl-CoA decarboxylation [31], leading to a decreased respiratory quotient (RQ). The MFA results were in agreement with this expectation. Experiment set 2, fed with propionic acid as a single substrate, had a lower RQ compared with Experiment set 1 (Table 3). However, for feeding strategies with mixed substrates, (Experiment set 3 and Experiment set 4), the RQ was very similar and remained at similar values throughout the accumulation experiments. The latter observations agreed with the composition data: similar molar fractions of propionic acid in the feed for Experiment set 3 and Experiment set 4 resulted in a constant molar fraction of propionic acid uptake relative to total carbon uptake flux (f_{Pr}).

Concerning active biomass synthesis, reaction stoichiometry indicates that propionyl-CoA gives higher theoretical growth yields compared with acetyl-CoA (1.06 mol PrCoA produces 1 mol X, and

1.27 mol AcCoA generates 1 mol X, see Supplementary Materials). According to this MFA analysis, propionyl-CoA was diverted to cell growth and PHA production during the exponential phase growth (0–8 h) in those accumulations that had propionic acid present at all times (see Experiment sets 2 and 3 in Figure 4b,d as examples), with propionyl-CoA having been shown to be the preferred substrate for active biomass growth by Lemos et al. [32] and Jiang et al.[16]. However, at the end of all experiments, when the decarboxylation rate was high, acetyl-CoA units were converted into 3HB monomers. As a consequence, 3HV monomers dominate in the early stages of accumulation, with their formation rate decreasing over time. To understand this result, there needs to be some consideration of the role that generation of reducing equivalents plays in controlling the pathways.

3.4. Pathways for Generation of Reducing Power

There are three important cofactors involved in PHA synthesis and regulation: coenzyme-A, $NADH/NAD^+$ and $NADPH/NADP^+$ [28,33]. The relative concentrations of acyl-CoA and free coenzyme-A are critical in controlling metabolic pathways, in particular PHA storage. NADH participates in catabolic reactions, while NADPH has an important role in reductive biosynthesis such as PHA biopolymers and active biomass [33]. In reduced metabolic networks, NADPH is not considered separately. The model used in the present study, developed by Pardelha et al. [17], works under the assumption that there exists free interchange between the reducing equivalents NADPH and NADH [34].

PHA production is favored when NADPH concentrations and $NADPH/NADP^+$ ratios are high [13]. In the model used in this current study, both PHA accumulation and energy production during oxidative phosphorylation require a source of reducing power. Therefore, the metabolic processes where reducing equivalents (NADH) are formed become key factors. Considering the processes outlined in Figure 1, these are reactions related to cellular growth (R_4 and R_5), the TCA cycle (R_6) and decarboxylation of propionyl-CoA to acetyl-CoA (R_3). In this sense, fluctuations in active biomass growth and PHA synthesis activities could be related to changes in the flux though the TCA cycle and/or decarboxylation of propionyl-CoA. On the other hand, VFA uptake and cell growth have an ATP requirement which is met by the TCA cycle and the electron transport of the respiratory chain [35].

One limitation for maximizing PHA yields is the regeneration of reducing equivalents (NADPH/NADH). MFA results showed that most of NADH was generated by the TCA cycle. According to the metabolic model, less reductive equivalents are necessary to produce 1 Cmol of propionyl-CoA* than 1 Cmol of acetyl-CoA* (0.167 vs. 0.25 mol NADH, respectively). Therefore, in order to produce 3HB monomer units, a greater amount of carbon must be directed to TCA cycle for NADH generation reducing equivalents (NADH).

To investigate the role of these pathways for generation of reducing equivalents in evolution of copolymer composition, metabolic flux analysis (MFA) was performed at different stages of the accumulations for all the feed regimes tested (Experiment sets 1, 2, 3, and 4) (see Figure 5).

In cultures fed with alternating substrates (Exp 4 and 4′), carbon fluxes for active biomass formation and 3HV monomer production remained constant during the accumulations (Figure 4). As shown in Figure 5b, NADH generation rate by propionyl-CoA decarboxylation (v_3) was kept low, to a level to cover energy production by TCA cycle requirements. On the other hand, in experiments with mixed acetic and propionic feeds (Exp 3 and 3′), the decarboxylation rates increased markedly when the active biomass growth rates attenuated, and thus more acetyl-CoA units became available and the 3HB production rate could increase as a result.

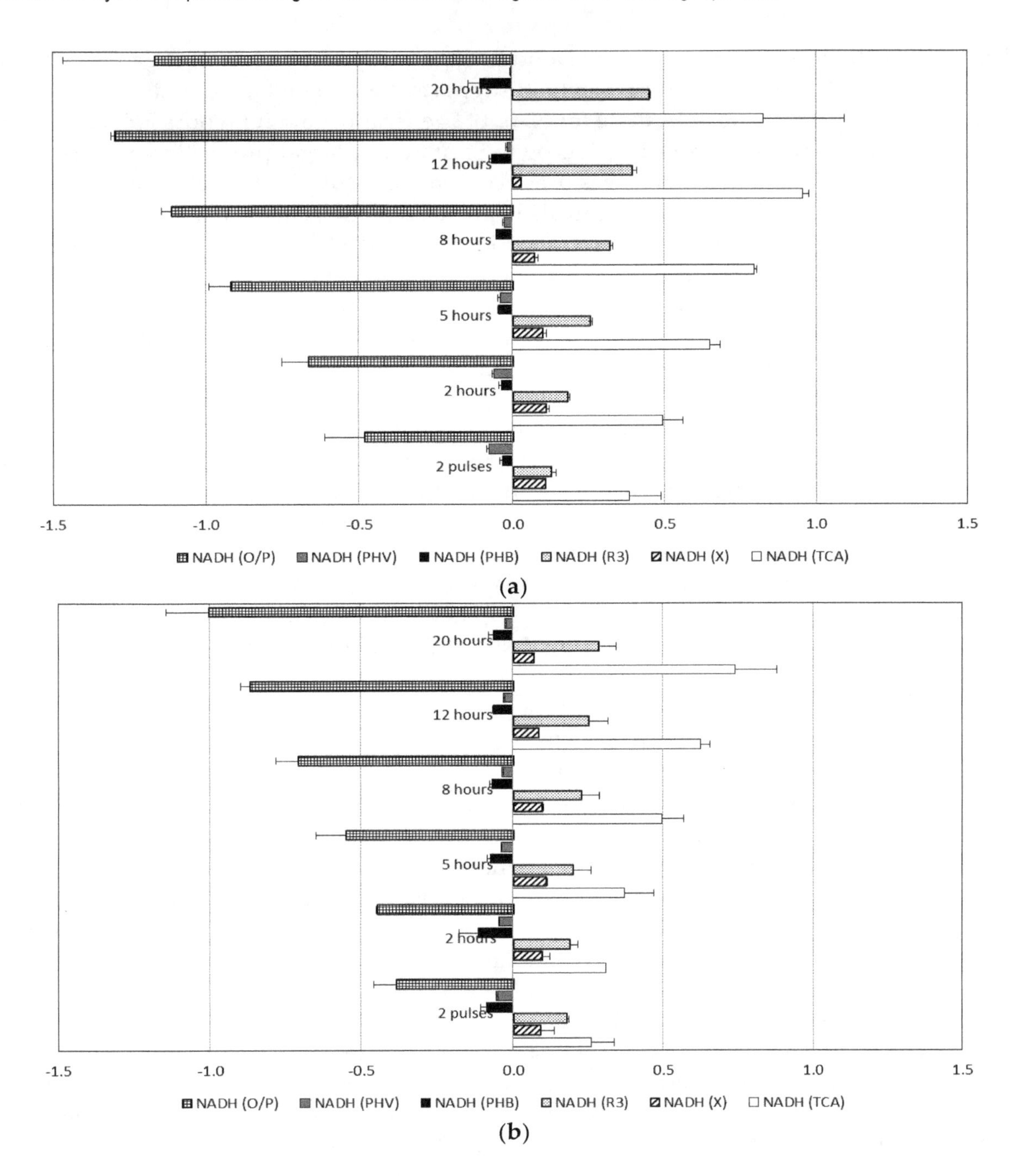

Figure 5. NADH generated and consumed at different stages of culture obtained by metabolic flux analysis (MFA): (**a**) Experiment set 3: 50% acetic acid and 50% propionic acid fed simultaneously; and (**b**) Experiment set 4: 100% acetic acid alternating with 100% propionic acid. (NADH was considered as internal metabolite).

As mentioned previously, propionic acid has been shown to be the preferred substrate for active biomass formation. Production of pure 3HV was not possible; however, it was close to 100% at the beginning of accumulations based on propionic acid alone (Experiment set 2). It has been demonstrated that keeping an optimal active biomass specific growth rate enhances PHBV copolymer synthesis in *Cupriavidus necator* [36]. However, at high specific cell growth rates, more substrate is used for active biomass formation and less is available for PHA production. The specific cell growth rate in the present study was relatively low compared to pure cell cultures, and it was found that the higher the specific growth rate, the higher the %3HV (fitted data for specific growth rate and monomer specific synthesis rates are in the Supplementary Materials). Grousseau et al. [6] found that a higher PHB yield on substrate was obtained when the Entner–Doudoroff pathway was active. This pathway produces NADPH and is linked to anabolic requirements. It supports the idea that maintaining cellular growth offers an alternative pathway to TCA cycle for NADPH generation and it favours PHA synthesis.

When cells experienced a degree of cell growth limitation, a larger proportion of 3HB monomer units compared to 3HV monomers units was produced. In those cultures fed continuously with acetic acid (Experiment sets 1 and 3), reductive equivalents can be directly generated by acetyl-CoA pathway through the TCA cycle, which favors PHA production. Similar PHA fluxes were obtained using acetic or propionic acid as single substrates (Figure 4). However, total PHA yield on substrate was higher for cultures fed exclusively with acetic acid than cultures fed with propionic acid as sole substrate (Table 2). However, compared with accumulations fed with propionic acid exclusively (Experiment set 2) or periodically (Experiment set 4), acetic acid as a feed did not stimulate concurrent active growth and storage as much as propionic acid. Previous MFA studies have suggested that when acetic and propionic acids are fed simultaneously, the catabolic activity (TCA) primarily depends on acetic acid uptake [4]. According to the metabolic model, active biomass synthesis has a higher demand of ATP and NADH when it is generated from acetic acid rather than propionic acid. In further MFA calculations, the rate of active biomass synthesis is not considered as being proportional to the consumption rate of acetic and propionic acids. This has resulted in a non-redundant MFA system, which showed that when acetic acid and propionic acids are fed simultaneously, most of the new active biomass is synthesized from propionic acid uptake. Future studies where more experimental rates are available (such as CO_2 evolution) would be required to test this hypothesis.

4. Conclusions

This study presented an analysis of PHA accumulation processes by mixed cultures when adopting different feeding strategies which favor concurrent cellular growth and carbon storage. Although higher maximum specific growth rates were achieved in cultures fed continuously with acetic acid as a sole substrate or as part of a mixture, constant cell growth was not achieved. Significant changes through time to the instantaneous 3HV content were observed under most accumulation conditions, and such changes cannot be adequately described by existing metabolic models. An alternating feeding strategy resulted in constant instantaneous 3HV content, despite the decarboxylation rate increasing with time. Overall, the 3HV monomer production rate is high. Finally, the incorporation of cellular metabolism in the evaluation of process performance for PHA production by mixed cultures offers an opportunity to help understand PHA polymer composition fluctuations and the carbon flux distribution in different cell physiological states. In this way, metabolic models can help improve the estimation of process final concentrations and yields. However, a better description of the 3HB:3HV fluctuations relative to active biomass growth needs a more detailed metabolic network to take into account the reactions in which NADPH and NADH are formed.

Supplementary Materials

Figure S1: Experimental data for PHA fraction and relative active biomass production; (**a**) Experiment set 1: 100% acetic acid; (**b**) Experiment set 2: 100% propionic acid. Figure S2: Fitted data for specific growth rate and monomer specific synthesis rate. Experiment set 1: 100% acetic acid; Experiment set 2: 100% propionic acid; Experiment set 3: 50% acetic acid and 50% propionic acid fed simultaneously; Experiment set 4: 100% acetic acid alternating with 100% propionic acid, Figure S3: NADH generated and consumed at different stages of culture obtained by metabolic flux analysis (MFA). (**a**) Experiment set 1: 100% acetic acid; (**b**) Experiment set 2: 100% propionic acid, Figure S4: Internal carbon flux distribution (v3, v6, v9, v10, vX) normalised respect to substrate uptake rate. (**a**) Experiment set 1: 100% acetic acid; (**b**) Experiment set 2: 100% propionic acid; (**c**) Experiment set 3: 50% acetic acid and 50% propionic acid fed simultaneously; (**d**) Experiment set 4: 100% acetic acid alternating with 100% propionic acid, Table S1: Metabolic network for PHA processes by microbial mixed cultures used in the present work, Table S2: Microstructure of PHA copolymer samples by 13C NMR analysis.

Acknowledgments: The authors would like to acknowledge Australian Research Council for funding this work through project LP0990917. The authors also thank AnoxKaldnes Sweden for funding through grant ARC LP0990917, and acknowledge gratefully the support provided by Monica Arcos-Hernandez, Lamija Karabegovic, Per Magnusson and Anton Karlsson in helping with the running of the pilot plant, accumulation studies and sample analysis. The authors confirm that there are no conflict of interests to declare.

Author Contributions: Liliana Montano-Herrera, Bronwyn Laycock, Steven Pratt and Alan Werker all conceived and designed the experiments; Liliana Montano-Herrera performed the experiments; Liliana Montano-Herrera, Bronwyn Laycock, Steven Pratt and Alan Werker all analyzed the data; Liliana Montano-Herrera wrote the paper, which was reviewed and edited by the other authors.

Abbreviations

%3HV	3HV fraction in total PHA (mol 3HV/mol PHA)
$\%3HV^{inst}$	Instantaneous 3HV content (mol 3HV/mol PHA)
%PHA	PHA intracellular content (gPHA/gVSS)
$\%PHA_0$	Initial PHA intracellular content (gPHA/gVSS)
f_{Pr}	Fraction of propionic acid uptake to total carbon uptake flux
3HB	3-Hydroxybutyrate
3HV	3-Hydroxyvalerate
AcCoA	Acetyl-CoA
ATP	Adenosine triphosphate
CO_2	Carbon dioxide
CoA	Coenzyme A
COD	Chemical oxygen demand
f_{PHA}	PHA fraction with respect to active biomass (PHA/X)
HAc	Acetic acid
HPr	Propionic acid
NAD^+	Nicotinamide adenine dinucleotide (oxidised)
NADH	Nicotinamide adenine dinucleotide (reduced)
$NADP^+$	Nicotinamide adenine dinucleotide phosphate (oxidised)
NADPH	Nicotinamide adenine dinucleotide phosphate (reduced)
PHA	Poly(3-hydroxyalkanoate) or PHA concentration at certain time
PHA_0	Initial PHA concentration
PHB	Poly(3-hydroxybutyrate)
PHV	Poly(3-hydroxyvalerate)
PrCoA	Propionyl-CoA
q_{HAc}	Acetic acid specific consumption rate (Cmol HAc/Cmol X·h)
q_{HB}	Specific 3HB monomer synthesis rate (Cmol 3HB/Cmol X·h)
q_{HPr}	Propionic acid specific consumption rate (Cmol HPr/Cmol X·h)
q_{HV}	Specific 3HV monomer synthesis rate (Cmol 3HV/Cmol X·h)
q_{PHA}	Specific PHA synthesis rate (Cmol PHA/Cmol X·h)
q_S	Specific VFA consumption rate (Cmol VFA/Cmol X·h)
RQ	Respiratory quotient
TCA	Tricarboxylic acid cycle
v	Reaction rate (Cmol/Cmol X·h)
VFA	Volatile fatty acid
VSS	Volatile suspended solids
X	Active biomass
X_0	Initial active biomass

References

1. Laycock, B.; Halley, P.; Pratt, S.; Werker, A.; Lant, P. The chemomechanical properties of microbial polyhydroxyalkanoates. *Prog. Polym. Sci.* **2013**, *38*, 536–583. [CrossRef]
2. Filipe, C.D.M.; Daigger, G.T.; Grady, C.P.L. A metabolic model for acetate uptake under anaerobic conditions by glycogen accumulating organisms: Stoichiometry, kinetics, and the effect of pH. *Biotechnol. Bioeng.* **2001**, *76*, 17–31. [CrossRef] [PubMed]
3. Taguchi, K.; Taguchi, S.; Sudesh, K.; Maehara, A.; Tsuge, T.; Doi, Y. Metabolic pathways and engineering of polyhydroxyalkanoate biosynthesis. *Biopolym. Online* **2005**, *3*. [CrossRef]

4. Dias, J.M.L.; Oehmen, A.; Serafim, L.S.; Lemos, P.C.; Reis, M.A.M.; Oliveira, R. Metabolic modelling of polyhydroxyalkanoate copolymers production by mixed microbial cultures. *BMC Syst. Biol.* **2008**, *2*, 59. [CrossRef] [PubMed]
5. Lemos, P.C.; Serafim, L.S.; Santos, M.M.; Reis, M.A.M.; Santos, H. Metabolic pathway for propionate utilization by phosphorus-accumulating organisms in activated sludge: C-13 labeling and in vivo nuclear magnetic resonance. *Appl. Environ. Microb.* **2003**, *69*, 241–251. [CrossRef]
6. Grousseau, E.; Blanchet, E.; Deleris, S.; Albuquerque, M.G.E.; Paul, E.; Uribelarrea, J.L. Impact of sustaining a controlled residual growth on polyhydroxybutyrate yield and production kinetics in *Cupriavidus necator*. *Bioresour. Technol.* **2013**, *148*, 30–38. [CrossRef] [PubMed]
7. Gurieff, N. Production of Biodegradable Polyhydroxyalkanoate Polymers Using Advanced Biological Wastewater Treatment Process Technology. Ph.D. Thesis, The University of Queensland, Brisbane, Australia, 2007.
8. Serafim, L.S.; Lemos, P.C.; Torres, C.; Reis, M.A.M.; Ramos, A.M. The influence of process parameters on the characteristics of polyhydroxyalkanoates produced by mixed cultures. *Macromol. Biosci.* **2008**, *8*, 355–366. [CrossRef] [PubMed]
9. Albuquerque, M.G.E.; Martino, V.; Pollet, E.; Averous, L.; Reis, M.A.M. Mixed culture polyhydroxyalkanoate (PHA) production from volatile fatty acid (VFA)-rich streams: Effect of substrate composition and feeding regime on pha productivity, composition and properties. *J. Biotechnol.* **2011**, *151*, 66–76. [CrossRef] [PubMed]
10. Ivanova, G.; Serafim, L.S.; Lemos, P.C.; Ramos, A.M.; Reis, M.A.M.; Cabrita, E.J. Influence of feeding strategies of mixed microbial cultures on the chemical composition and microstructure of copolyesters p(3HB-*co*-3HV) analyzed by NMR and statistical analysis. *Magn. Reson. Chem.* **2009**, *47*, 497–504. [CrossRef] [PubMed]
11. Arcos-Hernandez, M.V.; Laycock, B.; Donose, B.C.; Pratt, S.; Halley, P.; Al-Luaibi, S.; Werker, A.; Lant, P.A. Physicochemical and mechanical properties of mixed culture polyhydroxyalkanoate (PHBV). *Eur. Polym. J.* **2013**, *49*, 904–913. [CrossRef]
12. Johnson, K.; Van Loosdrecht, M.C.M.; Kleerebezem, R. Influence of ammonium on the accumulation of polyhydroxybutyrate (PHB) in aerobic open mixed cultures. *J. Biotechnol.* **2010**, *147*, 73–79. [CrossRef] [PubMed]
13. Pardelha, F.; Albuquerque, M.G.E.; Reis, M.A.M.; Dias, J.M.L.; Oliveira, R. Flux balance analysis of mixed microbial cultures: Application to the production of polyhydroxyalkanoates from complex mixtures of volatile fatty acids. *J. Biotechnol.* **2012**, *162*, 336–345. [CrossRef] [PubMed]
14. Serafim, L.S.; Lemos, P.C.; Oliveira, R.; Ramos, A.M.; Reis, M.A.M. High storage of PHB by mixed microbial cultures under aerobic dynamic feeding conditions. *Eur. Symp. Environ. Biotechnol.* **2004**, 479–482.
15. Valentino, F.; Karabegouic, L.; Majone, M.; Morgan-Sagastume, F.; Werker, A. Polyhydroxyalkanoate (PHA) storage within a mixed-culture biomass with simultaneous growth as a function of accumulation substrate nitrogen and phosphorus levels. *Water Res.* **2015**, *77*, 49–63. [CrossRef] [PubMed]
16. Jiang, Y.; Hebly, M.; Kleerebezem, R.; Muyzer, G.; van Loosdrecht, M.C.M. Metabolic modeling of mixed substrate uptake for polyhydroxyalkanoate (PHA) production. *Water Res.* **2011**, *45*, 1309–1321. [CrossRef] [PubMed]
17. Pardelha, F.; Albuquerque, M.G.E.; Reis, M.A.M.; Oliveira, R.; Dias, J.M.L. Dynamic metabolic modelling of volatile fatty acids conversion to polyhydroxyalkanoates by a mixed microbial culture. *New Biotechnol.* **2014**, *31*, 335–344. [CrossRef] [PubMed]
18. Murugan Janarthanan, O.; Laycock, B.; Montano-Herrera, L.; Lu, Y.; Arcos-Hernandez, M.V.; Werker, A.; Pratt, S. Fluxes in PHA-storing microbial communities during enrichment and biopolymer accumulation processes. *New Biotechnol.* **2016**, *33*, 61–72. [CrossRef] [PubMed]
19. Werker, A.G.; Bengtsson, S.O.H.; Karlsson, C.A.B. Method for Accumulation of Polyhydroxyalkanoates in Biomass with on-Line Monitoring for Feed Rate Control and Process Termination. WO 2011070544 A2, 16 June 2011.
20. Morgan-Sagastume, F.; Pratt, S.; Karlsson, A.; Cirne, D.; Lant, P.; Werker, A. Production of volatile fatty acids by fermentation of waste activated sludge pre-treated in full-scale thermal hydrolysis plants. *Bioresour. Technol.* **2011**, *102*, 3089–3097. [CrossRef] [PubMed]
21. American Public Health Association (APHA). *Standard Methods for the Examination of Water and Wastewater*; American Public Health Association: Washington, DC, USA, 1995.

22. Motulsky, H.J. *Prism5 Statistics Guide*; Graphpad Software Inc.: San Diego, CA, USA, 2007.
23. Zwietering, M.H.; Jongenburger, I.; Rombouts, F.M.; Vantriet, K. Modeling of the bacterial growth curve. *Appl. Environ. Microb.* **1990**, *56*, 1875–1881.
24. Third, K.A.; Newland, M.; Cord-Ruwisch, R. The effect of dissolved oxygen on PHB accumulation in activated sludge cultures. *Biotechnol. Bioeng.* **2003**, *82*, 238–250. [CrossRef] [PubMed]
25. Villadsen, J.; Nielsen, J.; Lidén, G. *Bioreaction Engineering Principles*, 3rd ed.; Springer: New York, NY, USA, 2011.
26. Klamt, S.; Saez-Rodriguez, J.; Gilles, E.D. Structural and functional analysis of cellular networks with cellnetanalyzer. *BMC Syst. Biol.* **2007**. [CrossRef] [PubMed]
27. Stephanopolulos, G.N.; Aristidou, A.A.; Nielsen, J. *Metabolic Engineering: Principles and Methodologies*; Academic Press: San Diego, CA, USA, 1998.
28. Escapa, I.F.; Garcia, J.L.; Buhler, B.; Blank, L.M.; Prieto, M.A. The polyhydroxyalkanoate metabolism controls carbon and energy spillage in *Pseudomonas putida*. *Environ. Microbiol.* **2012**, *14*, 1049–1063. [CrossRef] [PubMed]
29. Ren, Q.; de Roo, G.; Ruth, K.; Witholt, B.; Zinn, M.; Thony-Meyer, L. Simultaneous accumulation and degradation of polyhydroxyalkanoates: Futile cycle or clever regulation? *Biomacromolecules* **2009**, *10*, 916–922. [CrossRef] [PubMed]
30. Arias, S.; Bassas-Galia, M.; Molinari, G.; Timmis, K.N. Tight coupling of polymerization and depolymerization of polyhydroxyalkanoates ensures efficient management of carbon resources in *Pseudomonas putida*. *Microb. Biotechnol.* **2013**, *6*, 551–563. [CrossRef] [PubMed]
31. Lefebvre, G.; Rocher, M.; Braunegg, G. Effects of low dissolved-oxygen concentrations on poly(3-hydroxybutyrate-*co*-3-hydroxyvalerate) production by *Alcaligenes eutrophus*. *Appl. Environ. Microb.* **1997**, *63*, 827–833.
32. Lemos, P.C.; Serafim, L.S.; Reis, M.A.M. Synthesis of polyhydroxyalkanoates from different short-chain fatty acids by mixed cultures submitted to aerobic dynamic feeding. *J. Biotechnol.* **2006**, *122*, 226–238. [CrossRef] [PubMed]
33. Yu, J.; Si, Y.T. Metabolic carbon fluxes and biosynthesis of polyhydroxyalkanoates in *Ralstonia eutropha* on short chain fatty acids. *Biotechnol. Prog.* **2004**, *20*, 1015–1024. [CrossRef] [PubMed]
34. Kim, J.I.; Varner, J.D.; Ramkrishna, D. A hybrid model of anaerobic *E. Coli* GJT001: Combination of elementary flux modes and cybernetic variables. *Biotechnol. Prog.* **2008**, *24*, 993–1006. [CrossRef] [PubMed]
35. Zeng, A.P.; Ross, A.; Deckwer, W.D. A method to estimate the efficiency of oxidative-phosphorylation and biomass yield from ATP of a facultative anaerobe in continuous culture. *Biotechnol. Bioeng.* **1990**, *36*, 965–969. [CrossRef] [PubMed]
36. Shimizu, H.; Kozaki, Y.; Kodama, H.; Shioya, S. Maximum production strategy for biodegradable copolymer P(HB-*co*-HV) in fed-batch culture of *Alcaligenes eutrophus*. *Biotechnol. Bioeng.* **1999**, *62*, 518–525. [CrossRef]

12

Rheological Behavior of High Cell Density *Pseudomonas putida* LS46 Cultures during Production of Medium Chain Length Polyhydroxyalkanoate (PHA) Polymers

Warren Blunt [1,*,†], Marc Gaugler [2], Christophe Collet [2], Richard Sparling [3], Daniel J. Gapes [2], David B. Levin [1] and Nazim Cicek [1]

1 Department of Biosystems Engineering, University of Manitoba, Winnipeg, MB R3T 5V6, Canada; David.Levin@umanitoba.ca (D.B.L.); Nazim.Cicek@umanitoba.ca (N.C.)

2 Scion Research, Te Papa Tipu Innovation Park, 49 Sala Street, Private Bag 3020, Rotorua 3046, New Zealand; Marc.Gaugler@scionresearch.com (M.G.); Christophe.Collet@scionresearch.com (C.C.); Daniel.Gapes@scionresearch.com (D.J.G.)

3 Department of Microbiology, University of Manitoba, Winnipeg, MB R3T 2N2, Canada; Richard.Sparling@umanitoba.ca

* Correspondence: umbluntw@myumanitoba.ca or warren.blunt@nrc-cnrc.gc.ca

† Present address of corresponding author: National Research Council of Canada, 6100 Royalmount Avenue Montreal, QC H4P 2R2, Canada.

Abstract: The rheology of high-cell density (HCD) cultures is an important parameter for its impact on mixing and sparging, process scale-up, and downstream unit operations in bioprocess development. In this work, time-dependent rheological properties of HCD *Pseudomonas putida* LS46 cultures were monitored for microbial polyhydroxyalkanoate (PHA) production. As the cell density of the fed-batch cultivation increased (0 to 25 $g \cdot L^{-1}$ cell dry mass, CDM), the apparent viscosity increased nearly nine-fold throughout the fed-batch process. The medium behaved as a nearly Newtonian fluid at lower cell densities, and became increasingly shear-thinning as the cell density increased. However, shear-thickening behavior was observed at shearing rates of approximately 75 $rad \cdot s^{-1}$ or higher, and its onset increased with viscosity of the sample. The supernatant, which contained up to 9 $g \cdot L^{-1}$ soluble organic material, contributed more to the observed viscosity effect than did the presence of cells. Owing to this behavior, the oxygen transfer performance of the bioreactor, for otherwise constant operating conditions, was reduced by 50% over the cultivation time. This study has shown that the dynamic rheology of HCD cultures is an important engineering parameter that may impact the final outcome in PHA cultivations. Understanding and anticipating this behavior and its biochemical origins could be important for improving overall productivity, yield, process scalability, and the efficacy of downstream processing unit operations.

Keywords: PHA; viscosity; non-Newtonian fluid; fed-batch fermentation; oxygen transfer; *Pseudomonas putida*

---

1. Introduction

Recent concern over the accumulation of plastic waste in the natural environment (particularly micro-plastics) emphasizes the need to find alternative biodegradable polymers [1,2]. In this regard, PHA polymers are a promising replacement for petroleum-based plastic materials, being both renewable and completely biodegradable [3]. PHA polymers can have a variety of different monomer sub-unit compositions. This enables, to a large extent, a wide-range of physical and thermal properties and numerous potential applications [4,5]. Indeed, certain PHAs (depending on the composition and arrangement of the monomer subunits) have properties comparable to conventional petroleum-based plastics, like polyethylene and polypropylene.

PHA polymers are synthesized as intracellular reserves of carbon, energy, and reducing power by a wide-range of bacteria, and some archaea. While the cost of production currently limits applications for PHA to niche markets [6,7], development of more efficient bioprocesses may help to increase the economic viability and lessen the environmental impact of PHA production [8,9]. Currently, HCD cultures are widely seen as the best cultivation strategy to achieve high volumetric productivities [10]. Some HCD cultures for PHA production have reached cell densities in excess of 200 $g{\cdot}L^{-1}$ CDM [11–13]. Further details on HCD cultivations in PHA production are available in several recent reviews [14,15]. However, a common problem with HCD cultures in general is increasing medium viscosity [16]. This can lead to dead zones in the bioreactor and reduced heat and mass transfer capabilities, especially in large-scale bioreactors with inherently poor mixing capability [17].

Rheology is the study of the deformation of matter (in this case, the flow of liquid fermentation medium) under an applied stress. Previous studies have examined the rheology of cultivation medium for a variety of bioprocessing applications using different microorganisms and fungi. These include: xanthan gum production using *Xanthomonas* spp. [18], viscous mycelial (fungal) cultures for variety of bio-products [19–24], polyglutamic acid (PGA) production using *Bacillus subtilis* [25], mixtures of primary and secondary sewage sludge [26], and fermentation of sewage sludge [27,28], amongst others. Multiple studies have looked at rheological properties of extracellular polymeric substances (EPS) produced by *Pseudomonas* spp. [29–33]. However, many of these assessments examined rheological properties of an extracted polymer of interest, but did not directly quantify its effect on culture medium. Most studies show that fermentation medium behaves as a non-Newtonian fluid, meaning the apparent viscosity is dependent on the shear rate [19,23,34,35].

Since lack of adequate dissolved oxygen (DO) is a significant factor that limits productivity in HCD cultivations for PHA production [36], the effect of medium viscosity on the oxygen transfer rate could be important. Several previous studies have demonstrated inversely proportional relationships between viscosity and oxygen transfer in both model Newtonian fluids (glycerol, glucose solutions) as well as non-Newtonian fluids (xanthan gum, carboxymethylcellulose solutions) [17,37–42]. Such model fluids are often preferred to actual biological cultures because they are cheaper and easier to work with [17]. A few studies, however, have evaluated oxygen transfer characteristics in a biological medium [19,25]. In all cases, there is a consensus that the volumetric oxygen mass transfer coefficient, K_La, is inversely proportional to the medium viscosity.

Previous application of HCD cultivations in PHA production are numerous [15]. Yet, we can find no evidence that rheology of the culture medium has been studied to date; or at the very least, that information is not widely accessible. This includes both short-chain length (scl-) PHAs and medium chain length (mcl-) PHAs. Considering that PHAs are high molecular weight (M_w) polymers that can occupy up to 75–88% CDM [13,43] and be produced with relatively high titer [14,15], the examination of culture rheology and its effects on oxygen transfer could be an important contribution to process development, optimization, and scalability in PHA production. Furthermore, this could have significant impact on downstream processing, including pumping, filtration, centrifugation, or spray drying unit operations.

The objectives of this work were, therefore: 1) to examine time-dependent rheological behavior of HCD fed-batch cultures of *Pseudomonas putida* LS46 for production of medium chain length (mcl-) PHAs; 2) to gain understanding of the biochemical origins of these rheological changes; and 3) to further assess how viscosity impacts the oxygen mass transfer characteristics of the cultivation medium.

2. Materials and Methods

2.1. Micro-Organism, Medium, and Substrate

The strain used in this study was *Pseudomonas putida* LS46 [44], and strain maintenance procedures were as specified previously [45]. A slightly modified version of Ramsay's minimal medium used in all experimental studies [46]. However, the initial concentrations of $(NH_4)_2SO_4$, $MgSO_4$, $CaCl_2{\cdot}2H_2O$, and trace element solution were increased to 2 g·L^{-1}, 0.2 g·L^{-1}, 20 mg·L^{-1}, and 2 mL·L^{-1}, respectively. The $MgSO_4$, $CaCl_2{\cdot}2H_2O$, ferric ammonium citrate, and trace element solution were filter sterilized through a 0.2 μm filter after autoclaving. Octanoic acid was used as the substrate in these studies and was added through a sterile 0.2 μm filter after autoclaving to an initial concentration of 20 mM.

2.2. Reactor Setup and Operation

Most experiments for this work were conducted in a 7 L (total volume) bench-scale system with a 3 L working volume. This system was used to generate the meta-data supporting the rheological observations in the pilot-scale bioreactor, which is described below. The configuration and setup of the bench-scale bioreactor system has been described previously [45,47]. Aeration was maintained at a constant flow rate of 2 VVM (atmospheric air only), and a mixing cascade (350–1200 rpm) was used to control the DO signal at 40% (of saturation with atmospheric air at 30 °C) for as long as possible. A reactive pulse-feed strategy was applied in response to either a drop in the off-gas CO_2 signal or a rise in the DO signal, indicating carbon limitation. Sub-inhibitory pulses of octanoic acid (5–20 mM) and a 200 g·L^{-1} solution of $(NH_4)_2SO_4$ were added to the reactor via high-precision injector syringes automated by LabBoss software [48]. The bench-scale cultivation was performed three times.

Because of the larger sample volume (1 L) required for rheological analysis, the system used for generation of these samples was a pilot-scale stainless steel, sterilization in place (SIP) bioreactor with a 152 L total volume (Sartorius Stedim Biostat D-DCU, Göttingen, Germany). The bioreactor was equipped with three 160 mm diameter Rushton turbines, four baffles, pH and DO electrodes, and a ring-type sparger located underneath the impeller. The bioreactor was filled with an initial volume of 70 L medium, and sterilized at 121 °C for 20 min before cooling to 30 °C.

In the pilot-scale system, the DO was maintained at 40% (of air saturation at 30 °C) for as long as possible. The cascade for DO was maintained through: (1) incremental increases in pressure from 200 mbar to 1000 mbar; (2) incremental increases in stirring rate from 100 rpm to a maximum of 600 rpm; and (3) increasing aeration (atmospheric air only) from 10 litres per minute (LPM) up to 30 LPM (maximum of approx. 0.4 volumes of air per liquid volume per minute or VVM). At this scale, aeration was limited because of foaming and excessive gas holdup encountered at higher volumetric flow rates. The slight headspace overpressure was used to obtain similar growth rates and biomass production over time, as well as timing of the onset of oxygen-limited conditions, as compared to the bench-scale bioreactor. This implies less efficient mixing in the pilot-scale bioreactor. The pilot-scale experiment was also carried out using the above-described pulse-feed strategy, except feeding was done with calibrated peristaltic pumps. Due to time and resource constraints, the pilot scale cultivation was performed once.

In either bioreactor system, experiments were initiated with the addition of a 5% (vol/vol) inoculum, which was grown overnight in flask cultures. After 16–20 h, $(NH_4)_2SO_4$ was no longer fed because it was no longer being consumed rapidly due to DO limitation. The pH of the medium was generally maintained via the addition of NaOH with automated peristaltic pumps (4 M at bench-scale and 10 M at the pilot-scale).

2.3. Sample Treatment

Samples (20–40 mL) were periodically withdrawn from the bioreactor, generally in 1–3 h intervals. These were centrifuged for 10 min at 12,500× *g*. The pellet was washed once in PBS buffer, transferred into a pre-weighed 20 mL aluminum dish and dried at 60 °C until no further loss of mass was detected to determine the total biomass concentration ($[X_t]$, g·L^{-1} CDM). The PHA content of the biomass ($\%_{PHA}$) was determined by gas chromatography with a flame ionization detector (GC-FID) using the sample preparation, instrument, and operating parameters described previously [45]. The supernatant was decanted and stored at −20 °C for analysis of residual octanoic acid by GC-FID and ammonium was determined spectrophotometrically by the indophenol blue method. Further details of these analyses are available elsewhere [45].

2.4. *Viscosity Measurements (Pilot Scale)*

Periodic 1 L samples were withdrawn from the bioreactor for rheological analysis. For certain samples, a portion of the medium was centrifuged at 12,500× *g* for 15 min (Sorvall RC-6 Plus with an F12-6 × 500 LEX rotor) to investigate the cell-free supernatants. The medium viscosity was assessed using a DHR-2 Rheometer (TA Instruments, New Castle, DE, USA) equipped with a cup-and-bob measurement system (30 mm cup diameter; 28 mm bob diameter). The cup and bob geometry was chosen to mitigate effects from sample drying, but plate-plate and cup/vane geometry were also assessed. Although good results for all three measurement geometries were obtained, the cup/bob system was chosen because of a more defined flow in the measurement gap, lower end-effects compared to the vane geometry [49] and fewer artefacts due to sample drying during the test compared to the parallel plate geometry.

The samples were conditioned at 30 °C for 20 min prior to measuring. During this conditioning step, a constant shear of 1 s^{-1} was applied to avoid settlement of the samples. All samples were measured using a flow sweep between 2 and 1000 s^{-1}. The samples in the cup/bob assembly were inspected after completion of the rheological testing to ensure that no significant evaporation occurred that would have affected the viscosity results. All samples were measured in triplicate. The data analysis was done using TRIOS v4.1.0.31739 (TA Instruments, New Castle, DE, USA).

The measured shear stress at different shear rates during rotational rheology can be described using a variety of established rheological models. In this work, the fit was best described using the power law. A power-law fluid is an idealized fluid, and its shear stress is a function of shear rate as described by

$$\tau = \varphi \times \dot{\gamma}^{n} \tag{1}$$

where τ is the shear stress (mPa); φ is the Power law viscosity constant (mPa·s); $\dot{\gamma}$ is the shear rate (s^{-1}), and n is the rate index (dimensionless).

2.5. *Off-Line Measurement of the Volumetric Oxygen Mass Transfer Coefficient (Bench-Scale)*

The global volumetric oxygen mass transfer coefficient, K_La, was measured using the dynamic out-gassing method [50]. To avoid the impracticality of K_La determinations at scale (which would require a 3 L sample volume), a small-scale reactor with a 200 mL working volume was constructed to allow at-line K_La determination while using minimal (150 mL) sample volume taken at various points throughout the bench scale fed-batch cultivations. The goal was to show that, for a given reactor environment (with constant mixing, geometry, gas flow rates, etc.) the oxygen transfer performance of that system is reduced as the chemical matrix of the supernatant becomes increasingly complex and viscous over time. An unfortunate consequence or limitation, however, is that the determined K_La values are not representative of the actual reactor environment from which the samples were derived. Because of this, the results were expressed as a percent of the value measured using the 0 h sample.

This 200 mL reactor used for K_La determination was constructed from plexi-glass with height 7.9 cm and 5.4 cm in diameter. The reactor was equipped with compression fitting ports for a DO probe, gas inlet, and gas outlet. The reactor was stirred magnetically with a 2.5 cm stir bar at 1000 rpm, and either air or N_2 was delivered to the reactor at flow rates of 200 or 500 mL·min^{-1}, respectively. This was done using thermal mass flow controllers (Bronkhorst Hi-Tech, Ruurlo, the Netherlands), which were part of an off-gas sensor system previously described [48]. A minimum of three determinations was done for each sample, and this was replicated for three fed batch experiments. The unit was validated initially in trials using distilled water or Ramsay's medium, and K_La values of 34.3 ± 3.4 h^{-1} and 21.5 ± 2.1 h^{-1} were obtained, respectively. Not surprisingly, these were on the lower end of the values obtained previously in the bench-scale bioreactor system [45,47]. This is probably because: (1) lack of baffles in the miniature device; (2) a stir bar was used instead of a proper impeller in the miniature device; and (3) the point of release of the bubbles was above the stir bar in the miniaturized reactor (as opposed to underneath the impeller in the bioreactor).

2.6. Analysis of Organic Products in the Supernatant (Bench-Scale)

Soluble protein in the supernatant was determined spectrophotometrically at 595 nm using a modified Bradford Assay [51]. Briefly, 0.5 mL of supernatant was mixed with 0.5 mL of 0.4 M NaOH. The samples were boiled for 10 min, and centrifuged (12,500× *g* for 5 min). A 20 μL aliquot of each sample was then placed in triplicate wells of a 96-well plate with 200 μL of Bradford Reagent (obtained from Sigma-Aldrich, St Louis, MO, USA). Standards were prepared using bovine serum albumin (Sigma-Aldrich, St Louis, MO, USA) and diluted into 0.2 M NaOH at concentrations of 0–300 mg·L^{-1}. Samples outside this concentration range were diluted appropriately in distilled water and the analysis was redone.

Reducing sugars in the supernatant were determined by the Anthrone method adapted from a previous protocol [52]. Briefly, 0.5 mL of supernatant was added to glass reaction vials with sealed caps. Then 1 mL of 0.1% anthrone in concentrated H_2SO_4 was added to the vial (using filter tips) and sealed. Samples were placed in a water bath at 80 °C for 5 min, and then allowed to cool to room temperature. 200 μL of each sample was pipetted (again using filter tips) into triplicate wells in a 96-well plate and the color change (green-blue) was quantified spectrophotometrically at 620 nm. Standards were prepared using glucose at concentrations of 0–100 mg·L^{-1}.

DNA in the supernatant was quantified using the Qubit Fluorometer. A 5 to 20 μL volume of sample was diluted into a 200 μL total volume of the working solution and sample (working solution was the fluorescence dye diluted 1:200 in buffer). If further dilutions were required, the samples were diluted in distilled water. The broad range DNA standards were used, which had a concentration range of 0 to 5 ng·μL^{-1}.

Volatile solids in the supernatant were quantified using 50 mL crucibles. A known volume of supernatant was placed in pre-weighed crucibles that were kept in a desiccator. The crucibles were then oven-dried at 105 °C for 24 h and weighed again following an equilibration period in the desiccator, and then placed at 550 °C for at least 2 h. The final mass of the crucible was then measured following cooling and equilibration in a desiccator.

3. Results and Discussion

3.1. Growth and mcl-PHA Synthesis

The $[X_t]$, $\%_{PHA}$, and resulting PHA biomass ($[X_{PHA}]$, expressed in g·L^{-1}) are shown in Figure 1 for both bioreactor systems. The initially high mcl-PHA content at time zero is due to carry-over from the inoculum, which was grown in flasks for a sufficiently long period so as to induce oxygen limitation and mcl-PHA synthesis from octanoic acid [45]. At both scales, the onset of oxygen limitation occurred around 12–14 h post inoculation and caused carbon flux to shift from growth to mcl-PHA synthesis. Overall, growth and total biomass production were similar at both scales, although the final $\%_{PHA}$ in the pilot-scale bioreactor was slightly lower than at bench-scale.

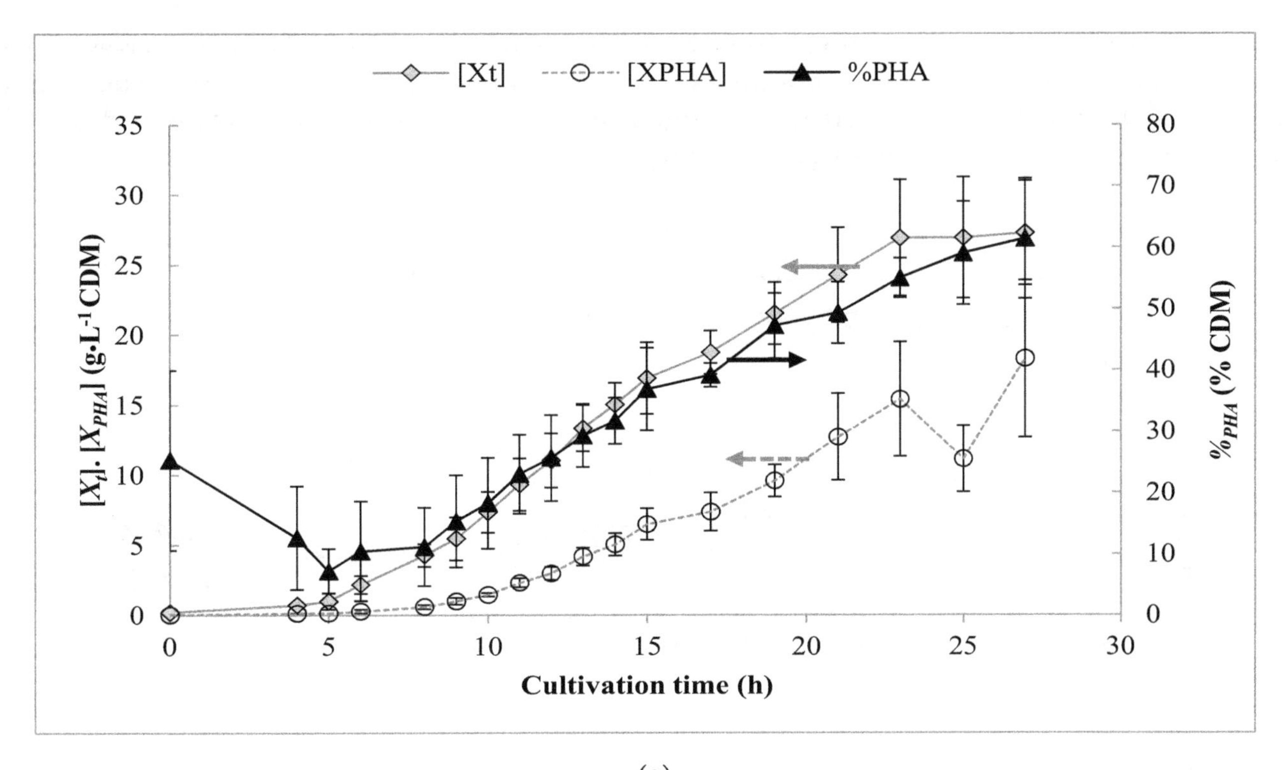

(**a**)

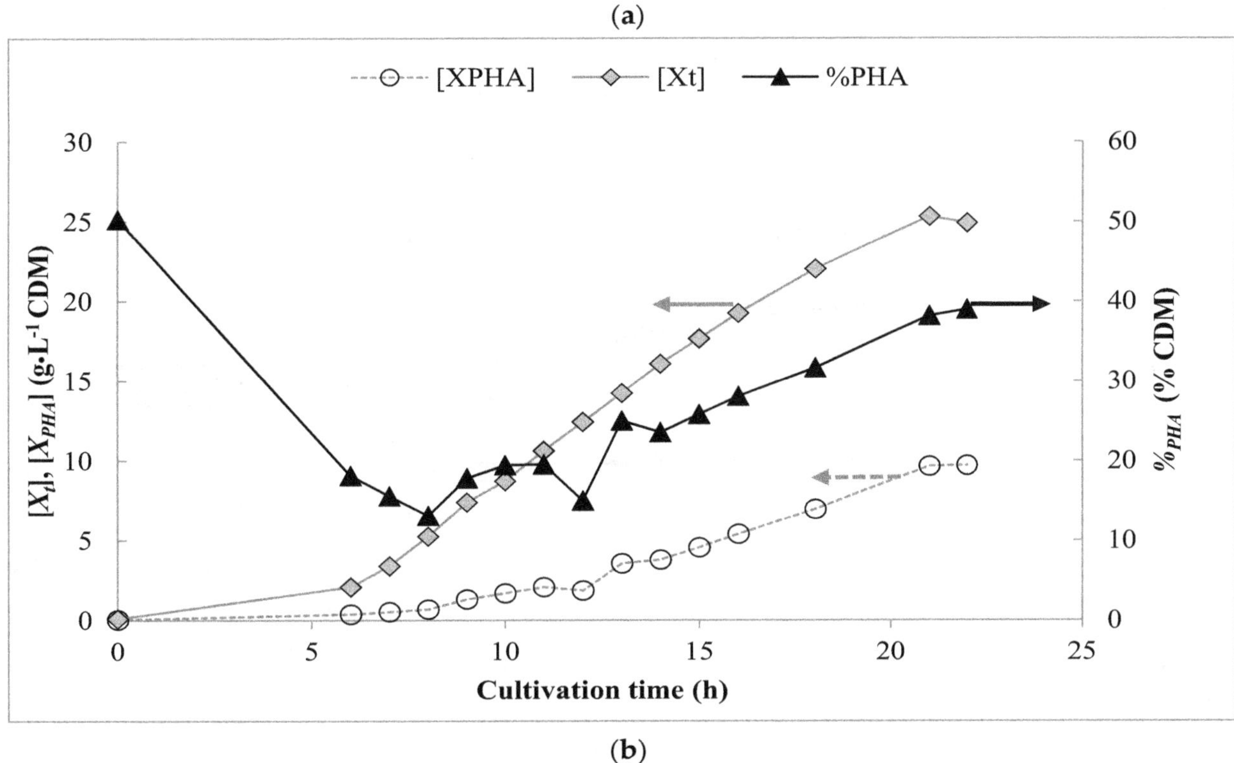

(**b**)

Figure 1. Results for biomass and PHA production obtained over the course of the pulse-feed fed-batch experiments (**a**) bench-scale bioreactor system (3 L initial working volume) and (**b**) pilot-scale bioreactor system (70 L initial working volume).

3.2. Rheological Characterization of the Cultivation Medium

Over the time course of the cultivation, the medium (even after centrifugation) became increasingly opaque and viscous. This appeared to significantly dampen the turbulence created for a given stirring input. The flow sweep curves for samples obtained from the pilot-scale system show viscosity as a function of shear rate for both cell suspensions (Figure 2a) and cell-free supernatants (Figure 2b) at various points in time in the bioreactor. In both cases, the medium appeared to behave as a Newtonian fluid up until 12 h for shear rates of approximately 75 s^{-1} or less. By 14 h (which corresponded to the onset of O_2 limitation in the bioreactor), a significant increase in viscosity was observed and the samples also became increasingly shear-thinning.

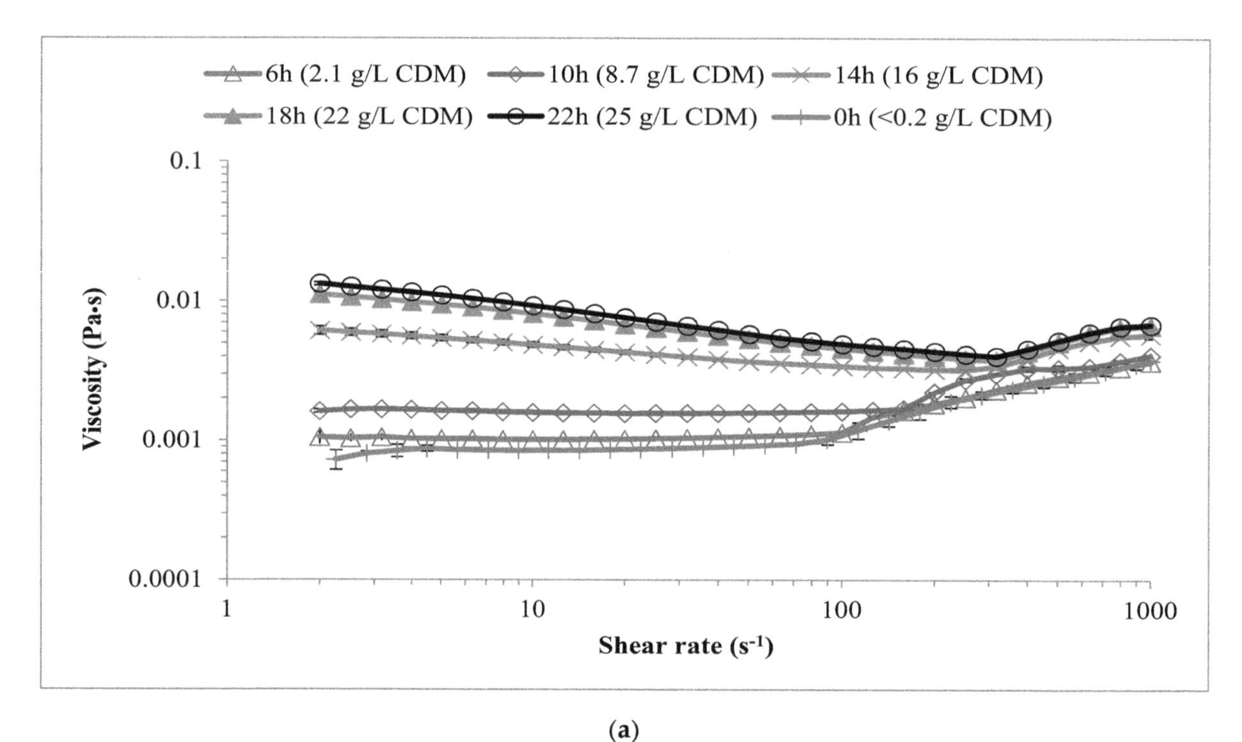

(a)

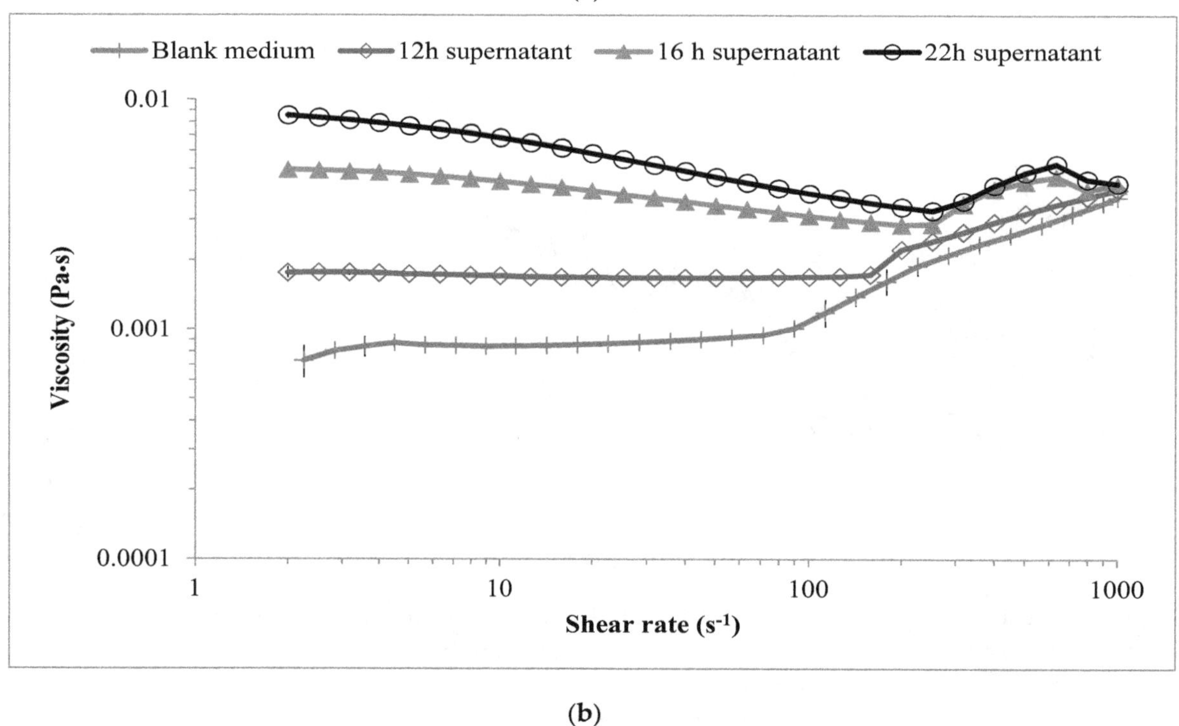

(b)

Figure 2. Flow sweep curves for samples of *P. putida* LS46 cultures obtained from the pilot-scale reactor. (**a**) Viscosity as a function of shear rate for 0–25 g·L^{-1} cell suspensions at various points in time and (**b**) viscosity as a function of shear rate for supernatant samples at various points in time. Error bars represent standard deviations of triplicate measurements for each sample.

This type of behavior has been described in previous studies of a variety of fermentation processes [34,35,53]. Similar rates were used in this study, but a unique attribute of this work is the viscosity of samples began to increase at a certain shear rate, which was consistent across the different measurements for the individual samples and increased with increasing sample viscosity (and hence cultivation time). This indicates a material property-related root cause rather than a measurement artefact.

In the initial (0 h) sample, this shear-thickening behavior was observed at shear rate of 76.5 s^{-1}, and increased with increasing sample viscosity up to nearly 300 s^{-1} in the final sample (25.3 g·L^{-1} CDM). This is an interesting observation because similar shear rates can easily be encountered in a bioreactor. This could suggest that increasing the shearing rate (bioreactor agitation rate) beyond this shear-thickening onset may actually cause a viscosity increase and reduce the oxygen transfer rate since viscosity is generally inversely proportional [41].

The relationship between shear-stress and shear rate was best described using the Power law ($R^2 > 0.99$ for all samples). A summary of model parameters for fitting the data from each sample with the power law is shown in Table 1, and in Figure 3 as a function of the corresponding total biomass of the sample. As shown, a strong linear relationship between the viscosity constant and the total biomass could be derived ($R^2 = 0.96$), and the slope was significantly different than zero ($p = 0.028$). However, the rate index did not seem to correlate with biomass in the culture ($R^2 = 0.63$), and the slope was not significantly different than zero ($p = 0.27$).

Table 1. Summary of model parameters ($n = 3$) for fitting the obtained data from each sample with the power law.

Sample	Power Law Constant: Viscosity (mPa·s)	Power Law Constant: Rate Index	Power Law: Regression
	Mean ± St. Dev.	Mean ± St. Dev.	Mean ± St. Dev.
0 h (<0.2 g·L^{-1})	0.16 ± 0.02	1.46 ± 0.02	1.00 ± 0.00
6 h (2.1 g·L^{-1})	0.22 ± 0.01	1.41 ± 0.01	1.00 ± 0.00
10 h (8.7 g·L^{-1})	0.45 ± 0.06	1.32 ± 0.02	1.00 ± 0.00
14 h (16.1 g·L^{-1})	0.55 ± 0.15	1.35 ± 0.04	1.00 ± 0.00
18 h (22 g·L^{-1})	0.62 ± 0.18	1.35 ± 0.05	1.00 ± 0.00
22 h (25.3 g·L^{-1})	0.80 ± 0.17	1.31 ± 0.03	0.99 ± 0.00
12 h supernatant	0.32 ± 0.02	1.37 ± 0.01	1.00 ± 0.00
16 h supernatant	1.91 ± 0.03	1.12 ± 0.00	0.99 ± 0.02
22 h supernatant	3.15 ± 0.13	1.05 ± 0.01	0.99 ± 0.00

The increase in viscosity of the culture as s function of $[X_t]$ is shown in Figure 3a. The viscosity at a shear rate of 10 s^{-1} was determined and compared from the flow sweep results to quantify the viscosity of the samples. Using a Tukey's range test, statistically significant (95%) differences in the rotational viscosities and shear-thickening onset across the different samples could be identified. These values are shown in Table 2, where it can be seen that over time, the apparent viscosity of the cell suspension (at a shear rate of 10 s^{-1}) increased from approximately 1.0 mPa·s to 9.2 mPa·s by 22 h. Interestingly, when the cells were removed by centrifugation, it was found that the viscosity of the 22 h supernatant was 6.8 mPa·s, which is nearly 75% of the value observed for the entire culture at 22 h (i.e., with $[X_t] = 25.3$ g·L^{-1}) suspended in that same matrix). From observations during the course of this work, when cells from 25.3 g·L^{-1} culture of *P. putida* LS46 (22 h) were re-suspended in fresh medium, the viscosity dropped to 1.7 mPa·s at shear rates of 10 s^{-1}, which was only slightly higher than the 0 h sample (1.01 mPa·s).

Table 2. Summary of average viscosity at a shearing rate of 10 rad·s^{-1} and onset of shear-thickening (n = 3).

Sample	Viscosity @ 10 s^{-1}, mPa·s	Shear Thickening Onset, s^{-1}
	Mean ± St.Dev	Mean ± St.Dev
0 h (<0.2 g·L^{-1})	1.01 ± 0.06	76.5 ± 13.0
6 h (2.1 g·L^{-1})	1.01 ± 0.02	94.8 ± 0.6
10 h (8.7 g·L^{-1})	1.59 ± 0.01	151.1 ± 0.6
14 h (16.1 g·L^{-1})	4.86 ± 0.21	258.9 ± 18.1
18 h (22 g·L^{-1})	8.10 ± 0.15	294.9 ± 9.1
22 h (25.3 g·L^{-1})	9.22 ± 0.11	293.3 ± 12.5
12 h supernatant	1.71 ± 0.01	156.3 ± 2.2
16 h supernatant	4.42 ± 0.07	218.9 ± 2.6
22 h supernatant	6.81 ± 0.09	256.6 ± 8.3

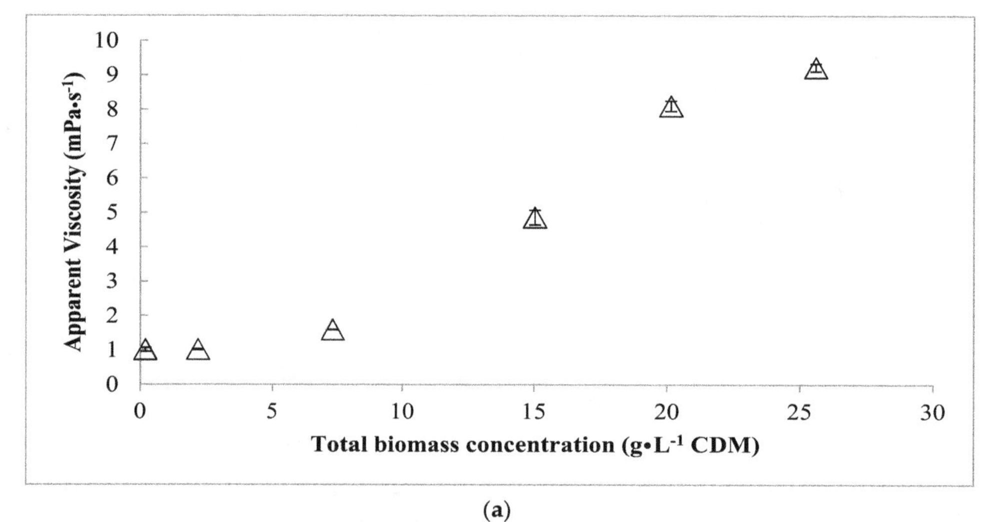

(a)

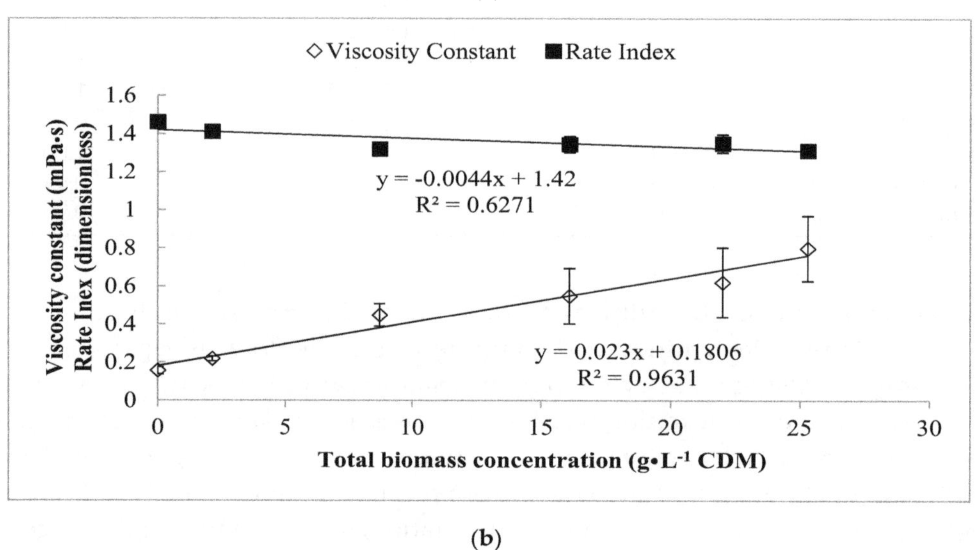

(b)

Figure 3. (**a**) Changes in apparent viscosity of the *P. putida* LS46 culture with increasing total biomass concentration over time and (**b**) changes in power-law constants describing culture rheology as a function of the total biomass in the (pilot-scale) fed batch cultivation at varying points over time. Error bars represent the standard deviations between technical replicate measurements (n = 3).

Typically, electrolyte solutions like microbial growth medium are slightly shear thinning [54]. In colloidal dispersions, shear-thinning is thought to be due to a more organized flow pattern of the molecules when subject to shear forces. This creates less stochastic (random) interactions, and results in reduced viscosity and decreased energy dissipation [55,56]. At higher shear rates, however, hydrodynamic forces can dominate over stochastic interactions, and the particle collisions are primarily due to shear forces rather than random thermal motions. This causes organization of the molecules into a more anisotropic state of so-called 'hydroclusters', and increases the difficulty by which molecules can flow around one another [55]. Although other theories exist (including order-disorder transition and dilatancy), this is perhaps the most commonly accepted mechanism for shear-thickening [56]. At the molecular level, the mechanism remains the subject of some debate, all theories essentially pertain to increased difficulty with particle-particle interactions in a flow path, and thus the volume fraction of particles is of importance [56,57]. The presence of high M_w polymers (particularly when suspended in a poor solvent), could further support the shear-thickening observations in this work [58]. In such situations, higher shear rates tend to cause high M_w macromolecules to extend in the flow path, breaking their intra-molecular associations and forming inter-molecular associations. This results in a gel network formation, which increases viscosity [59]. This is also a positive feedback mechanism in which the molecules of higher M_w extend first, and formation of gel networks causes the viscosity to increase. This, in turn, increases the shear stress, which then affects the molecules of lower M_w [59]. The intermolecular associations may include crosslinking, which is a known phenomenon with mcl-PHA [60–63].

3.3. *Quantifying Components of the Extracellular Matrix*

These data indicate that significant rheological changes to the culture medium occur over time, and much of this effect is not simply explained by the presence of cells. According to Newton et al. [53], in HCD *E. coli* cultures this behavior is the result of structural interactions between cells and cellular debris (high M_w nucleic acids, which can also from crosslinks) resulting from lysed cells. This could further contribute to a shear thickening effect. The following section describes the soluble organic material detected in the culture supernatant.

P. putida and other *Pseudomonas* spp. are known to produce significant quantities of extracellular polymeric substances (EPS) as precursors to biofilm formation. This is a particularly well-known phenomena with *P. aeruginosa*. Kachlany et al. [64] suggested that young *P. putida* G7 cells are encapsulated by an exopolysaccharide layer that is sloughed off as cells age. It is likely that high shear forces expedite the sloughing of this capsular material. That study also described the collapsed extracellular polymer from *P. putida* G7 as being a 'rope-like' material, which could certainly fit the proposed gel-formation theory for shear thickening behavior in polymer solutions.

Generally, the extracellular polymers associated with *Pseudomonas* spp. are composed predominantly of sugars, typically glucose, galactose, rhamnose and mannose [30,32,64–66]. Other extracellular secretions associated with *Pseudomonas* spp. include alginate [67], DNA [68–70], gellan [71], proteins [72], glycolipids and lipopolysaccharides [64,73,74], organic acids, [30,75], as well as acetylated sugars and uronic acids [65,66].

In this work, several of these putative EPS constituents and/or cell lysis products were monitored and quantified in the supernatants of cultivations performed at bench-scale to better understand the observed rheological behavior. These include proteins, reducing sugars, DNA, and extracellular PHA, as well as bulk measurement of carbonaceous products in the supernatant by volatile solids. These are shown over time in Figure 4.

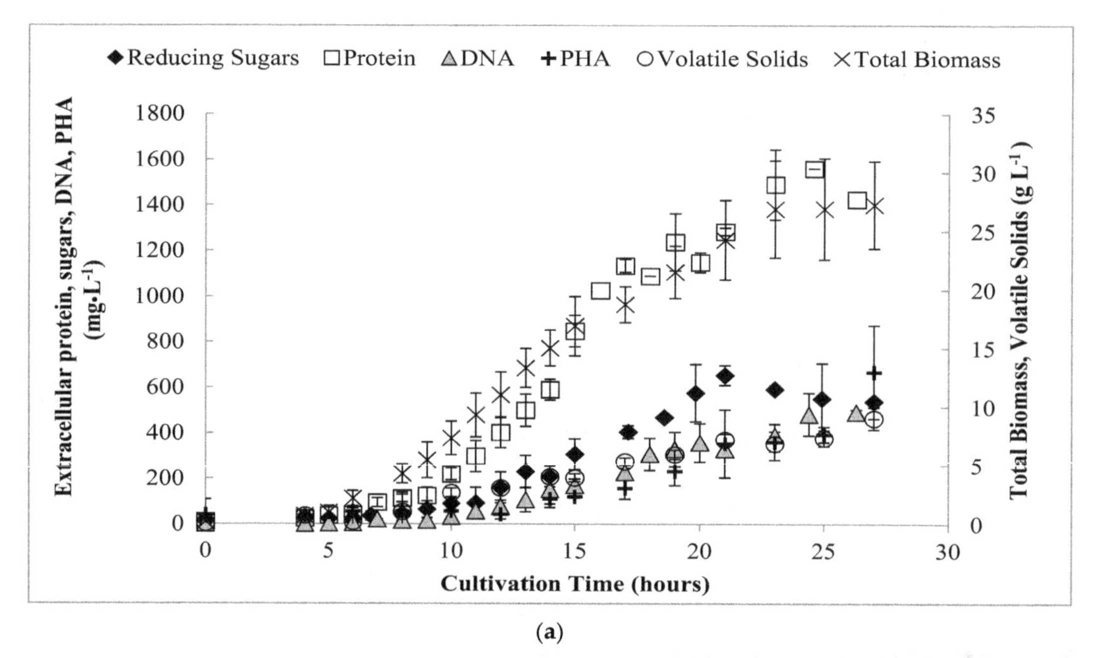

(**a**)

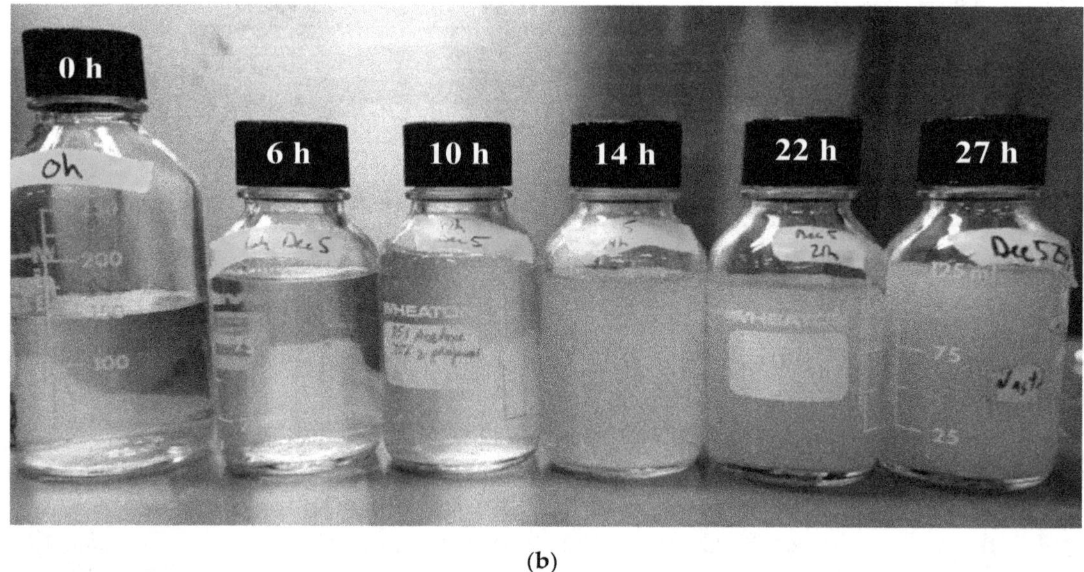

(**b**)

Figure 4. (**a**) Soluble (extracellular) organic material detected in the supernatant over the time course of the bench-scale cultivations, which is thought to contribute to the observed rheological behavior of the medium and (**b**) appearance of culture after centrifugation for fed-batch experiments at the cultivation time indicated. Error bars represent standard deviations between the mean values obtained from each of the biological replicate experiments.

In general, the concentrations of these components increased proportionally to $[X_t]$ with the exception of reducing sugars, which reached a maximum concentration of 0.65 g·L^{-1} at 21 h and then declined. The maximum concentrations (at 27 h) of proteins, PHA, and DNA in the medium were 1.56 g·L^{-1}, 0.67 g·L^{-1}, and 0.49 g·L^{-1}, respectively. Collectively, the components could account for at most 48% of the total VS detected in the supernatant, which reached a maximum of 9 g·L^{-1} by 27 h. Newton et al. [35] demonstrated that both protein (0–50 g·L^{-1}) and DNA (0–4 g·L^{-1}) contributed linearly to increased viscosity and the flow curves for solutions of protein and DNA exhibited shear thinning and Newtonian behavior, respectively. However, in that work a somewhat lesser increase in viscosity was noted (1.1 to 5.3 mPa·s) for a 48 g·L^{-1} *E. coli* culture, despite the presence of far more extracellular DNA (typically 3 g·L^{-1}) and protein (up to 40 g·L^{-1}).

In this work, we did not attempt to differentiate whether these components of the supernatant were due to the production of EPS or are simply cell lysis products from cultivation in a high-shear environment. We found evidence to make an argument in favor of either scenario, which likely implies the rheological behavior is due to a combination of physical and biochemical factors. Mg^{2+} is an intracellular metabolite that can leak from damaged cell membranes, which would precede cell lysis [76]. In this work, the Mg^{2+} concentration monitored in the supernatant began to increase after 14 h, which corresponds to the onset of O_2 limitation and maximum agitation rates. However, when the ratios of protein-to-DNA of the culture supernatant was compared to that of *P. putida* cell lysate, it was found that the cell lysate contained only about half the DNA fraction that was observed in the supernatant. This could suggest DNA release as a possible EPS component, and would be supported by previous studies using *Pseudomonas* spp. [68,72]. Furthermore, a similar experiment using lower shear rates using a maximum of 600 rpm mixing (compared to 1200 rpm) in the bench-scale bioreactor, but with pure O_2 to increase driving force for oxygen transfer. Although this method produced lower $[X_t]$, statistically indifferent ($p < 0.05$) yields of extracellular organic content (protein, sugars, DNA per unit $[X_t]$) were observed in comparison with the normal mixing condition of 1200 rpm. While this does not disprove the occurrence of significant cell lysis, it does show that this behavior is difficult to avoid, even at comparatively low bioreactor mixing rates.

3.4. *Engineering Significance: Effects on Oxygen Transfer Rate*

Knowledge of viscosity in bioprocesses is important for process scale up. Many empirical relationships (or dimensionless parameters like the Reynold's number, *Re*) describing the K_La, are inversely proportional to viscosity [50]. The K_La measured using the 200 mL bioreactor system with different supernatant samples obtained from the bench-scale bioreactor over time were assessed. The reduction in K_La over time is shown in Figure 5 as a function of the increasing amount of soluble organic material in the culture supernatant. As shown, K_La is expressed as a percent of the value measured at 0 h. The final values (obtained at 27 h) showed a significant ($p < 0.05$) reduction of 45–52% from the values measured at 0 h or 6 h. It was intended to perform a similar test using the entire culture, but that was not possible due to the high oxygen demand of the culture preventing observable changes in DO as well as excessive foaming when air was bubbled through the cell suspensions in the miniature reactor.

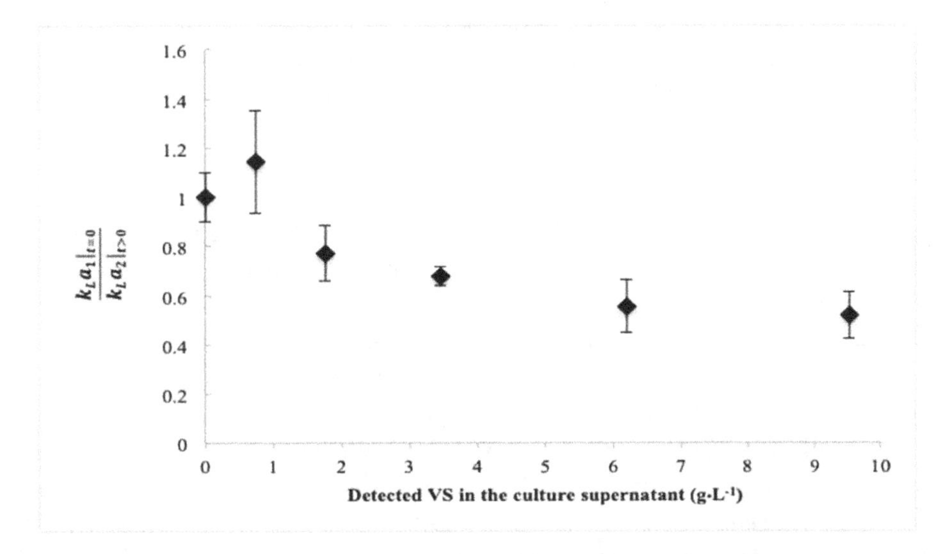

Figure 5. Shows the reduction in K_La over time (expressed as a fraction of the value measured at time zero) that might be anticipated from the increasing VS detected in the supernatant. The K_La values were measured in the described 200 mL reactor. Error bars represent standard deviation of the mean values determined for each biological replicate experiments.

According to Martin et al. [41], the reduction in oxygen transfer rates with increased fluid viscosity is due to: (1) reduced contact area between bubbles and a fluid because bubbles are more stable in viscous fluids; and (2) decreased liquid diffusivity due to reduced velocity profile in the liquid layer surrounding the bubble. Interestingly, the extracellular biosurfactants (glycolipids, lipopolysaccharides) known to be secreted by several *Pseudomonas* spp. can form a layer at the liquid–gas interface and decrease oxygen mass transfer [36].

Considering that other fed-batch strategies have achieved 100 $g{\cdot}L^{-1}$ CDM or more in PHA production [11,13,77,78], the reduction in K_La over time could be considerably more significant in those systems compared to that described in this work. This may often be neglected in bench-scale HCD PHA cultivations in which stirring and aeration are typically set as high as is realistically possible to maximize productivity. However, with increasing scale the required power consumption for mixing and aeration becomes significant [79], and so the bioreactor operation parameters must be carefully managed in order to save aeration costs. This aspect is emphasized in PHA production/waste treatment operations using enriched mixed cultures [80].

A characteristic of shear-thickening fluids is effective energy dissipation [56]. While this may be useful or interesting property for certain applications, for bioprocessing it is generally problematic, and would likely result in poor performance for the energy input to the bioreactor. The Reynold's number (*Re*) is often used as an dimensionless constant used to estimate power consumption required to mix the reactor contents, which can be a significant cost for aerobic processes like PHA production [79]. Using the obtained constants shown in Table 2, we estimated that for a given (un-gassed) power input to the stirrer, *Re* decreased by 44–55% as the cell density of the culture increased from approximately 0 to 25 $g{\cdot}L^{-1}$ following the approach of Gabelle et al. [17].

Shear thickening behaviour, although interesting, is problematic for the cultivation process as well as downstream operations, including pumping, centrifugation, filtration, or spray-drying [55,56]. In our production process using *P. putida* LS46, it certainly appears to be a difficult situation to avoid. Moving forward, efforts to alleviate such conditions might include: 1) investigation of lower-shear bioreactors such as air-lift configurations [81]; 2) modifying the medium with surface active molecules or flocculants to reduce or reverse shear-thickening [82]; 3) addition of extracellular enzymes to break up large macromolecules that may form gel networks and contribute to increased viscosity and shear-thickening; or 4) or engineering the bacterium to, or selection of strains that, avoid production of EPS. The latter could also help close the carbon balance and improve the overall PHA yield.

4. Conclusions

In moderately HCD fed-batch cultivations of *P. putida* LS46, significant changes in the rheological properties of the culture were observed. At lower shear rates the culture exhibited slight shear-thinning behavior, while the onset of shear-thickening was observed at shear rates that increased with sample viscosity (or increased $[X_t]$). A nearly nine-fold increase in viscosity (at 10 s^{-1}) was measured throughout the course of the cultivation process, approximately 75% of which was attributed to the supernatant rather than the presence of cells. Investigation of the culture supernatants revealed up to 9 $g{\cdot}L^{-1}$ VS being present in the supernatant, of which half was accounted for as extracellular proteins, sugars, DNA, and PHA. It was shown that this material could reduce the mass transfer coefficient associated with a given bioreactor system by up to 50% over the course of the cultivation process. Although difficulties in maintaining oxygen transfer are well known in HCD aerobic bioprocesses, this work has demonstrated that biochemically-induced changes in the medium composition played a significant role, rather than just the high oxygen demand associated with a HCD culture of strictly aerobic organisms.

Author Contributions: Individual contributions of authors are as follows: Design and conceptualization: W.B.; Methodology development: W.B. and C.C.; Acquisition of data: W.B. and M.G.; Formal analysis: W.B. and M.G.; Writing: W.B.; Review and editing: D.B.L., C.C., R.S., D.J.G., and N.C.; Supervision: C.C., R.S., D.J.G., D.B.L., and N.C.; Final approval: N.C.; Funding Acquisition: D.B.L. (GAPP and NSERC Discovery grants) and W.B. (NSERC CGS-D, NSERC-MSFSS, Sir Gordon Wu Scholarship, and Edward R. Toporeck fellowship).

Acknowledgments: Special thanks my colleagues and friends on the Biotransformation and Clean Technologies teams at Scion—including Tracey Bowers, Mark West, Kim McGrouther, Wajid Waheed Bhat, and Martin Cooke-Willis—for their support in ensuring a safe and positive experience in learning and operating the pilot-scale SIP bioreactors. At the University of Manitoba, thanks to Tanner Devlin in the Civil Engineering Department for his help constructing the bioreactor for K_La determinations. Thanks also to Scott Wushke (Department of Microbiology) for being interested in this project and providing helpful insights and constructive feedback.

References

1. Gasperi, J.; Wright, S.L.; Dris, R.; Collard, F.; Mandin, C.; Guerrouache, M.; Langlois, V.; Kelly, F.J.; Tassin, B. Microplastics in air: Are we breathing it in? *Curr. Opin. Environ. Sci. Health* **2018**, *1*, 1–5. [CrossRef]
2. Hale, R.C. Are the Risks from Microplastics Truly Trivial? *Environ. Sci. Technol.* **2018**, *52*, 931. [CrossRef]
3. Harding, K.; Dennis, J.; Vonblottnitz, H.; Harrison, S. Environmental analysis of plastic production processes: Comparing petroleum-based polypropylene and polyethylene with biologically-based poly-β-hydroxybutyric acid using life cycle analysis. *J. Biotechnol.* **2007**, *130*, 57–66. [CrossRef]
4. Masood, F.; Yasin, T.; Hameed, A. Polyhydroxyalkanoates—What are the uses? Current challenges and perspectives. *Crit. Rev. Biotechnol.* **2015**, *35*, 514–521. [CrossRef]
5. Koller, M. Biodegradable and Biocompatible Polyhydroxyalkanoates (PHA): Auspicious Microbial Macromolecules for Pharmaceutical and Therapeutic Applications. *Molecules* **2018**, *23*, 362. [CrossRef]
6. Kourmentza, C.; Plácido, J.; Venetsaneas, N.; Burniol-Figols, A.; Varrone, C.; Gavala, H.N.; Reis, M.A.M. Recent advances and challenges towards sustainable polyhydroxyalkanoate (PHA) production. *Bioengineering* **2017**, *4*, 55. [CrossRef]
7. Możejko-Ciesielska, J.; Kiewisz, R. Bacterial polyhydroxyalkanoates: Still fabulous? *Microbiol. Res.* **2016**, *192*, 271–282. [CrossRef]
8. Kaur, G.; Roy, I. Strategies for large-scale production of polyhydroxyalkanoates. *Chem. Biochem. Eng. Q.* **2015**, *29*, 157–172. [CrossRef]
9. Koller, M.; Maršálek, L.; de Sousa Dias, M.M.; Braunegg, G. Producing microbial polyhydroxyalkanoate (PHA) biopolyesters in a sustainable manner. *New Biotechnol.* **2017**, *37*, 24–38. [CrossRef]
10. Ienczak, J.L.; Schmidell, W.; de Aragão, G.M.F. High-cell-density culture strategies for polyhydroxyalkanoate production: A review. *J. Ind. Microbiol. Biotechnol.* **2013**, *40*, 275–286. [CrossRef]
11. Ryu, H.W.; Hahn, S.K.; Chang, Y.K.; Chang, H.N. Production of poly(3-hydroxybutyrate) by high cell density fed-batch culture of *Alcaligenes eutrophus* with phospate limitation. *Biotechnol. Bioeng.* **1997**, *55*, 28–32. [CrossRef]
12. Shang, L.; Jiang, M.; Chang, H.N. Poly(3-hydroxybutyrate) synthesis in fed-batch culture of *Ralstonia eutropha* with phosphate limitation under different glucose concentrations. *Biotechnol. Lett.* **2003**, *25*, 1415–1419. [CrossRef]
13. Wang, F.; Lee, S.Y. Poly(3-hydroxybutyrate) production with high productivity and high polymer content by a fed-batch culture of *Alcaligenes latus* under nitrogen limitation. *Appl. Environ. Microbiol.* **1997**, *63*, 3703–3706.
14. Koller, M. A Review on Established and Emerging Fermentation Schemes for Microbial Production of Polyhydroxyalkanoate (PHA) Biopolyesters. *Fermentation* **2018**, *4*, 30. [CrossRef]
15. Blunt, W.; Levin, D.; Cicek, N. Bioreactor Operating Strategies for Improved Polyhydroxyalkanoate (PHA) Productivity. *Polymers* **2018**, *10*, 1197. [CrossRef]
16. Riesenberg, D.; Guthke, R. High-cell-density cultivation of microorganisms. *Appl. Microbiol. Biotechnol.* **1999**, *51*, 422–430. [CrossRef]
17. Gabelle, J.C.; Augier, F.; Carvalho, A.; Rousset, R.; Morchain, J. Effect of tank size on kLa and mixing time in aerated stirred reactors with non-newtonian fluids. *Can. J. Chem. Eng.* **2011**, *89*, 1139–1153. [CrossRef]

18. Rottava, I.; Batesini, G.; Silva, M.F.; Lerin, L.; de Oliveira, D.; Padilha, F.F.; Toniazzo, G.; Mossi, A.; Cansian, R.L.; Di Luccio, M.; et al. Xanthan gum production and rheological behavior using different strains of Xanthomonas sp. *Carbohydr. Polym.* **2009**, *77*, 65–71. [CrossRef]
19. Badino, A.C.; Facciotti, M.C.R.; Schmidell, W. Volumetric oxygen transfer coefficients (kLa) in batch cultivations involving non-Newtonian broths. *Biochem. Eng. J.* **2001**, *8*, 111–119. [CrossRef]
20. Sinha, J.; Tae Bae, J.; Pil Park, J.; Hyun Song, C.; Won Yun, J. Effect of substrate concentration on broth rheology and fungal morphology during exo-biopolymer production by *Paecilomyces japonica* in a batch bioreactor. *Enzym. Microb. Technol.* **2001**, *29*, 392–399. [CrossRef]
21. Pedersen, A.G.; Bundgaard-Nielsen, M.; Nielsen, J.; Villadsen, J.; Hassager, O. Rheological characterization of media containing *Penicillium chrysogenum*. *Biotechnol. Bioeng.* **1993**, *41*, 162–164. [CrossRef]
22. Riley, G.L.; Tucker, K.G.; Paul, G.C.; Thomas, C.R. Effect of biomass concentration and mycelial morphology on fermentation broth rheology. *Biotechnol. Bioeng.* **2000**, *68*, 160–172. [CrossRef]
23. Dhillon, G.S.; Brar, S.K.; Kaur, S.; Verma, M. Rheological studies during submerged citric acid fermentation by *Aspergillus niger* in stirred fermentor using apple pomace ultrafiltration sludge. *Food Bioprocess Technol.* **2013**, *6*, 1240–1250. [CrossRef]
24. Rodríguez Porcel, E.M.; Casas López, J.L.; Sánchez Pérez, J.A.; Fernández Sevilla, J.M.; García Sánchez, J.L.; Chisti, Y. *Aspergillus terreus* Broth Rheology, Oxygen Transfer, and Lovastatin Production in a Gas-Agitated Slurry Reactor. *Ind. Eng. Chem. Res.* **2006**, *45*, 4837–4843. [CrossRef]
25. Richard, A.; Margaritis, A. Rheology, oxygen transfer, and molecular weight characteristics of poly(glutamic acid) fermentation by *Bacillus subtilis*. *Biotechnol. Bioeng.* **2003**, *82*, 299–305. [CrossRef]
26. Baroutian, S.; Eshtiaghi, N.; Gapes, D.J. Rheology of a primary and secondary sewage sludge mixture: Dependency on temperature and solid concentration. *Bioresour. Technol.* **2013**, *140*, 227–233. [CrossRef]
27. Brar, S.K.; Verma, M.; Tyagi, R.D.; Valéro, J.R.; Surampalli, R.Y. *Bacillus thuringiensis* fermentation of hydrolyzed sludge – Rheology and formulation studies. *Chemosphere* **2007**, *67*, 674–683. [CrossRef]
28. Verma, M.; Brar, S.K.; Tyagi, R.D.; Sahai, V.; Prévost, D.; Valéro, J.R.; Surampalli, R.Y. Bench-scale fermentation of Trichoderma viride on wastewater sludge: Rheology, lytic enzymes and biocontrol activity. *Enzym. Microb. Technol.* **2007**, *41*, 764–771. [CrossRef]
29. Sun, F.; Huang, Q.; Wu, J. Rheological behaviors of an exopolysaccharide from fermentation medium of a *Cordyceps sinensis* fungus (Cs-HK1). *Carbohydr. Polym.* **2014**, *114*, 506–513. [CrossRef]
30. Freitas, F.; Alves, V.D.; Pais, J.; Costa, N.; Oliveira, C.; Mafra, L.; Hilliou, L.; Oliveira, R.; Reis, M.A.M. Characterization of an extracellular polysaccharide produced by a *Pseudomonas* strain grown on glycerol. *Bioresour. Technol.* **2009**, *100*, 859–865. [CrossRef]
31. Freitas, F.; Alves, V.D.; Carvalheira, M.; Costa, N.; Oliveira, R.; Reis, M.A.M. Emulsifying behaviour and rheological properties of the extracellular polysaccharide produced by Pseudomonas oleovorans grown on glycerol byproduct. *Carbohydr. Polym.* **2009**, *78*, 549–556. [CrossRef]
32. Maalej, H.; Hmidet, N.; Boisset, C.; Bayma, E.; Heyraud, A.; Nasri, M. Rheological and emulsifying properties of a gel-like exopolysaccharide produced by *Pseudomonas stutzeri* AS22. *Food Hydrocoll.* **2016**, *52*, 634–647. [CrossRef]
33. Maalej, H.; Moalla, D.; Boisset, C.; Bardaa, S.; Ayed, H.B.; Sahnoun, Z.; Rebai, T.; Nasri, M.; Hmidet, N. Rhelogical, dermal wound healing and in vitro antioxidant properties of exopolysaccharide hydrogel from Pseudomonas stutzeri AS22. *Colloids Surf. B Biointerfaces* **2014**, *123*, 814–824. [CrossRef]
34. Goudar, C.T.; Strevett, K.A.; Shah, S.N. Influence of microbial concentration on the rheology of non-Newtonian fermentation broths. *Appl. Microbiol. Biotechnol.* **1999**, *51*, 310–315. [CrossRef]
35. Newton, J.M.; Vlahopoulou, J.; Zhou, Y. Investigating and modelling the effects of cell lysis on the rheological properties of fermentation broths. *Biochem. Eng. J.* **2017**, *121*, 38–48. [CrossRef]
36. Papapostolou, A.; Karasavvas, E.; Chatzidoukas, C. Oxygen mass transfer limitations set the performance boundaries of microbial PHA production processes—A model-based problem investigation supporting scale-up studies. *Biochem. Eng. J.* **2019**, *148*, 224–238. [CrossRef]
37. Arjunwadkar, S.J.; Sarvanan, K.; Kulkarni, P.R.; Pandit, A.B. Gas-liquid mass transfer in dual impeller bioreactor. *Biochem. Eng. J.* **1998**, *1*, 99–106. [CrossRef]
38. Buchholz, H.; Buchholz, R.; Niebeschütz, H.; Schügerl, K. Absorption of oxygen in highly viscous newtonian

and non-Newtonian fermentation model media in bubble column bioreactors. *Eur. J. Appl. Microbiol. Biotechnol.* **1978**, *6*, 115–126. [CrossRef]
39. García-Ochoa, F.; Gómez, E. Mass transfer coefficient in stirred tank reactors for xanthan gum solutions. *Biochem. Eng. J.* **1998**, *1*, 1–10. [CrossRef]
40. Herbst, H.; Schumpe, A.; Deckwer, W.D. Xanthan production in stirred tank fermenters: Oxygen transfer and scale-up. *Chem. Eng. Technol.* **1992**, *15*, 425–434. [CrossRef]
41. Martín, M.; Montes, F.J.; Galán, M.A. Mass transfer rates from bubbles in stirred tanks operating with viscous fluids. *Chem. Eng. Sci.* **2010**, *65*, 3814–3824. [CrossRef]
42. Puthli, M.S.; Rathod, V.K.; Pandit, A.B. Gas–liquid mass transfer studies with triple impeller system on a laboratory scale bioreactor. *Biochem. Eng. J.* **2005**, *23*, 25–30. [CrossRef]
43. Jiang, X.; Sun, Z.; Ramsay, J.A.; Ramsay, B.A. Fed-batch production of MCL-PHA with elevated 3-hydroxynonanoate content. *AMB Express* **2013**, *3*, 50. [CrossRef]
44. Sharma, P.K.; Fu, J.; Cicek, N.; Sparling, R.; Levin, D.B. Kinetics of medium-chain-length polyhydroxyalkanoate production by a novel isolate of *Pseudomonas putida* LS46. *Can. J. Microbiol.* **2012**, *58*, 982–989. [CrossRef]
45. Blunt, W.; Dartiailh, C.; Sparling, R.; Gapes, D.; Levin, D.B.; Cicek, N. Microaerophilic environments improve the productivity of medium chain length polyhydroxyalkanoate biosynthesis from fatty acids in *Pseudomonas putida* LS46. *Process Biochem.* **2017**, *59*, 18–25. [CrossRef]
46. Ramsay, B.A.; Lomaliza, K.; Chavarie, C.; Dubé, B.; Ramsay, J.A. Production of poly-(beta-hydroxybutyric-co-beta-hydroxyvaleric) acids. *Appl. Environ. Microbiol.* **1990**, *56*, 2093–2098.
47. Blunt, W.; Dartiailh, C.; Sparling, R.; Gapes, D.; Levin, D.B.; Cicek, N. Carbon flux to growth or polyhydroxyalkanoate synthesis under microaerophilic conditions is affected by fatty acid chain-length in *Pseudomonas putida* LS46. *Appl. Microbiol. Biotechnol.* **2018**, *102*, 6437–6449. [CrossRef]
48. Blunt, W.; Hossain, M.D.E.; Gapes, D.J.; Sparling, R.; Levin, D.B.; Cicek, N. Real-Time Monitoring of Microbial Fermentation End-Products in Biofuel Production with Titrimetric Off-Gas Analysis (TOGA). *Biol. Eng. Trans.* **2014**, *6*, 203–219.
49. Yang, H.; Wei, F.; Hu, K.; Zhou, G.; Lyu, J. Comparison of rheometric devices for measuring the rheological parameters of debris flow slurry. *J. Mt. Sci.* **2015**, *12*, 1125–1134. [CrossRef]
50. Garcia-Ochoa, F.; Gomez, E. Bioreactor scale-up and oxygen transfer rate in microbial processes: An overview. *Biotechnol. Adv.* **2009**, *27*, 153–176. [CrossRef]
51. Bradford, M.M. A rapid and sensitive method for the quantitation of microgram quantities of protein utilizing the principle of protein-dye binding. *Anal. Biochem.* **1976**, *72*, 248–254. [CrossRef]
52. Viles, F.J.; Silverman, L. Determination of Starch and Cellulose with Anthrone. *Anal. Chem.* **1949**, *21*, 950–953. [CrossRef]
53. Newton, J.M.; Schofield, D.; Vlahopoulou, J.; Zhou, Y. Detecting cell lysis using viscosity monitoring in *E. coli* fermentation to prevent product loss. *Biotechnol. Prog.* **2016**, *32*, 1069–1076. [CrossRef]
54. Doran, P.M. *Bioprocess Engineering Principles*, 2nd ed.; Elsevier/Academic Press: Amsterdam, The Netherlands; Boston, MA, USA, 2013; ISBN 978-0-12-220851-5.
55. Wagner, N.J.; Brady, J.F. Shear thickening in colloidal dispersions. *Phys. Today* **2009**, *62*, 27–32. [CrossRef]
56. Brown, E.; Jaeger, H.M. Shear thickening in concentrated suspensions: Phenomenology, mechanisms and relations to jamming. *Rep. Prog. Phys.* **2014**, *77*, 046602. [CrossRef]
57. Fernandez, N.; Mani, R.; Rinaldi, D.; Kadau, D.; Mosquet, M.; Lombois-Burger, H.; Cayer-Barrioz, J.; Herrmann, H.J.; Spencer, N.D.; Isa, L. Microscopic Mechanism for Shear Thickening of Non-Brownian Suspensions. *Phys. Rev. Lett.* **2013**, *111*, 108301. [CrossRef]
58. van Egmond, J.W. Shear-thickening in suspensions, associating polymers, worm-like micelles, and poor polymer solutions. *Curr. Opin. Colloid Interface Sci.* **1998**, *3*, 385–390. [CrossRef]
59. Ballard, M.J.; Buscall, R.; Waite, F.A. The theory of shear-thickening polymer solutions. *Polymer* **1988**, *29*, 1287–1293. [CrossRef]
60. Levine, A.C.; Sparano, A.; Twigg, F.F.; Numata, K.; Nomura, C.T. Influence of Cross-Linking on the Physical Properties and Cytotoxicity of Polyhydroxyalkanoate (PHA) Scaffolds for Tissue Engineering. *ACS Biomater. Sci. Eng.* **2015**, *1*, 567–576. [CrossRef]

61. Bassas, M.; Diaz, J.; Rodriguez, E.; Espuny, M.J.; Prieto, M.J.; Manresa, A. Microscopic examination in vivo and in vitro of natural and cross-linked polyunsaturated mclPHA. *Appl. Microbiol. Biotechnol.* **2008**, *78*, 587–596. [CrossRef]
62. Hazer, B.; Demirel, S.I.; Borcakli, M.; Eroglu, M.S.; Cakmak, M.; Erman, B. Free radical crosslinking of unsaturated bacterial polyesters obtained from soybean oily acids. *Polym. Bull.* **2001**, *46*, 389–394. [CrossRef]
63. Ashby, R.D.; Solaiman, D.K.Y.; Foglia, T.A.; Liu, C.K. Glucose/lipid mixed substrates as a means of controlling the properties of medium chain length poly(hydroxyalkanoates). *Biomacromolecules* **2001**, *2*, 211–216. [CrossRef]
64. Kachlany, S.C.; Levery, S.B.; Kim, J.S.; Reuhs, B.L.; Lion, L.W.; Ghiorse, W.C. Structure and carbohydrate analysis of the exopolysaccharide capsule of *Pseudomonas putida* G7. *Environ. Microbiol.* **2001**, *3*, 774–784. [CrossRef]
65. Celik, G.Y.; Aslim, B.; Beyatli, Y. Characterization and production of the exopolysaccharide (EPS) from *Pseudomonas aeruginosa* G1 and *Pseudomonas putida* G12 strains. *Carbohydr. Polym.* **2008**, *73*, 178–182. [CrossRef]
66. Wrangstadh, M.; Conway, P.L.; Kjelleberg, S. The production and release of an extracellular polysaccharide during starvation of a marine *Pseudomonas* sp. and the effect thereof on adhesion. *Arch. Microbiol.* **1986**, *145*, 220–227. [CrossRef]
67. Chang, W.S.; van de Mortel, M.; Nielsen, L.; Nino de Guzman, G.; Li, X.; Halverson, L.J. Alginate production by *Pseudomonas putida* creates a hydrated microenvironment and contributes to biofilm architecture and stress tolerance under water-limiting conditions. *J. Bacteriol.* **2007**, *189*, 8290–8299. [CrossRef]
68. Allesen-Holm, M.; Barken, K.B.; Yang, L.; Klausen, M.; Webb, J.S.; Kjelleberg, S.; Molin, S.; Givskov, M.; Tolker-Nielsen, T. A characterization of DNA release in *Pseudomonas aeruginosa* cultures and biofilms. *Mol. Microbiol.* **2006**, *59*, 1114–1128. [CrossRef]
69. Yang, L.; Hu, Y.; Liu, Y.; Zhang, J.; Ulstrup, J.; Molin, S. Distinct roles of extracellular polymeric substances in *Pseudomonas aeruginosa* biofilm development: EPS-mediated biofilm development. *Environ. Microbiol.* **2011**, *13*, 1705–1717. [CrossRef]
70. Steinberger, R.E.; Holden, P.A. Macromolecular composition of unsaturated *Pseudomonas aeruginosa* biofilms with time and carbon source. *Biofilms* **2004**, *1*, 37–47. [CrossRef]
71. Banik, R.M.; Kanari, B.; Upadhyay, S.N. Exopolysaccharide of the gellan family: Prospects and potential. *World J. Microbiol. Biotechnol.* **2000**, *16*, 407–414. [CrossRef]
72. Jahn, A.; Griebe, T.; Nielsen, P.H. Composition of *Pseudomonas putida* biofilms: Accumulation of protein in the biofilm matrix. *Biofouling* **1999**, *14*, 49–57. [CrossRef]
73. Wigneswaran, V.; Nielsen, K.F.; Sternberg, C.; Jensen, P.R.; Folkesson, A.; Jelsbak, L. Biofilm as a production platform for heterologous production of rhamnolipids by the non-pathogenic strain *Pseudomonas putida* KT2440. *Microb. Cell Factories* **2016**, *15*, 181. [CrossRef]
74. Gutiérrez-Gómez, U.; Servín-González, L.; Soberón-Chávez, G. Role of β-oxidation and de novo fatty acid synthesis in the production of rhamnolipids and polyhydroxyalkanoates by Pseudomonas aeruginosa. *Appl. Microbiol. Biotechnol.* **2019**, *103*, 3753–3760. [CrossRef]
75. Read, R.R.; Costerton, J.W. Purification and characterization of adhesive exopolysaccharides from *Pseudomonas putida* and *Pseudomonas fluorescens*. *Can. J. Microbiol.* **1987**, *33*, 1080–1090. [CrossRef]
76. Royce, L.A.; Liu, P.; Stebbins, M.J.; Hanson, B.C.; Jarboe, L.R. The damaging effects of short chain fatty acids on *Escherichia coli* membranes. *Appl. Microbiol. Biotechnol.* **2013**, *97*, 8317–8327. [CrossRef]
77. Cerrone, F.; Duane, G.; Casey, E.; Davis, R.; Belton, I.; Kenny, S.T.; Guzik, M.W.; Woods, T.; Babu, R.P.; O'Connor, K. Fed-batch strategies using butyrate for high cell density cultivation of *Pseudomonas putida* and its use as a biocatalyst. *Appl. Microbiol. Biotechnol.* **2014**, *98*, 9217–9228. [CrossRef]
78. Maclean, H.; Sun, Z.; Ramsay, J.A.; Ramsay, A.B. Decaying exponential feeding of nonanoic acid for the production of medium-chain-length poly(3-hydroxyalkanoates) by *Pseudomonas putida* KT2440. *Can. J. Chem.* **2008**, *86*, 564–569. [CrossRef]
79. Koller, M.; Sandholzer, D.; Salerno, A.; Braunegg, G.; Narodoslawsky, M. Biopolymer from industrial residues: Life cycle assessment of poly(hydroxyalkanoates) from whey. *Resour. Conserv. Recycl.* **2013**, *73*, 64–71. [CrossRef]
80. Coats, E.R.; Watson, B.S.; Brinkman, C.K. Polyhydroxyalkanoate synthesis by mixed microbial consortia

cultured on fermented dairy manure: Effect of aeration on process rates/yields and the associated microbial ecology. *Water Res.* **2016**, *106*, 26–40. [CrossRef]

81. Da Cruz Pradella, J.G.; Taciro, M.K.; Mateus, A.Y.P. High-cell-density poly (3-hydroxybutyrate) production from sucrose using *Burkholderia sacchari* culture in airlift bioreactor. *Bioresour. Technol.* **2010**, *101*, 8355–8360. [CrossRef]
82. Brown, E.; Forman, N.A.; Orellana, C.S.; Zhang, H.; Maynor, B.W.; Betts, D.E.; DeSimone, J.M.; Jaeger, H.M. Generality of shear thickening in dense suspensions. *Nat. Mater.* **2010**, *9*, 220–224. [CrossRef]

13

The Molecular Level Characterization of Biodegradable Polymers Originated from Polyethylene using Non-Oxygenated Polyethylene Wax as a Carbon Source for Polyhydroxyalkanoate Production

Brian Johnston [1,*], Guozhan Jiang [1], David Hill [1], Grazyna Adamus [2], Iwona Kwiecień [2], Magdalena Zięba [2], Wanda Sikorska [2], Matthew Green [3], Marek Kowalczuk [1,2] and Iza Radecka [1,*]

1 Wolverhampton School of Biology, Chemistry and Forensic Science, Faculty of Science and Engineering, University of Wolverhampton, Wolverhampton WV1 1LY, UK; Guozhan.jiang@wlv.ac.uk (G.J.); D.Hill@wlv.ac.uk (D.H.); M.Kowalczuk@wlv.ac.uk (M.K.)

2 Centre of Polymer and Carbon Materials, Polish Academy of Sciences, 41-800 Zabrze, Poland; Grazyna.Adamus@cmpw-pan.edu.pl (G.A.); ikwiecien@cmpw-pan.edu.pl (I.K.); mzieba@cmpw-pan.edu.pl (M.Z.); wsikorska@cmpw-pan.edu.pl (W.S.)

3 Recycling Technologies Ltd., South Marston Industrial Park, Swindon SN3 4WA, UK; Matt.Green@recyclingtechnologies.co.uk

* Correspondence: B.Johnston@wlv.ac.uk (B.J.); I.Radecka@wlv.ac.uk (I.R.)

Academic Editor: Martin Koller

Abstract: There is an increasing demand for bio-based polymers that are developed from recycled materials. The production of biodegradable polymers can include bio-technological (utilizing microorganisms or enzymes) or chemical synthesis procedures. This report demonstrates the corroboration of the molecular structure of polyhydroxyalkanoates (PHAs) obtained by the conversion of waste polyethylene (PE) via non-oxygenated PE wax (N-PEW) as an additional carbon source for a bacterial species. The N-PEW, obtained from a PE pyrolysis reaction, has been found to be a beneficial carbon source for PHA production with *Cupriavidus necator* H16. The production of the N-PEW is an alternative to oxidized polyethylene wax (O-PEW) (that has been used as a carbon source previously) as it is less time consuming to manufacture and offers fewer industrial applications. A range of molecular structural analytical techniques were performed on the PHAs obtained; which included nuclear magnetic resonance (NMR) and electrospray ionisation tandem mass spectrometry (ESI-MS/MS). Our study showed that the PHA formed from N-PEW contained 3-hydroxybutyrate (HB) with 11 mol% of 3-hydroxyvalerate (HV) units.

Keywords: polyhydroxyalkanoates; PHAs; non-oxidized PE wax; N-PEW; *Cupriavidus necator* H16

---

1. Introduction

Polyhydroxyalkanoates (PHAs) are a family of polyhydroxyesters with 3-, 4-, 5-, and 6-hydroxyalkanoic acids that are biodegradable, non-toxic, biocompatible organic polyesters synthesized by some species of bacteria [1]. Currently there is a great deal of interest in PHAs due to their far-reaching applications, which include being used as biological implants, pharmacological delivery systems, packaging materials, and many more, as they can provide unique properties that do not exist in some synthetic polymers [1]. Traditional plastics, however, originate from petrochemical sources and have been a vital business commodity since their first industrial production. They are lightweight, cheap to manufacture, strong, flexible, and adaptable, but petro-based polymers form a growing proportion of the municipal solid waste (MSW) produced world-wide annually [2] and their waste materials are extremely harmful on a microscopic scale. These persistent plastics infiltrate the food chain as they are mechanically broken down (but not completely) in our oceans. Microplastics, defined as fragments that are less than five millimetres in diameter, can range from the size of an insect down to the size of a virus [3]. These particles, and larger plastics, also have a deadly impact on sea-life through entanglement or ingestion. It has been stated that plastics now account for somewhere between 60 and 80% of all ocean and shoreline debris [3–6]. The challenge, for both economic and environmental reasons, is to develop novel strategies that can create alternative materials to match or surpass the physical and commercial appeal of current petro-based plastic, with no adverse environmental consequences.

Waste polyethylene (PE) is a potential carbon source that could be utilized to make value-added biopolymers, particularly as it is the most commonly-produced plastic, making up over 29% of worldwide plastic manufacture, while only 10% of it is recycled [7–10]. Unfortunately, the high molecular weight, hydrophobicity, and chemically-stable hydrocarbon covalent bonds of PE make it difficult for bacteria to metabolize effectively [11–15]. However, in the melt-form after oxidative degradation with free radical initiators and then sonication with an additional fermentation medium, PE waxes can be metabolized by bacteria and PHAs can be produced [16,17]. These polar PE waxes already have commercial uses; they can be used to produce PVC, as components of aqueous emulsions, lubricants, additives for varnishes, road marking paint, and adhesives [16,18–20]. The alternative non-polar, non-oxidized PE waxes (N-PEWs) have fewer industrial uses [7]. They are also cheaper and relatively easier to produce, which suggests that N-PEWs provide an alternative carbon source to O-PEWs.

We propose that N-PEWs could be a pragmatic application of waste PE as a carbon source for PHA synthesis using *Cupriavidus necator* H16. The pyrolysis of PE samples in anaerobic conditions yields a complex mixture of relatively low molecular weight hydrocarbons with a molecular mass of 200–1000 [7,16]. Due to the hydrophobic nature of the non-oxidized waxes, ultrasound sonication can be performed to breakdown and distribute the wax as an emulsion so that it will be readily metabolized by bacteria [16,21]. For alkane degradation (with oxygen), the most widely-used pathway is the oxidation of the terminal methyl group into a carboxylic acid, via an alcohol intermediate, followed by complete mineralization through β-oxidation [15,22]. The intercellularly-synthesized PHAs are then held within bacterial species like *Cupriavidus necator* (formerly *Ralstonia eutropha* or *Alcaligenes eutropha)* as granules. *C. necator* is a Gram-negative, hydrogen-oxidizing bacterium ("knallgas") that is able to grow at the interface of anaerobic and aerobic environments [12]. These microbes can synthesize PHA in the presence of excess amounts of carbon and limited nutrients [11] and they have been shown to metabolize a variety of carbon sources including fatty acids, hemicellulose, crude glycerol, methane, and even liquefied wood, for PHA production [11,23–25]. They are also able to accumulate up to 85% PHA per dry cell weight, often with 8–12 biopolymer granules per bacterium [16]. To extract the PHAs held within *C. necator*, their biomass is lyophilized and then hot solvent extraction is used, followed by precipitation in ethanol or hexane [16,23].

Here, we report on the production and molecular-level structural analysis of PHAs formed using *C. necator*, with non-oxidized wax as a carbon source in a nitrogen-rich tryptone soya broth (TSB) growth media. This is a novel use of N-PEWs with the aid of *C. necator*, which was selected due to its yield potential, limited resistance to copper metals (a possible contaminant from some wax production methods), growth rate at relatively low temperature, and its well-documented genetic profile and gene stability (for future modification and enhancement purposes) [16,25–27].

2. Materials and Methods

2.1. Microorganism

The bacterial species used for PHA production with N-PEWs and TSB or basal salts medium (BSM), was *C. necator* H16 (NCIMB 10442, ATCC 17699). This organism was obtained from the

University of Wolverhampton stock culture (freeze-dried and kept at −20 °C). Previous to the study, cultures were revived and grown overnight at 30 °C (optimum) in TSB at 150 rpm. The microorganism was then sub-cultured on tryptone soya agar (TSA) plates and incubated at 30 °C for 24 h.

2.2. Carbon Source and Chemicals

The non-oxidized wax was kindly provided by Recycling Technologies Ltd. (Swindon, UK) from waste plastics that were scanned and separated from contaminants (such as glass or stone) and then shredded and dried before chemical treatment [28]. These remnants were then passed to a thermal cracker, where the long hydrocarbon chains in the plastics were cracked into shorter chains. The hot hydrocarbon vapour leaving the reactor was then filtered and treated for impurities. At this stage the refined gas was condensed into a wax [28]. The wax was used as received without any further purification (more N-PEW structural details are found in Section 3.1).

2.3. Media

TSB and TSA were purchased from Lab M Ltd. (Lancashire, UK). Both of these growth media were prepared under aseptic conditions, using the instructions of the manufacturer. The basal salts medium (BSM) used contained distilled water, 1 g/L K_2HPO_4, 1 g/L KH_2PO_4, 1 g/L KNO_3, 1 g/L $(NH_4)_2SO_4$, 0.1 g/L $MgSO_4{\cdot}7H_2O$, 0.1 g/L NaCl, 10 mL/L trace elements solution. The trace element solution contained: 2 mg/L $CaCl_2$, 2 mg/L $CuSO_4{\cdot}5H_2O$, 2 mg/L $MnSO_4{\cdot}5H_2O$, 2 mg/L $ZnSO_4{\cdot}5H_2O$, 2 mg/L $FeSO_4$, 2 mg/L $(NH_4)_6Mo_7O_{24}{\cdot}4H_2O$. BSM salts were purchased from BDH Chemicals Ltd. (Poole, UK). Ringer's solution was also purchased from Lab M Ltd. (Lancashire, UK). A 1/4 strength tablet was used in 500 mL of deionized water; then it was completely dissolved. All media were sterilized by being autoclaved at standard conditions (121 °C for 15 min).

2.4. Fermentation Procedure

Starter cultures were prepared using 20 mL TSB (in a 50 mL flask) inoculated with a single *C. necator* colony from a TSA spread plate. That culture was then incubated (aerobically) for 24 h at 30 °C and 150 rpm in a rotary incubator (Incu-Shake MIDI, Shropshire, UK). After 24 h these cultures were checked for contamination by Gram staining and microscope observation.

Shake flask fermentation analysis was performed in triplicate using 500 mL wide neck Erlenmeyer flasks. Each flask used 1 g of N-PEW that was first put into a 50 mL beaker, covered with foil, and melted at 70 °C. Then 20 mL of sterile TSB or BSM was added to the melted wax, which caused the wax to become solid again. The temperature was then increased to re-melt the wax. The waxes were then sonicated (in TSB) for 8 min at 0.5 active and passive intervals with a power of 70% using a Bandelin Electronic sonicator, (Berlin, Germany) to form a TSB/wax emulsion or BSM/wax emulsion. This emulsion was tested for sterility by spread plating. The sterile emulsion was then added to 210 mL sterile TSB or BSM in a 500 mL flask, followed by 20 mL of the starter culture, giving a total volume of 250 mL. The experimental control was 230 mL TSB or BSM, inoculated with 20 mL of starter culture with no PE wax included. The flasks were incubated in a rotary incubator under the same conditions mentioned for 48 h.

Viable cell counts were done using the methodology of Miles and Misra [29]. Briefly, 5 mL samples were aseptically collected from the flask cultures at 0, 3, 24, and 48 h. Serial dilutions of each sample were performed to 10^{-8} and then 20 μL of each dilution was pipetted onto a standard TSA plate in triplicate. These plates were incubated at 30 °C for 48 h and the colonies were counted and expressed in $\log_{10}$ CFU mL^{-1}.

2.5. PHA Extraction Procedure

PHA extraction was performed after the 48 h fermentation time period had elapsed. The flasks were removed from their incubators and the flask contents were filtered using a sieve (to remove any conjugates of wax) and separated into centrifuge tubes, usually 35–40 mL per tube. The media was then centrifuged in a Sigma 6-16KS centrifuge for 10 min at 4500 rpm. At this point, there was only the pellet containing the biomass and the emulsion layer visible. The supernatant in each tube was discarded and the biomass was obtained and frozen overnight a −20 °C. This was followed by lyophilization using an Edwards freeze-drier (Modulyo, Crawley, UK) for 48 h at a temperature of −40 °C and at a pressure of 5 MBAR. The dry biomass was then weighed and recorded as cell dry weight (CDW) before being placed into an extraction thimble and, using Soxhlet extraction with HPLC grade chloroform for 48 h, the PHA was collected as a chloroform/biopolymer mixture. Then rotation evaporation was used (at 50 °C) to remove the chloroform. Polymer precipitation using n-hexane was performed in a round-bottom flask and then this was further separated by filtration (Watman No. 1 paper). If required, hexane was used to rinse the product further to remove any residue low molecular weight wax. The sample was then left in a fume cupboard to dry for up to five hours and the yield was recorded using:

$$\text{Percentage yield of PHA} = (\text{weight of extracted polymer})/(\text{cell dry weight}) \times 100$$

2.6. Characterization

2.6.1. FTIR

Fourier transform infrared spectroscopy (FTIR) was performed using a DuraScope (Genesis II) spectrophotometer (Smiths Detection Inc., Danbury, CT, USA). After a background scan, samples of N-PEW were placed under the lens. FTIR spectra were recorded in the transmittance mode with a resolution of 1 cm^{-1} in the range of 4000–400 cm^{-1}.

2.6.2. GPC

Gel permeation chromatography (GPC) was used to find the molecular weight and molecular weight distribution of the wax and polymers produced during this study. The analysis was conducted using a TOSOH EcoSec HLC/GPC 8320 (Tosoh Bioscience, Yamaguchi, Japan) system equipped with a RI and a UV detector (Tosoh Bioscience, Yamaguchi, Japan) operating at a temperature of 40 °C. The column used was TSKgel HZM-N (Tosoh Bioscience, Yamaguchi, Japan), calibrated against polystyrene standards. The UV detector was set at a wavelength of 254 nm. Chloroform was used as the eluent, at a flow rate of 0.25 mL/min. A sample size of 2 μL was injected into the system using an autosampler.

The wax GPC analysis included: (i) wax pre-sonication; (ii) post-sonication; (iii) post-fermentation (wax and TSB) and the control; (iv) post-shake flask N-PEW in TSB. The number-average molar mass (Mn) and the molecular mass distribution index (Mw/Mn) of the N-PEWs were determined in a $CHCl_3$ solution at 35–60 °C heating to melt the waxes, then 10 min of cooling, to ascertain whether the waxes were fully dissolved in the selected volume of HPLC-grade chloroform. This N-PEW/chloroform solution was carefully pipetted into a syringe and passed through a filter into a GPC ampoule.

2.6.3. NMR Analysis

Proton nuclear magnetic resonance (^{1}H-NMR) spectra were recorded with a Bruker Avance II (Bruker, Rheinstetten, Germany) operating at 600 MHz, with 64 scans, 2.65 s acquisition time, and an 11 μs pulse width. ^{13}C-NMR spectra were recorded with a Bruker Avance II operating at 150.9 MHz, with 20,480 scans, 0.9088 s acquisition times, and 9.40 μs pulse width. The ^{1}H-NMR and ^{13}C-NMR spectra were run in $CDCl_3$ at room temperature with tetramethylsilane (TMS) as an internal standard.

2.6.4. ESI-MS/MS Analysis and Identification of PHA at the Molecular Level

Electrospray mass spectrometry analysis was performed using a Finnigan LCQ ion trap mass spectrometer (Thermo Finnigan LCQ Fleet, San Jose, CA, USA). The oligomer samples, prepared as described by Kawalec et al. [30], were dissolved in a chloroform/methanol system (1:1 *v*/*v*), and the solutions were introduced into the ESI source by continuous infusion using the instrument syringe pump at a rate of 5 μL/min. The LCQ ESI source was operated at 4.5 kV, and the capillary heater was set to 200 °C; the nebulizing gas applied was nitrogen. For ESI-MS/MS experiments, the ions of interest were isolated monoisotopically in the ion trap and were activated by collisions. The helium damping gas that was present in the mass analyser acted as a collision gas. The analysis was performed in the positive-ion mode.

PHA with low molar mass was obtained via thermal degradation of bacterial PHA in the presence of potassium hydrogen carbonate ($KHCO_3$). The respective amount of polyester and salt, at a ratio PHA/$KHCO_3$ equal to 5, was introduced in a vial with ethanol. The mixture was then stirred for 30 min to provide a homogeneous suspension. The experiment was carried out in an oven at 180 °C. The length of time for the process was dependent on the molar mass plain PHA. For the PHAs with a mass bellow 100,000 it was 20 min, and for polymers with a mass between 200,000 and 400,000, it was 30 min. This time regime allowed for the obtaining of oligomers with a molar mass of around 2000. After the experiment, oligomers were protonated with the Dowex® 50WX4 in hydrogen form. The oligomer and Dowex® (1:1 wt%) was dissolved in chloroform and the mixture was stirred for four hours. Next, the Dowex® was removed by filtration. The oligomers obtained were characterized by GPC and ^{1}H-NMR spectrometry, as well as their structure at a molecular level using ESI-MS/MS. The ^{1}H-NMR analysis revealed the presence of characteristic signals corresponding to the protons of 3-hydroxybutyrate (HB) (and 3-hydroxyvalerate (HV)) repeating units, and also the signals attributed to the crotonate end groups.

3. Results

3.1. N-PEW Initial FTIR and NMR Analysis

The non-oxidized wax obtained for this study was tested to confirm its structure and chemical properties. The waxes were produced from waste plastic underwent thermal cracking to produce a refined gas that was then condensed into a wax [28]. FTIR was performed on O-PEW and N-PEW for comparison and to determine the presence or absence of functional groups associated with oxidation (Figure 1).

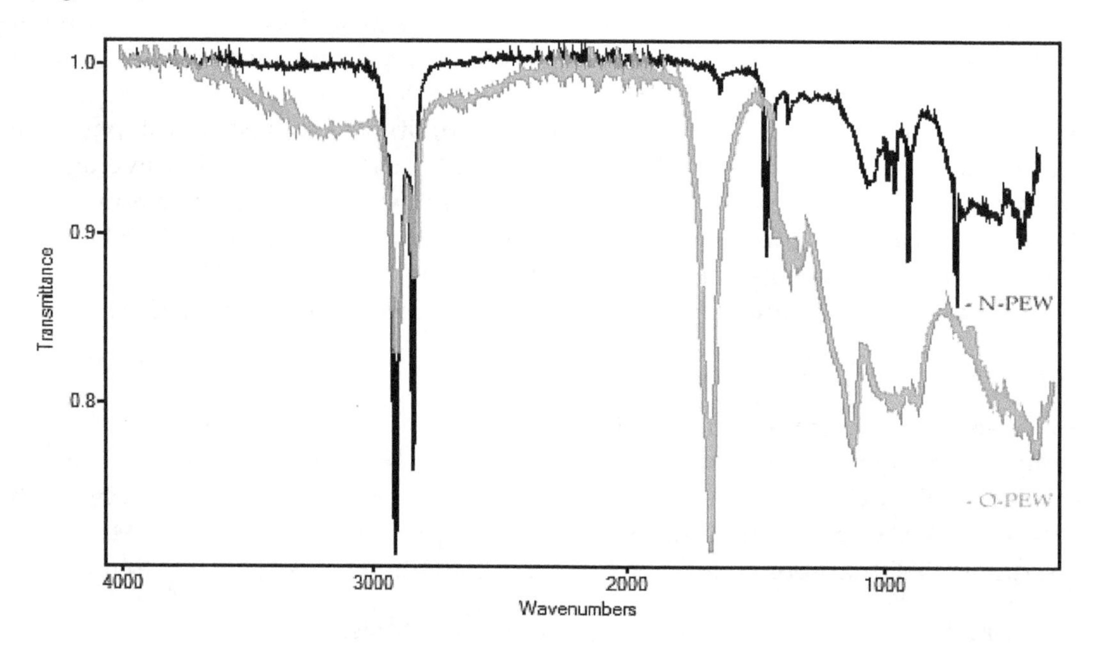

Figure 1. FTIR of O-PEW (blue trace) and N-PEW (black trace) used in this study.

The N-PEW (black) spectrum showed a strong absorption at 2800 cm^{-1}, an indication of a C-H stretch, which confirms saturated chains. There was no absorbance at 3500 cm^{-1} (C=O stretch overtone) and weak absorbance at 1790–1700 cm^{-1} (C=O stretch), as compared with the spectrum of the oxidized wax, further confirmation of hydrocarbon chains with a very small degree of oxidation. When these results are compared to the O-PEW (blue) reading (a wax with an acid number of 197), there was less absorbance at 2800 cm^{-1} than what is displayed in the N-PEW reading. There was also a greater absorbance at 1700 cm^{-1} and broad absorbencies at 1160–1060 cm^{-1} which are likely due to R-O-R aliphatic bonds.

The structure of the non-oxidized PE wax was further studied using ^{1}H-NMR and ^{13}C-NMR, as shown in Figure 2a,b. In the ^{1}H-NMR, there are strong signals at 0.88 and 1.25 ppm, which are attributed to CH_3 and CH_2. The signals at 1.6 and 2.0 ppm may be due to the allylic hydrogen or hydrogen adjacent to the C=O group. There are signals at 4.9 ppm, 5.4 ppm, and 5.8 ppm in the ^{1}H spectrum, and there are signals at 114 ppm and 139 ppm in the ^{13}C spectrum.

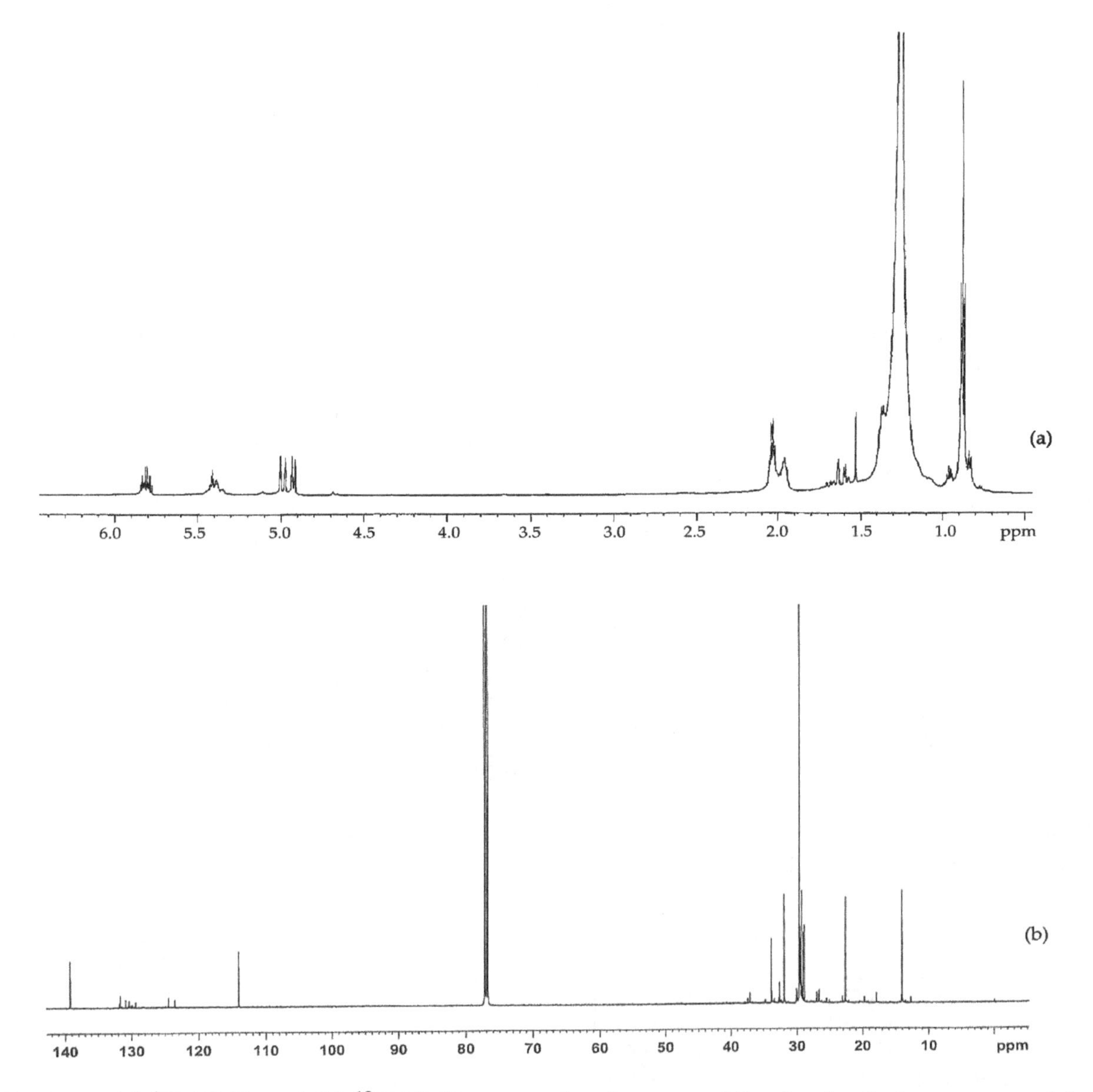

Figure 2. (**a**) ^{1}H-NMR and (**b**) ^{13}C-NMR spectra for the non-oxidized polyethylene wax used in this study.

These signals indicate that there are very small amounts of oxygenated functional groups in the non-oxidized wax. However, the amount of oxygenated functional groups are much less than those in the oxidized wax.

3.2. *Growth of Bacteria and PHA Yield*

The growth of *C. necator* H16 in TSB only (control), BSM only, and TSB/BSM supplemented with N-PEW is shown in Figure 3.

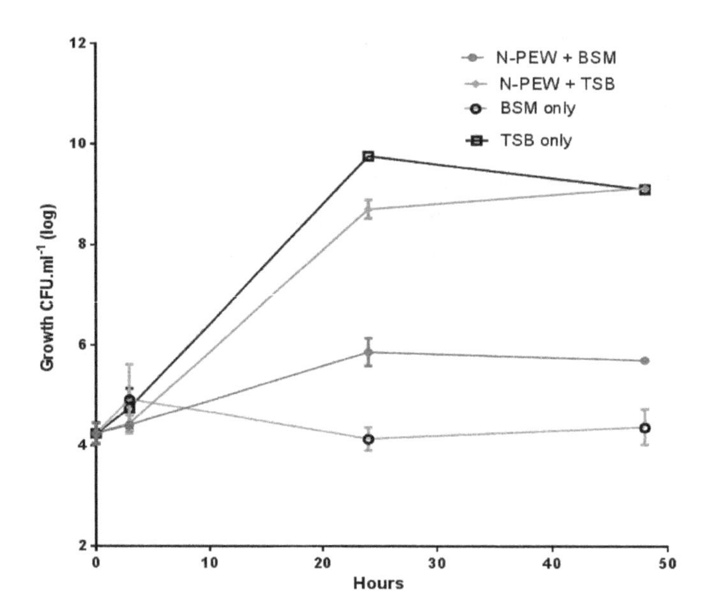

Figure 3. Small-scale fermentations of *C. necator* H16 with 4 g/L of N-PEW in TSB or BSM. Data points are arithmetic means of triplicates, while error bars denote the SE of the mean.

The bacteria were inoculated at approximately 4.1 $\log_{10}$ CFU mL^{-1} and grew to 9.1 $\log_{10}$ CFU mL^{-1} with TSB only and TSB with N-PEW as an additional carbon source. There was no significant difference ($p < 0.5$) between these growth curves, however, when the growth curves with and without N-PEW and BSM are examined, it is evident that there is a difference in final counts. There was little growth when BSM only was used, cell numbers were merely maintained at 4.1–4.3 $\log_{10}$ CFU mL^{-1}) over 48 h, but when N-PEW was added to BSM, there was more growth, an increase of 1.8 $\log_{10}$ CFU mL^{-1}, and the counts ended at 5.8 $\log_{10}$ CFU mL^{-1} rather than 4.3 $\log_{10}$ CFU mL^{-1} in BSM only. In addition, there was no product yield with BSM only and with BSM with N-PEW, suggesting that the wax, alone, is not sufficient for PHA production, but it can sustain cells (Table 1). When TSB only and TSB with N-PEW were used the product was visually different; the PHA produced from TSB only was paper-like in texture and white, while the PHA produced from the N-PEW with TSB was more brittle.

Table 1. The amount of PHA synthesized by *C. necator* using 4 g/L N-PEW. This experiment was conducted over a 48 h incubation period in TSB, TSB with N-PEW, and BSM with N-PEW at 30 °C (150 rpm).

Media	Average CDW (g/L)	Average PHA (g/L)	PHA (%*w*/*w*)
TSB only	0.98 ± 0.05	0.20 ± 0.05	20%
TSB with N-PEW	1.42 ± 0.20	0.46 ± 0.20	32%
BSM only	0.16 ± 0.06	ND	ND
BSM with N-PEW	0.20 ± 0.10	ND	ND

ND = non-detected; CDW = cellular dry weight.

3.3. *GPC of PHA and N-PEW*

GPC analysis of the PHAs obtained using N-PEW and TSB as growth medium showed that the number-average molar mass (Mn) was in the range of 64,000 g/mol with a molecular distribution of

9.7 (Mw/Mn). The Mn results for polyesters synthesized using TSB only were lower, in the range of 52,000 g/mol with a distribution of 5.7 (Mw/Mn).

GPC analysis of the waxes under different conditions during this investigation is reported in Table 2. The N-PEW results showed the number-average molar mass of normal N-PEWs used was (Mn) was 1600 g/mol with a weight average (Mw) of 5400 g/mol.

Table 2. GPC results of unprocessed, pre-sonicated, post-sonicated, post-fermentation, and N-PEWs after shake flask exposure for 48 h without bacterial inoculation.

N-PEW Conditions	Number-Average Molar Mass (Mn) in g/mol	Molecular Mass Distribution Index (Mw/Mn)
Normal N-PEW (pre-sonication)	1600	3.5
Post-sonication N-PEWs	1200	1.2
Post-fermentation	1100	1.3
Post-shake flask (no bacteria)	1100	1.3

The GPC results showed that before sonication there were two distinct types of N-PEW in the sample: high molecular weight (the first peak) and lower molecular weight chains (second peak). After the sonication procedure (Section 2.4) the higher molecular weight peak was removed as the high molecular weight chains were broken down to a Mn of 1200 g/mol with a molar mass index of 1.2. However, any further reduction in the Mn value cannot be attributed to bacterial activity as the flasks without bacteria produced similar Mn values of 1100 g/mol after 48 h.

3.4. PHA Identification and Characterization

The ^{1}H-NMR results of the PHAs formed from shake-flask fermentation experiments with TSB and N-PEW using *C. necator* is shown in Figure 4.

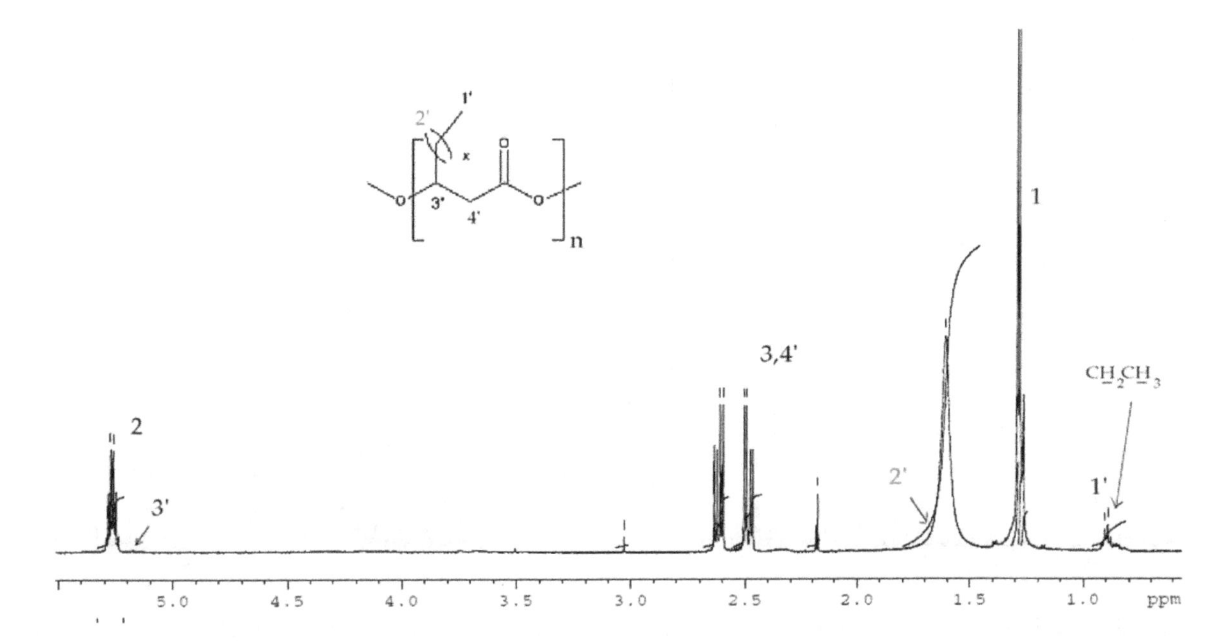

Figure 4. ^{1}H-NMR spectra of PHAs produced from small-scale fermentations with *C. necator* H16 using TSB media supplemented with N-PEW as a carbon source.

In the spectrum the signals corresponded to the protons of the HB repeating units (labelled 1, 2, and 3) and additional signals at $\delta = 0.9$ ($-CH_2\underline{CH_3}$) and $\delta = 5.17$ (-CH) correspond to HV-repeated units that were observed. After integrating these signals, the content of monomer units other than

3-hydroxybutyrate (3-HB) has been estimated at 11 mol% 3-hydroxyvalerate (HV). Due to some of overlapping of signals, the protons of 2′ are not visible.

3.5. ESI-MS/MS

The controlled thermal degradation of PHAs synthesized from N-PEW, induced by $KHCO_3$, was performed according to the procedure referred to in Section 2.6.4 [30]. This kind of E1cB degradation leads to PHA oligomers with unsaturated and carboxylic end groups. The results of PHAs formed from TSB with N-PEW are displayed in Figures 5–7. Scheme 1 shows the general structure of the ions presented in the ESI-MS spectra.

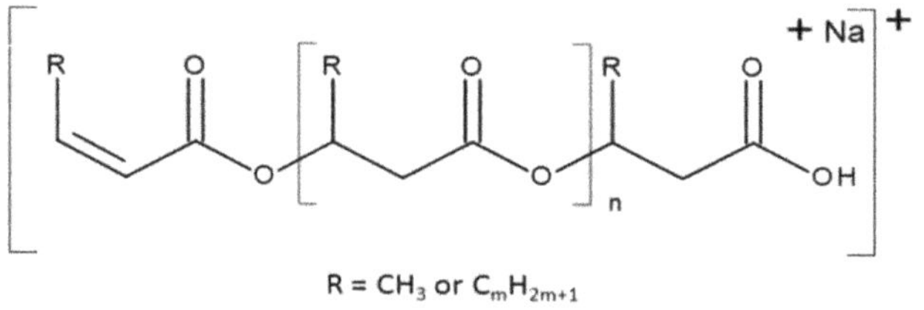

R = CH_3 or C_mH_{2m+1}

Scheme 1. The general formula of ions observed in the ESI-MS spectra.

There are two series of ions visible in the mass spectrum. The main series of the ions at *m/z* 1571, 1485, 1399, 1313, 1277, 1141, 1055, 969, 883, and 797 corresponds to the sodium adduct of oligomers with crotonate and carboxyl end groups, calculated according to the formula m/z 86 + (86 × n) + 23, as shown in Scheme 1. There are also a small series of ions besides the main series. The m/z difference is 14 between the main series and the small series, which corresponds to a CH_2 group.

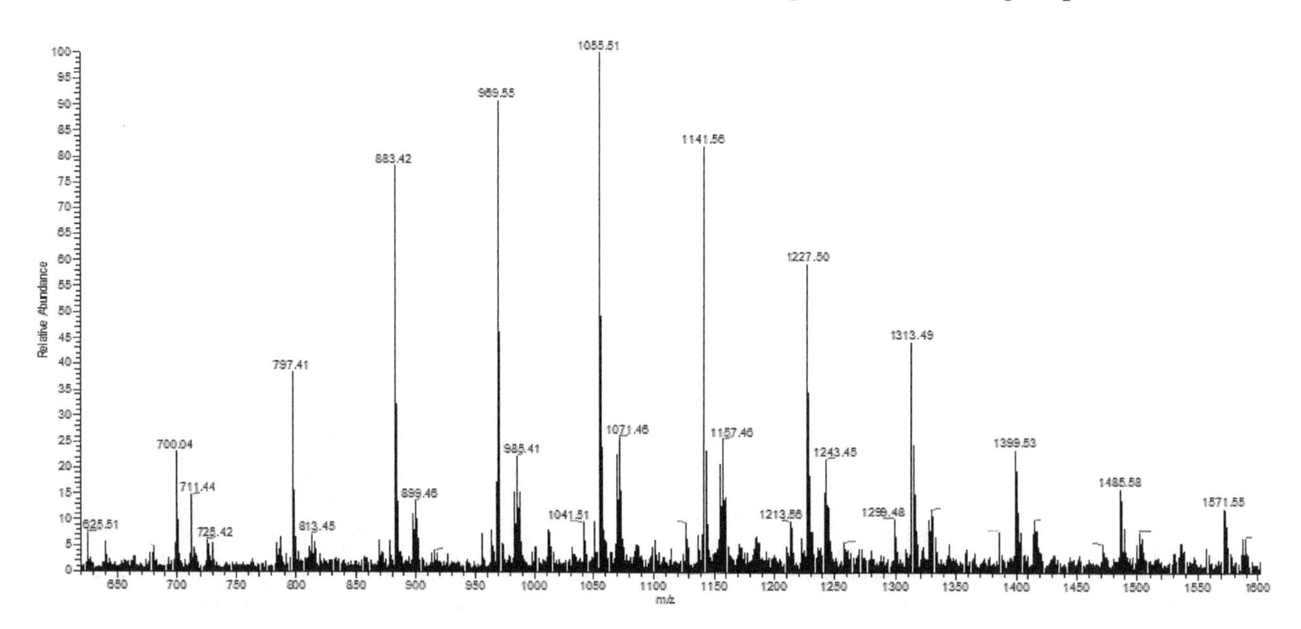

Figure 5. The ESI-MS spectrum (positive-ion mode) of the PHA oligomers, obtained via partial thermal degradation of the biopolyester produced by *C. necator* H16 utilizing TSB and N-PEW as an additional carbon source.

The ESI-MS^2 spectrum of the sodium adduct of the biopolyester oligomers at m/z 1055 is presented in Figure 6. The product ions at m/z 969, 883, 797, 711, 625, 539, 453, and 367 correspond to the oligo (3-hydroxybutyrate) terminated with carboxyl end groups.

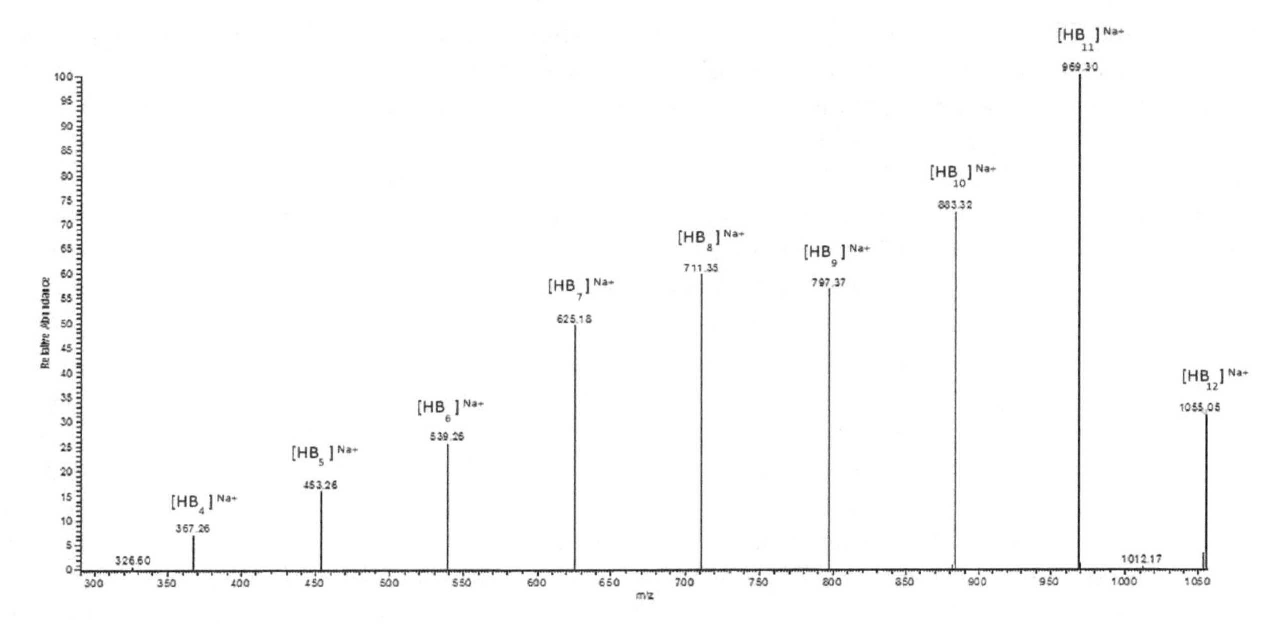

Figure 6. ESI-MS/MS spectrum for the selected sodium adduct ion of oligomers of $[HB_{12} + Na]^+$ at 1055 m/z.

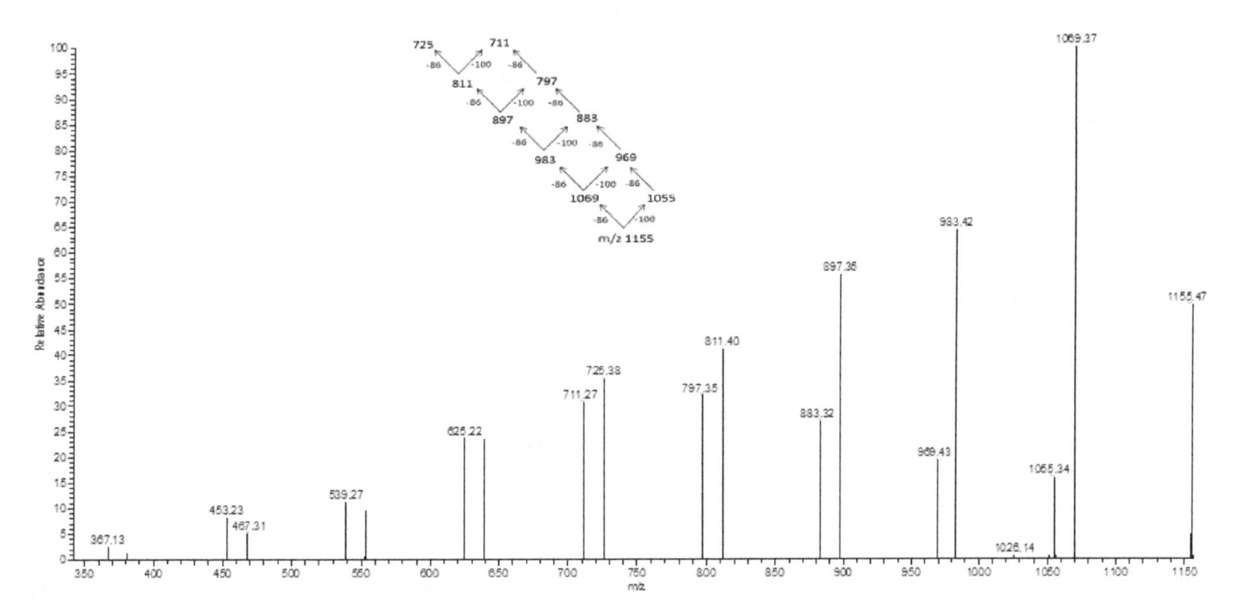

Figure 7. ESI-MS/MS spectrum obtained for the selected sodium adduct of oligomers of $[HB_{13}HV + Na]^+$ at m/z at 1155 m/z.

In Figure 7 the fragmentation of the sodium adduct ion of oligomer of $[HB_{13}HV + Na]^+$ at m/z at 1155 m/z is presented. This parent ion produces fragment ions of subsequent oligomers according to the theoretical fragmentation pathway diagram. The loss of 86 Da equates to the loss of crotonic acid (e.g., 1155–1069 m/z), and the loss of 100 Da equates to an HV unit. The fragmentation spectrum of these ions confirms that the most intensive ions in the clusters correspond to sodium adducts of 3-hydroxybutyrate oligomers. Therefore, the fragmentation spectrum for the precursor ion at m/z 1155 confirms the presence of 3-hydroxybutyrate and randomly-distributed 3-hydroxyvalerate co-monomer units in polyester chains produced from N-PEW and TSB.

4. Discussion

The FTIR spectra showed that the N-PEWs received were non-polar, saturated hydrocarbons (Figure 1). However, the NMR spectra revealed that there were traces of oxygenated functional groups

(Figure 2). The GPC analysis of the N-PEWs showed that they have a lower molecular mass than O-PEWs, which could make it a more accessible carbon source [16]. Although the hydrocarbon chains are hydrophobic, their shorter chains (after sonication [21]) would be able to be distributed within the TSB or BSM media, and from the growth curves in Figure 3, the addition of the wax made a difference (+1.8 log10 CFU mL^{-1}) when it was added to BSM fermentations. The cell growth analysis also showed that N-PEWs do not have an adverse effect upon the growth rate of *C. necator* H16 in either TSB (not significant) or BSM. This was also reported by Radecka et al. [16] with O-PEW. In an experiment conducted by Zhang et al. [31] it was stated that PE has very low antimicrobial properties, limiting growth in Gram-positive and Gram-negative bacterial cells. However, in this study there was no evidence of antimicrobial activity. The low antimicrobial activity reported by Zhang et al. could have been due to the presence of antimicrobial moieties, such as chlorine or fluorine held within the polymer chains, rather than the PE itself. These contaminants could have been freed during the oxidation process and then dispersed into the medium to inhibit microbial growth. It is, therefore, likely that, in a similar way to O-PEW, N-PEWs are being successfully metabolized as fatty acids via β-oxidation. However, the GPC data was not able to fully support this theory as any reduction in Mn or Mw occurred in the controls without bacteria. Those smaller hydrocarbon chains that were not successfully recovered after the 48 h fermentation could have either been metabolized (in the inoculated flask) or they were fully dissolved in the medium. Another possibility is that some of the wax attached to the sides of the flasks as a thin film. For these reasons it was not possible to give a definitive value of the wax metabolized by the bacteria; a possible answer to this would be use a labelled variant of the wax. The presence of accessible carbon sources, such as fatty and carboxylic acids from O-PEW means that more carbon is present in the wax-supplemented cultures in a form that promotes the synthesis of PHAs [16]. It is possible that some of the larger alkane chains from N-PEWs that are more inaccessible for bacterial metabolism could be acting as oxygen inhibitors, limiting distribution within the system. This could then contribute to the stress conditions that push the growth culture towards the production of PHA [11,16].

The PHA yield of TSB cultures supplemented with N-PEW was 32% with a cell dry weight 1.42 g/L compared to shake flask experiments using TSB exclusively, which produced a lower yield of 20% and a dry weight of 0.98 g/L. There was growth of *C. necator* with N-PEW in BSM (shown by cell counts), however, no PHAs were recovered after 48 h. This is an indication that the addition of N-PEW into TSB has an effect upon the PHA production and the structure, with a Mn of 64,000 g/mol with a molecular distribution of 9.7 (Mw/Mn). The Mn results for PHAs produced with TSB only were lower, in the range of 52,000 g/mol with a distribution of 5.7 (Mw/Mn). These dispersity values indicate that N-PEW with TSB produces a more varied range of polyesters, and it is further evidence that N-PEWs are being metabolized.

^{1}H-NMR was used to determine the structure of the copolymers produced. It was found that PHB was produced in the TSB only fermentation (data not shown) [16]. In the spectrum (Figure 4) the signals corresponded to the protons of the HB repeating units (labelled 1, 2, and 3) and additional signals at $\delta = 0.9$ (-CH_2CH_3) and $\delta = 5.17$ (-CH) correspond to HV repeated units that were observed. These signals are typical for PHAs with longer aliphatic groups in the β-position [32,33]. After integrating these signals, the content of monomer units other than 3-hydroxybutyrate (3-HB) has been estimated at 11 mol% 3-hydroxyvalerate (HV). However, due to some overlapping of certain signals, the protons of $2'$ were not visible, although they have been reported in that location in PHBV previously [33].

In the N-PEW supplemented cultures, electrospray mass spectrometry (ESI-MS) determined copolymers were present. This method is known to provide more detail than GC and NMR when used for PHBV characterization. This is because it can corroborate whether a copolymer is a co-block polymer or randomly distributed. Therefore, it is feasible to find the precise sequence distribution of the oligomers generated and construct their sequences [34]. The oligomers that were obtained by partial depolymerisation with the controlled thermal degradation of the PHAs produced were induced by $KHCO_3$ (Figures 5–7). This procedure was conducted following the methodology of Kawalec et al. [30]. The E1cB degradation method displayed end groups with both unsaturated and carboxylic acid properties. Based on the mass assignment of single charged ions in the mass spectrum, the structure of the end groups and repeating units can be presumed to be sodium adducts of individual co-oligoester chains of 3-hydroxybutyrate (HB) and 3-hydroxyvalerate (HV). The fragmentation of these ions resulted from the random breakage of ester bonds along both sides of the oligomer chains and the results were consistent with previous studies [16,32,34]. The fragmentation data also suggests that the presence of N-PEW as a carbon source produces different PHAs to those synthesized from media supplemented with O-PEW, which has been reported to produce hydroxyhexanoate (HH) comonomer units, in addition to PHB and PHBV [16,32–36]. This is significant because an increase in the ratio of 3HV to 3HB can decrease the melting point and improve the mechanical strength of the polymers produced [37].

5. Conclusions

N-PEW, a substance produced from waste PE materials is a potential carbon source for PHA production with *C. necator*. The N-PEWs did not mix with growth media, such as TSB, due to their hydrophobic nature; however, with ultrasound sonication, these waxes formed an emulsion with reduced-length hydrocarbon chains that did provide an increased cell count, clearly acting as a viable carbon source. The structural and thermal characterization of the biopolymers produced showed that by adjusting the carbon source, the types of polymer synthesized can be changed. This variation affects the molecular masses and properties of the PHB or PHBV biopolmyers. N-PEW does not limit the growth rate of bacterial cells; in fact the cultures with N-PEW as an additional carbon source produced a higher yield of PHAs. BSM supplemented with N-PEW produced no PHAs, while TSB and N-PEW can produce yields of up to 40% (*w*/*w*), TSB only experiments gave a yield of 20% (*w*/*w*).

The ^{1}H-NMR and ESI-MS/MS analysis indicated that in the presence of TSB/N-PEW the chains of the resulting biopolyester contained 3-HB repeating units with 11 mol% of 3-HV units randomly distributed along the copolymer chain. This is different to the PHAs reported from O-PEW/TSB fermentations, that produced PHBV and 3HH repeating units. The N-PEW/TSB media also yielded polymers with greater polymer diversity. Ultimately this analytical technique has enabled the characterization of PHAs synthesized from N-PEW/TSB media using *C. necator* H16, at the molecular level even when the contents of 3HB is high and the levels of other PHA oligomers (such as HV) are limited. This study demonstrates that plastic manufacturing companies could recycle more and utilize a wider range of materials for the production of value-added bioplastics and possibly direct the type of polyesters formed by selecting the level of wax oxidation. By using N-PEWs, food-based materials previously used for biopolymer synthesis, such as corn seed or sugar cane, could be reserved. These N-PEWs are a viable carbon source that can yield truly biodegradable, bio-compatible, non-toxic bioplastics for diverse applications.

Acknowledgments: This research was funded by the Research Investment Fund, University of Wolverhampton, Faculty of Science and Engineering, UK. This work was also partially supported under the EU 7FP BIOCLEAN Project, Contract No. 312100, "New biotechnological approaches for biodegrading and promoting the environmental biotransformation of synthetic polymeric materials".

Author Contributions: Matthew Green was responsible for N-PEW production. Brian Johnston, Iza Radecka, and David Hill were responsible for the bacterial production of PHA and PHA initial characterization. Brian Johnston and Guozhan Jiang were responsible for further GPC and NMR analysis. Grazyna Adamus, Iwona Kwiecień, Magdalena Zięba, Wanda Sikorska, and Marek Kowalczuk were responsible for PHA characterization using ESI-MS/MS. Iza Radecka, Marek Kowalczuk, and Brian Johnston were the main persons involved in the planning of the experiments and interpretation of the data of PHA characterization.

References

1. Chen, G.Q. A microbial polyhydroxyalkanoates (PHA) based bio- and materials industry. *Chem. Soc. Rev.* **2009**, *38*, 2434–2446. [CrossRef] [PubMed]
2. Al-Salem, S.M.; Lettieri, P.; Baeyens, J. Recycling and recovery routes of plastic solid waste (PSW): A review. *Waste Manag.* **2009**, *29*, 2625–2643. [CrossRef]
3. Bondareff, J.M.; Carey, M.; Lyden-Kluss, C. Plastics in the Ocean: The Environmental Plague of Our Time. *Roger Williams UL Rev.* **2017**, *22*, 360–506.
4. Derraik, J.G.B. The pollution of the marine environment by plastic debris: A review. *Mar. Pollut. Bull.* **2002**, *44*, 842–852. [CrossRef]
5. Barnes, D.K.A.; Galgani, F.; Thompson, R.C.; Barlaz, M. Accumulation and fragmentation of plastic debris in global environments. *Philos. Trans. R. Soc. Lond. B* **2009**, *364*, 1985–1998. [CrossRef] [PubMed]
6. Hopewell, J.; Dvorak, R.; Kosior, E. Plastics recycling: Challenges and opportunities. *Philos. Trans. R. Soc. Lond. B* **2009**, *364*, 2115–2126. [CrossRef]
7. Guzik, M.; Kenny, S.; Duane, G.; Casey, E.; Woods, T.; Babu, R.; Nikodinovic-Runic, J.; Murray, M.; O'Connor, K. Conversion of post consumer polyethylene to the biodegradable polymer polyhydroxyalkanoate. *Appl. Microbiol. Biotechnol.* **2014**, *98*, 4223–4232. [CrossRef] [PubMed]
8. Miskolczi, N.; Bartha, L.; Deak, G.; Jover, B. Thermal degradation of municipal plastic waste for production of fuel-like hydrocarbons. *Polym. Degrad. Stabil.* **2004**, *86*, 357–366. [CrossRef]
9. Global Demand for Polyethylene to Reach 99.6 Million Tons in 2018. Available online: https://pgjonline.com/2014/12/10/global-demand-for-polyethylene-to-reach-99-6-million-tons-in-2018/ (accessed on 12 July 2017).
10. Wei, R.; Zimmermann, W. Biocatalysis as a Green Route for Recycling the Recalcitrant Plastic Polyethylene Terephthalate. Available online: http://onlinelibrary.wiley.com/doi/10.1111/1751-7915.12714/full (accessed on 12 July 2017).
11. Verlinden, R.A.J.; Hill, D.J.; Kenward, M.A.; Williams, C.D.; Radecka, I. Bacterial synthesis of biodegradable polyhydroxyalkanoates. *J. Appl. Microbiol.* **2007**, *102*, 1437–1449. [CrossRef]
12. Shah, A.A.; Hasan, F.; Hameed, A.; Ahme, S. Biological degradation of plastics: A comprehensive review. *Biotechnol. Adv.* **2008**, *26*, 246–265. [CrossRef] [PubMed]
13. Philip, S.; Keshavartz, T.; Roy, I. Polyhydroxyalkanoates: Biodegradable polymers with a range of applications. *J. Chem. Technol. Biotechnol.* **2007**, *82*, 233–247. [CrossRef]
14. Verlinden, R.A.J.; Hill, D.J.; Kenward, M.A.; Williams, C.D.; Radecka, I. Production of polyhydroxyalkanoates from waste frying oil by *Cupriavidus necator*. *AMB Express* **2011**, *1*, 1–11. [CrossRef] [PubMed]
15. Leja, K.; Lewandowicz, G. Polymer biodegradation and biodegradable polymers—A review. *Pol. J. Environ. Stud.* **2010**, *19*, 255–266.
16. Radecka, I.; Irorere, V.; Jiang, G.; Hill, D.; Williams, C.; Adamus, G.; Kwiecień, M.; Marek, A.A.; Zawadiak, J.; Johnston, B.; et al. Oxidized Polyethylene Wax as a Potential Carbon Source for PHA Production. *Materials* **2016**, *9*, 367. [CrossRef] [PubMed]
17. Zawadiak, J.; Orlinska, B.; Marek, A.A. Catalytic oxidation of polyethylene with oxygen in aqueous dispersion. *J. Appl. Polym. Sci.* **2013**, *127*, 976–981. [CrossRef]
18. Benefield, R.E.; Boozer, C.E. Thermoplastic Polyolefin and Ethylene Copolymers with Oxidized Polyolefin. Patent US 4,990,568 A, 5 February 1991.
19. Seven, M.K. Process for Oxidizing Linear Low Molecular Weight Polyethylene. Patent WO 2005066219 A1, 21 July 2005.
20. Basstech International. Available online: http://basstechintl.com/products/polyethylene-wax/ (accessed on 12 July 2017).
21. Malykh, N.V.; Petrov, V.M.; Mal'tzev, L.I. Ultrasonic and hydrodynamic cavitation and liquid hydrocarbon cracking. *XX Sess. Russ. Acoust. Soc.* **2008**, *10*, 345–348.
22. Gautam, R.; Bassi, A.S.; Yanful, E.K.; Cullen, E. Biodegradation of automotive waste polyester polyurethane foam using Pseudomonas chlororaphis ATCC55729. *Int. Biodeterior. Biodegrad.* **2007**, *60*, 245–249. [CrossRef]
23. Jiang, G.; Hill, D.J.; Kowalczuk, M.; Johnston, B.; Adamus, G.; Irorere, V.; Radecka, I. Carbon Sources for Polyhydroxyalkanoates and an Integrated Biorefinery. *Int. J. Mol. Sci.* **2016**, *17*, 1157. [CrossRef] [PubMed]

24. Chaijamrus, S.; Udpuay, N. Production and Characterization of Polyhydroxybutyrate from Molasses and Corn Steep Liquor produced by Bacillus megaterium ATCC 6748. *Agric. Eng. Int. CIGR J.* **2008**, *X*, Manuscript FP 07 030.
25. Koller, M.; Sousa Dias, M.M.; Rodríguez-Contreras, A.; Kunaver, K.; Žagar, E.; Kržan, A.; Braunegg, G. Liquefied Wood as Inexpensive Precursor-Feedstock for Bio-Mediated Incorporation of (*R*)-3-Hydroxyvalerate into Polyhydroxyalkanoates. *Materials* **2015**, *8*, 6543–6557. [CrossRef] [PubMed]
26. Casida, L.E., Jr. Response in soil of *Cupriavidus necator* and other copper-resistant bacterial predators of bacteria to addition of water, soluble nutrients, various bacterial species, or Bacillus thuringiensis spores and crystals. *Appl. Environ. Microbiol.* **1988**, *54*, 2161–2166. [PubMed]
27. Schwartz, E.; Henne, A.; Cramm, R.; Eitinger, T.; Friedrich, B.; Gottschalk, G. Complete nucleotide sequence of pHG1: A Ralstonia eutropha H16 megaplasmid encoding key enzymes of H_2-based lithoautotrophy and anaerobiosis. *J. Mol. Biol.* **2003**, *332*, 369–383. [CrossRef]
28. Recycling Technologies. Available online: http://recyclingtechnologies.co.uk/technology/the-rt7000/#thermal-cracking (accessed on 27 July 2017).
29. Miles, A.A.; Misra, S.S.; Irwin, J.O. The estimation of the bactericidal power of the blood. *Epidemiol. Infect.* **1938**, *38*, 732–749. [CrossRef]
30. Kawalec, M.; Sobota, M.; Scandola, M.; Kowalczuk, M.; Kurcok, P. A convenient route to PHB macromonomers via anionically controlled moderate-temperature degradation of PHB. *J. Polym. Sci. Polym. Chem.* **2010**, *48*, 5490–5497. [CrossRef]
31. Zhang, W.; Luo, Y.; Wang, H.; Jiang, J.; Pu, S.; Chu, P.K. Ag and Ag/N2 plasma modification of polyethylene for the enhancement of antibacterial properties and cell growth/proliferation. *Acta Biomater.* **2008**, *4*, 2028–2036. [CrossRef] [PubMed]
32. Zagar, E.; Krzan, A.; Adamus, G.; Kowalczuk, M. Sequence distribution in microbial poly(3-hydroxybutyrate-co-3-hydroxyvalerate) co-polyesters determined by NMR and MS. *Biomacromolecules* **2006**, *7*, 2210–2216. [CrossRef] [PubMed]
33. Adamus, G.; Sikorska, W.; Janeczek, H.; Kwiecien, M.; Sobota, M.; Kowalczuk, M. Novel block copolymers of atactic PHB with natural PHA for cardiovascular engineering: Synthesis and characterization. *Eur. Polym. J.* **2012**, *48*, 621–631. [CrossRef]
34. Wei, L.; Guho, N.M.; Coats, E.R.; McDonald, A.G. Characterization of Poly(3-hydroxybutyrate-co-3-hydroxyvalerate) Biosynthesized by Mixed Microbial Consortia Fed Fermented Dairy Manure. *J. Appl. Polym. Sci.* **2014**. [CrossRef]
35. Adamus, G.; Sikorska, W.; Kowalczuk, M.; Noda, I.; Satkowski, M.M. Electrospray ion-trap multistage mass spectrometry for characterisation of co-monomer compositional distribution of bacterial poly(3-hydroxybutyrate-co-3-hydroxyhexanoate) at the molecular level. *Rapid Commun. Mass Spectrom.* **2003**, *17*, 2260–2266. [CrossRef] [PubMed]
36. Adamus, G. Aliphatic polyesters for advanced technologies structural characterization of biopolyesters with the aid of mass spectrometry. *Macromol. Symp.* **2006**, *239*, 77–83. [CrossRef]
37. Rudnik, E. Biodegradable Polymers from Renewable Sources. In *Composite Polymer Materials*, 1st ed.; Elsevier: Amsterdam, The Netherlands, 2008; p. 21.

14

Biodegradable Polymeric Substances Produced by a Marine Bacterium from a Surplus Stream of the Biodiesel Industry

Sourish Bhattacharya [1,†], Sonam Dubey [2,†], Priyanka Singh [3], Anupama Shrivastava [4] and Sandhya Mishra [2,*]

1 Process Design and Engineering Cell, CSIR-Central Salt and Marine Chemicals Research Institute, Bhavnagar 364002, India; sourishb@csmcri.org
2 Salt and Marine Chemicals, CSIR-Central Salt and Marine Chemicals Research Institute, Bhavnagar 364002, India; sonamdubey20@gmail.com
3 DTU BIOSUSTAIN, Novo Nordisk Foundation Center for Biosustainability, Technical University of Denmark, Lyngby 2800, Denmark; prnksingh254@gmail.com
4 Research & Product Development, Algallio Biotech Private Limited, Vadodara 390020, India; anupamashrivastav@gmail.com
* Correspondence: smishra@csmcri.org
† These authors contributed equally to this work.

Academic Editor: Martin Koller

Abstract: Crude glycerol is generated as a by-product during transesterification process and during hydrolysis of fat in the soap-manufacturing process, and poses a problem for waste management. In the present approach, an efficient process was designed for simultaneous production of 0.2 g/L extracellular ε-polylysine and 64.6% (w/w) intracellular polyhydroxyalkanoate (PHA) in the same fermentation broth (1 L shake flask) utilizing *Jatropha* biodiesel waste residues as carbon rich source by marine bacterial strain (*Bacillus licheniformis* PL26), isolated from west coast of India. The synthesized ε-polylysine and polyhydroxyalkanoate PHA by *Bacillus licheniformis* PL26 was characterized by thermogravimetric analysis (TGA), differential scanning colorimetry (DSC), Fourier transform infrared spectroscopy (FTIR), and ^{1}H Nuclear magnetic resonance spectroscopy (NMR). The PHA produced by *Bacillus licheniformis* was found to be poly-3-hydroxybutyrate-co-3-hydroxyvalerate (P3HB-co-3HV). The developed process needs to be statistically optimized further for gaining still better yield of both the products in an efficient manner.

Keywords: crude glycerol; polyhydroxyalkanoate; ε-polylysine; *Bacillus licheniformis*; fermentation

---

1. Introduction

Most of the global economy is driven by petroleum fuels as the main source of energy. However, due to market fluctuation, it is moving towards a sustainable bio based economy as fossil reserves are projected to decline completely by 2050 [1–3]. In addition to over exploitation of petroleum deposits, climate change and other negative environmental effects from exhaust gases lead researchers for the search of renewable alternatives such as biodiesel [4]. Biodiesel is an appealing alternative [5,6], which is clean burning, non-toxic and biodegradable [7]. It is a fatty acid methyl ester compound produced by a transesterification process of animal or plant oils with methanol in the presence of a catalyst [8–10]. Generally, glycerol is obtained in huge amount as a by-product in production of biodiesel [11,12]. With every 100 lbs of biodiesel produced by transesterification of vegetable oils or animal fats, 10 lbs of crude glycerol is generated [13,14]. However, the tremendous growth of the biodiesel industry has created a glycerol surplus that resulted in a dramatic 10-fold decrease in crude glycerol prices over the last few years. This decrease in prices resulted a problem for the glycerol producing and refining industries and also the economic viability of the biodiesel industry has also been greatly affected [15,16].

For sustainable development and commercialization of biodiesel production, effective utilization of crude glycerol into value added products are desired. However, conversion of crude glycerol will also promote the accretion of integrated biorefineries. 1,3-propanediol, citric acid, PHA's, ε-polylysine, butanol, hydrogen, ethanol, phytase, lipase, succinic acid, docosahexaenoic acid, eicosapentanoic acid, monoglycerides, lipids and syngas can be produced from crude glycerol through microbial strains [17–31]. However, many of the available technologies need further optimization and development in the form of efficient and sustainable form for its incorporation in bio-refineries.

The utilization of low quality glycerol obtained as by-product of biodiesel production is a big challenge as this glycerol cannot be used for direct food and cosmetic uses. An effective usage for conversion of crude glycerol to specific products may cut down the biodiesel production costs. The process for biodiesel preparation and generating value added products simultaneously through the use of waste generated during the biodiesel production is a very effective approach [32]. Clearly, the development of processes to convert crude glycerol into higher-value products is an urgent need.

Various studies on tuning the material properties of the polyhydroxyalkanoate (PHA) polymer are carried out for its higher applicability in diverse areas [33]. Nerve 2010 reported production of ε-polylysine from *Streptomyces albulus* (CCRC 11814) utilizing crude glycerol as carbon source through aerobic fermentation yielding 0.2 g/L ε-polylysine. However, growth rate of *Streptomyces albulus* (CCRC 11814) was slow due to other impurities such as methanol and other salts present in crude glycerol.

ε-polylysine and polyhydroxyalkanoates are important biopolymers which can be conjugated with other biopolymers for its various applications e.g., ε-polylysine may be utilized as water absorbable hydrogels, drug carriers and anticancer agents. Simultaneously, PHA may be used in drug delivery systems, atrial septal defect repair and cardiovascular stents. ε-polylysine and polyhydroxyalkanoates are important biopolymers which are synthesized through microbe in an efficient and eco-friendly manner. However, these biopolymers may be conjugated with other biopolymers for its application as water absorbable hydrogels, drug carriers and anticancer agents. The hydrogels prepared from these complex biopolymers may be used for its application in quick peritoneal repair and prevention of post-surgical intraabdominal adhesions.

In the present study, for sustainable development of biodiesel production, efforts have been made for effective utilization of crude glycerol as the carbon source for simultaneous production of ε-polylysine and PHA. The present bioconversion route will effectively convert the waste stream of biodiesel production into value added products. In addition, metabolic engineering may used in future for improving product yield of such strains.

2. Materials and Methods

2.1. *Materials*

2.1.1. Chemicals

Peptone, yeast extract, iron(III) citrate, NaCl, $MgCl_2$, Na_2SO_4, $CaCl_2$, KCl, $NaHCO_3$, KBr, $SrCl_2$, H_3BO_3, Na_2O_3Si, NaF, $(NH_4)(NO_3)$ and Na_2HPO_4 were purchased from M/S Hi-Media Limited Mumbai, and were of the highest purity available. ε-PL was procured from Handary SA, Brussels, Belgium, Polyhydroxybutyrate and 3-hydroxyvalerate from Sigma, Bangalore, India.

2.1.2. Growth and Maintenance of *Bacillus licheniformis*

Bacillus licheniformis was isolated from sea brine of experimental salt farm, Bhavnagar, India. It was maintained on Zobell marine agar plates containing (g/L) peptone 5.0; yeast extract 1.0; iron(III) citrate 0.1; NaCl 19.45; $MgCl_2$ 8.8; Na_2SO_4 3.24; $CaCl_2$ 1.8; KCl 0.55; $NaHCO_3$ 0.16; KBr 0.08;

$SrCl_2$ 0.034; H_3BO_3 0.022; Na_2O_3Si 0.004; NaF 0.0024; $(NH_4)(NO_3)$ 0.0016; Na_2HPO_4 0.008; agar, 1.5, at pH 7.6 ± 0.2. 5 mL of glycerol was added to the above medium. The slants were incubated at 37 °C for 4 days and then stored at 4 °C.

Experiments were carried out in 250 mL Erlenmeyer flasks with 100 mL of production medium with following components (g/L): yeast extract, 10; glucose, 50; $(NH_4)_2SO_4$, 15; $MgSO_4$, 0.5; K_2HPO_4, 0.8; KH_2PO_4, 1.4; $FeSO_4$, 0.04; $ZnSO_4$, 0.04. The pH of the medium was adjusted to 6.8 with 1 N NaOH before sterilization [34]. 10% (v/v) of a 48-h-old culture (approximately 8.9×10^8 cells/mL) was used as inoculum. Shake flask cultures of the organism were incubated at temperature 37 ± 2 °C with continuous agitation at 150 rpm for 96 h. These fermentation parameters were kept uniform for all the studies conducted. All experiments were carried out in triplicates.

2.2. *Fermentation*

2.2.1. Culture Media

The strain *Bacillus licheniformis* PL26 was cultivated in Zobell marine broth, Himedia, Mumbai, India. The media was adjusted to pH 7.6 ± 0.2. The plates were incubated at 37 °C temperature for 48 h.

2.2.2. Inoculum Development

The seed culture inoculated with loopful of *Bacillus licheniformis* was incubated overnight at 30 °C in an incubator shaker at 120 rpm. The inoculum for the production batch was prepared by using a single colony of *B. licheniformis* PL26 having 100 mL working volume.

2.2.3. Simultaneous Production of ε-Polylysine and PHA

The marine bacteria was cultured in Zobell marine broth to obtain seed culture having $O.D._{600}$ of 2.3. Zobell marine medium comprising (g/L) peptone 5.0; yeast extract 1.0; iron(III) citrate 0.1; NaCl 19.45; $MgCl_2$ 8.8; Na_2SO_4 3.24; $CaCl_2$ 1.8; KCl 0.55; $NaHCO_3$ 0.16; KBr 0.08; $SrCl_2$ 0.034; H_3BO_3 0.022; Na_2O_3Si 0.004; NaF 0.0024; $(NH_4)(NO_3)$ 0.0016; Na_2HPO_4 0.008 in one liter of the medium maintained at pH 7.6 ± 0.2. 20% seed culture was inoculated in the production medium which contained 20 g crude glycerol, yeast extract 5 g, $(NH_4)_2SO_4$ 10 g, K_2HPO_4 0.8 g, KH_2PO_4 1.36 g, $MgSO_4$ 0.5 g, $ZnSO_4$ 0.04 g, $FeSO_4$ 0.03 g in one litre of the medium maintained at pH 8.9 ± 0.2.

2.2.4. Analysis of ε-Polylysine

The culture broth was harvested after fermentation and cells were separated by centrifugation at 15,296× *g* rcf for 10 min in refrigerated centrifuge. 1 mL of supernatant was added to 1 mL of 1 mM methyl orange, mixed thoroughly under shaking condition along with incubating it at 37 °C for 60 min [35]. Further, the solution was centrifuged at 15,296× *g* rcf for 10 min in refrigerated centrifuge and absorbance of the supernatant was measured at 465 nm on UV-vis spectrophotometer (Varian, Palo Alto, CA, USA). A standard curve was derived from measurements with known amounts (0.1–2 mg/mL) of standard ε-PL procured from Handary S.A. [36].

2.2.5. Percentage Carbon Utilization of *Bacillus licheniformis* PL26

Percentage carbon utilized by *Bacillus licheniformis* PL26 was calculated a

$$\text{Carbon utilization (\%)} = \frac{\text{Total utilized carbon by bacteria}}{\text{Total carbon present in the medium}} \times 100 \quad (1)$$

Glycerol estimation was carried out by using "Waters Alliance" high performance chromatographic system equipped with RI detector (Waters 2414 model, Waters India Ltd., Bangalore, India) and separation module (Waters 2695 model, Waters India Ltd., Bangalore, India). Chromatographic separations were performed on an "Aminex HPX-87H" column (300 × 7.8 mm)

(Bio-Rad Laboratories, Richmond, CA, USA) with a precolumn (30 × 4.6 mm) of the same stationary phase (DVB-S, hydrogen form, Richmond, CA, USA). Isocratic elution at a flow rate of 0.6 mL/min was carried out using a mixture of 5 mM sulfuric acid. Peak detection was made by keeping the cells of the RI detector at 30 °C. The samples were appropriately degassed, twice diluted with double-distilled water, filtered through a "Whatman" 0.45-μm filter membrane (GE Healthcare Life Sciences, Little Chalfont, Buckinghamshire, UK), and then injected (50-μL loop volume). Data were obtained and processed by using "Waters EMPOWER" software (waters Corporation India, Bangalore, India). Peak identification was carried out by spiking the sample with pure standards and comparing the retention times with those of pure compounds.

2.2.6. Extraction and Purification of PHA

After completion of 96 h fermentation, culture broth was centrifuged at 15,296× *g* rcf for 10 min in refrigerated centrifuge. The cell pellets were oven dried overnight at 60 °C. Cellular digestion of dried cell pellet was carried out by re-suspending it in 6% (v/v) sodium hypochlorite solution followed by centrifugation at 10,000 rpm for 5 min. Further, the digested cell pellets were washed twice with methanol followed by distilled water to remove the traces of impurities resulting in a purified product, which was further dissolved in chloroform and weighed after air drying [37].

2.3. Purification of ε-Polylysine

2.3.1. Precipitation of Polycationic ε-PL with TPB^- Anion from the Supernatant

After removal of cells through centrifugation, the supernatant obtained was treated with sodium tetraphenylborate for precipitating ε-PL as a polyelectrolyte salt with the TPB^- anion. The polycationic ε-PL salt with the TPB^- anion was further purified by washing the mixed precipitate with acetone to remove triphenylborate and benzene. Thereafter, the precipitate reacted with 1 M HCl for obtaining ε-PL hydrochloride.

2.3.2. Analytical Methods

FT-IR spectra of obtained PHA were recorded on a Perkin-Elmer Spectrum GX (FT-IR System, Waltham, MA, USA) instrument. ^{1}H Nuclear magnetic resonance spectroscopy of PHA was determined on Bruker Avance-II 500 (Ultra shield) spectrometer, Bangalore, India, at 500 MHz, in $CDCl_3$. Proton ^{1}H NMR spectroscopy was also used to determine copolymer composition through running standards of 3HB and 3HV. Differential scanning calorimetry (DSC) of PHA was carried out using a DSC 204 F1 phoenix instrument with Netzsch software (NETZSCH Technologies India Pvt. Ltd., Chennai, India). The PHA samples were scanned from −20 °C to 500 °C with the heating rate of 10 °C/min. Glass transition temperature and onset melting points were determined in the scan between −20 °C and 500 °C in DSC analysis. Thermo-gravimetric analysis (TGA) of PHA was carried out in temperature range of 27–500 °C using TG 209 F1 instrument (NETZSCH Technologies India Pvt. Ltd., Chennai, India).

3. Results and Discussions

3.1. Simultaneous ε-Polylysine and PHA Production by B. licheniformis PL26

After complete submerged fermentation of 96 h at an agitation of 220 rpm, fermentation broth was centrifuged to obtain supernatant containing ε-polylysine and biomass for PHA extraction.

Simultaneous production of ε-polylysine and PHA by *B. licheniformis* PL26 was obtained utilizing crude glycerol as the carbon source. However, *B. licheniformis* is able to produce 0.2 g/L ε-polylysine extracellularly in the fermentation broth along with 64.59% PHA with respect to dry cell weight i.e., 1.1 g/L P(3HB-co-3HV) having 96 h production age at 37 °C (Figures S1–S3).

In the present case, *B. licheniformis* PL26 is able to produce both ε-polylysine and PHA in the same fermentation broth, which is not reported till date. However, *Streptomyces albulus* (CCRC 11814) is reported to produce ε-polylysine utilizing crude glycerol [3]. *S. albulus* being an Actinomycetes, it possesses a relatively slower growth rate as compared to *Bacillus* sp. As previously reported, *S. albulus* produces ε-polylysine after 120 h in M3G medium containing glucose as the carbon source, 0.2 g/L of ε-polylysine was produced by *Streptomyces albulus* (CCRC 11814) after 168 h [3], but in present study, ε-polylysine was produced after 96 h by *B. licheniformis* at a concentration of 0.2 g/L. In addition, fast growth rate and less production age of *B. licheniformis* which is producing 64.59% PHA with respect to dry cell weight i.e., 1.1 g/L P(3HB-co-3HV), which may be considered as the additional advantage of the process. Simultaneously, crude glycerol was replaced with analytical grade pure glycerol with similar concentration of the carbon source and it was found that pure glycerol yielded 0.06 g/L ε-polylysine and 0.4 g/L P(3HB-co-3HV).

0.2 g/L of ε-Poly-L-lysine produced from *Bacillus licheniformis* PL26 is in similar range with respect to its production from a wild strain *Streptomyces albulus* CCRC 11814 as reported in the literature [3]. However, similar sort of system developed by Moralejo-Ga'rate 2013, wherein using microbial community, simultaneous production of PHA and polyglucose was done [37]. Similarly, using halobacterium *Haloferax mediterranei*, simultaneously poly(3-hydroxybutyrate-*co*-3-hydroxyvalerate and extracellular polysaccharide (EPS) was produced [38]. Few microbes or microbial consortium have potential to produce two different polymers simultaneously using single carbon and nitrogen source.

3.2. Percentage Carbon Utilization of Bacillus licheniformis PL26

Carbon utilization percentage of isolated *Bacillus licheniformis* PL26 is shown in tab:bioengineering-03-00034-t001. *Bacillus licheniformis* isolated from salt pan has a 30% total carbon utilization percentage as it utilizes 0.21% total carbon from 0.7% total carbon present in the production medium.

Table 1. Percentage carbon utilization by *Bacillus licheniformis.*

Parameter	Concentration
Total Carbon content in fermentation medium	0.7%
Total Carbon left in the supernatant after complete fermentation (96 h production age)	0.41%
Carbon present in the biomass	0.21%
Percentage carbon utilized	30%

3.3. Characterization of Purified PHA

The polymer extracted from *B. licheniformis* PL26 grown in production media was characterized through TGA, DSC, NMR and *Fourier transform infrared spectroscopy* (FTIR).

Figure 1 indicates TGA analysis to analyze the thermal decomposition of the extracted polymer through the thermogravimetric analyzer. The extracted polymer showed 0.28 g weight loss out of 0.35 g till 500 °C temperature. Mass change of 0.21 g PHA out of 0.35 g PHA was found in the temperature range of 225–325 °C.

The DSC analysis shown in Figure 2 indicates the melting temperature of the standard sample and the extracted polymer through sodium hypochlorite treatment. The thermal degradation was obtained at 268 °C for the obtained polymer and 250 °C of standard PHA from Sigma Aldrich.

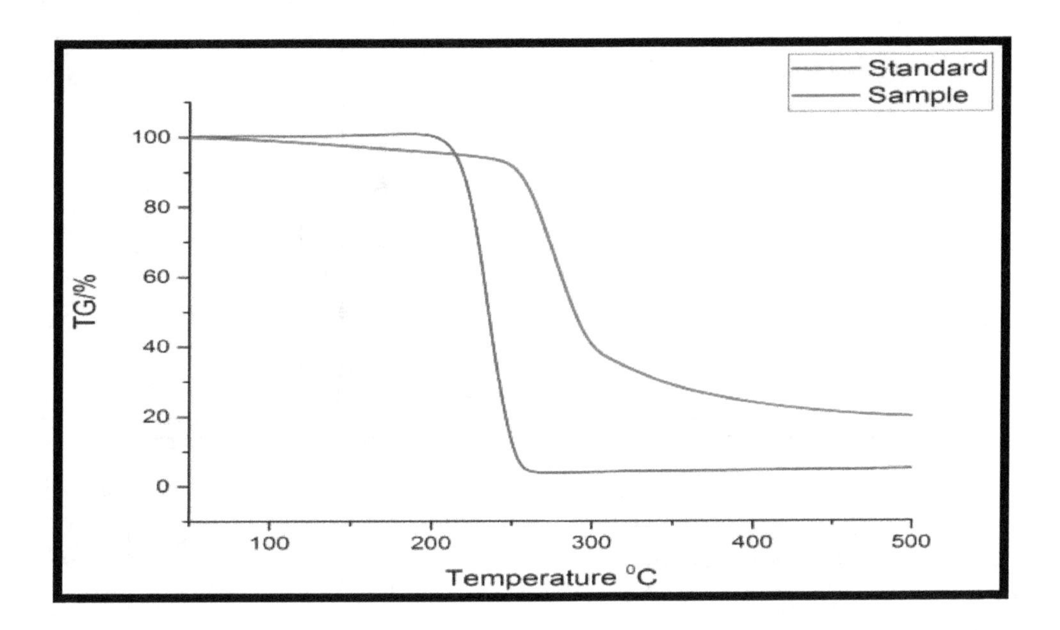

Figure 1. Thermogravimetric analysis (TGA) of purified PHA obtained from *Bacillus licheniformis*.

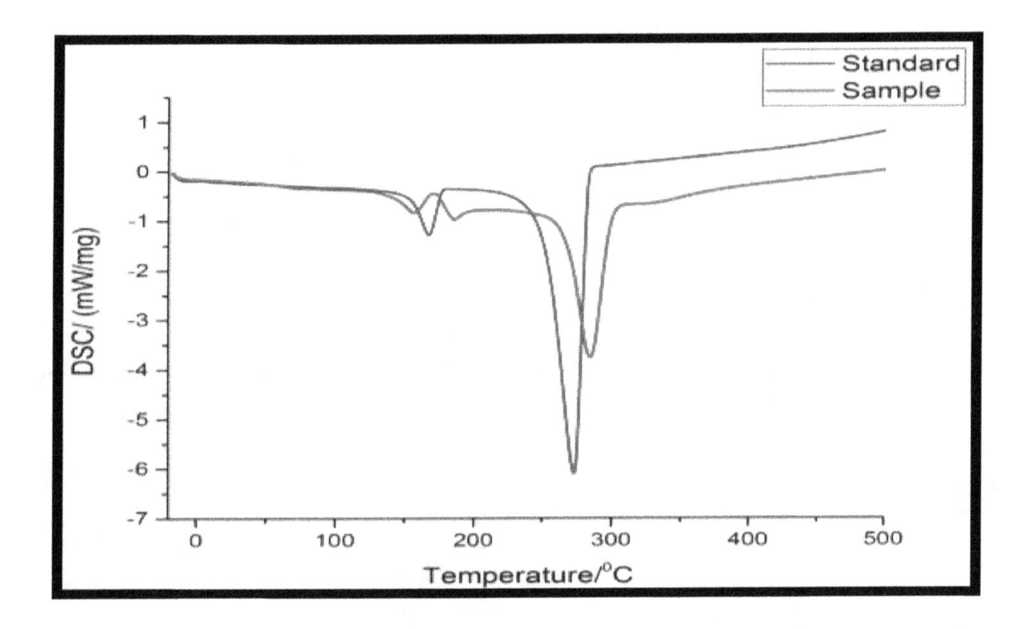

Figure 2. Differential scanning calorimetry (DSC) of purified PHA from *Bacillus licheniformis*.

^{1}H NMR spectra of the extracted polymer, standard PHB and standard 3 hydroxy valerate were found to be comparable with respect to each other. Prominent peaks were observed at $\delta = 1.6$ ppm for CH_3, $\delta = 2.4$ ppm for CH_2 and $\delta = 5.2$ ppm for CH group (Figure 3).

Infra red (IR) spectra (Figure 4) showed intense peaks at 1724 and 1283 cm^{-1} corresponding to –C=O and –CH group which are in correlation of peaks of standard PHA. Peaks at 1380, 1456 and 2932 correspond to $-CH_3$, $-CH_2$ and –CH group which are in correlation of peaks of standard PHA as shown in Figure 5.

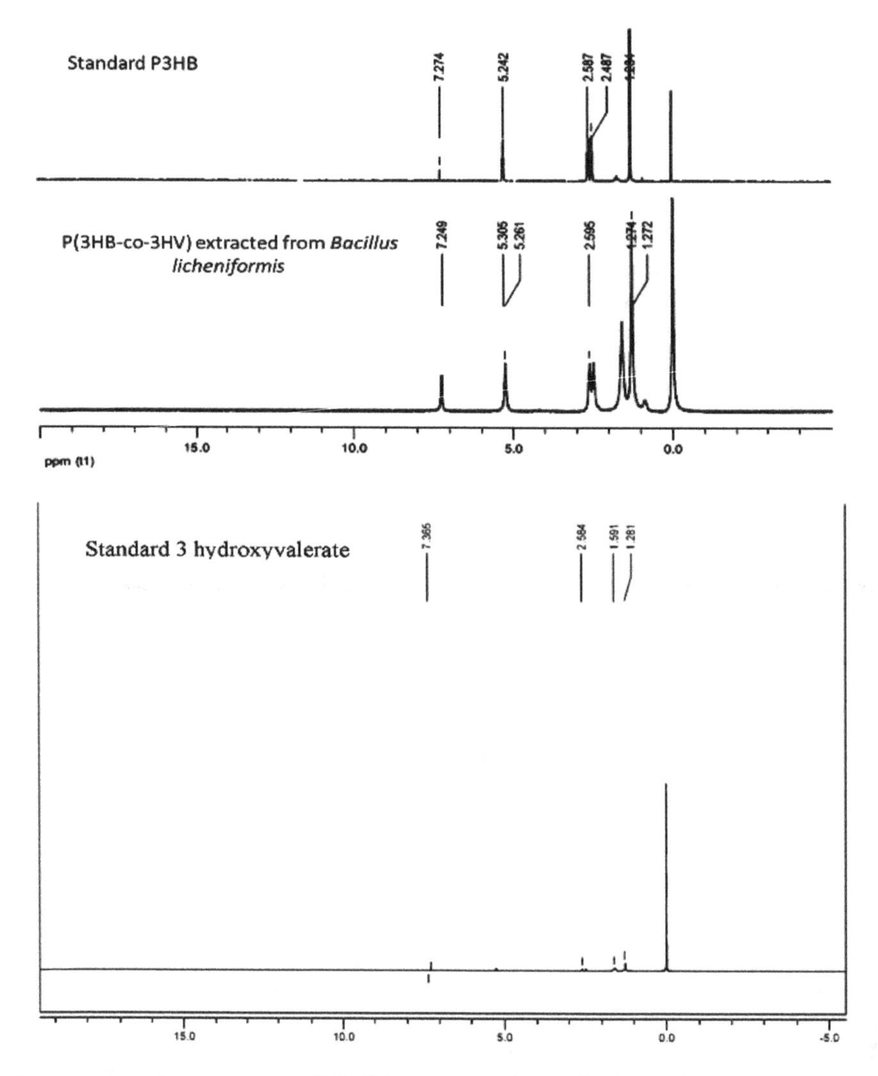

Figure 3. Nuclear magnetic resonance (NMR) spectra of purified product along with standard PHB and standard 3 hydroxy valerate.

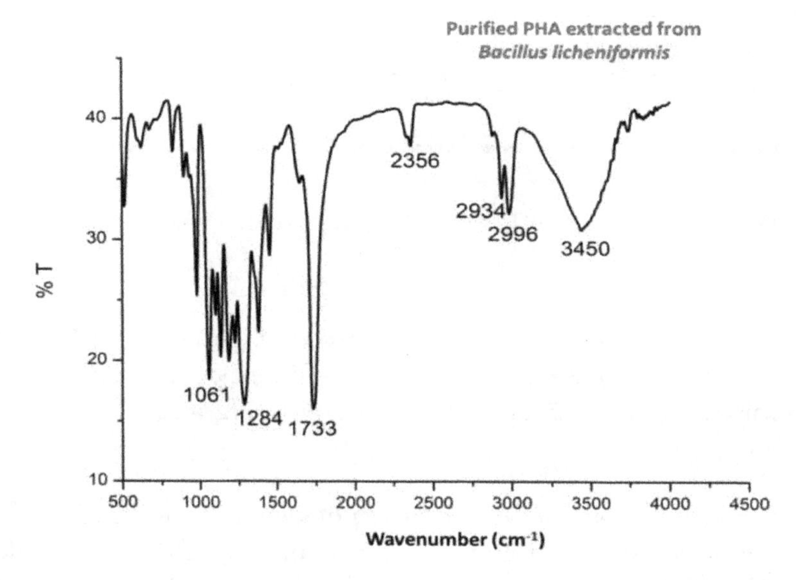

Figure 4. Fourier transform infrared spectroscopy (FTIR) spectra of purified PHA recovered from *Bacillus licheniformis*.

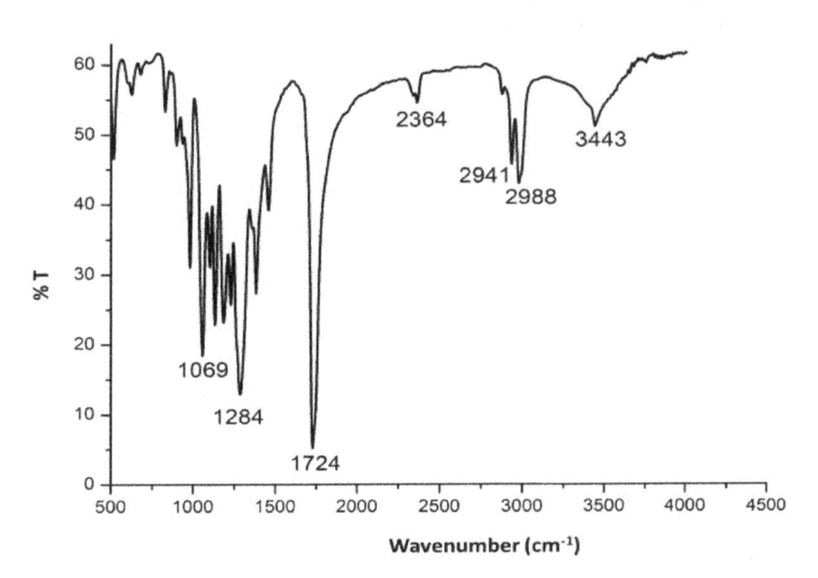

Figure 5. Fourier transform infrared spectroscopy (FTIR) spectra of standard PHA procured from Sigma Aldrich.

3.4. Characterization of ε-Polylysine

^{1}H NMR of ε-Polylysine in D_2O Isolated from *B. licheniformis* PL26

Protons (H^a, H^c) attached to α-amino groups arrived together as broad singlet at δ 3.76 ppm and protons (H^b, H^d) attached to ε-amino groups arrived together as broad singlet at δ 3.14 ppm. Protons attached to β and β′ carbons come at δ 1.75 ppm as broad singlet. While the other protons attached to carbons come at δ 1.47 ppm (4H, ε and ε′) and δ 1.30 ppm (4H, γ and γ′ respectively (Figure 6). Overall, the peaks showing peptide linkage between α-carboxyl group and the ε-amino group, confirming the structure as ε-polylysine.

As per Jia et al., 2010, chemical shift of ε-H in the ε-polylysine units $\delta_{\varepsilon H}$ and at the N-terminal $\delta'_{\varepsilon H}$ are 3.097 and 2.863 of 5 KDa ε-polylysine protein [39]. However, similar results were obtained in the present case, wherein chemical shift of ε-H in the ε-polylysine units, $\delta_{\varepsilon H}$ is at δ 3.14 ppm, which are in similar range with respect to the mentioned reports.

The described process showed the potential of utilizing *Bacillus licheniformis* for the production of PHA (64.59% w/w w.r.t cell dry mass i.e., 1.1 g/L P(3HB-co-3HV)) as well as ε-polylysine (0.2 g/L) using crude glycerol as the carbon source. In addition, there are no such reports as per our knowledge wherein two polymers are produced at a time in the same fermentation broth. Previously, Bera et al., 2014 reported microbial synthesis of such polymer (P(3HB-co-3HV)) by *Halomonas hydrothermalis* (MTCC accession no. 5445) from seaweed derived levulinic acid at a concentration of 57.5% PHA/dry cell weight and Ghosh et al., 2011 reported microbial synthesis of PHA from biodiesel by-products i.e., from crude glycerol and *Jatropha* deoiled cake hydrolysate at a concentration of 75% PHA/dry cell weight. However, further optimization will be required for efficient production of PHA and ε-polylysine at larger scale.

The most important parameter responsible for cost of the production in PHA is substrate for carbon and energy source. The economic feasibility of PHA would depend on few important factors like growth rate of microbe for generation of biomass, substrate cost and recovery process including the solvent involved. However, production of other biopolymer along with PHA in the fermentation medium will have additional advantage in further reducing the production cost. PHA production from biodiesel waste stream can reduce production cost and at the same time solves the problem of waste disposal. In order to increase the PHA yield, the concentration of carbon source and other nutrient source desired for the microbial production may be optimized.

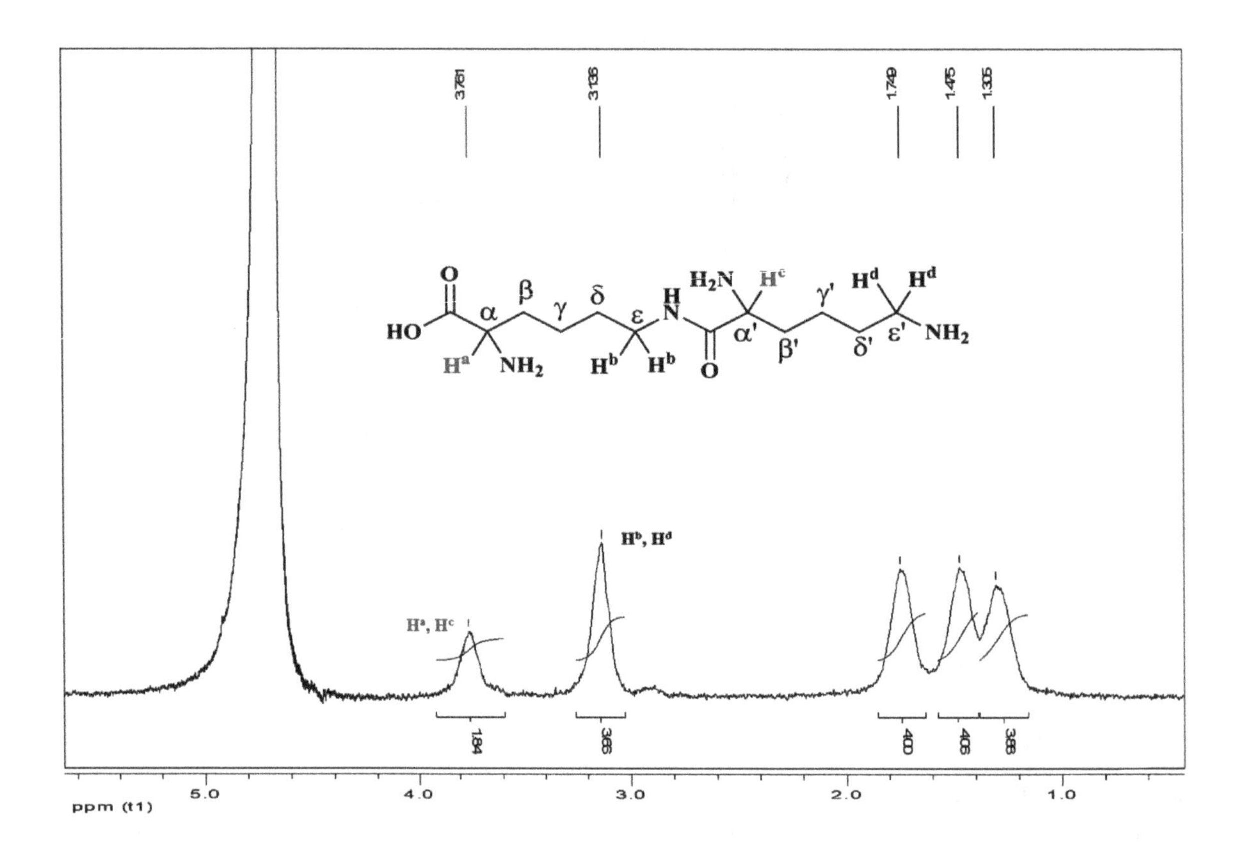

Figure 6. ^{1}H NMR of ε-PL in D_2O isolated from *Bacillus licheniformis*.

4. Conclusions

In the present study, an integrated process for simultaneous production of extracellular ε-polylysine and intracellular P(3HB-co-3HV) developed through marine bacterial strain (*Bacillus licheniformis*) isolated from west coast of India utilizing Jatropha biodiesel waste residues as carbon rich source. A maximum of 0.2 g/L ε-polylysine content and 64.6% (w/w) P(3HB-co-3HV) production with respect to dry biomass was obtained in the fermentation broth using *Bacillus licheniformis*. ε-polylysine and PHA are important class of biopolymers which have various applications in food, agriculture, medicine, pharmacy, controlled drug release, tissue engineering, etc. Although, the present approach provides a solution for the effective utilization of biodiesel by-product, still, the developed process needs to be optimized further for gaining still better yield of both the products for its acclamation as cost effective and sustainable process.

Acknowledgments: Financial assistance from CSIR as a part of EMPOWER scheme is gratefully acknowledged. We also acknowledge analytical science division of CSIR-CSMCRI for their timely support and cooperation. Authors are also grateful to CSC 0105, CSC 0203, OLP 0077 for providing financial support. We extend our gratitude to BDIM for providing PRIS no. CSIR-CSMCRI—043/2016 registration number.

Author Contributions: Sourish Bhattacharya developed the protocol for isolation and screening of potential ε-polylysine halophilic bacteria, followed by optimizing medium composition for simultaneous production of ε-polylysine and polyhydroxyalkanoate in the medium. Further, the extraction, purification and characterization of ε-polylysine was also conducted by him. Priyanka Singh assisted in isolation of bacteria from west coast of India and screening for identifying potential ε-polylysine halophilic bacterial isolate. Anupama Shrivastava and Sonam Dubey optimized the process for extraction and purification of polyhydroxyalkanoate. Sandhya Mishra monitored the overall experiments being performed and helped in the interpretation of results and polishing the manuscript.

References and Notes

1. Campbell, C.J.; Laherrère, J.H. The end of cheap oil. *Sci. Am.* **1998**, *3*, 78–83. [CrossRef]
2. Sheehan, J.; Camobreco, V.; Duffield, J.; Graboski, M.; Shapouri, H. Life Cycle Inventory of Biodiesel and Petroleum Diesel for Use in an Urban Bus. Available online: http://www.nrel.gov/docs/legosti/fy98/24089.pdf (accessed on 17 January 2016).
3. Nerve, Z. Upgrading of Biodiesel-Derived Glycerol in the Biosynthesis of ε-Poly-L-Lysine: An Integrated Biorefinery Approach. Available online: http://wiredspace.wits.ac.za/handle/10539/9253?show=full (accessed on 17 January 2016).
4. Vasudevan, P.T.; Briggs, M. Biodiesel production—Current state of the art and challenges. *J. Ind. Microbiol. Biotechnol.* **2008**, *35*, 421–430. [CrossRef] [PubMed]
5. Adhikari, S.; Fernando, S.D.; To, S.D.F.; Bricka, R.M.; Steele, P.H.; Haryanto, A. Conversion of glycerol to hydrogen via a steam reforming process over nickel catalysts. *Energy Fuels* **2008**, *22*, 1220–1226. [CrossRef]
6. Wirawan, S.S.; Tambunan, A.H. The current status and prospects of biodiesel development in Indonesia: A review. In Proceedings of the Third Asia Biomass Workshop, Tsukuba, Japan, 16 November 2006.
7. Marchetti, J.M.; Miguel, V.U.; Errazu, A.F. Possible methods for biodiesel production. *Renew. Sustain. Energy Rev.* **2007**, *11*, 1300–1311. [CrossRef]
8. Johnson, D.T.; Taconi, K.A. The glycerin glut: Options for the value-added conversion of crude glycerol resulting from biodiesel production. *Environ. Prog.* **2007**, *26*, 338–348. [CrossRef]
9. Thompson, J.C.; He, B.B. Characterization of crude glycerol from biodiesel production from multiple feedstock. *Appl. Eng. Agric.* **2006**, *22*, 261–265. [CrossRef]
10. Dasari, M.A.; Kiatsimkul, P.P.; Sutterlin, W.R.; Suppes, G.J. Low-pressure hydrogenolysis of glycerol to propylene glycol. *Appl. Catal. A Gen.* **2005**, *281*, 225–231. [CrossRef]
11. Lemke, D. Volumes of Versatility. Auri Ag Innovation News.
12. Gallan, M.; Bonet, J.; Sire, R.; Reneaume, J.; Plesu, A.E. From residual to useful oil: Revalorization of glycerine from the biodiesel synthesis. *Bioresour. Technol.* **2009**, *100*, 3775–3778. [CrossRef] [PubMed]
13. McCoy, M. Glycerin surplus. *Chem. Eng. News* **2006**, *84*, 7–8. [CrossRef]
14. Yazdani, S.S.; Gonzalez, R. Anaerobic fermentation of glycerol: A path to economic viability for the biofuels industry. *Curr. Opin. Biotechnol.* **2007**, *18*, 213–219. [CrossRef] [PubMed]
15. Ghosh, P.K.; Mishra, S.C.P.; Gandhi, M.R.; Upadhyay, S.C.; Paul, P.; Anand, P.S.; Popat, K.M.; Shrivastav, A.V.; Mishra, S.K.; Ondhiya, N.; et al. Integrated Process for the Production of *Jatropha* Methyl Ester and by Products. EU Patent 2,475,754, 18 July 2012.
16. Yang, F.; Hanna, M.A.; Sun, R. Value-added uses for crude glycerol—A byproduct of biodiesel production. *Biotechnol. Biofuels* **2012**, *5*, 1–10. [CrossRef] [PubMed]
17. Ito, T.; Nakashimada, Y.; Senba, K.; Matsui, T.; Nishio, N. Hydrogen and ethanol production from glycerol-containing wastes discharged after biodiesel manufacturing process. *J. Biosci. Bioeng.* **2005**, *100*, 260–265. [CrossRef] [PubMed]
18. Shrivastav, A.; Mishra, S.K.; Shethia, B.; Pancha, I.; Jain, D.; Mishra, S. Isolation of promising bacterial strains from soil and marine environment for polyhydroxyalkanoates (PHAs) production utilizing *Jatropha* biodiesel byproduct. *Int. J. Biol. Macromol.* **2010**, *47*, 283–287. [CrossRef] [PubMed]
19. Rymowicz, W.; Rywińska, A.; Marcinkiewicz, M. High-yield production of erythritol from raw glycerol in fed-batch cultures of *Yarrowia lipolytica*. *Biotechnol. Lett.* **2009**, *31*, 377–380. [CrossRef] [PubMed]
20. Tang, S.; Boehme, L.; Lam, H.; Zhang, Z. *Pichia pastoris* fermentation for phytase production using crude glycerol from biodiesel production as the sole carbon source. *Biochem. Eng. J.* **2009**, *43*, 157–162. [CrossRef]
21. Volpato, G.; Rodrigues, R.C.; Heck, J.X.; Ayub, M.A.Z. Production of organic solvent tolerant lipase by *Staphylococcus caseolyticus* EX17 using raw glycerol as substrate. *J. Chem. Technol. Biotechnol.* **2008**, *83*, 821–828. [CrossRef]
22. Scholten, E.; Renz, T.; Thomas, J. Continuous Cultivation Approach for Fermentative Succinic Acid Production from Crude Glycerol by *Basfia succiniciproducen* DD1. *Biotechnol. Lett.* **2009**, *31*, 1947–1951.

[CrossRef] [PubMed]

23. Ethier, S.; Woisard, K.; Vaughan, D.; Wen, Z.Y. Continuous culture of the microalgae *Schizochytrium limacinum* on biodiesel-derived crude glycerol for producing docosahexaenoic acid. *Bioresour. Technol.* **2011**, *102*, 88–93. [CrossRef] [PubMed]
24. Athalye, S.K.; Garcia, R.A.; Wen, Z.Y. Use of biodiesel-derived crude glycerol for producing eicosapentaenoic acid (EPA) by the fungus *Pythium Irregular*. *J. Agric. Food. Chem.* **2009**, *57*, 2739–2744. [CrossRef] [PubMed]
25. Choi, W.J.; Hartono, M.R.; Chan, W.H.; Yeo, S.S. Ethanol Production from Biodiesel-Derived Crude Glycerol by Newly Isolated *Kluyvera Cryocrescen*. *Appl. Microbiol. Biotechnol.* **2011**, *89*, 1255–1264. [CrossRef] [PubMed]
26. Oh, B.R.; Seo, J.W.; Heo, S.Y.; Hong, W.K.; Luo, L.H.; Joe, M.; Park, D.H.; Kim, C.H. Enhancement of ethanol production from glycerol in a *Klebsiella pneumoniae* mutant strain by the inactivation of lactate dehydrogenase. *Bioresour. Technol.* **2011**, *102*, 3918–3922. [CrossRef] [PubMed]
27. Taconi, K.A.; Venkataramanan, K.P.; Johnson, D.T. Growth and solvent production by *Clostridium pasteurianum* ATCC® 6013™ utilizing biodiesel-derived crude glycerol as the sole carbon source. *Environ. Prog. Sustain. Energy* **2009**, *28*, 100–110. [CrossRef]
28. Poblete-Castro, I.; Binger, D.; Oehlert, R.; Rohde, M. Comparison of mcl-Poly (3-hydroxyalkanoates) synthesis by different *Pseudomonas putida* strains from crude glycerol: Citrate accumulates at high titer under PHA-producing conditions. *BMC Biotechnol.* **2014**, *14*, 962. [CrossRef] [PubMed]
29. Moita, R.; Freches, A.; Lemos, P.C. Crude glycerol as feedstock for polyhydroxyalkanoates production by mixed microbial cultures. *Water Res.* **2014**, *58*, 9–20. [CrossRef] [PubMed]
30. Hermann-Krauss, C.; Koller, M.; Muhr, A.; Fasl, H.; Stelzer, F.; Braunegg, G. Archaeal production of polyhydroxyalkanoate (PHA) co-and terpolyesters from biodiesel industry-derived by-products. *Archaea* **2013**, *2013*, 129268. [CrossRef] [PubMed]
31. González-Pajuelo, M.; Andrade, J.C.; Vasconcelos, I. Production of 1,3-propanediol by *Clostridium butyricum* VPI 3266 using a synthetic medium and raw glycerol. *J. Ind. Microbiol. Biotechnol.* **2004**, *31*, 442–446. [CrossRef] [PubMed]
32. Bera, A.; Dubey, S.; Bhayani, K.; Mondal, D.; Mishra, S.; Ghosh, P.K. Microbial synthesis of polyhydroxyalkanoate using seaweed-derived crude levulinic acid as co-nutrient. *Int. J. Biol. Macromol.* **2015**, *72*, 487–494. [CrossRef] [PubMed]
33. Chheda, A.H.; Vernekar, M.R. Improved production of natural food preservative ε-poly-L-lysine using a novel producer *Bacillus cereus*. *Food Biosci.* **2014**, *30*, 56–63. [CrossRef]
34. Itzhaki, R.F. Colorimetric method for estimating polylysine and polyarginine. *Anal. Biochem.* **1972**, *50*, 569–574. [CrossRef]
35. Dhangdhariya, J.H.; Dubey, S.; Trivedi, H.B.; Pancha, I.; Bhatt, J.K.; Dave, B.P.; Mishra, S. Polyhydroxyalkanoate from marine *Bacillus megaterium* using CSMCRI's Dry Sea Mix as a novel growth medium. *Int. J. Biol. Macromol.* **2015**, *76*, 254–261. [CrossRef] [PubMed]
36. Chheda, A.H.; Vernekar, M.R. Enhancement of ε-poly-l-lysine (ε-PL) production by a novel producer *Bacillus cereus* using metabolic precursors and glucose feeding. *3 Biotech* **2015**, *5*, 839–846. [CrossRef]
37. Koller, M.; Chiellini, E.; Braunegg, G. Study on the Production and Re-use of Poly (3-hydroxybutyrate-co-3-hydroxyvalerate) and Extracellular Polysaccharide by the Archaeon *Haloferax mediterranei* Strain DSM 1411. *Chem. Biochem. Eng. Q.* **2015**, *29*, 87–98. [CrossRef]
38. Moralejo-Gárate, H.; Palmeiro-Sánchez, T.; Kleerebezem, R.; Mosquera-Corral, A.; Campos, J.L.; van Loosdrecht, M. Influence of the cycle length on the production of PHA and polyglucose from glycerol by bacterial enrichments in sequencing batch reactors. *Biotechnol. Bioeng.* **2013**, *110*, 3148–3155. [CrossRef] [PubMed]
39. Jia, S.; Fan, B.; Dai, Y.; Wang, G.; Peng, P.; Jia, Y. Fractionation and Characterization of ε-poly-L-lysine from *Streptomyces albulus* CGMCC 1986. *Food Sci. Biotechnol.* **2010**, *19*, 361–366. [CrossRef]

Permissions

All chapters in this book were first published in MDPI; hereby published with permission under the Creative Commons Attribution License or equivalent. Every chapter published in this book has been scrutinized by our experts. Their significance has been extensively debated. The topics covered herein carry significant findings which will fuel the growth of the discipline. They may even be implemented as practical applications or may be referred to as a beginning point for another development.

The contributors of this book come from diverse backgrounds, making this book a truly international effort. This book will bring forth new frontiers with its revolutionizing research information and detailed analysis of the nascent developments around the world.

We would like to thank all the contributing authors for lending their expertise to make the book truly unique. They have played a crucial role in the development of this book. Without their invaluable contributions this book wouldn't have been possible. They have made vital efforts to compile up to date information on the varied aspects of this subject to make this book a valuable addition to the collection of many professionals and students.

This book was conceptualized with the vision of imparting up-to-date information and advanced data in this field. To ensure the same, a matchless editorial board was set up. Every individual on the board went through rigorous rounds of assessment to prove their worth. After which they invested a large part of their time researching and compiling the most relevant data for our readers.

The editorial board has been involved in producing this book since its inception. They have spent rigorous hours researching and exploring the diverse topics which have resulted in the successful publishing of this book. They have passed on their knowledge of decades through this book. To expedite this challenging task, the publisher supported the team at every step. A small team of assistant editors was also appointed to further simplify the editing procedure and attain best results for the readers.

Apart from the editorial board, the designing team has also invested a significant amount of their time in understanding the subject and creating the most relevant covers. They scrutinized every image to scout for the most suitable representation of the subject and create an appropriate cover for the book.

The publishing team has been an ardent support to the editorial, designing and production team. Their endless efforts to recruit the best for this project, has resulted in the accomplishment of this book. They are a veteran in the field of academics and their pool of knowledge is as vast as their experience in printing. Their expertise and guidance has proved useful at every step. Their uncompromising quality standards have made this book an exceptional effort. Their encouragement from time to time has been an inspiration for everyone.

The publisher and the editorial board hope that this book will prove to be a valuable piece of knowledge for researchers, students, practitioners and scholars across the globe.

List of Contributors

Dana I. Colpa, Wen Zhou, Jan Pier Wempe, Janneke Krooneman and Gert-Jan W. Euverink
Products and Processes for Biotechnology Group, Engineering and Technology Institute Groningen, University of Groningen, Nijenborgh 4, 9747 AG Groningen, The Netherlands

Jelmer Tamis
Paques Technology B.V., Tjalke de Boerstrjitte 24, 8561 EL Balk, The Netherlands

Marc C. A. Stuart
Groningen Biomolecular Sciences and Biotechnology Institute, University of Groningen, Nijenborgh 7, 9747 AG Groningen, The Netherlands

Diana Gomes Gradíssimo and Agenor Valadares Santos
Post Graduation Program in Biotechnology, Institute of Biological Sciences, Universidade Federal do Pará, Augusto Corrêa Street, Guamá, Belém, PA 66075-110, Brazil
Laboratory of Biotechnology of Enzymes and Biotransformations, Institute of Biological Sciences, Universidade Federal do Pará, Augusto Corrêa Street, Guamá, Belém, PA 66075-110, Brazil

Luciana Pereira Xavier
Laboratory of Biotechnology of Enzymes and Biotransformations, Institute of Biological Sciences, Universidade Federal do Pará, Augusto Corrêa Street, Guamá, Belém, PA 66075-110, Brazil

Dan Kucera, Pavla Benesova, Peter Ladicky, Miloslav Pekar, Petr Sedlacek and Stanislav Obruca
Faculty of Chemistry, Brno University of Technology, Purkynova 118, 612 00 Brno, Czech Republic

Mohsen Moradi and Soheil Rezazadeh Mofradnia
Department of Chemical Engineering, Faculty of Engineering, Islamic Azad University North Tehran Branch, Tehran 1651153311, Iran

Hamid Rashedi
Biotechnology Group, School of Chemical Engineering, College of Engineering, University of Tehran, Tehran 11155-4563, Iran

Kianoush Khosravi-Darani
Department of Food Technology Research, Faculty of Nutrition Sciences and Food Technology/National Nutrition and Food Technology Research Institute, Shahid Beheshti University of Medical Sciences, Tehran 19395–4741, Iran

Reihaneh Ashouri
Department of Environment, Faculty of Environment and Energy, Science and Research Branch, Islamic Azad University, Tehran 1477893855, Iran

Fatemeh Yazdian
Department of Life Science Engineering, Faculty of New Science & Technology, University of Tehran, Tehran 1417466191, Iran

Clemens Troschl
Institute of Environmental Biotechnology, Department of Agrobiotechnology, IFA-Tulln, University of Natural Resources and Life Sciences, Vienna, Tulln 3430, Austria

Katharina Meixner and Bernhard Drosg
Bioenergy2020+ GmbH, Tulln 3430, Austria

Grazia Licciardello
Consiglio per la Ricerca in agricoltura e l'analisi dell'Economia Agraria-Centro di ricerca Olivicoltura, Frutticoltura e Agrumicoltura (CREA), Corso Savoia 190, 95024 Acireale, Italy

Antonino F. Catara
Formerly, Science and Technologies Park of Sicily, ZI Blocco Palma I, Via V. Lancia 57, 95121 Catania, Italy

Vittoria Catara
Dipartimento di Agricoltura, Alimentazione e Ambiente, Università degli studi di Catania, Via Santa Sofia 100, 95130 Catania, Italy

Constantina Kourmentza and Maria A. M. Reis
UCIBIO-REQUIMTE, Department of Chemistry, Faculdade de Ciências e Tecnologia/Universidade Nova de Lisboa, 2829-516 Caparica, Portugal

Jersson Plácido
Centre for Cytochrome P450 Biodiversity, Institute of Life Science, Swansea University Medical School, Singleton Park, Swansea SA2 8PP, UK

Nikolaos Venetsaneas
Faculty of Engineering and the Environment, University of Southampton, Highfield, Southampton SO17 1BJ, UK
European Bioenergy Research Institute (EBRI), Aston University, Aston Triangle, Birmingham B4 7ET, UK

Anna Burniol-Figols, Cristiano Varrone and Hariklia N. Gavala
Department of Chemical and Biochemical Engineering, Center for Bioprocess Engineering, Søltofts Plads, Technical University of Denmark, Building 229, 2800 Kgs. Lyngby, Denmark

Alejandra Rodriguez-Contreras
Department of Materials Science and Metallurgical Engineering, Universitat Politècnica de Catalunya (UPC), Escola d'Enginyeria de Barcelona Est (EEBE), Eduard Maristany 10-14, 08930 Barcelona, Spain

Miguel Miranda de Sousa Dias
Université Pierre et Marie Curie UPMC, Institut national de la santé et de la recherche médicale INSERM, Centre national de la recherche scientifique CNRS, Institut de la Vision, Sorbonne Universités, 17 rue Moreau, 75012 Paris 06, France

Martin Koller
Institute of Chemistry, University of Graz, NAWI Graz, Heinrichstrasse 28/III, 8010 Graz, Austria
ARENA—Association for Resource Efficient and Sustainable Technologies, Inffeldgasse 21b, 8010 Graz, Austria

Gerhart Braunegg
ARENA—Association for Resource Efficient and Sustainable Technologies, Inffeldgasse 21b, 8010 Graz, Austria

Dario Puppi, Andrea Morelli and Federica Chiellini
BIOLab Research Group, Department of Chemistry & Industrial Chemistry, University of Pisa, UdR INSTM Pisa, Via Moruzzi, 13, 56124 Pisa, Italy

Yolanda González-García, Janessa Grieve, Juan Carlos Meza-Contreras and José Antonio Silva-Guzman
Department of Wood, Cellulose and Paper, University of Guadalajara, 45020 Zapopan, Mexico

Berenice Clifton-García
Department of Chemical Engineering, University of Guadalajara, 44430 Guadalajara, Mexico

Liliana Montano-Herrera, Bronwyn Laycock and Steven Pratt
School of Chemical Engineering, University of Queensland, St Lucia QLD 4072, Australia

Alan Werker
Veolia Water Technologies AB—AnoxKaldnes, Klosterängsvägen 11A SE-226 47 Lund, Sweden

Warren Blunt, David B. Levin and Nazim Cicek
Department of Biosystems Engineering, University of Manitoba, Winnipeg, MB R3T 5V6, Canada

Marc Gaugler, Christophe Collet and Daniel J. Gapes
Scion Research, Te Papa Tipu Innovation Park, 49 Sala Street, Rotorua 3046, New Zealand

Richard Sparling
Department of Microbiology, University of Manitoba, Winnipeg, MB R3T 2N2, Canada

Brian Johnston, Guozhan Jiang, David Hill and Iza Radecka
Wolverhampton School of Biology, Chemistry and Forensic Science, Faculty of Science and Engineering, University of Wolverhampton, Wolverhampton WV1 1LY, UK

Grazyna Adamus, Iwona Kwiecień, Magdalena Zięba and Wanda Sikorska
Centre of Polymer and Carbon Materials, Polish Academy of Sciences, 41-800 Zabrze, Poland

Marek Kowalczuk
Wolverhampton School of Biology, Chemistry and Forensic Science, Faculty of Science and Engineering, University of Wolverhampton, Wolverhampton WV1 1LY, UK
Centre of Polymer and Carbon Materials, Polish Academy of Sciences, 41-800 Zabrze, Poland

Matthew Green
Recycling Technologies Ltd., South Marston Industrial Park, Swindon SN3 4WA, UK

Sourish Bhattacharya
Process Design and Engineering Cell, CSIR-Central Salt and Marine Chemicals Research Institute, Bhavnagar 364002, India

Sonam Dubey and Sandhya Mishra
Salt and Marine Chemicals, CSIR-Central Salt and Marine Chemicals Research Institute, Bhavnagar 364002, India

Priyanka Singh
DTU BIOSUSTAIN, Novo Nordisk Foundation Center for Biosustainability, Technical University of Denmark, Lyngby 2800, Denmark

Anupama Shrivastava
Research & Product Development, Algallio Biotech Private Limited, Vadodara 390020, India

Index